FUNDAMENTALS OF STRESS ANALYSIS

MATRIX SERIES IN MECHANICAL AND AERONAUTICAL ENGINEERING

Francis J. Hale, *Editor*

Fundamentals of Stress Analysis
McKinley

Energy: A Crisis, A Dilemma or Just Another Problem
Doolittle

FUNDAMENTALS OF STRESS ANALYSIS

JOE W. McKINLEY
Professor of Aerospace Engineering
California State Polytechnic University, Pomona

MATRIX PUBLISHERS, INC.
Portland, Oregon

10 9 8 7 6 5 4 3 2 1

Library of Congress catalog card number: 78-71525

ISBN: 0-916460-24-X

Matrix Publishers, Inc.
30 NW 23rd Place
Portland, OR 97210

Production: Patricia Miller
Illustrations: Scientific Illustrators
Printing: Pantagraph Printing
Editor: Merl K. Miller

PREFACE

This text is written for undergraduate and beginning level graduate students in the area of stress analysis. It is also intended for engineers who are practicing or learning stress analysis in their jobs. It is equally suitable for aerospace, mechanical or civil engineers or, in fact, for any engineer who has need for an understanding of stress analysis.

This book stresses the essential fundamental skills of stress analysis so that any person who completes it is prepared for engineering practice and for further formal study or self-study.

As an exceptional feature in the area of stress analysis, material is included from the areas of classical stress analysis, finite element stress analysis and experimental stress analysis so that a solid fundamental base in these three areas is established.

As a textbook, enough material is contained for three to four quarters of junior to senior to beginning graduate level instruction. The initial prerequisites are statics and calculus. For chapter 6 some knowledge of FORTRAN is required. The learning objectives for each chapter are stated. Thus, the statement of expected student progress, stated in behavioral terms, will be an aid to learning.

The numerous problems provide adequate practice as an aid to learning and the inclusion of a computer program listing permits the student to prepare and run his or her own finite element stress analyses.

A foundation in classical stress analysis is given in chapters 1 through 5 which include the areas of: uniaxial stress-strain, basic material properties, torsional stress and deflection, beam stress and deflection, biaxial stresses and the theory of elasticity.

Subsequently, one may choose to pursue the finite element method and will find that other chapters will provide advanced work as desired. Most engineers will benefit from the chapters on virtual work and statically indeterminate structures. Additional areas of general interest are stress concentration and metal fatigue and the theory of elastic stability.

Aerospace engineering students will find the chapter on flange-web structures of special interest.

Those interested in experimental stress analysis will find chapters on electrical resistance strain gages and photo-elasticity.

The author would like to express his appreciation to the following people who have significantly guided his professional development: Professor George Forman, Dr. Roger Anderson, Dr. Harvey Christensen, Dr. Hussein Kamel, Dr. Ralph Richard, Professor R. D. Sutherland and Dean Beaumont Davison, and most especially his junior high school algebra teacher, Jewel Johnson. In addition, the author has been encouraged and inspired by his parents and by his children, Frank and Celia McKinley. Special thanks are also due for the help and encouragement of Glenda Menges, including editorial assistance.

The typing assistance of the following persons is acknowledged with thanks: Margie Glantz, Barbara Fuller, Deborah Germain and Sandra Pinedo.

The support and encouragement of many students is acknowledged and special thanks for the diligent pursuit and reporting of errors in the original manuscript goes to Mr. Don Durston. Mr. Don Hilsen wrote the finite element computer program and his contribution is gratefully acknowledged. Valuable editorial assistance was given by Dr. Francis J. Hale of North Carolina State University. The author would like to extend a special thanks to Colonel Robert M. Wilson, Ph.D., Professor and Head, Department of Mechanics, United States Military Academy for his valuable editorial assistance.

The support of California State Polytechnic University, Pomona, California, especially in the form of a sabbatical leave for the purpose of finalizing the text, is gratefully acknowledged.

Joe W. McKinley
Pomona, California
February, 1979

CONTENTS

Chapter 1

INTRODUCTION

1.1 THE NATURE OF STRESS ANALYSIS

The job of the Stress Analyst, or Structural Engineer, whether he or she be in the aerospace, civil or mechanical field, is to assure that structures do not fail. The structure may be an airplane, building, freeway overpass, machine component, automobile, missile, spacecraft, bridge or any other object which carries a load. Failure may mean one or more of several things. Among these are: fracture, permanent deformation, excessive deflection and buckling.

Among the many tasks which the Stress Analyst might perform in the course of his career, we can list the following:

- Analyzing a building to assure that it will survive an earthquake
- Designing an aircraft wing of minimum weight
- Determining the cause of failure of an automobile drive shaft and redesigning it so it will not fail in the future
- Analyzing each component of a missile fuselage to be sure it will carry all the loads required for a mission
- Designing a helicopter blade to last a specified minimum lifetime when it is subjected to repeating loads
- Checking allowable stresses and deflections in the track structure of an amusement ride in order to assure that no failures will occur.

There are, of course, a very wide variety of assignments which the Stress Analyst may undertake; some of them are very challenging. Some of them are more routine.

The Stress Analyst is concerned with the calculation of <u>stress</u> in the structure. Stress is defined as the force per unit area in the structure and will be defined in much greater detail later. Quite frequently, the Stress Analyst is given the structure and the loading as his input and is asked to calculate the stress in the structure. Once this stress is calculated, the analyst must then compare it to the <u>allowable stress</u> of the material from which the structure is made. The comparison of stress in the part to the allowable stress of the material is of primary concern insofar as whether or not the part will fail. This whole matter can be a rather complicated one, as will be seen in later pages.

The allowable stress in a material is generally a material property, and is found by testing samples of the material in various ways. The stress analyst must be familiar with the various materials and their properties.

A second concern of the Stress Analyst is with the question of how much a given structure will deflect under a given load. Many structures are <u>deflection-governed</u>

designs and are much stiffer than they would need to be from the standpoint of stress alone. A common example of this is a building, whose members are stressed relatively lightly, but whose deflections are kept quite small for the occupants' peace of mind.

Other than stress and deflection, there are several areas of concern to the Stress Analyst. One is the question of buckling or instability of a structure. A student may relate to this phenomena by recalling the behavior of a long slender member (fishing pole, welding rod, yardstick, thin plastic ruler) when it is compressed. It will spring out of shape, but will recover its original configuration when the load is released. (If the load is kept below a certain limit.) This phenomenon governs the design of many structural elements: building columns and aircraft skin panels are two examples.

An additional area of concern is with structural fatigue. This concerns the behavior of a structure which is subjected to repeated, or cyclic, loading. A special behavior then governs the crack initiation and propagation, up to the ultimate fracture of the part. Fatigue is an area that Stress Analysts in all fields must be concerned with because it is responsible for a large number of the structural failures which actually occur. Going onward from these areas, there are innumerable areas of concern: creep, composite materials, dynamic buckling and flutter to name a few, which may be considered to be specialist areas for those pursuing careers in stress analysis, but which are not discussed in this book. The objective of this book is to take the beginning stress analysis student through all of the fundamentals which will prepare him for employment and also prepare him for further specialized study.

Like most engineers today, the Stress Analyst finds that the use of the digital computer is inevitable in his work. This is true in large firms and is becoming increasingly true in smaller firms, to the point where unfamiliarity with the computer is equivalent to illiteracy insofar as the Stress Analyst is concerned. In stress analysis, the major impact of the computer has been in the finite element method. This topic is discussed in Chapter 6 of this book in order to introduce it to stress analysis students early in their training.

Many structural problems are so complex, for one reason or another, that they defy mathematical analysis. For this reason, the Stress Analyst is frequently concerned with the subject of experimental stress analysis. Here we find structures, structural components or models tested under actual service conditions (as in flight testing) or in the laboratory. The primary methods for accomplishing experimental stress analysis are the bonded electrical resistance strain gage and photoelasticity. These two methods are described in this book.

Quite often, the subject of stress analysis is begun with the study of the behavior of structural elements: i.e., uniaxial elements, torsion elements, beam elements. Later, the behavior of assembled structures is considered. With the increasing use of the finite element method, this approach makes even more sense, and that is the approach which is followed in this book. Usually, the stress analyst functions in the midst of many conflicts. People are generally hassling him for less weight in the structure, or for more space. He may be under pressures to minimize the cost of the structure. He may be asked to utilize a warehousefull of a particular material which the company would otherwise be stuck with. The analyst must possess a number of skills in engineering, mathematics, physics, materials science, etc., in order to do this. Hopefully, he will possess and develop a degree of artistry and creativity as well.

Objectives

The learning objectives for each section will be stated in advance. This serves to prepare the student to read the material for maximum learning. Although the list of learning objectives is not exhaustive (the list may be supplemented by the instructor and the student) it will serve to highlight the most important parts.

As a result of "learning" the material in this section, the student will be able to do the following:

1. State or recognize the definition of an elastic material.
2. Write and illustrate his own definition of normal stress.
3. Draw free-body diagrams of structural "slices" to obtain the internal forces.
4. Calculate the normal stresses in a uniaxial element subjected to tension or compression.
5. Recognize uniaxial elements. State the definition of uniaxial elements. Produce examples of uniaxial elements.
6. State the simplified criterion for failure (fracture) in a uniaxial element. Determine if such an element will fail.
7. State the definition of ultimate stress. Find the numerical value of ultimate stress in a table.
8. Write and illustrate his own definitions of shear stress.
9. Calculate the stresses in an element subjected to pure shear.
10. State the simplified failure criteria for the failure (fracture) of elements in pure shear. Determine if an element will fail.
11. Calculate factors and margins of safety for normal and shear stress situations.
12. State the assumptions and limitations for the formulas in this section. Recognize whether or not a particular element meets them.

1.2 THE NATURE OF STRESS

The subject of statics, which is a prerequisite to this text, deals primarily with the equilibrium of external forces on a rigid body. The subject of stress analysis deals with the equilibrium of internal forces on an elastic (or at least deformable) body. The term elastic means that there is a linear relation between stress and strain. This will be discussed in greater detail later.

A simple example of the relationship of statics to stress analysis, and a definition of normal stress, can be obtained by considering a uniform rod in tension as depicted in Figure 1.2.1(a).

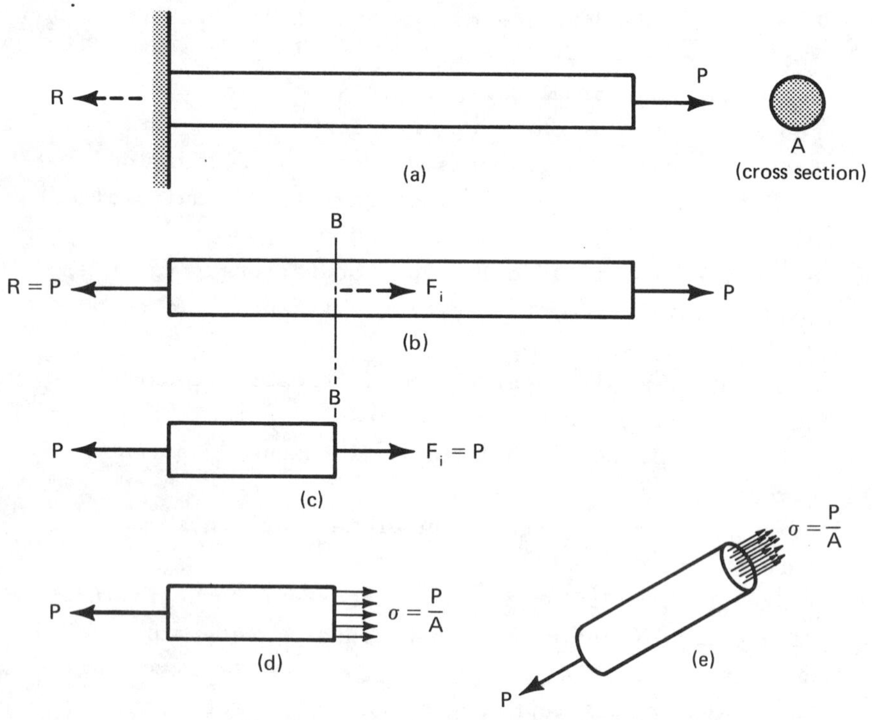

Figure 1.2.1

Consider, for simplicity, that the rod has uniform cross section. A static analysis, which is essentially the result of writing the equilibrium equations between the applied load P and the wall reaction R, determines that R = P, in the direction shown, and results in the free body diagram of Figure 1.2.1(b).

In Figure 1.2.1(c), we examine the internal equilibrium of this structure by asking what <u>internal force</u>, F_i, is present at section B. It is the presence of this internal force which causes stress in the rod. This is done by actually slicing or cutting the structure through at section B, throwing one part of it away and examining the <u>equilibrium</u> of the remaining section. As shown in Figure 1.2.1(c), this results in $F_i = P$, which is a static equilibrium of forces.

The <u>normal stress</u> is <u>defined</u> as the normal force divided by the cross sectional area. In the example we have used, the stresses are distributed uniformly over the cross section. Therefore the normal stress, as shown in Figures 1.2.1(d) and (e) is

$$\sigma = \frac{P}{A} \quad \frac{\text{lb}}{\text{in}^2} \text{ or } \frac{\text{N}}{\text{m}^2} \tag{1.1}$$

We have used tensile stress in our example. The definition applies equally well to compressive stress.

Stress describes the intensity of the force, or the amount of force carried by each unit of area. Speaking in a simplified manner for a moment (there are many important exceptions to be discussed later) it is quite often true that a rod such as the one used in our example, of <u>any size</u>, will fail (fracture) when its normal stress reaches some value which is unique to the material from which the rod is made. This value of stress is called the <u>ultimate stress</u> or <u>ultimate strength</u> of the material.

A similar observation may be made when the loading on a part is parallel to the cut plane, rather than normal to it. This is illustrated by considering a structure which could represent a bolt or a rivet in shear as shown in Figure 1.2.2.

We shall ignore the small amount of bending which occurs as a result of the depicted loading. We can see from statics that the resultant R equals the applied load P. By examining a slice, or section, of the structure we see

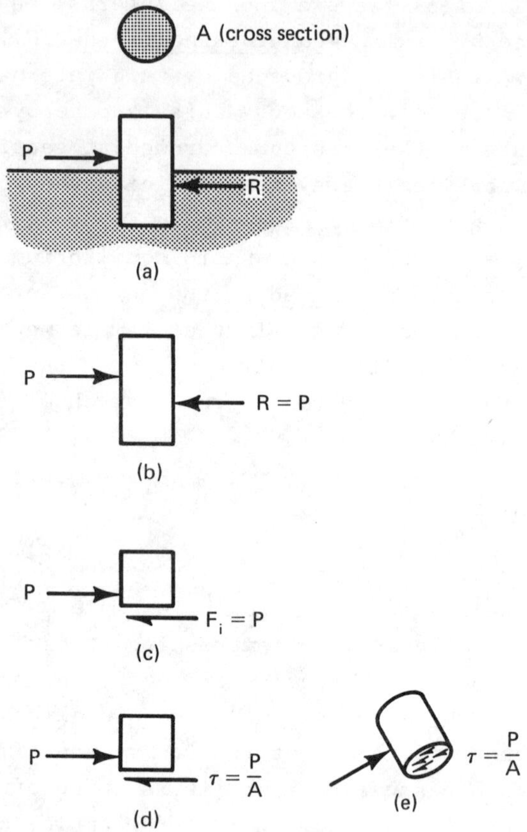

Figure 1.2.2

that the internal force F_i equals P. The force per unit area in the plane of the cut section, the <u>tangential force per unit area is defined as the shear stress</u> τ.

$$\tau = \frac{P}{A} \; \frac{\text{lb}}{\text{in}^2} \text{ or } \frac{\text{N}}{\text{m}^2} \tag{1.2}$$

The situation depicted in Figure 1.2.2 is referred to as pure shear, or direct shear. We again find that shear stress defines the intensity of the force, or the amount of force carried by each unit area. Thus we find that (with some important exceptions) many structures such as bolts and rivets will fail (fracture) when the shear stress equals a material property known as the ultimate shear stress or ultimate shear strength, regardless of their size.

The general statement can be made that a structure will fail when the stress due to the applied loading exceeds the stress the material from which it is made is capable of carrying. This general statement needs much further elaboration, as we shall see in subsequent chapters. If we denote the normal stress at which a material will fail as S_{tu} and that for the shear ultimate as S_{su}, we may define the factor of safety as

$$\text{Factor of Safety}_{\text{Normal}} = FS_N = \frac{\text{ultimate stress}}{\text{applied stress}} = \frac{S_{tu}}{\sigma} \tag{1.3}$$

$$\text{Factor of Safety}_{\text{Shear}} = FS_s = \frac{\text{ultimate stress}}{\text{applied stress}} = \frac{S_{su}}{\tau} \tag{1.4}$$

In the aerospace industries, the term margin of safety is used. This is defined as

$$MS = FS - 1 \tag{1.5}$$

Structure such as those we have discussed can be assumed to fail when their factor of safety equals one or their margin of safety equals zero. Important <u>assumptions</u> and <u>limitations</u> exist for the formulas we have developed thus far. It is extremely important to be aware that throughout the practice of stress analysis, at all levels, the formulas used are subject to such assumptions and limitations. Errors in stress analysis, and subsequent failures, may be caused by attempting to apply a formula in violation of the assumptions and limitations under which it was developed.

Equation 1.1 may be used to find the normal stress subject to the following assumptions and limitations:

1. The cross sectional area of the rod must be uniform, or nearly so. Although smooth, gradual changes in cross section do not cause appreciable errors in Equation 1.1, any abrupt change in cross section will cause errors of large magnitude. Abrupt changes in cross section cause <u>stress concentrations</u>, which will be discussed in Chapter 10.
2. The localized effects in the area where the load is applied are not given by Equation 1.1.
3. The line of action of the force must pass through the centroid of the cross section. If this is not true, bending will exist as well as tension or compression.
4. If the rod is in compression, it must be "stocky" enough so that it does not buckle.

It is usually not obvious to the beginner, but it is true that the stress at any point on the interior face of a structure may be resolved into one of the two types we have discussed: normal stress and shear stress. If we consider Figure 1.2.3, we note that the external forces acting on the most general elastic solid can be equilibrated at any

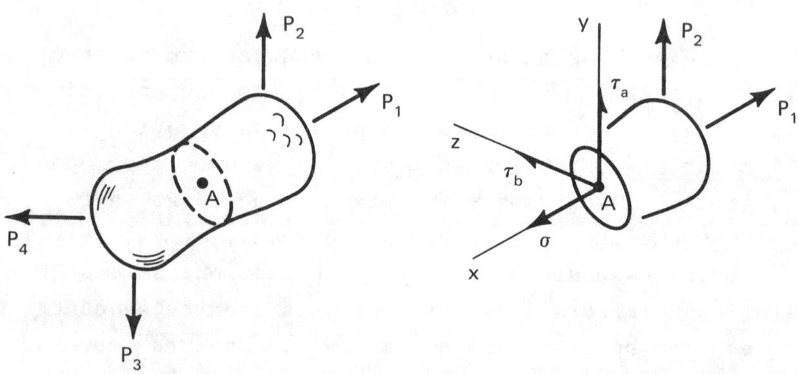

Figure 1.2.3

internal point by stresses normal and tangential to the plane of the slice.

The stress at point A can be represented by a normal stress σ and two in-plane shear stresses τ_a and τ_b.

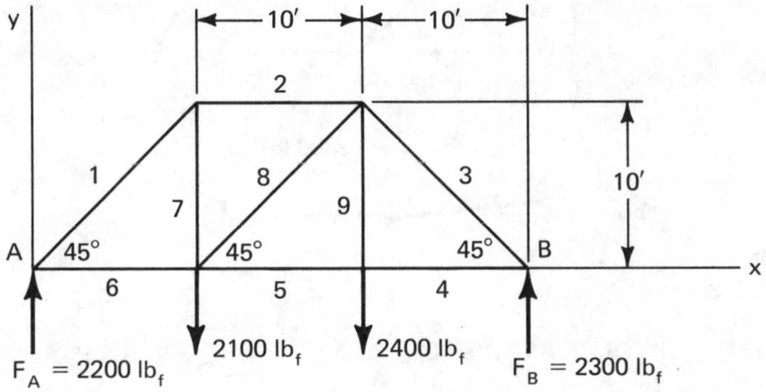

Example 1.1. For the truss shown, the area of each member is 1.75 in². Find the following and tabulate your results:

(a) Force in each member

(b) Stress in each member

Solution:

Calculate reactions:

(a) $+\uparrow \Sigma F_y = 0$

$$F_A + F_B - 2100 - 2400 = 0$$

$$F_A + F_B = 4500 \text{ lb}_f$$

$$\circlearrowleft\!+ \ \Sigma M_A = 0$$

$$(2100 \text{ lbs})(10 \text{ ft}) + (2400 \text{ lbs})(20 \text{ ft}) - F_B(30 \text{ ft}) = 0$$

$$21{,}000 \text{ ft} - \text{lb}_f + 48{,}000 \text{ ft} - \text{lb}_f = F_B(30 \text{ ft})$$

$$69{,}000 \text{ ft} - \text{lb}_f = F_B(30 \text{ ft})$$

$$F_B = 2300 \text{ lb}_f$$

$$F_A + F_B = 4500 \text{ lb}_f,\ F_A = 4500 \text{ lb}_f - 2300 \text{ lb}_f$$

$$F_A = 2200 \text{ lb}_f$$

Consider:

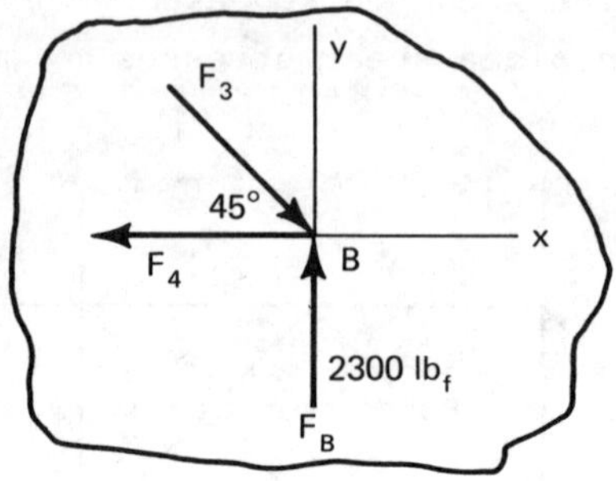

$+\uparrow \Sigma F_y = 0$

$2300\ lb_f - F_3(\text{SIN } 45°) = 0$

$2300\ lb_f = F_3(.707)$

$F_3 = 3260\ lb_f(C)$

$\overset{+}{\rightarrow} \Sigma F_x = 0$

$F_3(\text{COS } 45°) - F_4 = 0$

$F_3(.707) - F_4 = 0$

$(3260\ lb_f)(.707) = F_4$

$F_4 = 2300\ lb_f(T)$

Consider:

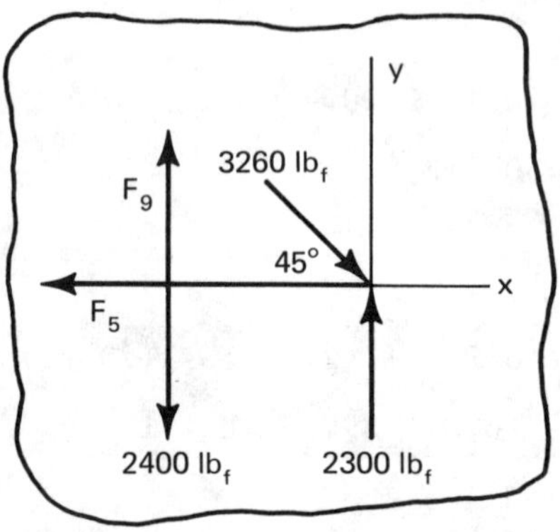

$\overset{+}{\rightarrow} \Sigma F_x = 0$

$(3260\ lb_f)(\text{COS } 45°) - F_5 = 0$

$(3260\ lb_f)(.707) = F_5$

$F_5 = 2300\ lb_f(T)$

$+\uparrow \Sigma F_y = 0$

$F_9 + (2300\ lb_f) - (3260\ lb_f)(SIN\ 45°) - (2400\ lb_f) = 0$

$F_9 + (2300\ lb_f) - (3260\ lb_f)(.707) - 2400\ lb_f = 0$

$F_9 - 2400\ lb_f = 0$

$F_9 = 2400\ lb_f(T)$

Now Consider:

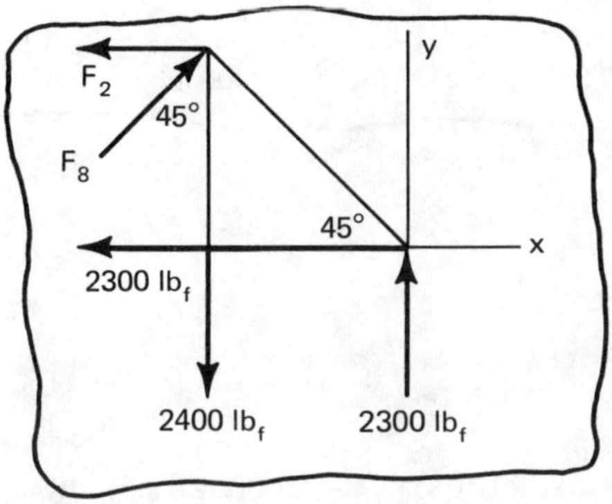

$\overset{+}{\rightarrow} \Sigma F_x = 0$

$F_8(COS\ 45°) - F_2 - 2300\ lb_f = 0$

$F_8(.707) - F_2 = 2300\ lb_f$

$+\uparrow \Sigma F_y = 0$

$F_8(SIN\ 45°) + 2300\ lb_f - 2400\ lb_f = 0$

$F_8(.707) = 100\ lb_f$

$F_8 = 142\ lb_f\ (C)$

$F_8(.707) - F_2 = 2300\ lb_f$

$(142\ lb_f)(.707) - F_2 = 2300\ lb_f$

$100\ lb_f - F_2 = 2300\ lb_f$

$-F_2 = 2200\ lb_f$

...The force acts in a direction opposite to the direction assumed.

$F_2 = 2200\ lb_f\ (C)$

Consider:

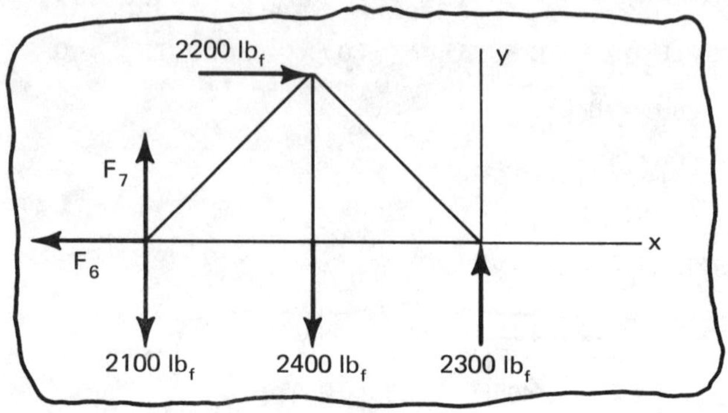

$\overset{+}{\rightarrow}\Sigma F_x = 0$

$2200\ lb_f - F_6 = 0$

$F_6 = 2200\ lb_f(T)$

$+\uparrow \Sigma F_y = 0$

$F_7 + 2300\ lb_f - 2100\ lb_f - 2400\ lb_f = 0$

$F_7 + 2300\ lb_f = 4500\ lb_f$

$F_7 = 2200\ lb_f(T)$

Consider:

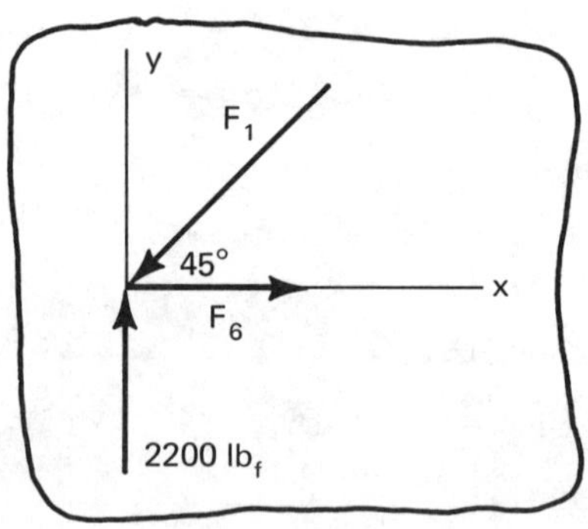

$\overset{+}{\rightarrow}\Sigma F_x = 0$

$F_6 - F_1(COS\ 45°) = 0$

$F_6 - F_1(.707) = 0$

$+\uparrow \Sigma F_y = 0$

$(2200\ lb_f) - F_1(SIN\ 45°) = 0$

$2200\ lb_f - F_1(.707) = 0$

$F_1(.707) = 2200\ lb_f$

$\underline{F_1 = 3110\ lb_f(C)}$

$F_6 = (3110\ lb_f)(.707) = 2200\ lb_f(T)$: Check

(b) Stress $(\sigma) = \frac{F}{A_o}$, $A_o = 1.75\ in^2$

$$\sigma_1 = \frac{F_1}{A_o} = \frac{3110\ lb_f}{1.74\ in^2} = 1780\ lb_f/in^2(C)$$

$$\sigma_2 = \frac{F_2}{A_o} = \frac{2200\ lb_f}{1.75\ in^2} = 1260\ lb_f/in^2(C)$$

$$\sigma_3 = \frac{F_3}{A_o} = \frac{3260\ lb_f}{1.75\ in^2} = 1860\ lb_f/in^2(C)$$

$$\sigma_4 = \frac{F_4}{A_o} = \frac{2300\ lb_f}{1.75\ in^2} = 1310\ lb_f/in^2(T)$$

$$\sigma_4 = \sigma_5 = 1310\ lb_f/in^2(T)$$

$$\sigma_6 = \sigma_7 = 1260\ lb_f/in^2(T)$$

$$\sigma_8 = \frac{F_8}{A_o} = \frac{142\ lb_f}{1.75\ in^2} = 81\ lb_f/in^2(C)$$

$$\sigma_9 = \frac{F_9}{A_o} = \frac{2400\ lb_f}{1.75\ in^2} = 1370\ lb_f/in^2(T)$$

Problems

Problem 1. Calculate the normal stress and the shear stress on the structural members as shown below. Find the normal stress in the tension members and the shear stress in the bolt. Discuss anything which you feel is an approximation or is not quite true.

Note: Draw free bodies of the structure and structural pieces in this and all other problems.

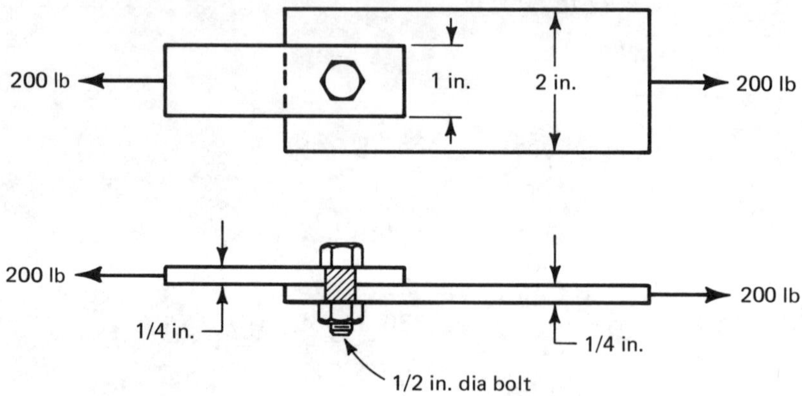

Problem 2. A rod of cross sectional area A = 0.785 in^2 hangs vertically under its own weight. Determine the location of maximum tensile stress. If the specific weight is 0.100 lb/in^3, calculate the value of the maximum tensile stress.

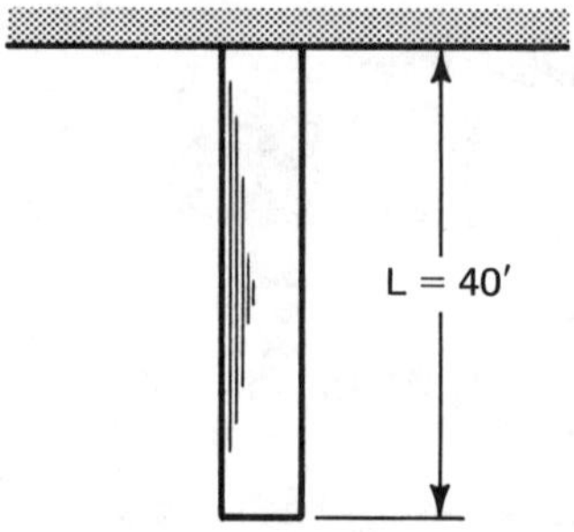

Problem 3. Repeat problem 2 in SI units. This may be done for all problems.

Problem 4. A rod of square cross section is loaded as shown. Calculate the normal stress in each of the sections. In what areas would you expect your answers to be invalid? Draw free-body diagrams of each section.

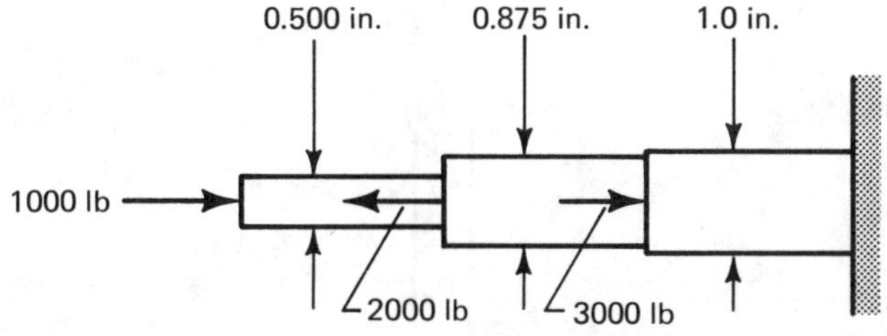

Problem 5. A rod is loaded by a concentrated axial force and by its own weight. It is of circular cross section and its diameter varies linearly from top to bottom. The bottom diameter is 0.50 in and the top diameter is 1.0 in. Find the location and magnitude of the maximum normal stress. Assume the rod is made of steel.

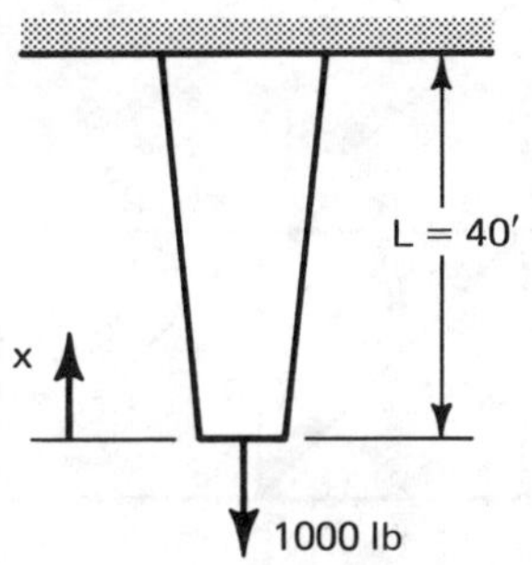

Problem 6. Define normal stress in your own words. Use sketches.

Problem 7. From the table of properties in Appendix A, find the ultimate tensile strength of:

Steel: AISI 4130
Aluminum: 7075-T6
Magnesium: AZ61A

Problem 8. A connecting rod is to be made from AISI 4130 steel. Assume that its compressive ultimate strength is equal to its tensile ultimate strength. If the maximum compressive force on the rod is 4,000 lb, find the required area. Use a factor of safety of 3.0.

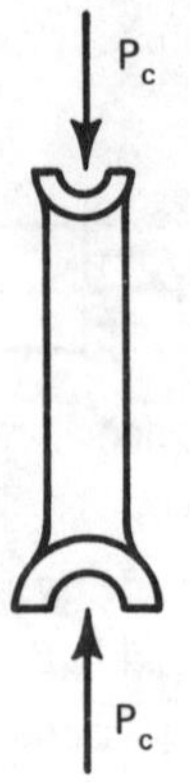

Problem 9. A truss is to be designed of AZ61A magnesium. Assume that the compressive members of the truss will not buckle. Determine the required area of each member. Use a factor of safety of 2.5.

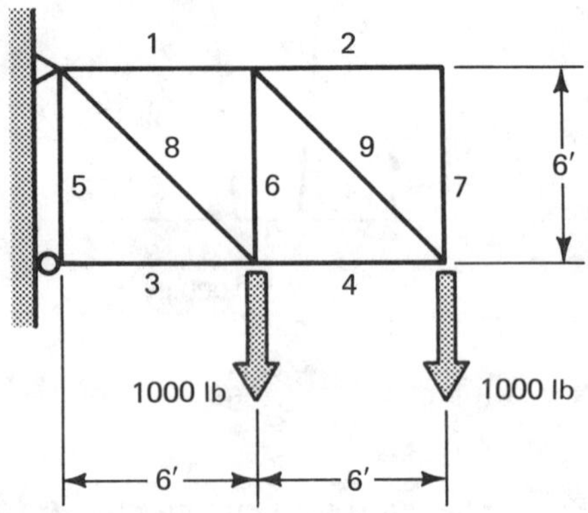

Problem 10. An aircraft landing gear (simplified) can be represented in side view as a plane 2-D truss. For the loads shown, size the members with margin of safety of 0.5 if they are to be made from 7075-T6 aluminum. See figure at top of page 19.

Problem 11. Two 1/4 in. dia. bolts are in shear as shown in second figure on page 19. Calculate the shear stress in each bolt. Draw complete free body diagrams.

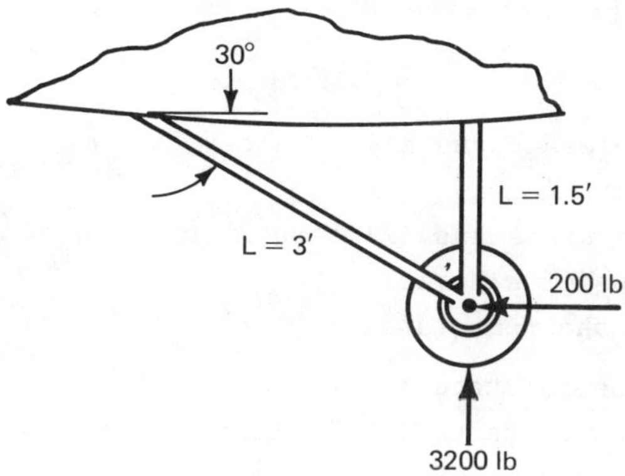

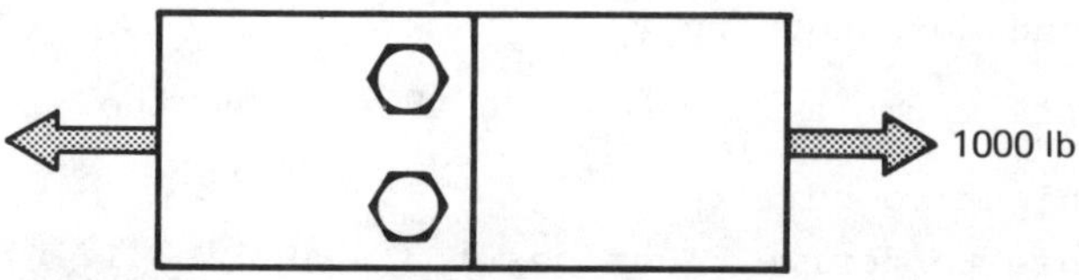

Problem 12. A 1/2 in. dia. bolt is in "double shear" as shown. Determine the shear stress at interfaces a-a and b-b. You will need to draw complete free-body diagrams.

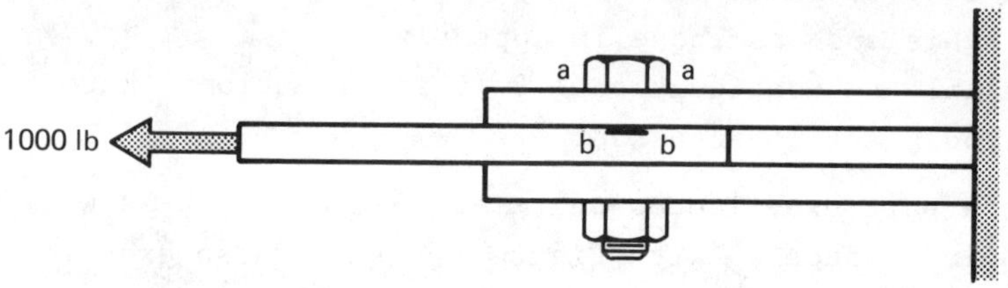

Problem 13. Write the assumptions and limitations for the formula for normal stress, Eq. 1.1, from memory. Use your own words.

Problem 14. Write and illustrate the definition of shear stress in your own words.

Problem 15. Write your own list of assumptions and limitations for the equation for shear stress, Eq. 1.2.

1.3 UNIAXIAL STRAIN, STRESS-STRAIN AND DEFLECTION

Objectives

1. Be able to state and illustrate your own definition of uniaxial strain.
2. Be able to state and illustrate your own definition of the modulus of elasticity.
3. Regarding the formula $\Delta L = \frac{PL}{AE}$:
   a. be able to state it from memory
   b. be able to state its limitations and assumptions from memory
   c. be able to recognize the situations where it does and does not apply
4. For the formula $\Delta L = \frac{wL^2}{2E}$, be able to do the a, b and c of objective number 3.
5. Be able to derive Equations 1.3.9 and 1.3.12. (Closed book.)
6. Be able to state and explain the principle of superposition. Be able to state and recognize the conditions under which it can be used.
7. Be able to find values of the modulus of elasticity in a table such as those in Appendix A.
8. Be able to use Equation 1.3.10 in situations where P, A and E are not constant.

When a part experiences uniaxial loading, such as was discussed in connection with Figure 1.2.1, it also experiences a uniaxial stretching, or deflection. All structural elements, and all structures, exhibit this behavior although many civil and mechanical structures are so stiff (rigid) that their deformations are too small to be discernable by human senses alone. In Figure 1.3.1, we depict the behavior of a uniaxial element. Tensile behavior is depicted; the basic concept for compression is the same.

The entire rod is stretched by amount ΔL. A differential length dx is stretched by amount $d\delta$. Strain is the elongation per unit length of the bar, and is defined as

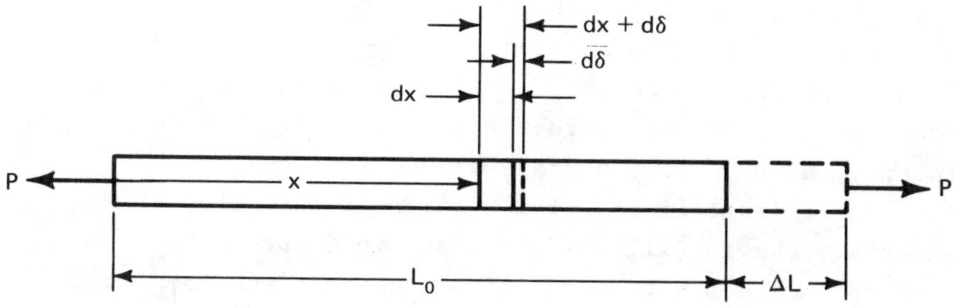

Figure 1.3.1

$$\varepsilon = \frac{\Delta L}{L_o} \qquad (1.3.1)$$

For the infinitesmal length dx, we can define the local strain as

$$\varepsilon = \frac{d\delta}{dx} \qquad (1.3.2)$$

The definition of strain as the elongation per unit length is the counterpart of the definition of stress as the force per unit area. One of the most important observations in stress analysis was that of Robert Hooke (1635-1703) who published in 1678 the results of a number of investigations. Hooke's law states that the deflection of a body is proportional to the force applied.

In our case, the example of Figure 1.3.1, we can state that

$$P = K\Delta L \qquad (1.3.3)$$

where K is a constant of proportionality. We can go a step further and state that stress is proportional to strain. This is because $\sigma = P/A$ and $\varepsilon = \Delta L/L_o$. In other words if we divide both sides of Equation 1.3.3 by constants, we will still have a linear equation.

The proportionality between stress and strain is written as

$$\sigma = E\varepsilon \quad \frac{lb}{in^2} \text{ or } \frac{N}{m^2} \qquad (1.3.4)$$

where E is a material constant known as the modulus of elasticity or Young's modulus.

Equation 1.3.4 describes a linear relation between stress and strain. For most structural materials, this relationship is only valid up to a certain value of stress known as the proportional limit. The details of this kind of material behavior will be discussed in Chapter 2.

We can now derive the equation for the elongation (or contraction) of a *uniform*, *elastic* rod which is loaded by a *concentrated force* at each *end*. The preceding italics summarize the assumptions and limitations which apply to the resulting formula.

Equation 1.3.2 can be written

$$d\delta = \varepsilon \, dx \tag{1.3.5}$$

Using Equation 1.3.4

$$d\delta = \frac{\sigma}{E} \, dx \tag{1.3.6}$$

Since $\sigma = P/A$ is constant throughout a uniform rod, we can write

$$d\delta = \frac{P}{AE} \, dx \tag{1.3.7}$$

Integrating over the total length of the rod

$$\int_{0}^{L} d\delta = \Delta L = \int_{0}^{L} \frac{P}{AE} \, dx \tag{1.3.8}$$

and

$$\Delta L = \frac{PL}{AE} \text{ in or m} \tag{1.3.9}$$

Example. A rod is suspended vertically as shown. If we neglect the weight of the rod, what is the deflection due to the 1000 lb weight on the end?

Material: Aluminum 7075-T6

$L = 6$ ft or 1.8288 m

$D = 1/4$ in or 6.35 mm

$E = 10 \times 10^6$ psi or 6.8948×10^{10} N/m^2

$P = 1000$ lb or 4.4482×10^3 N

Solution.

Calculate area:

$$A = (\pi/4)D^2 = \pi/4(.25)^2$$
$$= 4.9087 \times 10^{-2} \text{ in}^2$$
$$= \pi/4\ (6.35)^2 = 31.6692 \text{ mm}^2$$

Calculate elongation:

$$\Delta L = \frac{PL}{AE} = \frac{1000(6 \times 12)}{4.9087 \times 10^{-2} \times 10 \times 106} = \underline{\underline{0.14688 \text{ in}}}$$

$$\Delta L = \frac{PL}{AE} = \frac{4.4482 \times 10^3 \times 1.8288}{31.6692 \times 10^{-6} \times 6.8948 \times 10^{10}} = \underline{\underline{3.7256 \times 10^{-3}\text{m}}}$$

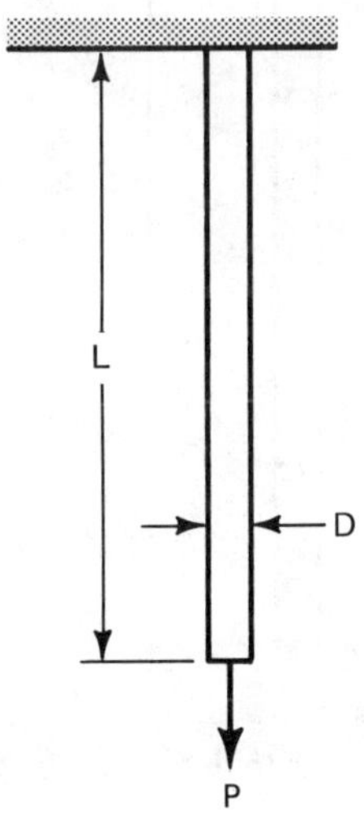

Figure 1.3.2

When we recall that one of the assumptions underlying Equation 1.3.9 is the assumption of a linear material, we are alerted to check for the possibility of the stress in the rod exceeding the yield stress of the material.

Checking stress: $\sigma = \frac{P}{A} = \frac{1,000}{4.9087 \times 10^{-2}} = 20,372 \text{ lb/in}^2$

Since the proportional limit of 7075-T6 aluminum is approximately 72000 psi (see Appendix A), we are indeed in the elastic range and our use of Equation 1.39 is correct.

Let us now consider the elongation of a bar which is suspended under its own weight.

L = length; in or m
A = area of cross section; in^2 or m^2
E = modulus of elasticity; psi or N/m^2
w = weight density; lb_f/in^3 or N/m^3

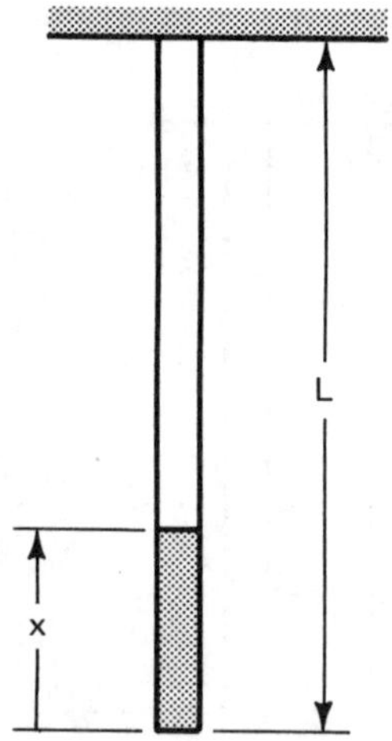

Figure 1.3.3

To solve this problem, we refer to the coordinate system of Figure 1.3.3. We also note that Equation 1.3.8 is not <u>necessarily</u> restricted to situations where P, A and E are constants with respect to x. If the derivation of Equation

1.3.8 is reviewed at this time, we will note that we can relax the restrictions for the uniform rod with concentrated loads on the end to allow:

P varying continuously as a function of x,
A varying gradually and continuously as a function of x,
E varying continuously as a function of x, and still end up with Equation 1.3.8 in the form

$$\Delta L = \int_0^L \frac{P(x)}{E(x)\,A(x)}\,dx \tag{1.3.10}$$

In our present problem, we have constant A and E and the <u>internal</u> <u>force</u> at any variable position x is just the weight of the portion of the rod being supported, i.e., the shaded portion shown in Figure 1.3.3.

Thus we have

$$P(x) = wAx \tag{1.3.11}$$

and the use of Equation 1.3.11 in Equation 1.3.10 gives

$$\Delta L = \int_0^L \frac{wAx}{AE}\,dx = \int_0^L \frac{wxdx}{E} = \frac{wL^2}{2E} \tag{1.3.12}$$

This equation shows, interestingly enough, that the answer is independent of area.

In the result of Equation 1.3.12, we are still restricted to linear problems, i.e., $\sigma = E\varepsilon$ is still restricted to the linear range.

A useful and simple technique for the solution of <u>linear</u> problems is the technique of <u>superposition</u>. This technique is used widely throughout stress analysis and basically it says that you can add up all the separate effects to get the total result.

To illustrate this, consider Figure 1.3.4. In this figure we consider a uniform rod which is loaded on the ends by concentrated forces P_1 and P_2. We neglect the deflection of the rod due to its own weight. Furthermore, we state that the sum of $P_1 + P_2 = P_T$. Now we note that the final deflection of the rod is the same whether we consider P_1 to be applied first and then P_2; P_2 applied first and then P_1; or the final deflection being just due to P_T. This is illustrated below.

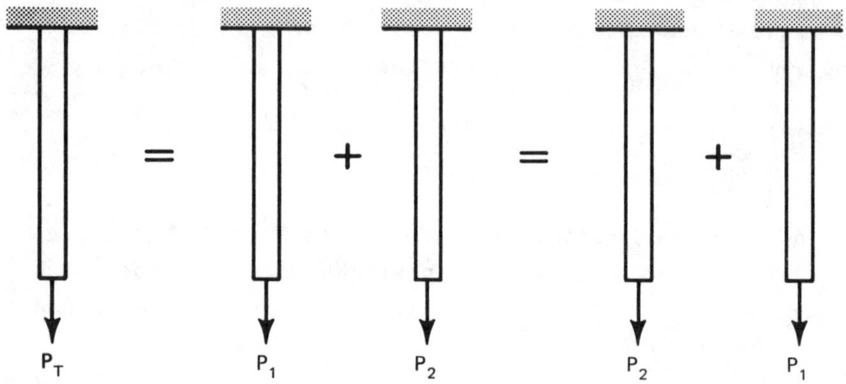

$$\Delta L_T = \Delta L_1 + \Delta L_2 = \Delta L_2 + \Delta L_1$$

$$\Delta L_T = \frac{P_T L}{AE} = \frac{P_1 L}{AE} + \frac{P_2 L}{AE} = \frac{(P_1 + P_2)L}{AE} = \frac{P_T L}{AE}$$

$$= \frac{P_2 L}{AE} + \frac{P_1 L}{AE} = \frac{(P_2 + P_1)L}{AE} = \frac{P_T L}{AE}$$

Figure 1.3.4

In Figure 1.3.5, we observe the graphical interpretation of the principle of superposition and note further the fact that separate parts can be added in any order to make the whole, provided the force-deflection equations are linear.

Note that the slope of the line is the constant in the equation $P = K\Delta L$ and as such, this slope is the measure of the stiffness of the rod. To further illustrate the principle of superposition, consider the uniform rod of Figure 1.3.6, which is loaded by a concentrated load and also its own weight.

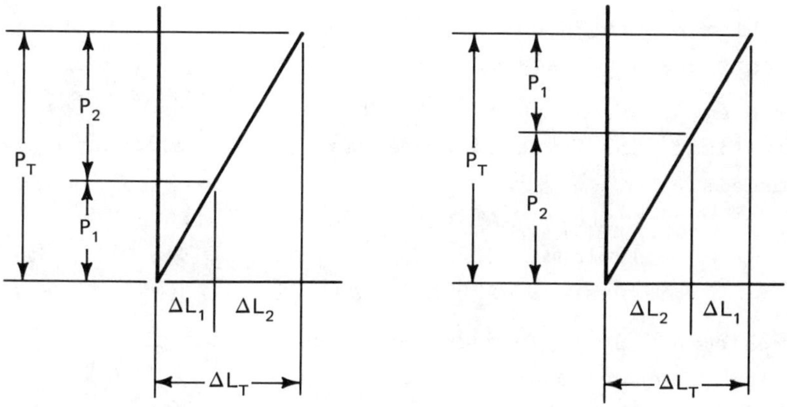

Figure I.3.5

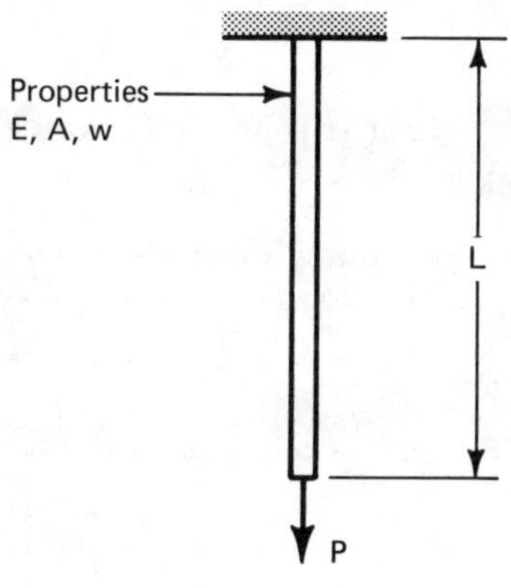

Figure I.3.6

By <u>superimposing</u> Equations 1.3.9 and 1.3.12, we can obtain the total deflection of this rod as

$$\Delta L = \frac{PL}{AE} + \frac{wL^2}{2E}$$

Problems

Problem 1. Derive Equation 1.3.12 from the beginning. Explain what you are doing in each step. Use a derivation similar to the derivation of Equation 1.3.9.

Problem 2. For problem 9, page 18, calculate the extension or contraction of each of the truss members.

Problem 3. From Appendix A, find the modulus of elasticity and estimate the limit of linearity for the following materials:

Steel:	AISI 1025; AISI 4130
Aluminum:	2024-T3; 7075-T6
Magnesium:	AZ61A

Problem 4. State in your own words the definition of strain. Use your own illustrations.

Problem 5. State and illustrate your own definition of the modulus of elasticity.

Problem 6. List the limitations and assumptions which apply to Equation 1.3.9.

Problem 7. List the limitations and assumptions which apply to Equation 1.3.12.

Problem 8. The correct equation for the deflection of the rod due to its own weight is:

a. $\delta = \dfrac{wL}{AE}$

b. $\delta = \dfrac{wL}{2AE}$

c. $\delta = \dfrac{2wL}{AE}$

d. δ = None of these

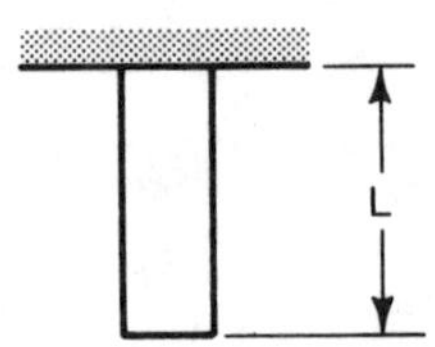

Problem 9. Find ΔL due to its own weight and the load P.

Weight = 500 lb.

Area = 0.1 in^2

$E = 10^6$ psi

L = 10 ft

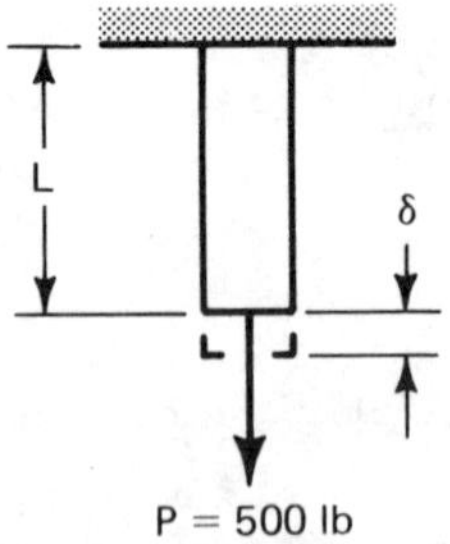

Problem 10. Write the algebraic equation to solve for the total deflection of the rod in terms of A_1, A_2, E_1, E_2, P_1, P_2 and L. Neglect the weight of the rod.

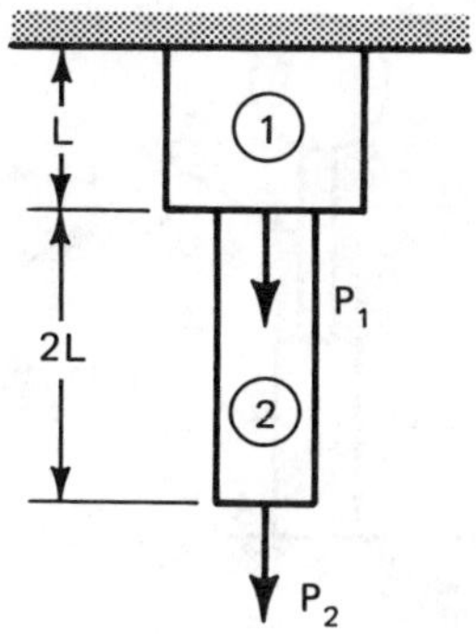

Problem 11. Write the equation for the deflection of the rod due to P. Assume symbolic values for any variables you need. (Such as E.)

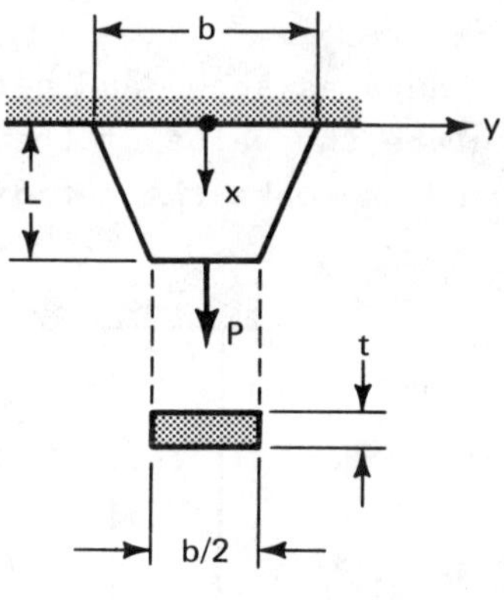

Problem 12. A water tower is constructed as shown. It consists of a heavy sphere on top of two pipe columns. Show the following in symbolic form:

a. The stress in each column as a function of length along the column.

b. The net shortening of the columns as assembled compared to their free length. Include the weight of the structure.

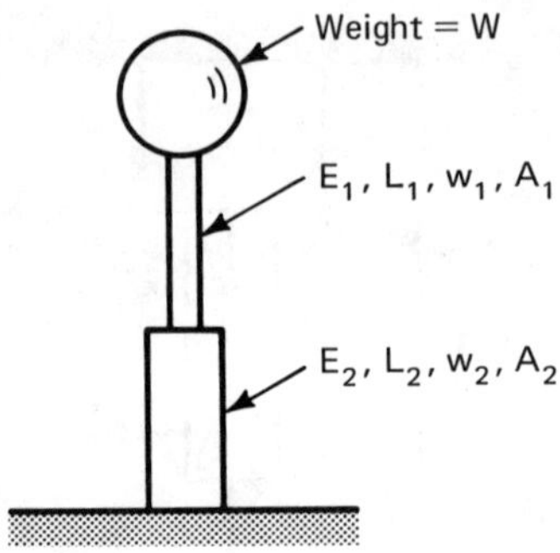

Problem 13.

<u>Section 1</u>:	E_1	^ psi	L_1	^ in
	w_1	^ lb/in^3	A_1	^ in^2
<u>Section 2</u>.	E_2	^ psi	L_2	^ in
	w_2	^ lb/in^3	A_2	^ in^2

A composite bar hangs as shown and deflects under its own weight. Calculate the value of the load P, which needs to be applied upward, to make the net displacement ΔL = ZERO.

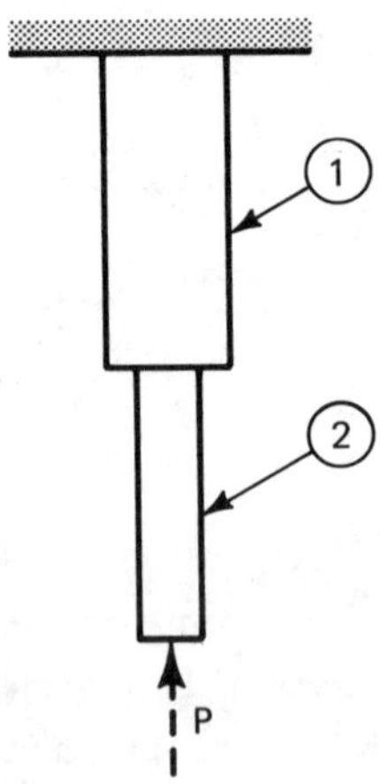

Problem 14.

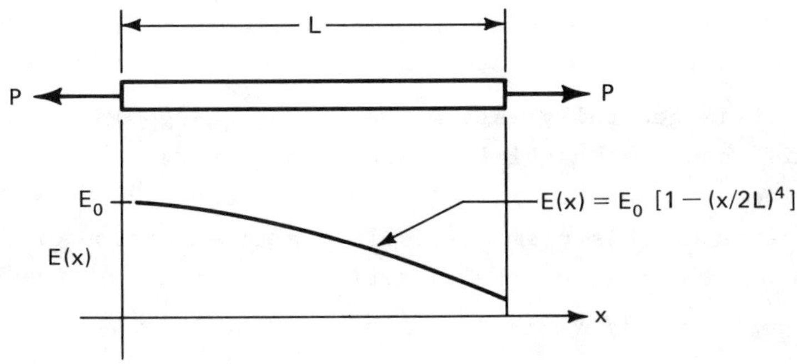

A bar of uniform cross section has area = A in^2. It is loaded by constant load P, and is of length L. The modulus of elasticity, E, varies with length as shown. Calculate the elongation.

1.4 THERMAL STRAIN, POISSON'S RATIO, STRAIN ENERGY, STIFFNESS AND FLEXIBILITY MATRIX NOTATIONS

Objectives

1. Be able to state and use Equation 1.4.1 from memory.
2. Be able to apply the logical steps of the example of Figure 1.4.2 to solve similar problems.
3. Be able to include the thermal deflection equations in the general method of superposition.
4. Be able to write and illustrate your own definition of Poisson's ratio from memory.
5. Be able to look up values for coefficient of thermal expansion and Poisson's ratio in tables.
6. Be able to calculate the lateral deflections accompanying longitudinal deflections.
7. Be able to state and illustrate your own definition of elastic strain energy.
8. Be able to state and use Castigliano's theorem to calculate structural deflections. Be aware of the appropriate assumptions, limitations and application of the theorem.

9. Be able to state Equation 1.4.11 from memory. Be able to use it to calculate uniaxial rod element stiffness matrices.

It is generally well known that a change in temperature will cause a material to expand or contract. The rod element is a uniaxial element. The effects of temperature change on this element results in an expansion or contraction along the axis of the element. This is illustrated in Figure 1.4.1, where an increase in temperature is shown.

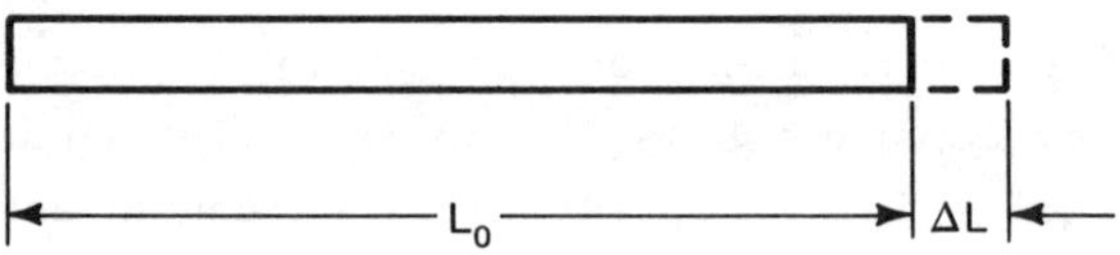

Figure 1.4.1

We then have

$$\Delta L = \alpha \Delta T L_o \tag{1.4.1}$$

Where:

ΔL = the change in length; in or m

α = the coefficient of thermal expansion; strain/°F or strain/°C (almost always a positive quantity; i.e., temperature increase means expansion.)

ΔT = temperature change; °F or °C (temperature increase is positive).

L_o = original length; in or m.

Notice that when we have thermal strains and the structure or structural element is unconstrained, we have a situation of strain without stress. If the structure is constrained, then a change in temperature will cause stress without strain. Consider the following examples:

Example: A rod is constrained against motion of the ends as shown. Consider Lo = 8 ft; α = 6.3 10^{-6} strain/°F let the temperature decrease by amount $\Delta T = -200°F$. The rod is steel and has a cross sectional area of 0.1 in^2. What will the stress in the rod be?

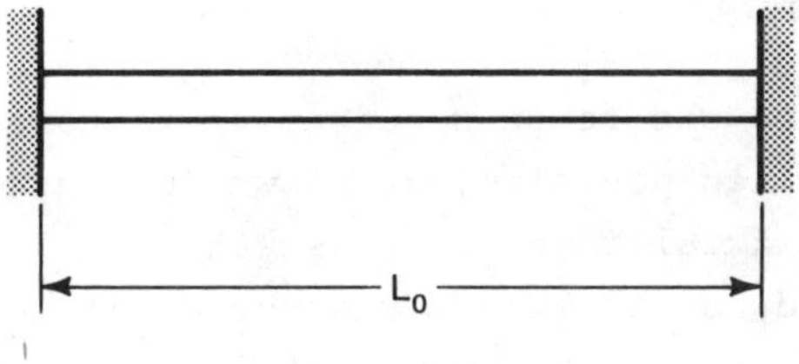

Figure 1.4.2(a)

The solution to this problem illustrates an important technique for the solution of structural problems of this type. The procedure is:

a. Relax (remove) the constraint and let the structure expand or contract freely.
b. Calculate the force necessary to restore the structure to the constrained position.

In accordance with this procedure, we remove (for example) the right constraint of our structure and allow it to contract as shown in Figure 1.4.2(b).

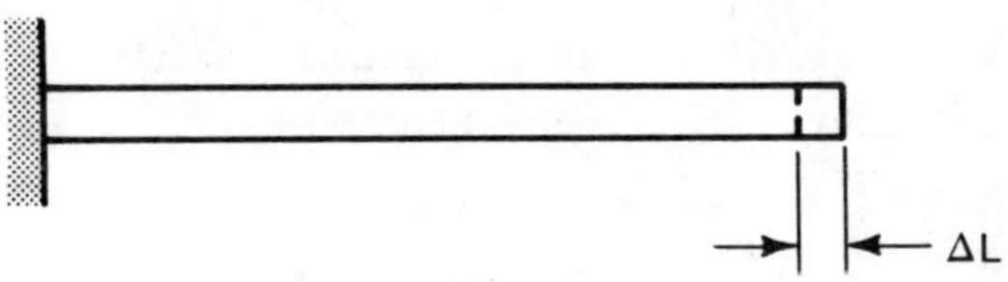

Figure 1.4.2(b)

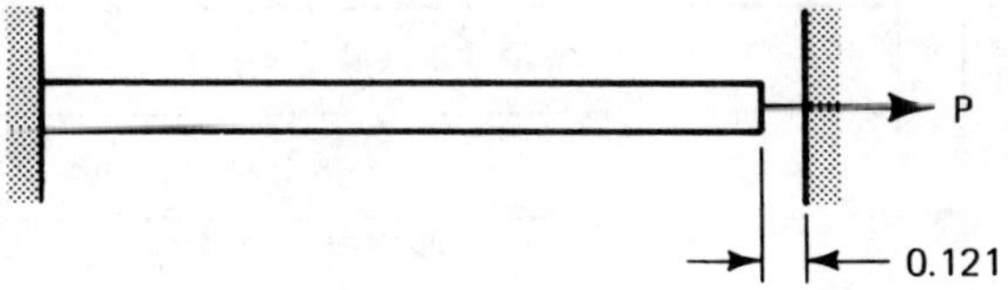

Figure 1.4.2(c)

Then from Equation 1.4.1

$$\Delta L = \alpha \Delta T\ L_o$$
$$= 6.3 \times 10^{-6} \times (-200) \times 8 \times 12$$
$$= -0.121 \text{ in}$$

Next calculate the force required to restore the structure to the constrained position, as shown in Figure 1.4.2(c). We need enough axial force to close the fictitious gap we have created, or to extend the rod 0.121 in. We use Equation 1.3.9.

$$\Delta L = \frac{PL}{AE}$$

$$P = \frac{\Delta L\ AE}{L} = \frac{0.121 \times 0.1 \times 29 \times 10^6}{8 \times 12}$$

$P = 3655$ lb (tension required to close the gap)

and the stress in the rod due to the temperature change is, from Equation 1.1

$$\sigma = \frac{P}{A} = \frac{3655}{0.1} = 36{,}550 \text{ psi} \quad \text{(tension)}$$

In a uniaxial member, longitudinal strain is accompanied by lateral strain. This is illustrated in Figure 1.4.3.

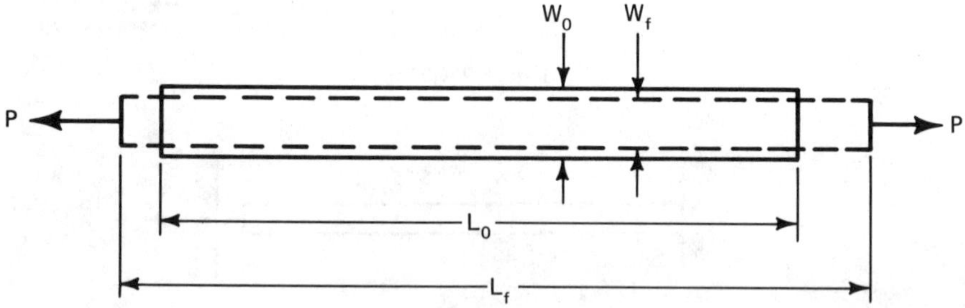

Figure 1.4.3

We see that the change in length $\Delta L = L_f - L_o$ is accompanied by a change in width $\Delta W = W_f - W_o$. Usually an extension of the rod is accompanied by a reduction in width. The longitudinal strain, as previously defined, is

$$\varepsilon_{long} = \frac{\Delta L}{L_o} \tag{1.4.1}$$

The lateral strain has a similar definition

$$\varepsilon_{lat} = \frac{\Delta W}{W_o} \tag{1.4.2}$$

Then Poisson's ratio is defined as

$$\nu = \left| \frac{\varepsilon_{lat}}{\varepsilon_{long}} \right| \tag{1.4.3}$$

Another important structural concept which is conveniently introduced by considering a uniaxial rod is the concept of elastic strain energy. The force-deflection curve for a uniaxial rod is shown in Figure 1.4.4.

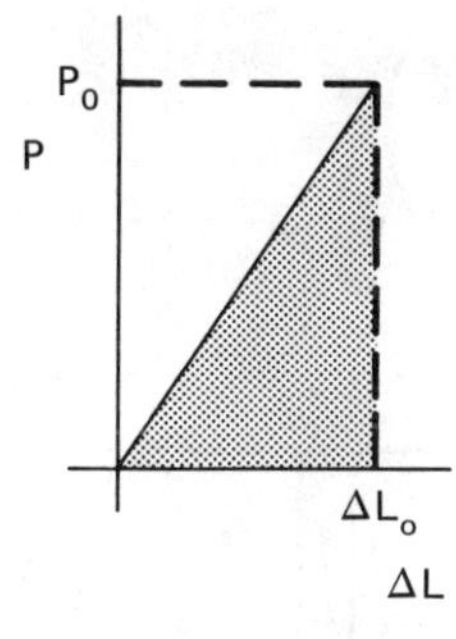

Figure 1.4.4

The <u>elastic strain energy</u> required to reach state P_o; Δ_{L_o} is defined as the shaded portion under the P vs. ΔL curve to that point.

$$U = 1/2\ P_o\ \Delta L_o \tag{1.4.4}$$

since $\Delta L_o = P_o L/AE$, we can write

$$U = 1/2 \frac{P_o^2 L}{AE} \qquad (1.4.5)$$

An interesting application of Equation 1.4.5 is Castigliano's theorem. Castigliano's theorem says that the deflection of a structure at the point where the load is applied and in the sense of the applied load is equal to the first partial derivative of the strain energy with respect to the load.

$$\Delta L = \frac{\partial U}{\partial P_o} \qquad (1.4.6)$$

(By sense we mean that the deflection will be linear, because we have a linear force and that it will be in the direction of the applied load.)

As an illustration, consider the rod loaded axially as in Figure 1.4.5. We neglect the weight of the rod. Castigliano's theorem can be used to calculate ΔL. Of course, we have solved this problem already but this illustrates a new method.

$$\Delta L = \frac{\partial U}{\partial P_o} = \frac{\partial}{\partial P_o}\left(\frac{1}{2}\frac{P_o^2 L}{AE}\right)$$

$$\Delta L = \frac{P_o L}{AE}$$

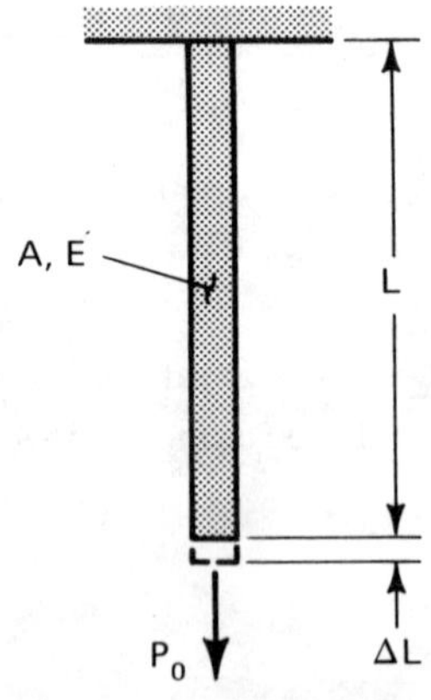

Figure 1.4.5

As previously mentioned, in the linear equation

$$P = K\Delta L \tag{1.4.7}$$

the term K(lb/in or n/m) is defined as the <u>stiffness</u> of the element. This equation can be written as

$$\Delta L = (1/K)P$$
$$\Delta L = FP \tag{1.4.8}$$

where: F = the flexibility of the element.

In preparation for future work in the finite element area (Chapter 6), let us state the basic equation for the stiffness of a rod element in matrix form. We need to consider a slightly different situation for the rod element than we have before. The two end forces and the two end displacements of the element must be related. This is shown in Figure 1.4.6.

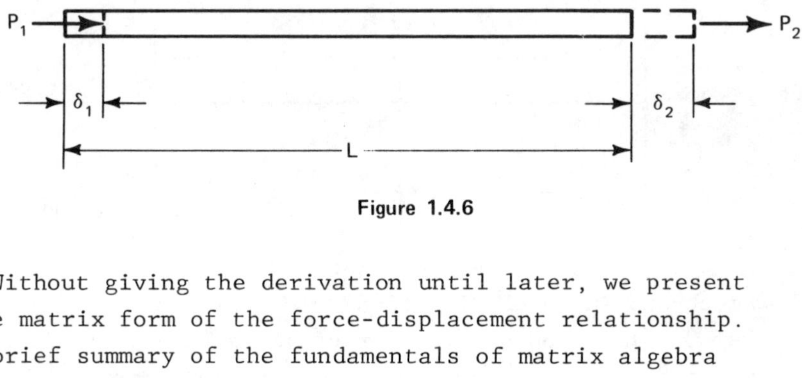

Figure 1.4.6

Without giving the derivation until later, we present the matrix form of the force-displacement relationship. A brief summary of the fundamentals of matrix algebra is given in Appendix B.

As

$$\begin{Bmatrix} P_1 \\ P_2 \end{Bmatrix} = \frac{AE}{L}\begin{bmatrix} 1 & -1 \\ -1 & 1 \end{bmatrix}\begin{Bmatrix} \delta_1 \\ \delta_2 \end{Bmatrix} \tag{1.4.9}$$

or, in condensed form

$$\{P\} = [K]\{\delta\} \tag{1.4.10}$$

The matrix $[K] = \frac{AE}{L}\begin{bmatrix} 1 & -1 \\ -1 & 1 \end{bmatrix}$ is the element (1.4.11)

<u>stiffness matrix</u> for the uniaxial rod element.

Problems

Problem 1. For problem 12, page 29, if the coefficient of thermal expansion of both columns is given by α, calculate (symbolically) the temperature rise which would restore the structure to its original free length.

Problem 2. The structure shown is made of two abutting uniaxial rods of cross sectional area 0.25 in^2. Rod 1 is made of steel and rod 2 is made of aluminum. The temperature is increased by 100°F. (See Appendix A for properties.) What is the stress in each rod?

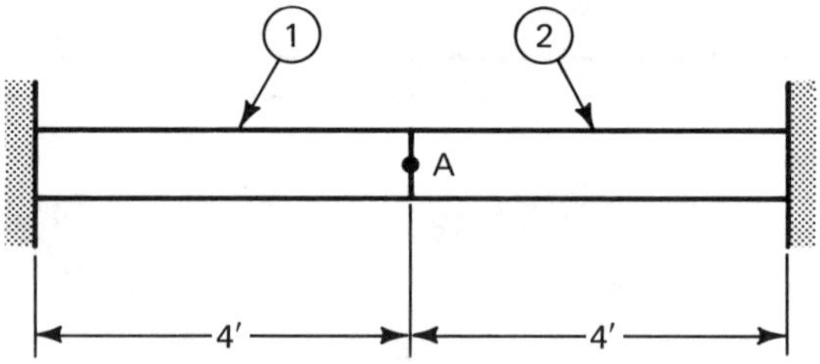

Will the point at the interface, labeled A, move? If so, by how much and in which direction?

Problem 3. If a rod of circular cross section is loaded as shown in the figure, what is the change in diameter of the rod? Neglect the weight of the rod.

Material: Aluminum 7075-T6

$L = 10$ ft.

$P = 1000$ lb

$D = 0.50$ in

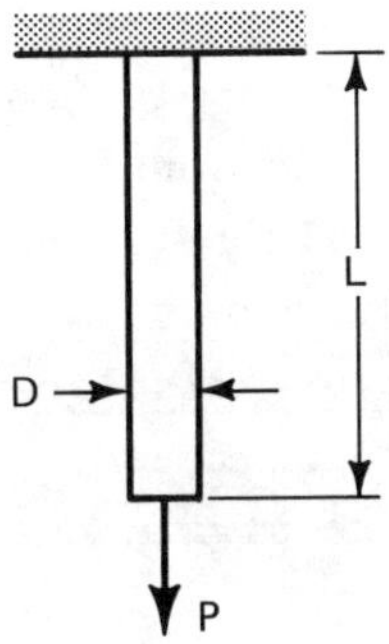

Problem 4. Castigliano's theorem may be extended to the calculation of the vertical deflection of point A of the truss shown.

	Member 1	Member 2
Area	0.1 in^2	0.15 in^2
Material	Al;7075-T6	Steel;AISI 4130

$L_1 = 4'$

$L_2 = 3'$

$P = 500$ lb

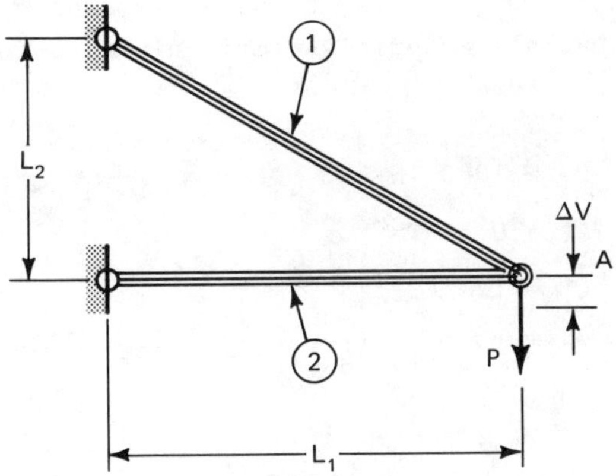

We can write the strain energy of the assembled structure as the sum of the strain energy of the two individual elements.

$$U_T = U_1 + U_2 = \frac{1}{2}\frac{f_1^2 L_1}{A_1 E_1} + \frac{1}{2}\frac{f_2^2 L_2}{A_2 E_2}$$

Where f_1 and f_2 are the internal forces in member one and two respectively, and are found from statics. Now f_1 and f_2 are functions of P and can be found from statics. (We must do all this symbolically first and then put in the numbers.)

The <u>structure displacement</u>, the vertical displacement at point A, can be found by taking the partial derivative of the structure strain energy with respect to the applied load P.

$$v = \frac{\partial u_T}{\partial p} = \frac{\partial u_1}{\partial p} + \frac{\partial u_2}{\partial p}$$

Use this method to calculate Δv for this structure.

Problem 5. Write the element stiffness matrix for each of the elements in problem 4.

Problem 6. Calculate the total vertical deflection of the endpoint of the structure. Include the weight of the structure and the temperature change. Is the assumption that all materials remain elastic violated anywhere in the structure?

P_1 = 1000 lb

A_1 = 0.50 in^2

L_1 = 12 ft.

Material $_1$ = AISI 1025 steel

ΔT_1 = 175°F

P_2 = 500 lb

A_1 = 0.25 in^2

L_2 = 10 ft

Material$_2$ = 2024-T3 Aluminum

ΔT_2 = 100°F

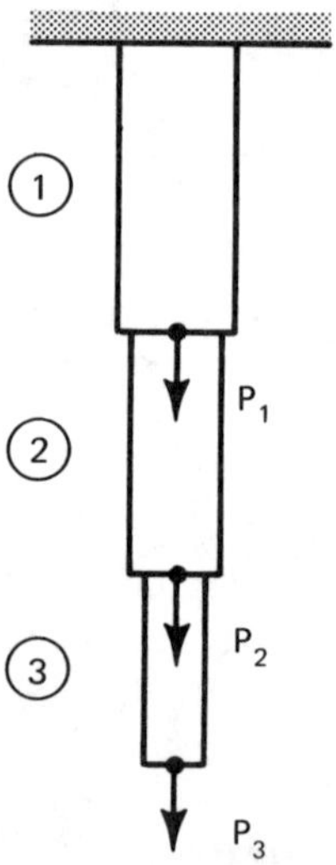

$P_3 = 200$ lb

$A_3 = 0.125$ in^2

$L_3 = 8$ ft.

Material_3 = MIA magnesium

$\Delta T_3 = 50°F$

Problem 7. Calculate the strain energy stored in each of the members of problem 5.

Chapter 2

BASIC PROPERTIES OF MATERIALS BASED ON THE UNIAXIAL STRESS-STRAIN CURVE

2.1 INTRODUCTION

One of the most important factors in determining whether or not a given structure is safe lies in how much stress a given material will take before failure. The concept of failure is actually a complicated one and will be discussed more fully in Chapter 5.

Generally, materials are considered to have failed when: 1) they reach their yield strength and thus suffer permanent elongation, or 2) they reach their ultimate strength and thus suffer fracture.

A great deal can be learned about properties of materials from a study of the stress-strain properties of a uniaxial tension (or compression) test specimen. This is another reason for beginning our study of structures with the uniaxial rod element. Most of the time, the stress analyst will find himself making decisions about materials based on information from the uniaxial test data. This is because this particular test is relatively easy to run and the resulting data are widely published.

The uniaxial test leads to a stress-strain curve which in turn permits the identification of several interesting properties:

a. The yield strength
b. The ultimate strength
c. The modulus of elasticity
d. The ductility

We must be sure that we clearly understand the meaning of these terms and their application.

One difficult skill, but an essential one, is how to find needed material properties. We shall discuss this to a degree.

It is also important to understand what variables (size, temperature, etc.) affect the material properties and under what circumstances. This will be discussed briefly.

The uniaxial tension test and the properties resulting from it are simple, yet they form the basic essential background on which nearly all further work on material properties and material failure is based.

The selection of the "best" material for a given structure is a complicated process, and depends on many factors such as cost and availability. One thing remains basic: the structure must carry the load without failing.

2.2 OBJECTIVES FOR PROPERTIES OF MATERIALS

1. The student will be able to write and illustrate a definition of the following:
   a. Ultimate tensile stress
   b. Proportional limit
   c. Yield stress
   d. Modulus of elasticity

   e. Resilience
   f. Toughness
   g. Ductility
   h. Tangent modulus
   i. Secant modulus

2. The student will be able to sketch a typical stress-strain curve and label the nine properties mentioned in part 1.
3. When given the designation of a structural material, the student will be able to look up the appropriate properties in part 1 in a table. The tables in the text will be used. Successful performance includes finding the right table, material condition, number and units.
4. Be able to sketch typical stress-strain curves for materials such as: steel, aluminum, magnesium.
5. Be able to define, discuss or recognize ductile vs. brittle behavior.
6. Be able to state or recognize the effects of alloying and heat treating on: E, hardness, S_{tu}, S_{ty}, ductility.
7. Be able to state from memory nominal values of modulus of elasticity and Poisson's ratio for steel and aluminum.
8. Be able to explain in words and with a diagram the difference between elastic and inelastic behavior of a material.
9. Be able to discuss the effects of elevated temperature (short-time) on the modulus of elasticity, yield point, ultimate tensile strength and ductility of a material.
10. Presented with curves such as Figures 2.5.1 to 2.5.5, be able to determine the effect of temperature on the material.
11. Be able to explain and illustrate the general effect of lowered temperature on the tensile strength of a material.
12. Be able to explain and show by a sketch of stress-strain curves the difference between materials which show a definite yield point and those which do not. Be able to explain and illustrate how the yield point is

defined for the latter material. Be able to find the yield point or yield stress by the offset method.

13. Be able to define and distinguish the difference between the proportional limit, the elastic limit and the yield point or yield stress.

14. Be able to state the difference between stress and strength.

2.3 THE STRESS-STRAIN CURVE AND THE BASIC PROPERTIES

The stress analyst basically has the task of: a) calculating the stresses in a structure or part and b) determining if the structure or part is adequate. The designer basically has the task of utilizing his knowledge of stress analysis, material properties and manufacturing techniques to synthesize (create) a structure to perform a given task. In either case, a knowledge of the properties of materials is essential to these engineers.

We begin with a study of the uniaxial stress-strain properties for two reasons:

1. These are the most readily available of all the properties.
2. The understanding of these properties forms the basis for advanced work.

Usually these properties are found by taking a standard tensile specimen and testing it in a standard way. See, for example, the ASTM Standard E8-57T which is shown in part in Appendix C. The specimen is pulled in tension in a large hydraulic or mechanical machine such as shown in Figure 2.3.1. The load and elongation may be recorded by hand and the resulting stress and strain calculated, or, in more modern equipment, the stress-strain curve may be automatically recorded. In any event, the result is the production of a stress-strain curve.

Two stress-strain curves are shown in Figures 2.3.2 and 2.3.3. It is important to distinguish between the two

Figure 2.3.1 A typical tensile-test machine set up

characteristic shapes of the curves. The curve of Figure 2.3.2 is typical of the curves for steels and is characterized by the feature labeled "A." This feature shows a pronounced hooking and flattening of the curve. This is known as the yield point, yield stress or yield strength of the material. The term <u>yield point</u> is reserved for materials which show this clearly defined "point" when a

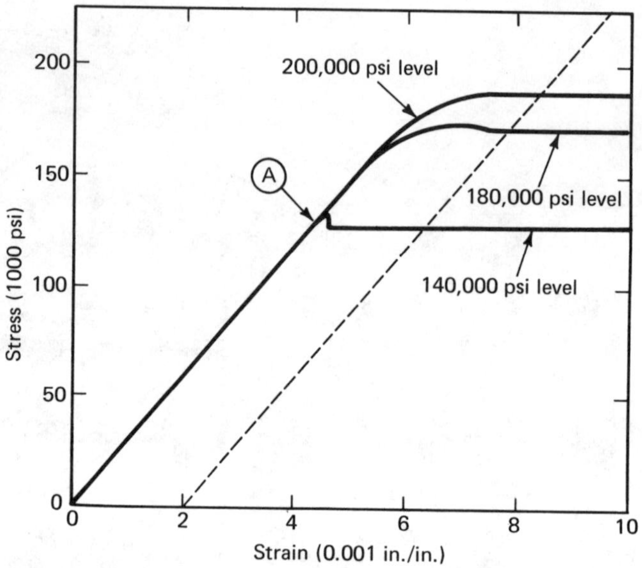

Figure 2.3.2 Typical tensile stress-strain curves at room temperature for heat-treated AISI 4340 alloy steel

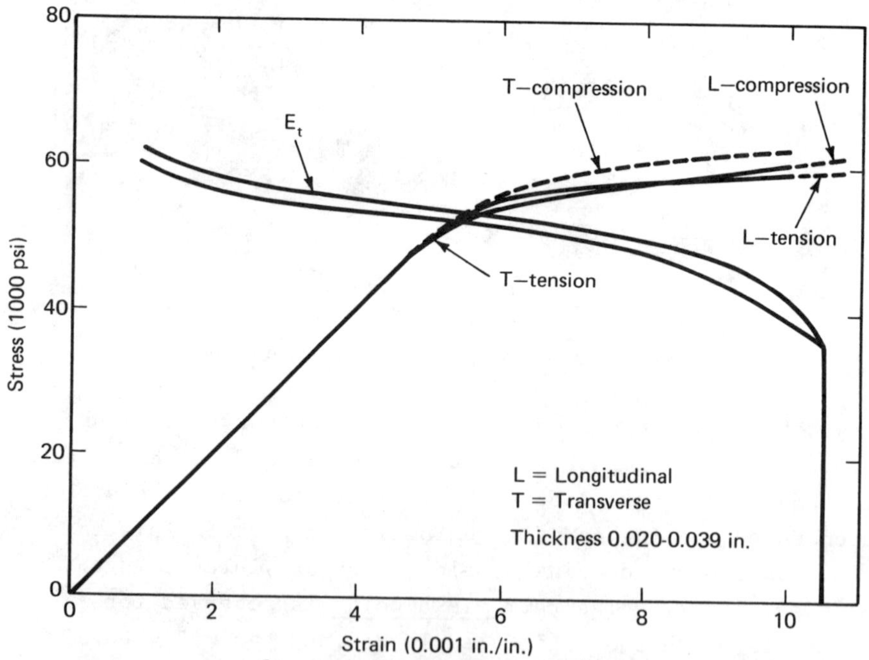

Figure 2.3.3 Typical stress-strain and tangent-modulus curves for Clad 2014-T6 aluminum alloy (sheet) at room temperature. Ref. (1)

dramatic extension of the specimen is taking place with little or no increase in load. We can also see that for steels, this is the point where the linear relation between stress and strain ends. (For all practical purposes.)

In contrast to this behavior, we see the curve of Figure 2.3.3. In this curve we have a gradual change of slope and there is no clear point where the material yields or the linearity ends. This curve is typical of many non-ferrous metals--especially Aluminum.

The concept of a yield point is quite useful. In steels, it provides a demarcation between linear and non-linear behavior. The yield point concept also provides one clear definition of failure. For these reasons, it has been agreed that a yield stress will be <u>defined</u> for materials which do not have a true yield point. The way this is done for Aluminum is shown in Figure 2.3.4. (This figure is not drawn to scale.)

We can also note from Figure 2.3.3 that there are some minor differences in the tensile and compressive properties of aluminum. There are also slight differences depending on whether or not the material is used with the load axis

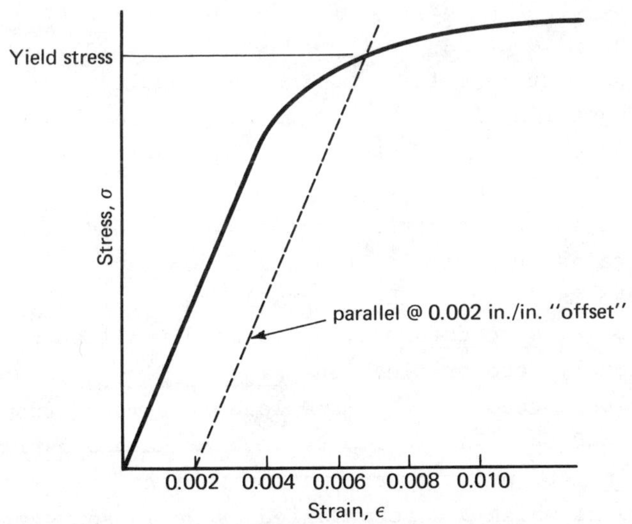

Figure 2.3.4 The offset method for defining the yield strength of Aluminum

oriented parallel (longitudinal) or transverse to the grain axis of the metal. (Rolled Aluminum has pronounced elongation of the metal grains in the direction of rolling.)

Some materials, such as concrete, will have markedly different properties in tension as compared to compression.

The maximum height obtained by the stress-strain curve in tension is called the ultimate tensile strength. This is the stress at which a specimen of the material will fracture in static tension.

Note that steel and aluminum both have a range where the stress-strain curve is a straight line. The material does satisfy the equation $\sigma = E\varepsilon$, and the modulus of elasticity E is the slope of the line. The end of the linear portion of the stress-strain curve is called the proportional limit.

Some materials do not have this initial linear portion of the stress-strain curve. For example, typical stress-strain curves for cast magnesium alloy are shown in Figure 2.3.5.

When the stress-strain curve is not linear, several local moduli are defined. The two principal ones are the tangent modulus and the secant modulus. These are illustrated in Figure 2.3.6. At a particular point on the stress-strain curve such as point "A" in Figure 2.3.6, the local tangent line to the curve provides the tangent modulus, E_t; the slope of a straight line from the origin to point A provides the secant modulus, E_s. A plot of tangent modulus is included in the curves of Figure 2.3.3.

In the range of the stress-strain curves below the yield point, strains are small. They are also recoverable, i.e., if the load is removed, the part will return to its original length. By elastic deformation, we mean small strains which are completely recoverable. The elastic limit is the limit of dimensional recovery when the load is removed from the part. Beyond the elastic limit, we will have plastic deformation. By plastic deformation we mean relatively large strains which are accompanied by permanent set. When the load is removed, the part will not recover its original

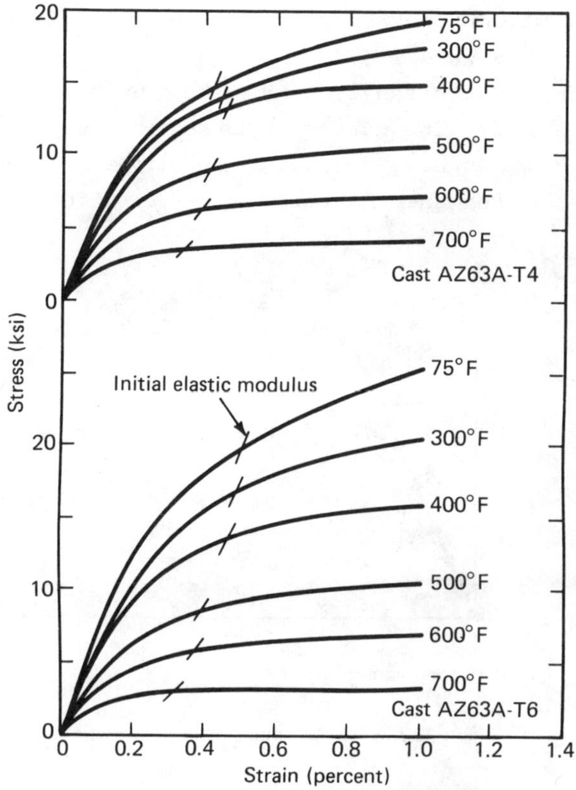

Figure 2.3.5 Tension stress-strain relationships for cast AZ63A-T6 and AZ63A-T4. Ref. 6

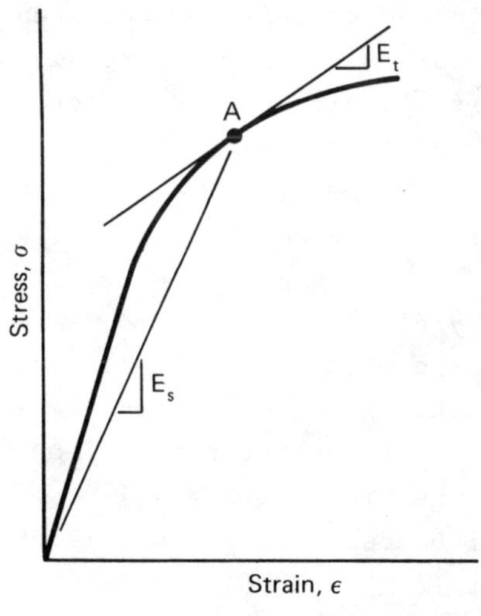

Figure 2.3.6 The tangent and secant modulus

length, but will be permanently stretched. This is illustrated in Figure 2.3.7.

Within the elastic range, loading and unloading occur along line 0-A-0. The part returns to its original length. In the plastic range, the loading and unloading occurs along line 0-A-B-C. Permanent strain 0-C is left in the part.

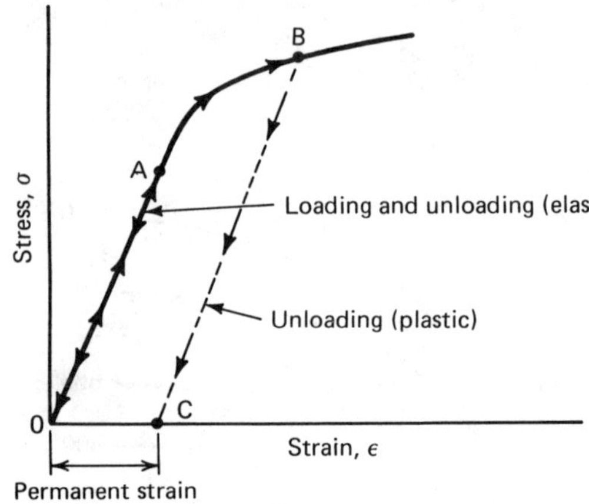

Figure 2.3.7 Elastic and plastic deformation

The question of whether or not a part experiences permanent strain is quite significant. In most structural applications a part which has yielded or suffered permanent strain is considered to have failed. Permanent set is not allowed in: aircraft, automobiles, civil structures, etc. We have actually defined three terms which have distinct definitions, namely:

1. Proportional limit
2. Elastic limit
3. Yield point or yield stress

Although these three terms are different by formal definition, for practical purposes they are assumed to occur at the yield strength of the material. Thus, of the three above values, only the yield strength is tabulated. The yield strength will be reported as the yield point if the material has a definite yield point or as an arbitrarily defined yield stress otherwise.

It is appropriate at this point to clarify the use of the terms stress and strength. In much of the literature today, these two terms have specific meanings as follows:

1. Stress is the force per unit area that the part experiences due to its shape and the loads applied to it. Stress is not a function of the material from which the part is made; only load and shape terms appear in stress formulas; i.e., $\sigma = P/A$.
2. Strength is the amount of stress which a particular material will take in some specified fashion before failure occurs; i.e., yield strength. Strength is a material property. In some literature, however, strength values may be given names such as "yield stress."

Figures 2.3.8, 2.3.9, and 2.3.10 are included to represent typical data which might appear in the literature. Figure 2.3.8 shows a comparison of several different materials, ranging from magnesium to steel, plotted on the same scale. Figure 2.3.9 shows the affects of alloying and heat treating on steel. This figure is not comprehensive, but it does provide comparison. Figure 2.3.10 shows the scatter bands which may develop in the various tension tests. The stress analyst is well advised to determine whether the values he is using represent minimum, average or some other value for the property in question.

The stress-strain curves also provide a measure of another important material property-ductility. Ductility is the ability of the material to elongate in the plastic range. Ductility is usually reported as a percentage elongation at failure. The failed tensile specimen is put back together and the permanent set of the original gage length is measured. Then the ductility is

$$D = \frac{L_f - L_o}{L_o} \tag{2.3.1}$$

The numerical value for ductility will depend on the length of the specimen which was used in the test. Therefore the gage length should be reported when ductility values

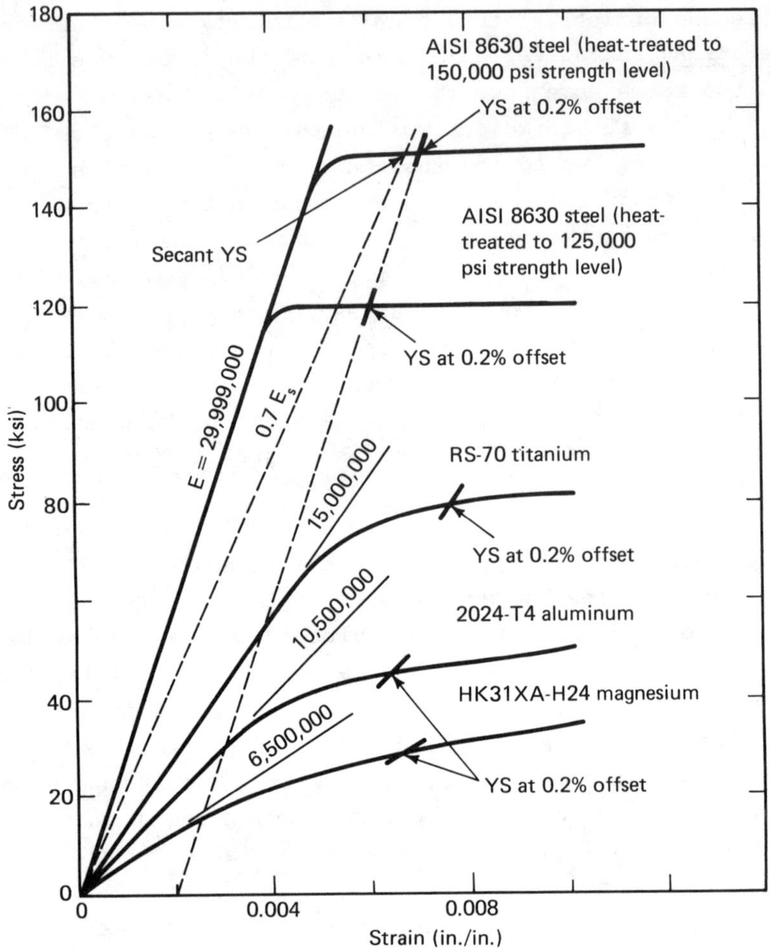

Figure 2.3.8 Comparison of stress-strain curves for several common materials. Ref. (3)

are given. Figure 2.3.11 shows a comparison between a low-strength, ductile material and a high-strength, brittle material. The general affect of alloying and heat treating materials is to increase yield and tensile strength and to decrease ductility. An increase in tensile strength is accompanied by an increase in hardness. Hardness is a measure of the resistance of a material to penetration.

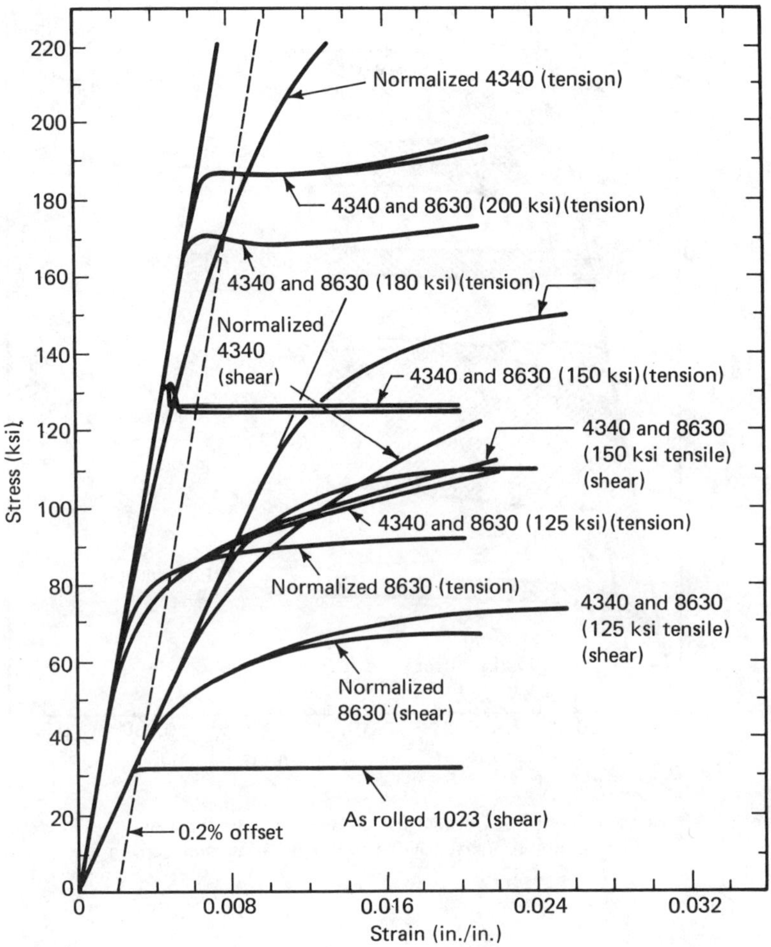

Figure 2.3.9 Stress-strain curves in tension and shear for AISI 4340 and AISI 8630 steels heat-treated to 125, 150, 180, and 200 ksi tensile strength and normalized. Ref. (3)

The more plastic strain to failure, i.e., the more the stress-strain curve goes to the right before failure, the more ductile the material will be.

Two terms which relate to the energy absorbing ability of the material are the <u>modulus of resilience</u> and the <u>modulus of toughness</u>. The modulus of resilience is the strain

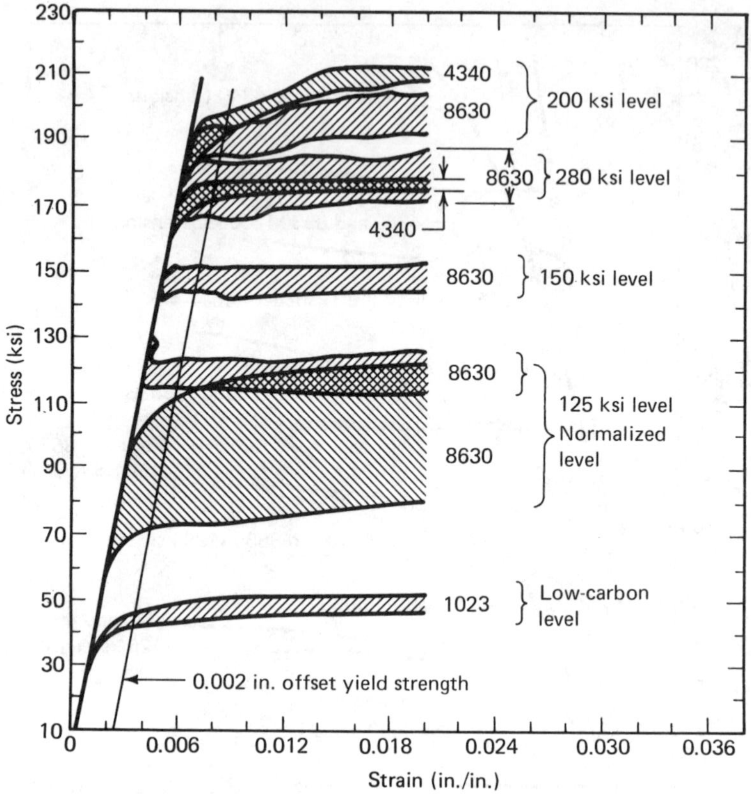

Figure 2.3.10 Stress-strain curves drawn as scatter bands to illustrate the spread in results of numerous tensile tests of the alloy and carbon steels. Ref. (3)

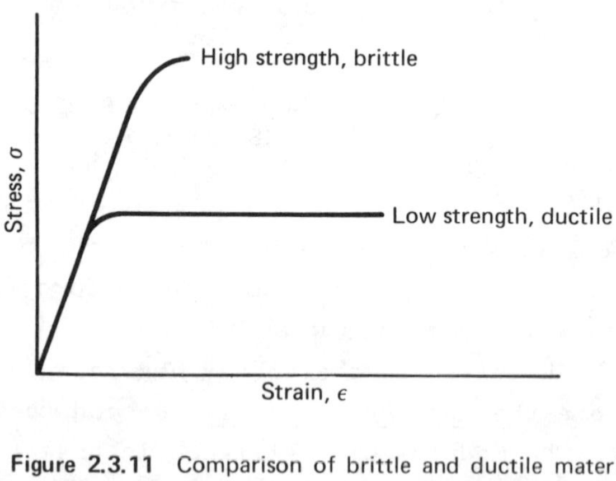

Figure 2.3.11 Comparison of brittle and ductile materials

energy per unit volume to the proportional limit. The modulus of toughness is the strain energy per unit volume to fracture.

These two terms are represented by areas under the stress-strain curve as illustrated in Figure 2.3.12. (The area under a stress-strain curve has units of $lb/in^2 \times in/in = in\text{-}lb/in^3$.)

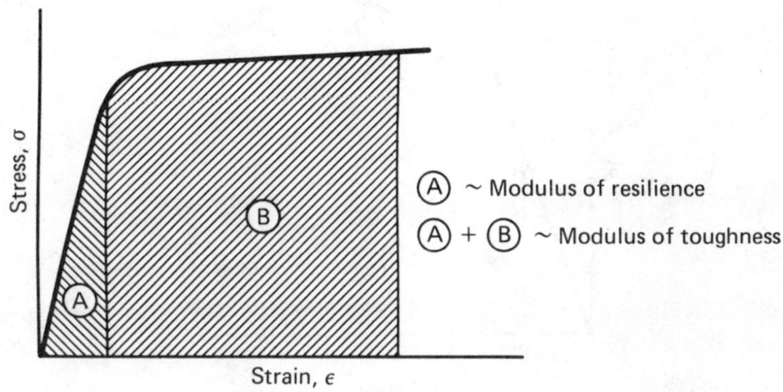

Figure 2.3.12 Modulus of resilience and modulus of toughness

2.4 SHEAR STRESS-STRAIN PROPERTIES

A portion of material which is undergoing direct shear is illustrated in Figure 2.4.1. The material is said to be in pure shear. An element of material subjected to pure shear will undergo an angular distortion, as illustra-

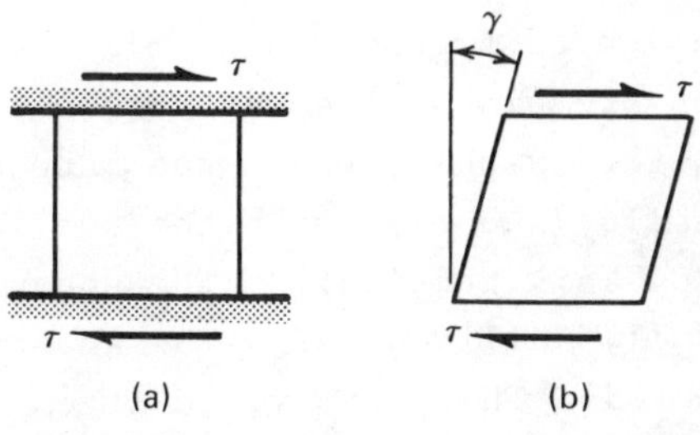

Figure 2.4.1 (a) material in direct shear; (b) distortion due to shear

ted, while the length of the sides does not change. Therefore there is shear distortion but no axial or normal strain. The angle γ, in radians, which measures the change in angle from the original right angle, is called the shear strain.

If shear stress is plotted vs. shear strain, a curve such as Figure 2.4.2 is obtained.

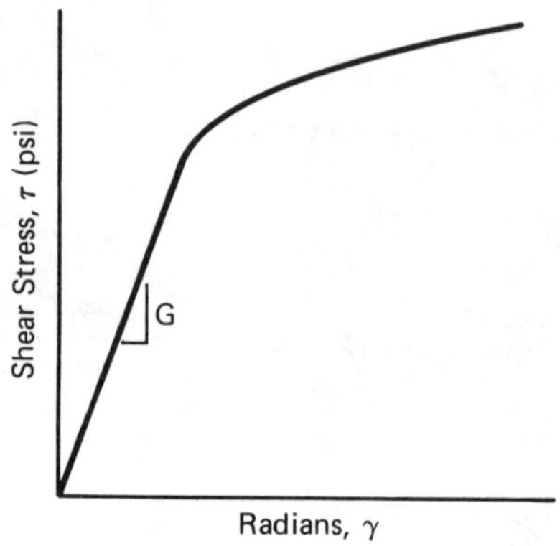

Figure 2.4.2

We usually observe a linear portion of the curve. In this region, $\tau = G\gamma$. The term G is an elastic constant called the shear modulus of elasticity.

Three elastic constants for a material have been defined. They are E, G and ν. These three constants are related by

$$G = \frac{E}{2(1 + \nu)} \tag{2.4.1}$$

A close examination of Figure 2.3.8 will reveal that some of the curves are plots of shear stress vs. shear strain. Compare, for example, the curve labeled "4340 and 8630 (200 ksi tensile)(shear)," with the one labeled "4340 and 8630 (200 ksi)(tension)." The shear curve has a linear slope of 11.5 million psi. The shear yield is approximately 130 ksi. The shear yield strength and the shear ultimate strength are properties which frequently can be found in tables.

2.5 THE EFFECT OF TEMPERATURE ON MATERIAL PROPERTIES

The general effect of elevated temperature on the properties of metals is to decrease E, F_{tu} and F_{ty}[1] and to increase ductility. The general effect of elevated temperatures on 2014-T6 Aluminum is shown in Figures 2.5.1-5.

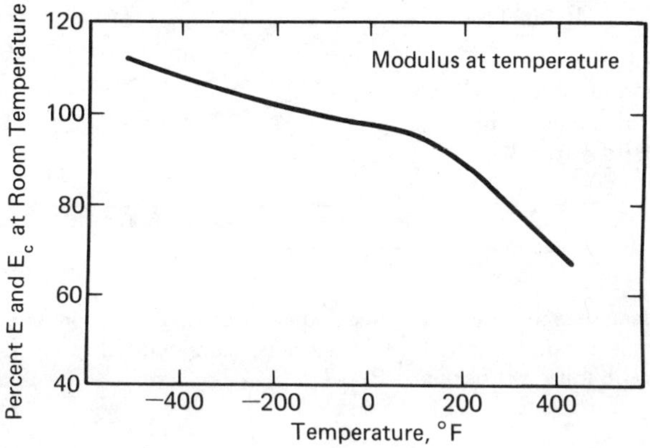

Figure 2.5.1 Effect of temperature on the tensile and compressive modulus (E and E_c) of 2014 and 2017 aluminum alloys. Ref. (1)

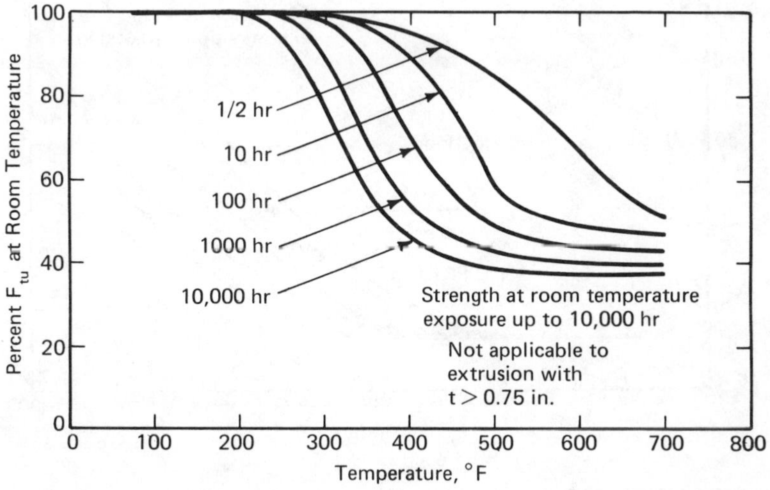

Figure 2.5.2 Effect of exposure at elevated temperatures on the room-temperature ultimate tensile strength (F_{tu}) of 2014-T6 aluminum alloy (all products except thick extrusions). Ref. (1)

---

[1] F_{tu} and F_{ty} are frequently used for tensile ultimate strength and tensile yield strength, respectively, especially in MIL-HBK-5B.

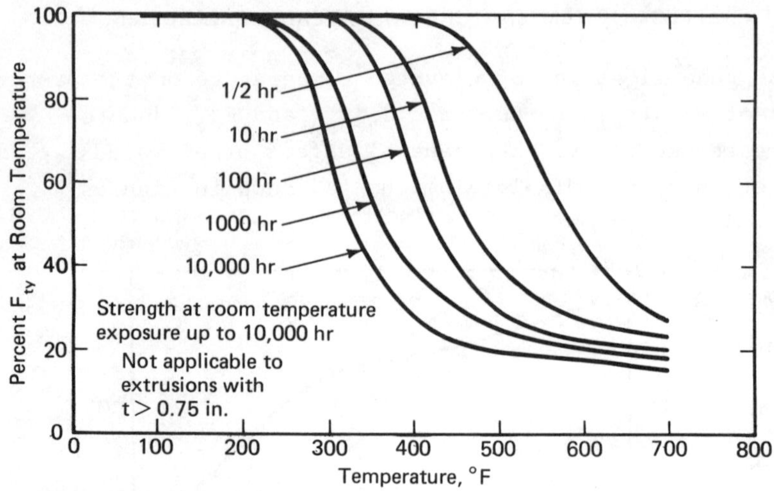

Figure 2.5.3 Effect of exposure at elevated temperatures on the room-temperature tensile yield strength (F_{ty}) of 2014-T6 aluminum alloy (all products except thick extrusions). Ref. (1)

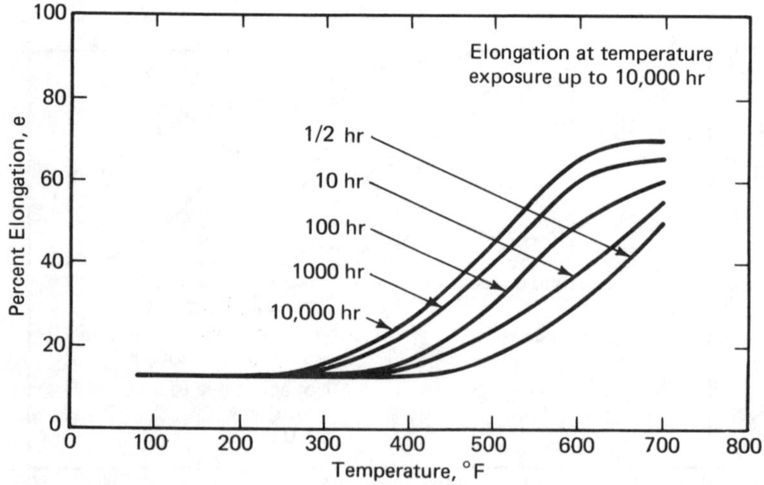

Figure 2.5.4 Effect of temperature on the elongation of 2014-T6 aluminum alloy (all products except thick extrusions). Ref. (1)

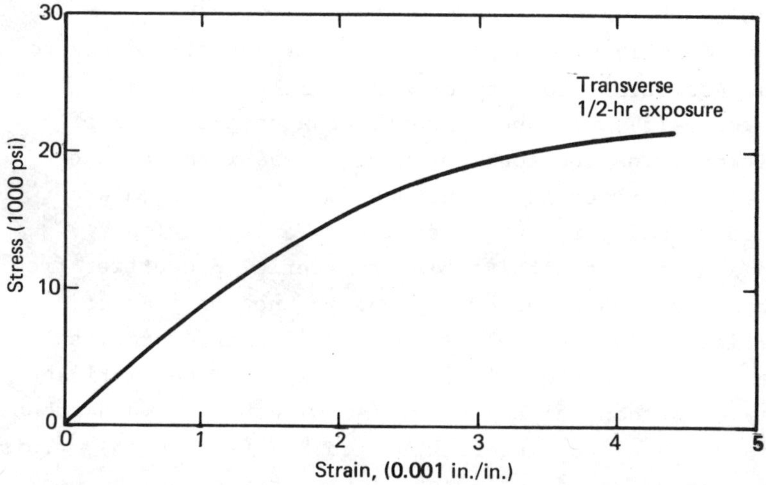

Figure 2.5.5 Typical compressive stress-strain curve for clad 2014-T6 aluminum alloy (sheet) at 500° F. Ref. (1)

Figures 2.5.5 and 2.3.3 should be compared for an overall view of the effect of temperature on the uniaxial stress-strain curve.

2.6 SOURCES AND REFERENCES FOR MATERIAL PROPERTIES

The stress analyst must have access to the properties of materials if he is to make an intelligent decision about the suitability of a material for a given situation.

In some cases the source for material properties may be specified. For example, most aerospace vehicles are designed from Mil-Hbk-5B (1)*; civil structures are often designed to the American Institute of Steel Construction (AISC) Handbook (2).

In other situations the source of materials and material properties may not be so closely specified. In these situations a number of sources for material properties

*Numbers in parentheses refer to references at the end of each chapter.

are available. A brief review of these will be given. It is frequently necessary to search rather diligently to find a particular property of a particular material.

Reference (1), Mil-Hbk-5B contains a complete set of definitions, instructions for use, stress formulas and material properties for structural materials which are produced to military specification. Table A.1 (in Appendix A) is a typical table of material properties from this handbook. Reference (2) contains not only material properties but a wide selection of tables and formulas for design application of standard shapes. The American Society of Mechanical Engineers (ASME) publishes a handbook entitled, "Metals Engineering Design," Reference (3). This is a very good source of material properties and the book has a comprehensive discussion of the significance of material properties and the applicability of material to given design situations. Figures 2.3.8-10 are from Reference (3). The American Society for Testing and Materials (ASTM) publishes a many-volumed series of standards, standard test method and properties. Reference (4) includes stainless steel. One page of a typical standard which is co-sponsored by the American National Standards Institute, and some tensile properties for representative stainless steels are shown in Appendix A.

An excellent source of properties for a wide range of materials and conditions is entitled, "Metals Reference Book," Reference (5). Three representative tables: A.4,5; A.12,13 show the presentation for pure Titanium and Titanium alloys. CRC publishes the "Handbook of Materials Science," Reference (6). This is an excellent source for a wide variety of materials and includes both metallic and non-metallic materials. Sample tables for metals are shown in Tables A.6-11. Reference (6) also shows some representative strength properties for wood. See for example Table A.14.

Another excellent source of metallic and non-metallic material properties is, "Engineering Materials Handbook,"

Reference (7). Table A.15 is taken from Reference (7) and gives some typical properties for Magnesium. Another publication of the American Society of Mechanical Engineers which provides an excellent source of material properties is entitled, "Metals Properties," Reference (8). A typical listing of properties for AISI 8640 steel is shown in Table A.16. The Society of Automotive Engineers (SAE) publishes a handbook, Reference (9), which gives materials properties and also gives a broad discussion of suitable usages for materials. Table A.17 gives some representative properties for Copper and Copper alloys. Probably the most comprehensive collection of properties and the most complete discussion of metals and their uses is contained in the eight-volume "Metals Handbooks," of the American Society for Metals (ASM), Reference (10). Some additional tables are included in Appendix A; Tables A.18-20.

The notations used in this text will be as follows:

Modulus of elasticity - E
Shear modulus of elasticity - G
Ultimate tensile strength - S_{tu}
Tensile yield strength - S_{ty}
Poisson's ratio - ν

This useage of symbols is not universal and each Table and Handbook must be checked for its own useage.

A final word about terminology is the use of limit load and ultimate load in a special sense in aircraft and missile analysis as follows:

Limit Load: The load factor that the airplane can sustain without objectionable permanent deformation of any part. (Associated with S_{ty}.)

Ultimate Load: The load factor that the airplane can sustain without fracture of any part. (Associated with S_{tu}.)

Problems

Problem 1. Given a typical tensile stress strain diagram of an aluminum alloy, label the axes and show the following:

a. Proportional limit
b. Yield stress
c. Ultimate tensile stress
d. Modulus of elasticity
e. Elastic region
f. Plastic region

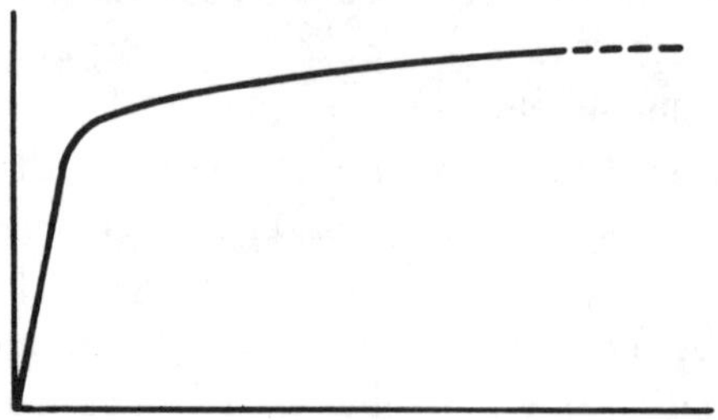

Problem 2. Define the following in your own words, closed book. Use sketches and illustrations.

a. Ultimate Tensile Stress
b. Proportional limit
c. Yield Stress
d. Modulus of Elasticity

Problem 3. Check yourself on these questions - closed book.

a. Materials which undergo a lot of plastic deformation before fracture are (ductile, brittle).
b. Alloying and heat treating, generally:
 (raise, lower, do not effect) modulus of elasticity
 (raise, lower, do not effect) hardness
 (raise, lower, do not effect) ultimate tensile strength
 (raise, lower, do not effect) yield strength
 (raise, lower, do not effect) ductility
c. Sketch a stress-strain curve and label it "BASIC." If a material is alloyed, sketch the effects, and indicate these on a stress-strain curve, drawn to the same scale, labeled "ALLOYED." Indicate general trends.
d. Give typical values for ultimate tensile strength, yield strength modulus of elasticity and Poisson's ratio for steel, aluminum magnesium and titanium.

Problem 4. What are nominal values of the following properties for 7075-T6 aluminum? Give units.

a. E ______________

b. F_{tu} ______________

c. F_{ty} ______________

d. density __________

e. α ______________

Problem 5. Give approximate values of toughness for 7075-T6 aluminum at room temperature and at 400°F.

Problem 6. A uniaxial rod one foot long must carry a load of 1,000 lb without fracturing. What material will give a rod of least weight?

a. AISI 4130 steel

b. 2024-T3 Aluminum

c. Titanium 318

Problem 7. A designer has a choice between AISI 1025 steel and 2024-T3 aluminum. Which material will give a minimum weight structure based on

a. Limit load

b. Ultimate load

Explain why.

Problem 8. What are the yield strengths of the materials of Figure 2.3.2 and 2.3.4 in the text? Explain how you obtain the answers.

Problem 9. An engineer is designing a part which is not to experience any permanent elongation while in service. The material property he is most concerned with is:

a. The resilience

b. The toughness

c. The tensile yield strength

d. The modulus of elasticity

e. The tensile ultimate strength

Problem 10. An engineer who is designing a part which is not to experience fracture while in service would be most concerned with which of the properties in question number 9?

a. The resilience

b. The toughness

c. The tensile yield strength
d. The modulus of elasticity
e. The tensile ultimate strength

Problem 11. Referring to Figure 1, the proportional limit is given by point

a. 0
b. A
c. B
d. C
e. F

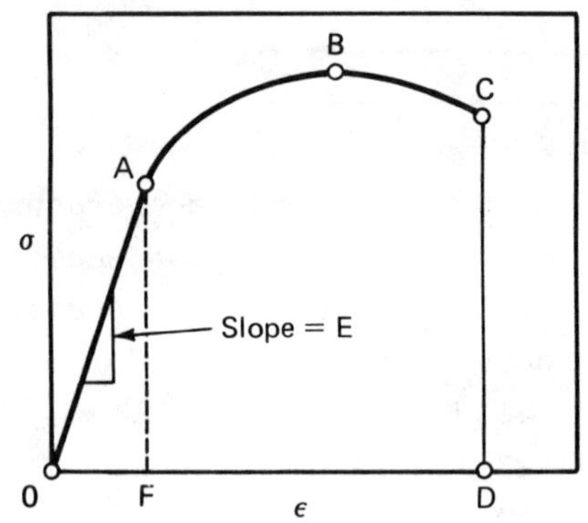

Figure 1

Problem 12. In Figure 1, the elastic limit is most nearly given by point

a. O
b. A
c. B
d. C
e. F

Problem 13. In Figure 1, the ultimate tensile strength is given by point

a. O
b. A
c. B
d. C
e. F

Problem 14. In Figure 1, the modulus of elasticity is given by point

a. O
b. A
c. B
d. C
e. E

Problem 15. In Figure 1, the toughness is given by area

a. OAF

b. OABCD

c. Neither of these

Problem 16. In Figure 1, the resilience is given by area

a. OAF

b. OABCD

c. Neither of these

Problem 17. A 2 inch length specimen of material is tested to failure. After it fails its ends are matched together and it is found to be 2.125 inches long. Its ductility is

a. .0625%

b. 2.125 in.

c. 0.125 in.

d. 0.625%

e. 6.25%

Problem 18. Give your definition of toughness.

Problem 19. Sketch a "typical" stress-strain curve for aluminum. Label the axes. Label the yield strengh and ultimate strength.

Problem 20. Repeat question 19 for a "typical" mild steel.

Problem 21. Check yourself on these questions - closed book.

a. Define yield stress. Why is it important?

b. Typical values of yield stress for a high-strength aluminum alloy is:
   1. 200 ksi
   2. 70-80 ksi
   3. 5 ksi
   4. 1 ksi

c. Modulus of elasticity, E, for aluminum is about
   1. 10×10^6 psi
   2. 30×10^6 psi
   3. 8×10^{12} psi
   4. 10^4 psi

d. Typical values of Poisson's ratio for structural metals is:
   1. 0.01
   2. 0.03
   3. 0.98
   4. 0.30

Problem 22. For the material of Figure 2.3.5 (Cast AZ63A-T4; 75°F), find the yield stress by the offset method. Find the tangent and secant moduli for this stress point.

Problem 23. A 1/4 inch diameter bolt is to carry 1,000 lb. in tension at 325°F. If the bolt must not yield, select from the available sources in the text the materials which will be successful. (Use 1/4 inch nominal diameter to estimate the bolt area and ignore the effects of the threads.)

Problem 24. A hollow circular tube is used in a truss configuration as shown below. Determine the area of AISI 4130 tubing required for member A. Use a factor of safety of 2.0 based on S_{tu}. (F_{tu} in the table.)

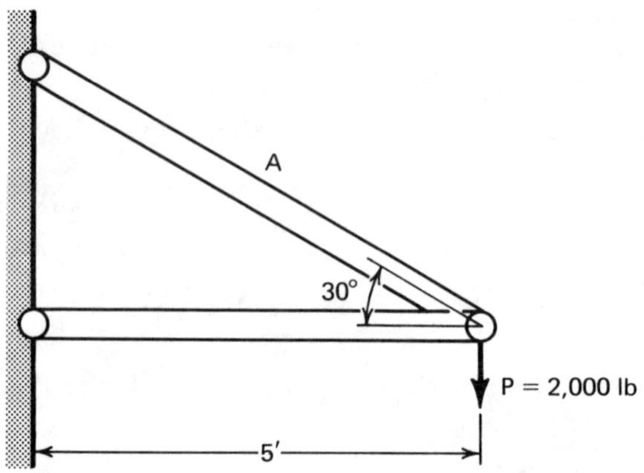

Problem 25. A rod is to carry 1000 lb. in tension for an extended period at 250°F. If the rod is 2 feet long, calculate the following quantities for the materials below. (Use a table for the results.) A) The area required for no yielding. B) The area required for no

fracture. C) The weight of material for A) and B). D) The elongation of the rod. Materials to be used:

Magnesium AZ63A-T4
Aluminum 2014-T6
Titanium 230
Aluminum 2219-T6
Monel® Alloy 400 Annealed

References

1. "Metallic Materials for Aerospace Structures," (MIL-HBK-5B), United States Government Printing Office. (Note: MIL-HBK-5C is now available.)
2. "Manual of Steel Construction," (AISC Handbook), American Institute of Steel Construction, 7th edition, New York, 1970.
3. "Metals Engineering Design," (ASME Handbook), Oscar J. Horger, Editor, 2nd edition, McGraw-Hill, New York, 1965.
4. "1977 Annual Book of ASTM Standards," Part 5, American Society for Testing and Materials, Philadelphia, PA, 1977.
5. "Metals Reference Book," Colin J. Smithells, Editor, 5th edition, Butterworths, London, 1976.
6. "CRC Handbook of Materials Science," Volume I, General Properties, Volume II, Metals, Composites and Refactory Metals, CRC Press, Cleveland, Ohio, 1975, 1976.
7. "Engineering Materials Handbook," Charles L. Mantell, Editor-in-Chief, 1st edition, McGraw-Hill, 1958.
8. "Metals Properties," (ASME Handbook), Samuel L. Hoyt, Editor, McGraw-Hill, 1954.
9. "SAE Handbook 1977," Part 1, Materials, Parts and Components, Society of Automotive Engineers, Warrendale, PA, 1977.
10. "Metals Handbook," (in eight volumes), Taylor Lymann, Editor, American Society for Metals, 8th Edition, Metals Park, Ohio, 1961.

Chapter 3

TORSION AND THE TORSION ELEMENT

3.1 INTRODUCTION

A structural element which is being twisted with respect to its longitudinal axis is in torsion. This is illustrated in Figure 3.1.1, where the double arrowheads indicate twisting in accordance with the right hand rule as shown.

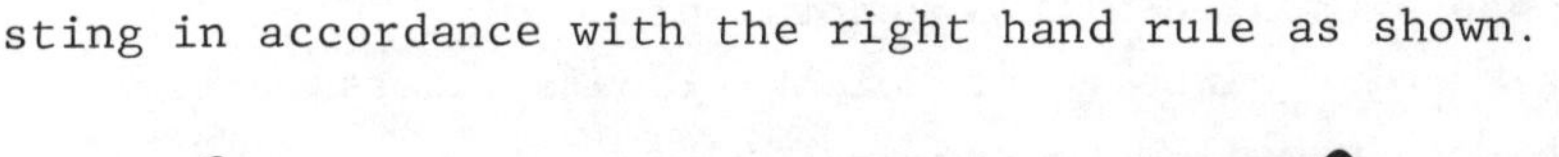

Figure 3.1.1 A structural element in torsion. The double arrowheads indicate twisting according to the right hand rule.

Elements may be in torsion due to the direct application of torque; due to their presence in a structure where they are twisted due to the deformation of surrounding structure; or due to the fact that they are being used to transmit horsepower.

A shaft which is being used to transmit horsepower has internal torque given by the following formula, which is a handy one to have memorized:

$$T = \frac{63,000 \text{ HP}}{N} \tag{3.1.1}$$

where: T is the torque in lb. - in.
HP is the horsepower transmitted
N is the rpm of the shaft

In analyzing elements in torsion, it is quite useful to be able to classify them into one of three categories and then proceed in a definite method of analysis which depends upon which category the <u>cross section</u> falls in.

The three categories are:

1. Circular (either solid or hollow)
2. Hollow, closed, thin wall
3. Other

We shall proceed to discuss the classification of the cross section in torsion and the method of analysis for each.

Learning Objectives for Torsion

1. Be able to distinguish between three cases in torsion when presented with a general problem.
   a. Solid circular or hollow circular cross section
   b. Thin walled closed section
   c. Other
2. Having made the proper identification in Part 1, above, be able to use the proper formula to calculate the stress and deflection. Also be able to use the proper tables and formulas from the text.
3. Be able to state the five assumptions associated with a circular cross section.

4. Be able to calculate the torque on a shaft given the shaft rpm and the horsepower transmitted.
5. Know from memory the following formulas. Know the meaning and units of each term. Be able to solve problems using these formulas. Know to which of the cases in objective 1 above the formulas apply.

$$\tau = \frac{Tr}{J}$$

$$\theta = \frac{TL}{JG}$$

6. Be able to derive the formulas in objective 5.
7. Know from memory the definition of the polar moment of inertia, J, and know the formula for J for a circular solid or hollow cross section.
8. Be able to give a brief written discussion on the difference in behavior (and difference in resulting equations) when the cross section of a solid non-circular torsional member is being discussed as compared to a circular one.
9. Be able to use tables in the text for torsion problems (stress and deflection) for solid non-circular cross sections.
10. Be able to use tables in the text to solve for torsional stresses and deflections of members made up of thin plates (thin wall open sections).
11. Be able to use equations from the text to calculate the shear stress and angular rotation of thin wall closed sections.
12. Recognize at least a rule-of-thumb figure for r/t as a criteria for treating a torsional member as a thick-wall or thin-wall cross section.

3.2 CLASSIFYING THE SHAPE OF THE CROSS SECTION IN TORSION

The shape of the cross section in torsion is the basis for the method of analysis to be used. We shall distinguish three categories:

1. Circular (solid or hollow)
2. Thin wall, closed, hollow
3. Other

The classification into one of these three categories is not difficult. However, some comments are in order so that doubtful areas may be overcome.

The choice of a circular cross section is obvious. The circular cross section may be solid or hollow as indicated in Figure 3.2.1. If the cross section is hollow, the circles must be concentric.

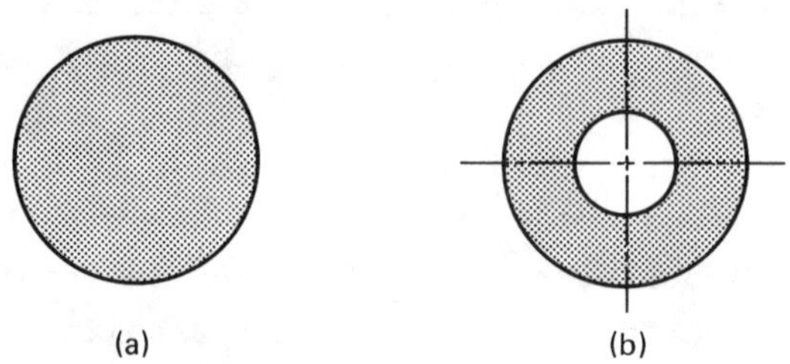

Figure 3.2.1 (a) solid circular cross section; (b) hollow circular cross section

The thin-wall, hollow, closed cross section may be of any shape. The wall thickness does not have to be constant. The question of what a thin wall is frequently arises as a guide, the ratio of thickness, t, to some significant size parameter D (see Figure 3.2.2) should meet the requirement

$$t/D < 0.1 \qquad (3.2.1)$$

The hollow circular thin wall tube may be treated under this classification, or it may be treated as a category one.

The hollow, thin-wall tube must be completely closed. If there is a break in the cross section, caused by a longitudinal cut, the section does not qualify under category two. Some examples are shown in Figure 3.2.3. They are category three, "other."

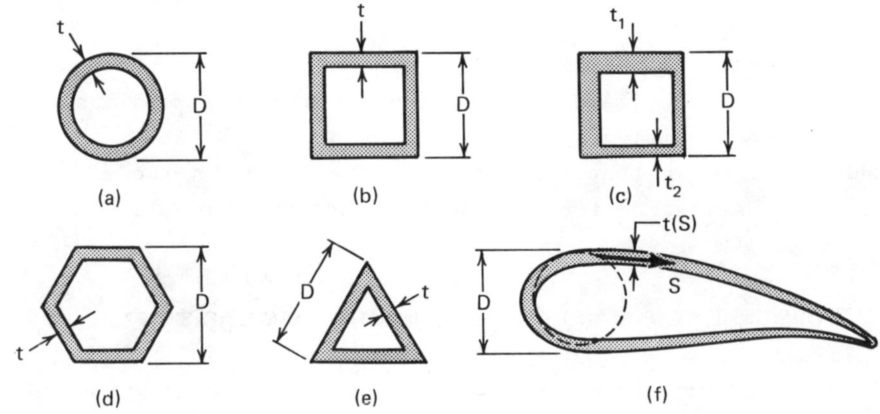

Figure 3.2.2

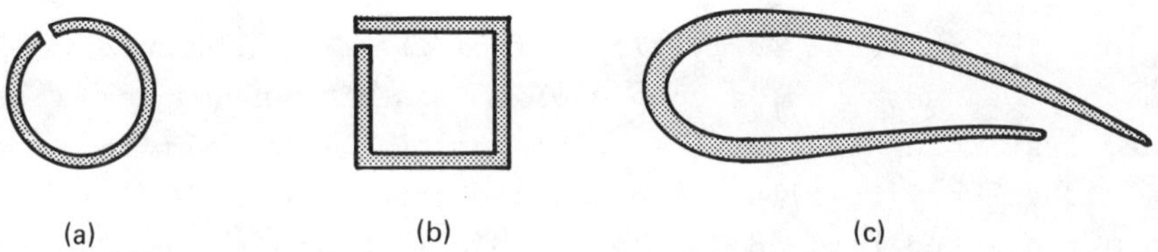

Figure 3.2.3

Any section which is not classified as a category one or two, will be (automatically) category three.

Some examples of category three, in addition to those shown in Figure 3.2.3, are shown in Figure 3.2.4.

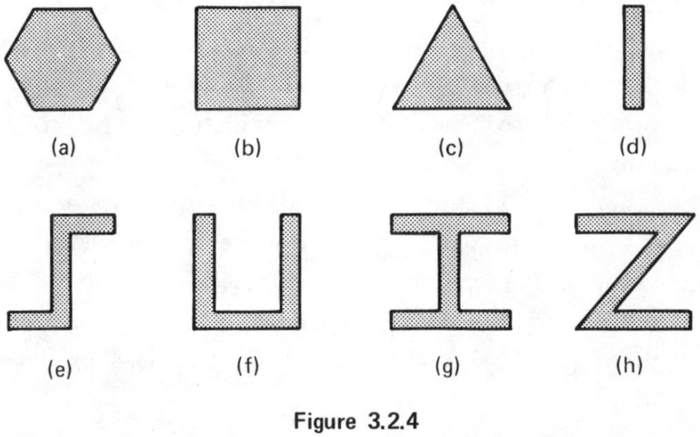

Figure 3.2.4

Once the process of categorizing the cross sections is completed, the stress analyst proceeds by the appropriate method to solve for the angular twist and torsional stress in the element.

We shall begin the discussion of the analysis methods with the derivation for the equations for the circular shaft.

3.3 TORSIONAL STRESS AND DEFLECTION IN THE CIRCULAR SHAFT

There is a nearly universal question form students when they read derivations of stress formulas or when they hear the derivation presented in lecture. They ask, "Where in the world did anyone get the idea for this in the first place?"

In all structural derivations, from the most elementary to the most advanced, a couple of things always happen. The first thing is that the person doing the derivation is either very familiar with how the structure is behaving when subjected to load, or he is able to make some good guesses. He is very curious about the behavior of the structure. (I call this "knowing how the structure feels.") The first assumptions or observations about the structure are usually about the deformation shape or geometry.

The second thing that happens is that the old familiar principles are applied. Principles which we have already learned. These are: equilibrium, stress-strain and compatibility. (We haven't quite learned compatability yet.) If the student will approach derivations with this in mind, perhaps the derivations will not seem so overwhelming. We shall follow this approach with the torsion of the circular shaft.

The discussion and illustrations will apply to a solid circular shaft; however, all of this section applies as well to a hollow circular shaft. The deformation of a circular shaft is unique in several ways. Consider the shaft of Figure 3.3.1(a) which is being twisted by torque T. One of the first observations is that a cross section,

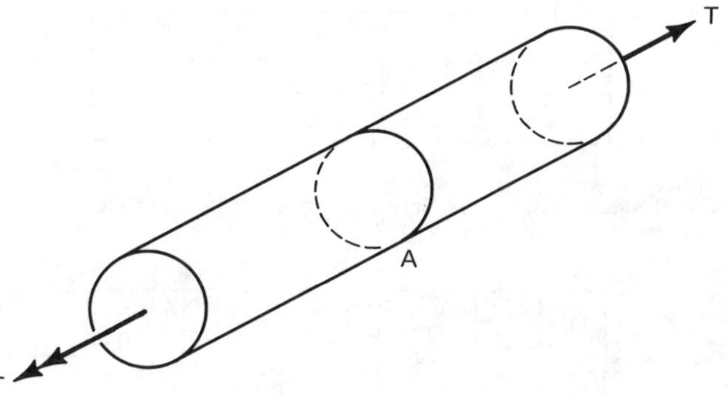

Figure 3.3.1 (a) The cross-section A remains circular

such as A, remains a circle after deformation. A second observation is that cross section A remains flat, from a side view, after deformation has taken place. In other words, there is no movement of the material in this cross section along the longitudinal directions. This is illustrated in Figure 3.3.1(b). We can then observe that

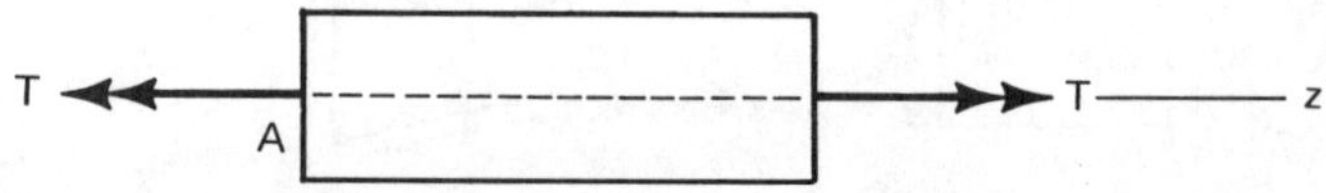

Figure 3.3.1 (b) The cross-section A remains flat in a side view

if we look at a side view such as is shown in Figure 3.3.1 (c), and we observe small squares which are formed by circumferential lines and by lines which are generators of the cylinder, then after deformations these lines will appear as in Figure 3.3.1(d). The twist of one end of the shaft with respect to the other has caused distortion of the squares into diamonds. During the distortion, the generators remain straight for small strains. The circumferential lines remain flat, and remain perpendicular to the longitudinal axis of the shaft. All of the above effects can be seen by twisting a foam rubber circular shaft with these lines inscribed on them.

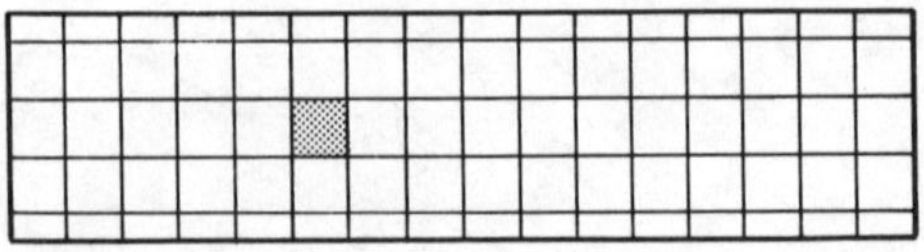

Figure 3.3.1 (c) Side view of shaft before twisting

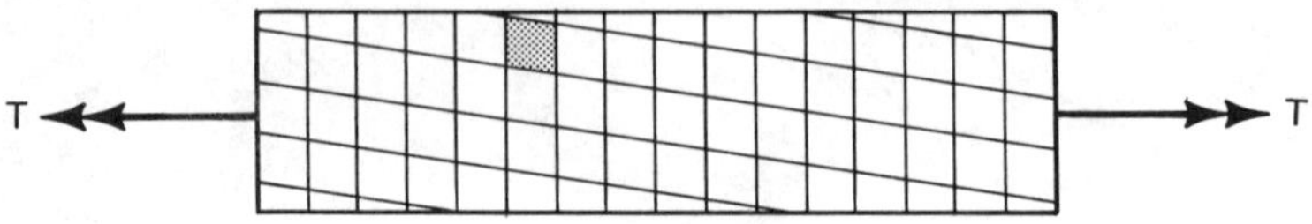

Figure 3.3.1 (d) Side view of shaft after twisting

If we consider a small square, such as the shaded one, we can postulate that the change from square to diamond shape is caused by shear stress, as shown by Figure 3.3.2.

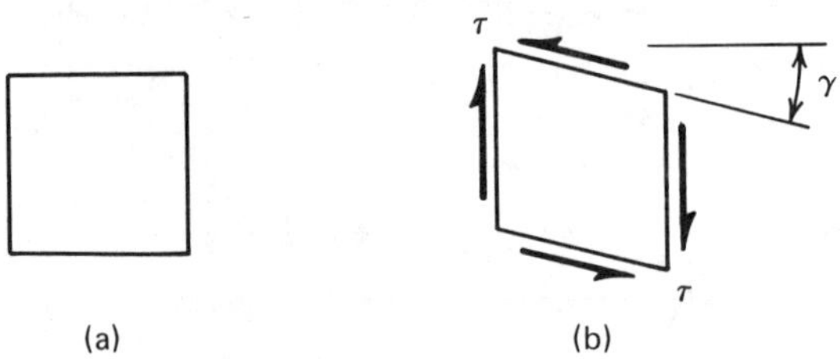

Figure 3.3.2 Distortion of square inscribed on the side of the shaft. (a) before twisting; (b) after twisting

The shear stress arrows act as shown in Figure 3.3.2. A more complete depiction of the shear stress arrows is shown in Figure 3.3.3.

In Figure 3.3.3, a cutaway view shows the behavior of an element on a cross section such as A. The effect of torsion is to try to wring, or shear, one face of the cross section with respect to the other. This results in stresses on the small element as shown.

If we now look at a view of cross section A, as shown in Figure 3.3.4, we can see that a shear stress in the face of cross section A contributes to a torque about 0,

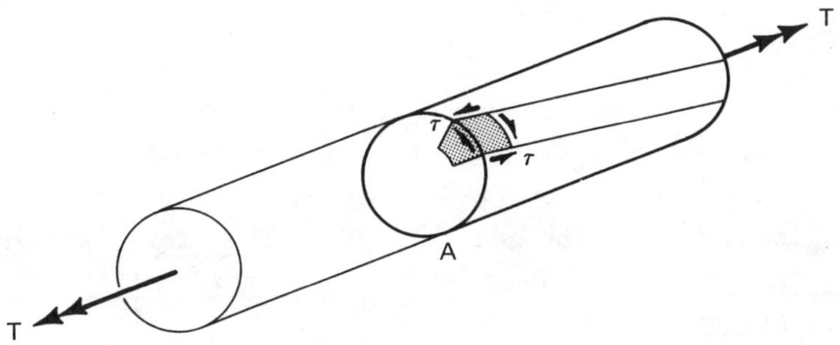

Figure 3.3.3 Distribution of shear stress on an element of the bar in torsion

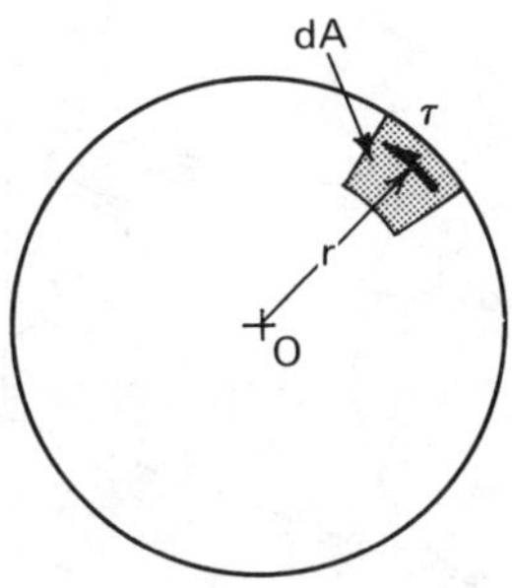

Figure 3.3.4 The shear stress τ on element dA contributes to torque about 0

because τdA is a force and $\tau \cdot dA \cdot r$ is a moment. We now apply the concept of <u>static equilibrium</u> of external forces. From this we know that

$$\int_A \tau \cdot r \cdot dA = T \tag{3.3.1}$$

A final picture will serve to depict the geometry of deformation. In Figure 3.3.5, we see that the generators of the cylinder deform into a straight line with inclination of γ, the shear strain angle. (Refer also to Figure 3.3.2.) The angle of twist of one face relative to the other in length dz is given by $d\theta$. We observe that $d\theta/dz$ is the twist per unit length and is a constant. From Figure 3.3.5, we observe, for small angles γ that

$$r d\theta = \gamma dz \tag{3.3.2}$$

$$\gamma = r \frac{d\theta}{dz} \tag{3.3.3}$$

Therefore, the same picture as Figure 3.3.5 will hold true for any radius. Equation (3.3.3) says that the shear strain γ in an element such as dA in Figure 3.3.4 varies linearly from the neutral axis. Therefore, strain varies linearly from the center to the outside of the circular cross section.

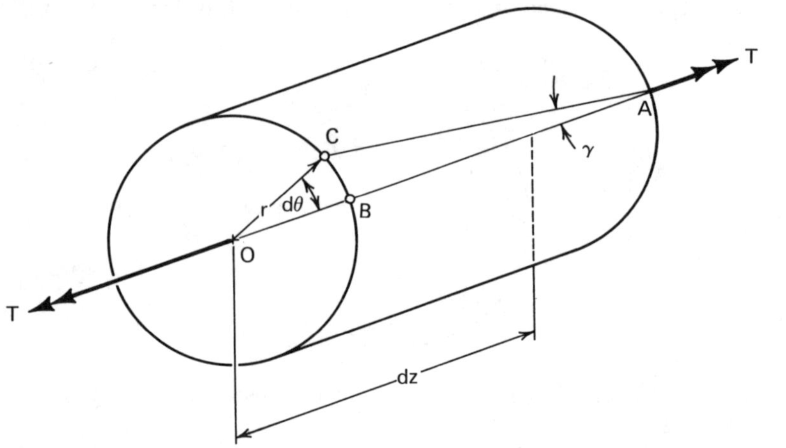

Figure 3.3.5 Generator AB deforms to line AC through shear strain angle γ

Now we use the applicable stress-strain relation

$$\tau = G\gamma \tag{3.3.4}$$

Using the relationship restricts us to elastic behavior of the material.

Returning to equation (3.3.1) and substituting equations (3.3.3) and (3.3.4), we have

$$T = \int_A G\gamma \; r \; dA \tag{3.3.5}$$

$$T = \int_A Gr^2 \frac{d\theta}{dz} \, dA \tag{3.3.6}$$

Since G and $\frac{d\theta}{dz}$ are constant with respect to integration over dA

$$T = G \frac{d\theta}{dz} \int_A r^2 dA \tag{3.3.7}$$

We recognize $\int_A r^2 dA$ as the polar moment of inertia, J. This is a tabulated area property.

So

$$T = G \frac{d\theta}{dz} J \tag{3.3.8}$$

and

$$\boxed{\frac{d\theta}{dz} = \frac{T}{GJ}} \tag{3.3.9}$$

The angle of twist per unit length is related to the torque, the shear modulus of elasticity and the polar moment of inertia by equation (3.3.0).

Using equations (3.3.3) and (3.3.4) once again, we have

$$\frac{\gamma}{r} = \frac{T}{GJ} \tag{3.3.10}$$

$$\boxed{\tau = \frac{Tr}{J}} \tag{3.3.11}$$

The shear stress as a function of radius is given in terms of applied torque and polar moment of inertia of equation (3.3.11).

The distribution of γ and τ with respect to radius is shown in Figure 3.3.6. This same distribution applies all the way around the circumferential coordinate of the circle.

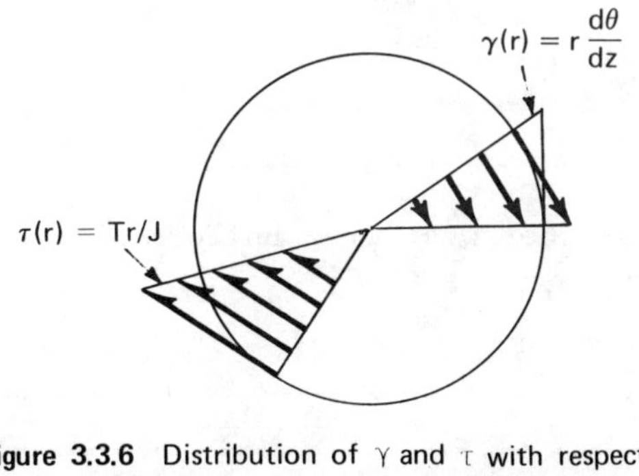

Figure 3.3.6 Distribution of γ and τ with respect to r

Example: 3.3.1. Circular shaft in torsion.

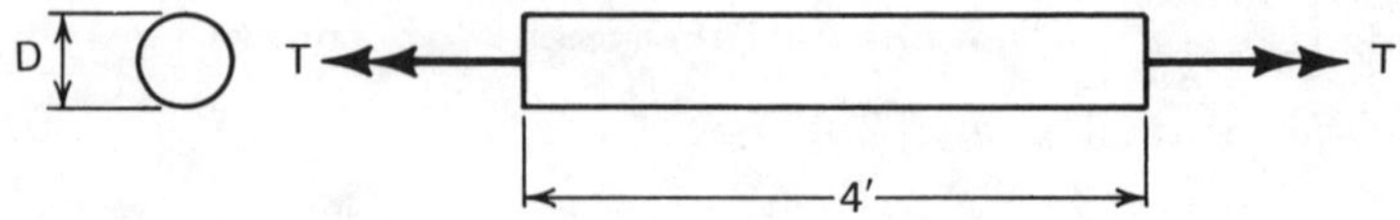

A four foot length of 7075-T6 Aluminum bar has a diameter of 1.25 in and carries a torque of 1,000 in-lb. What is the angle of twist? What is the maximum shear stress?

Solution:

The angle of twist per unit length is

$$\frac{d\theta}{dz} = \frac{T}{JG}$$

So:

$$d\theta = \frac{Tdz}{JG}$$

$T = 1{,}000$ in-lb

G (Appendix A) $= 3.9 \times 10^6$ psi

J for a circle $= \frac{\pi D^4}{32}$ (Appendix C)

$$= \frac{\pi (1.25)^4}{32} = 0.2397 \text{ in}^4$$

$$\theta = \int_0^L \frac{T}{JG}\, dz = \frac{TL}{JG} = \frac{1{,}000\ (4 \times 12)}{0.2397 \times 3.9 \times 10^6} = \underline{\underline{0.0513 \text{ rad}}}$$

$$= \underline{\underline{2.94°}}$$

Note that

$$\boxed{\theta = \frac{TL}{JG}} \qquad (3.3.12)$$

is the solution for the twist of a uniform shaft of length L.

The shear stress is given by

$$\tau_{max} = \frac{Tr_o}{J}$$

$T = 1,000$ in-lb

r_o = outside radius = 0.625 in

$J = 0.2397 \text{ in}^4$

$$\tau_{max} = \frac{1,000\ (0.625)}{0.2397} = \underline{\underline{2607\ \text{lb/in}^2}}$$

Example 3.3.2. Hollow Circular Shaft.

If a 0.5 in dia. hole is drilled in the shaft of Example 3.3.1, calculate:

1. The % reduction of volume (and thus weight).
2. The % increase in angle of twist.
3. The % increase in stress.

Solution:

The % reduction in area is

$$\frac{A_o - A_f}{A_o} \times 100$$

where: A_o = original area

A_f = final area

and

$$A_o = \frac{\pi D_o^{\ 2}}{4} = \frac{\pi(1.25)^2}{4} = 1.227 \text{ in}^2$$

$$A_i = \frac{\pi D_i^{\ 2}}{4} = \frac{\pi(0.5)^2}{4} = 0.196 \text{ in}^2$$

$$A_f = 1.227 - 0.196 - 1.031 \text{ in}^2$$

And % red. in area $= \dfrac{1.227 - 1.031}{1.227} \times 100 =$ $\underline{\underline{15.9\%\ \text{reduction}}}$.

The increase in angle of twist and shear stress is governed by the new polar moment of inertia

$$J = \frac{\pi}{32}(D_o^4 - D_i^4)$$

$$= \frac{\pi}{32}(1.25^4 - 0.5^4) = 0.2335 \text{ in}^4$$

The new angle of twist is

$$\theta = \frac{TL}{JG} = \frac{1,000\ (48)}{0.2335 \times 3.9 \times 10^6} = \underline{.0527 \text{ rad}}$$

The % increase in angle of twist is

$$\% \text{ increase} = 100 \times \frac{\theta_{new} - \theta_{old}}{\theta_{new}} = \frac{0.0527 - 0.0513}{.0527} \times 100$$

$$= \underline{2.66\%}$$

The new stress is given by:

$$\tau_{max} = \frac{Tr_o}{J}$$

(The stress distribution for a hollow shaft is shown in the sketch.)

$$\tau_{max} = \frac{1,000\ (0.625)}{0.2335} = \underline{2676 \text{ psi}}$$

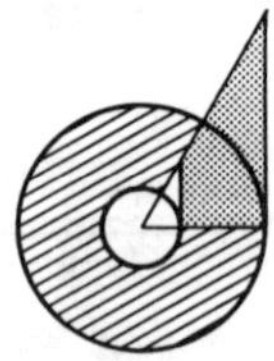

The % increase in stress is:

$$\frac{\tau_{new} - \tau_{old} \times 100}{\tau_{new}}$$

$$= \frac{2676 - 2607}{2676} \times 100 = \underline{2.58\%}$$

The above numbers show why shafts in actual machinery are often hollow. A significant weight saving can be obtained with a small deflection and stress penalty.

3.4 TORSION OF THIN-WALL, CLOSED HOLLOW TUBES

In Equation (3.2.1) the guideline $t/D < 0.1$ was given for a thin wall tube. A more accurate description of a thin wall tube is that the shear stress variation across the wall is negligible. Consider Figure 3.4.1 which shows a hollow circular tube with a thick wall compared to a thin wall. In the thick wall tube the stress variation from inside to outside is significant. In the thin wall the stress variation is small.

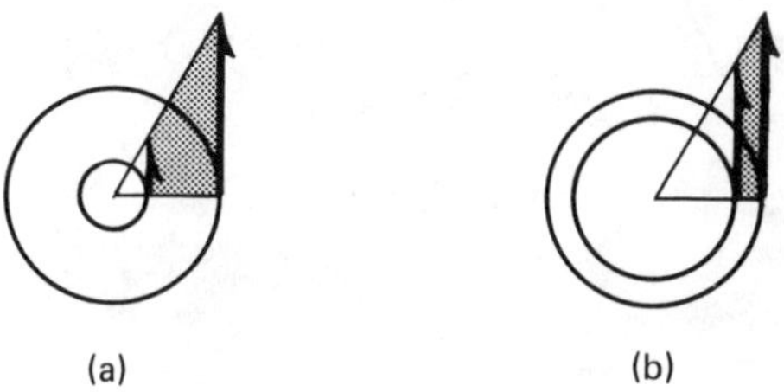

Figure 3.4.1 Hollow circular tubes in torsion

The decision as to whether or not something is thin wall may require some judgement, but basically, the wall is thin enough with respect to the rest of the structure so that the shear stress variation across the wall thickness is essentially constant as shown in Figure 3.4.2.

For thin wall structures, it is customary to work with the shear flow in lb per in, defined by

$$q = \tau \cdot t \ \frac{\text{lb}}{\text{in}} \tag{3.4.1}$$

The shear flow is assumed to act at the mid-line of the wall. The coordinate S runs around the mid line of the wall as shown.

Imposition of <u>equilibrium</u> of the <u>in-plane</u> <u>moments</u> gives:

$$q \cdot ds \times r = T \tag{3.4.2}$$

(force) (arm) = (applied torque)

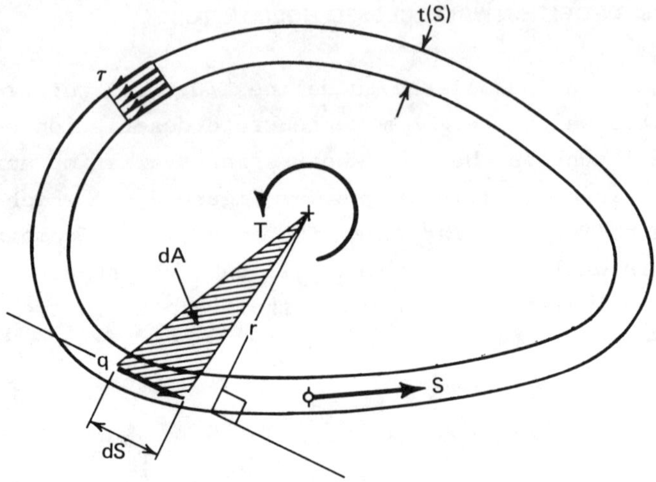

Figure 3.4.2

Now note that

$$rds = 2 \; (dA) \tag{3.4.3}$$

where (dA) is the shaded area in Figure 3.4.2. Letting ds cover section gives

$$q = \frac{T}{2 \; Ⓐ} \tag{3.4.4}$$

where Ⓐ is the <u>entire</u> area enclosed by the centerline of the tube. (<u>Not</u> the area of actual material.) Note that even though the thickness t (and thus the shear stress) may vary around the circumferential coordinate of the tube, the shear flow q <u>remains constant</u>.

<u>Example 3.4.1</u>

For the tube shown, calculate the shear flow. Calculate the shear stress in each leg. T = 1,000 in-lb.

Solution:

The area Ⓐ is total area enclosed by the centerline of the tube. This is shown as shaded in the figure.

Ⓐ = 4 x 4 = 16 in^2

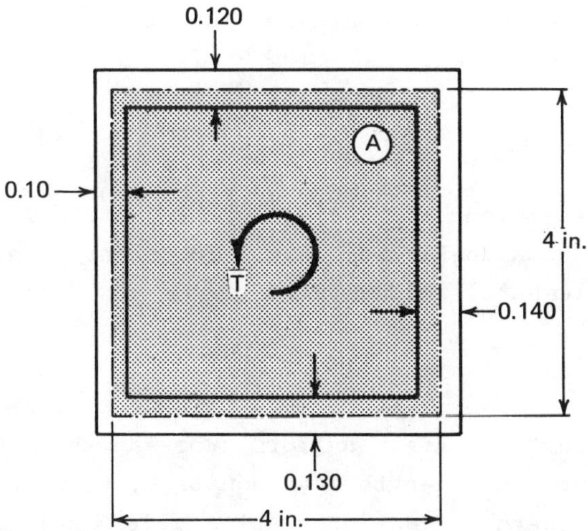

Then

$$q = \frac{T}{2\,Ⓐ} = \frac{1000}{(2)16} = 31.25 \text{ lb/in}$$

In the 0.10 wall: $\tau = q/t = 31.25/0.10 = 312.5$ psi

In the 0.120 wall: $\tau = q/t = 31.25/0.120 = 260.4$ psi

In the 0.140 wall: $\tau = q/t = 31.25/0.140 = 223.2$ psi

In the 0.130 wall: $\tau = q/t = 31.25/0.130 = 240.4$ psi

When we consider the angular deflection of a hollow, closed, thin wall structure, it is convenient to use Castigliano's theorem. The strain energy of a structure in shear, per unit volume, is, see Figure 3.4.3.

$$U = \frac{1}{2}\tau \cdot \gamma \tag{3.4.5}$$

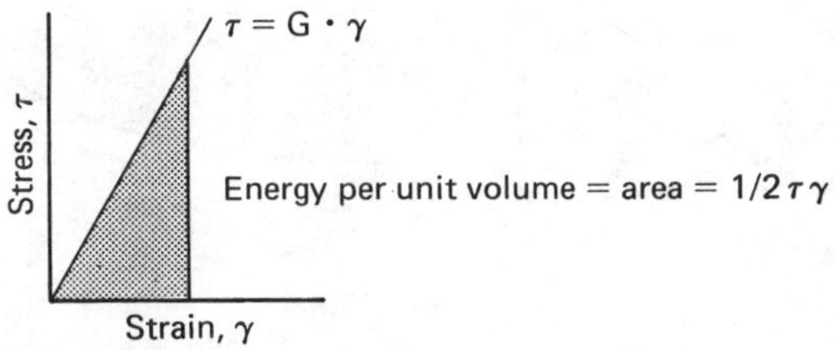

Figure 3.4.3

Since we will restrict ourselves, as usual, to the elastic range, we have $\tau = G\gamma$, and

$$U = \frac{\tau^2}{2G} \tag{3.4.6}$$

Consider now the element of tube along the Z axis (longitudinal axis) as shown in Figure 3.4.4. The shear strain energy on the elemental volume is

$$dU = \frac{\tau^2 t}{2G}\, ds \cdot dz \tag{3.4.7}$$

If we assume that we have a uniform tube and that the angle of twist per unit length is a constant, then we can write the total energy as an integration over the length and around the circumference.

$$U = \int_0^L \oint_s \frac{\tau^2 t}{2G}\, ds \cdot dz \tag{3.4.8}$$

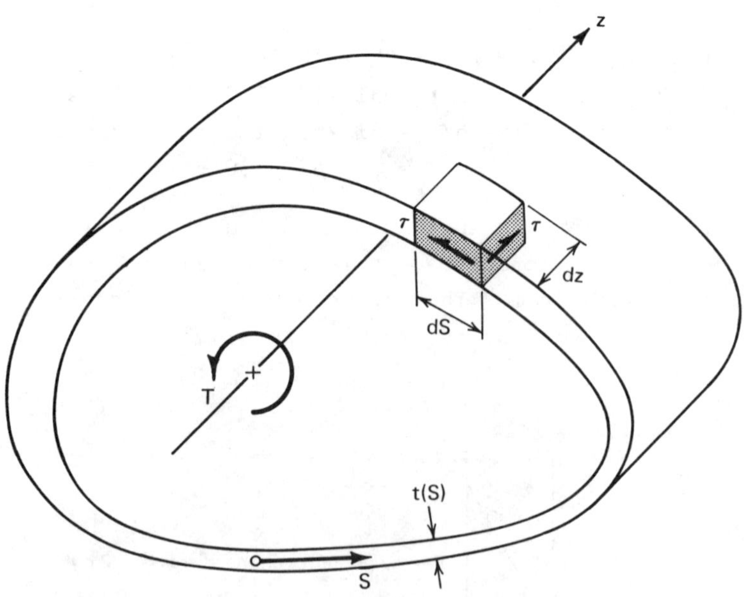

Figure 3.4.4

Using equations (3.4.1) and (3.4.4) in equation (3.4.8) gives

$$U = \int_0^L \oint_s \frac{T^2}{8 \textcircled{A}^2 t(s) G} \, ds \cdot dz \tag{3.4.9}$$

Bring out the variables which are not functions of S.

$$U = \int_0^L \frac{T^2}{8 \textcircled{A}^2 G} \, dz \oint_s \frac{ds}{t(s)} \tag{3.4.10}$$

By extending our discussion of Castigliano's theorem given in Chapter 1, we can infer that the derivative of the torsional strain energy with respect to the torque gives the angle of twist. Therefore:

$$\theta = \frac{\partial U}{\partial T} = \int_0^L \frac{T}{4 A^2 G} \, dz \oint_s \frac{ds}{t(s)} \tag{3.4.11}$$

and

$$\theta = \frac{TL}{4 \textcircled{A}^2 G} \oint_s \frac{ds}{t(s)} \tag{3.4.12}$$

If we compare this with

$$\theta = \frac{TL}{JG} \tag{3.3.12}$$

Then we see that the area property for this tube is given by

$$J_H = \frac{4 \textcircled{A}^2}{\int_s \frac{ds}{t(s)}} \tag{3.4.13}$$

<u>Example 3.4.2</u>

For the tube of example 3.4.1, find the angle of twist if the length is 6 ft. and the tube is made of AISI 1025 steel.

Solution:

$$\theta = \frac{TL}{GJ_H}$$

$T = 1{,}000$ in. lb

$L = 6 \times 12 = 72$ in

$G = 11 \times 10^6$ psi (Appendix A)

To calculate J_H, we can replace the integration by a summation along each leg of the tube, since t is constant in each leg.

$$J_H = \frac{4 Ⓐ^2}{\sum_{i=1}^{4} \frac{S_i}{t_i}}$$

$Ⓐ^2 = 16^2 = 256 \text{ in}^4$

$s_1/t_1 = 4/0.10 = 40$

$s_2/t_2 = 4/0.120 = 33.3$

$s_3/t_3 = 4/0.140 = 28.6$

$s_4/t_4 = 4/0.130 = \underline{30.8}$

$$\sum \frac{si}{ti} = 132.7$$

$$J_H = \frac{4 Ⓐ^2}{\sum \frac{si}{ti}} = \frac{4 \times 256}{132.7} = 7.71 \text{ in}^4$$

$$\theta = \frac{1000 \times 72}{11 \times 10^6 \times 7.71} = 0.000848 \text{ radians}$$

$$= 0.04864°$$

3.5 TORSION OF THE "OTHER" CROSS SECTIONS

For practical purposes, the method for stress analysis of cross-sections classified as "other" is simply stated: find the proper formula in a handbook or other reference. Reference (1), listed at the end of the chapter, is an excellent source.

In this section, we will list a few of the more common other shapes which are used in practice. One common shape is the rectangle, which is shown in Figure 3.5.1.

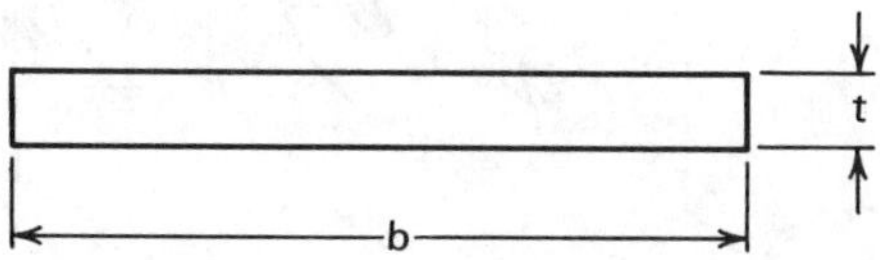

Figure 3.5.1

The ratio of b/t is unity for a square cross section and approaches infinity for a long, thin rectangular cross section. The appropriate formulas, presented without derivation, are:

$$\tau = \frac{T}{\alpha b t^2} \tag{3.5.1}$$

$$\theta = \frac{TL}{\beta b t^3 G} \tag{3.5.2}$$

Where α and β are given in Table 3.5.1.

Table 3.5.1

b/t	1.00	1.50	1.75	2.00	2.50	3.00	4	6	8	10	∞
α	0.208	0.231	0.239	0.246	0.258	0.267	0.282	0.299	0.307	0.313	0.333
β	0.141	0.196	0.214	0.229	0.249	0.263	0.281	0.299	0.307	0.313	0.333

Equations 3.5.1 and 3.5.2 can be used for rectangular shapes which are not flat, as shown in Figure 3.5.2. The dimension b is now taken to be the running length of the centerline.

For shapes made up of thin rectangles with legs of different thickness, such as shown in Figure 3.5.3, the formula

$$\theta = \frac{TL}{J_{eq} G} \tag{3.5.3}$$

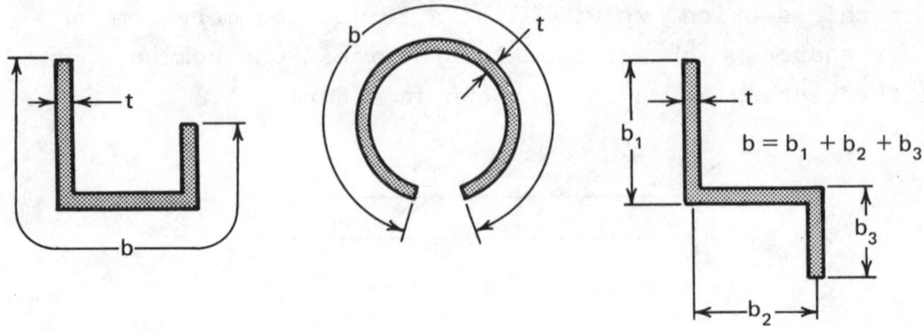

Figure 3.5.2

is used, with

$$J_{eq} = \beta_1 b_1 t_1{}^3 + \beta_2 b_2 t_2{}^3 + \beta_3 b_3 t_3{}^3 \qquad (3.5.4)$$

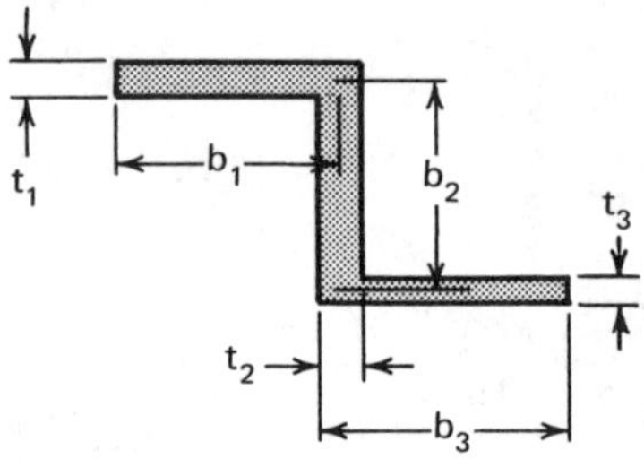

Figure 3.5.3

Sections of other shapes are infinite in variety. Table 3.5.2, taken from Reference (1) provides a sample. For more complex shapes, additional tables may be found in various handbooks and publications. References (2) thru (5) contain useful information.

Table 3.5.2

Any elongated section with axis of symmetry OX. U = length, A = area of section, I_x = moment of inertia about axis of symmetry.

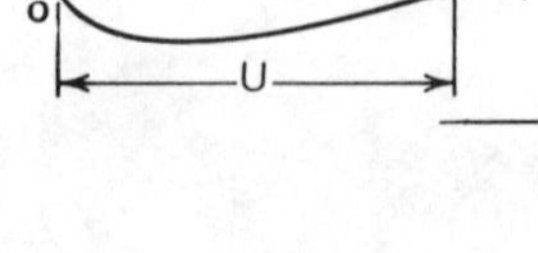

$$K = \frac{4I_x}{\left(1 + 16\,\dfrac{I_x}{AU^2}\right)}$$

Any elongated section or thin open tube. dU = elementary length along median line, t = thickness normal to median line, A = area of section.

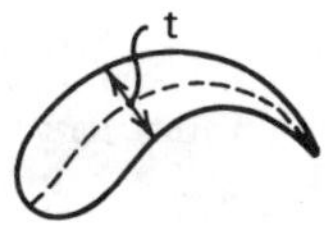

$$K = \frac{\frac{F}{3}}{\left(1 + \frac{4}{3}\frac{F}{AU^2}\right)}$$

where $F = \int_0^U t^3 dU$

Any solid, fairly compact section without reentrant angles. J = polar moment of inertia about centroidal axis; A = area of section.

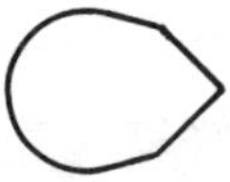

$$K = \frac{A^4}{40J}$$

I section, flange thickness uniform. r = fillet, radius, D = diameter largest inscribed circle, t = b if b < d, t = d if d < b, t_1 = b if b > d, t_1 = d if d > b.

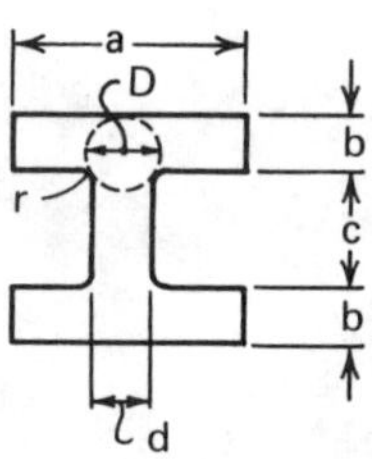

$$K = 2K_1 + K_2 + 2\alpha D^4$$

where

$$K_1 = ab^3\left[\frac{1}{3} - 0.21\frac{b}{a}\left(1 - \frac{b^4}{12a^4}\right)\right]$$

$$K_2 = \frac{1}{3}cd^3$$

$$\alpha = \frac{t}{t_1}\left(0.15 + 0.1\frac{r}{b}\right)$$

For all solid section of irregular form the max shear stress occurs at or very near one of the points where the largest inscribed circle touches the boundary,* and of these, at the one where the curvature of the boundary is algebraically least. (Convexity represents positive, concavity negative, curvature of the boundary.) At a point where the curvature is positive (boundary of section straight or convex) this max stress is given approximately by: $s = G\frac{\theta}{L}c$ or $s = \frac{T}{K}c$ where

$$c = \frac{D}{1 + \frac{\pi^2 D^4}{16A^2}}\left[1 + 0.15\left(\frac{\pi^2 D^4}{16A^2} - \frac{D}{2r}\right)\right],$$

where D = diameter of largest inscribed circle
r = radius of curvature of boundary at the point (positive for this case)
A = area of the section

At a point where the curvature is negative (boundary of section concave, or reentrant) this max stress is given approximately by $s = G\frac{\theta}{L}c$ or $s = \frac{T}{K}c$ where

$$c = \frac{D}{1 + \frac{\pi^2 D^4}{16A^2}}\left[1 + \left\{0.118 \log_e\left(1 - \frac{D}{2r}\right) - 0.238\frac{D}{2r}\right\}\tanh\frac{2\phi}{\pi}\right]$$

where D, A, and r have same meaning as before and ϕ = angle through which a tangent to the boundary rotates in turning or traveling around the reentrant portion, measured in radians. (Here r is negative.)

*Unless at some other point on boundary there is a sharp reentrant angle, causing high local stress.

Note: $\theta = \frac{TL}{KG}$

Circular sector

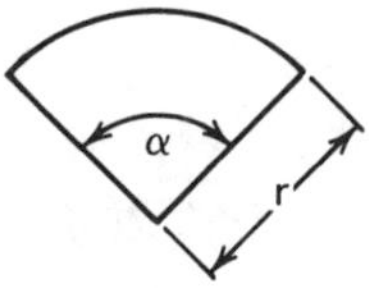

$K = Cr^4$

where C depends on α and has values as follows:

α	45°	60°	90°	120°
C	0.0181	0.0349	0.0825	0.148
α	180°	270°	300°	360°
C	0.296	0.528	0.686	0.878

Max $s = \frac{T}{Q}$ on radial boundary. Here $Q = Cr^3$, where C has values as follows:

α	60°	120°	180°
C	0.0712	0.227	0.35

Isosceles triangle

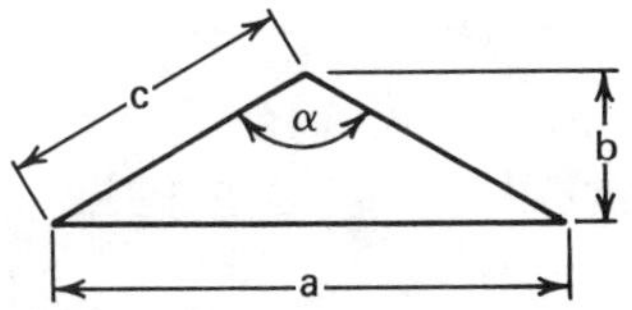

$$K = \frac{a^3b^3}{15a^2 + 20b^2}$$

For $\alpha = 90°$, $K = 0.0261c^4$

For $\alpha = 60°$, $K = 0.0216c^4$

$Q = 0.0666ab^2$

For $\alpha = 90°$, $Q = 0.0554c^3$

For $\alpha = 60°$, $Q = 0.050c^3$

Max S_s at center longest side

Trapezoid

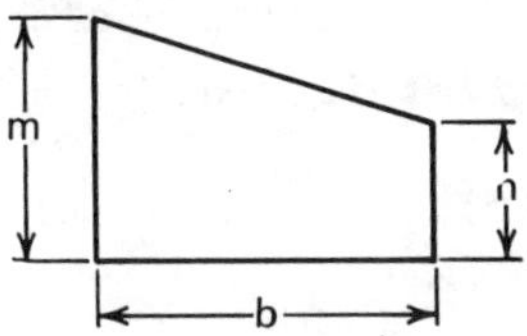

$$K = \frac{1}{12}b(m + n)(m^2 + n^2) - V_L m^4 - V_s n^4$$

where

$$V_L = 0.10504 - 0.10s + 0.0848s^2 - 0.06746s^3 + 0.0515s^4$$

$$V_s = 0.10504 + 0.10s + 0.0848s^2 + 0.0674s^3 + 0.0515s^4$$

and $s = \frac{m - n}{b}$

Max S as for elongated section cases.

Circular shaft with opposite sides flattened

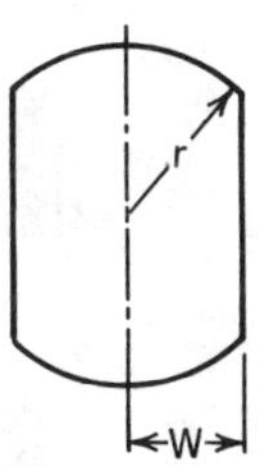

$K = cr^4$

where c depends on ratio w/r and has values as follows:

$\frac{w}{r}$	$\frac{7}{8}$	$\frac{3}{4}$	$\frac{5}{8}$	$\frac{1}{2}$
c	1.357	1.076	0.733	0.438

$Q = cr^3$

where c depends on ratio w/r and has values as follows:

$\frac{w}{r}$	$\frac{7}{8}$	$\frac{3}{4}$	$\frac{5}{8}$	$\frac{1}{2}$
c	1.155	0.912	0.638	0.471

Closed Book Problems

You should be able to answer these questions without referring to the text.

1. Give the formula for torsional shear flow in <u>thin-wall</u> members. Define all terms. Use a sketch.
2. T F This formula is restricted to circular thin-wall members.
3. T F The member must be a closed tube.
4. T F The wall of the member must be of constant thickness.
5. T F Shear stress is assumed to be constant across the thickness of the wall.
6. Sketch the distribution of shear stress magnitude along line OA as a function of radius r for a circular torsion bar.

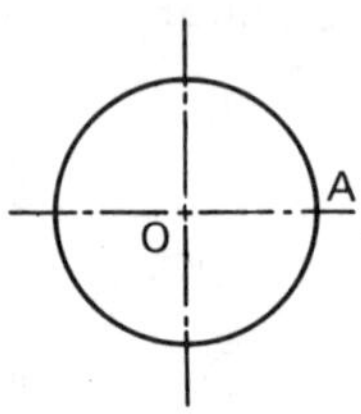

7. Classify the cross sections according to the following categories: (1) circular (solid or hollow); (2) hollow thin wall, closed; (3) other.

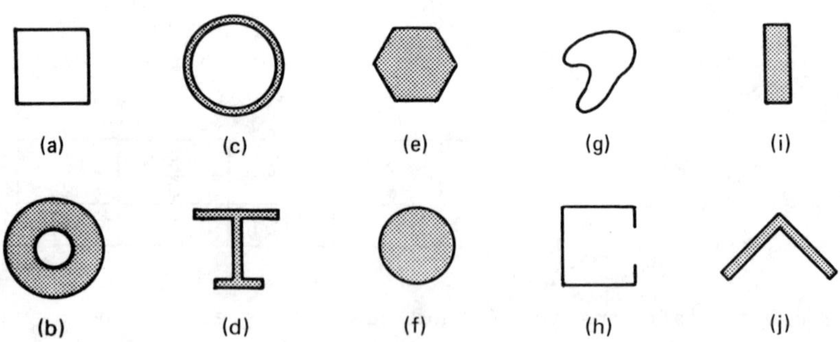

8. Derive the equation of: (a) deflection, and (b) shear strain of a circular cross section (in terms of: τ, θ, G, r, γ, z).

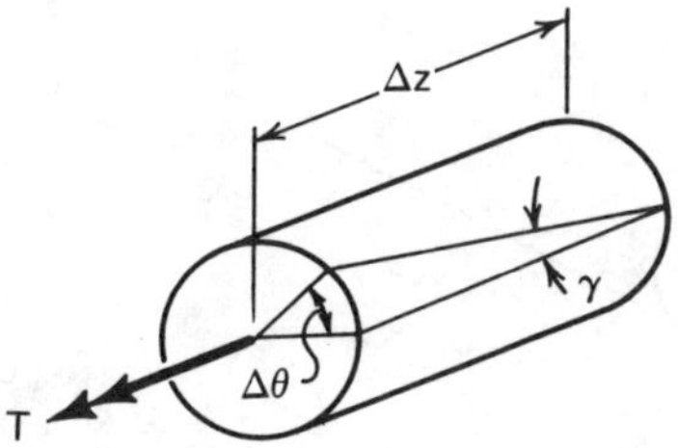

9. State the meaning, units, and physical significance of the terms of the answers to problem 8. (τ, θ, G, r, γ, z.)
10. State the equation which describes the shear stress in terms of (r, J, τ, M) for circular cross sections.
11. State the meaning and units of each term: (as in problem 10: r, J, τ, M).
12. A solid round steel bar is to be used to replace a torsion bar in a Porsche 914 (a car, for those who are not so inclined). The bar is 4 feet long, $G = 10 \times 10^6$. At a torque of 10,000 in-lb, the bar may deflect only .2 rad. Find the minimum diameter which may be used.
13. What is the shear stress under the conditions stated in problem 12?
14. State the formula for shear flow in a thin-walled, closed member in torsion.
15. Define the terms and state the units associated with the answer to problem 14.

Problems

Problem 1. The torque on a shaft is distributed along the length in a linear fashion such that

$$T(x) = T_o + \frac{x}{L}(T_L - T_o)$$

Derive an expression for the total angle of twist θ.

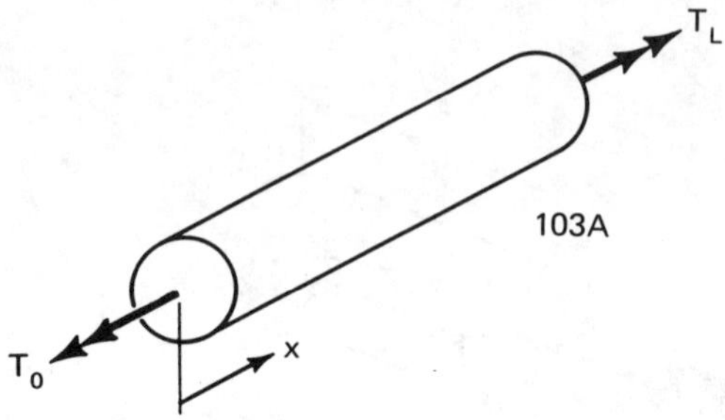

Problem 2. A prop is arranged as shown with a drive shaft length of 7.2'. For an engine HP of 265 and an RPM of 900, calculate the size of circular shaft required if the max shear stress is to be 25,000 psi. Calculate θ for this case. Select a material of strength of at least 50,000 psi. $F_{sy} \simeq 0.5\ F_{ty}$.

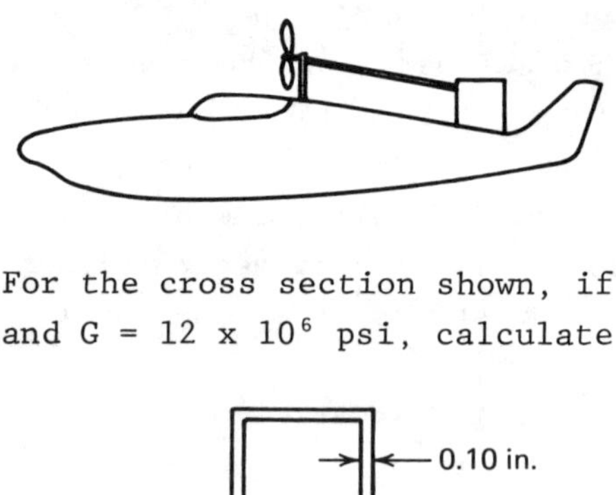

Problem 3. For the cross section shown, if L = 3', T = 10^3 in-lb and G = 12×10^6 psi, calculate τ_{max}, θ.

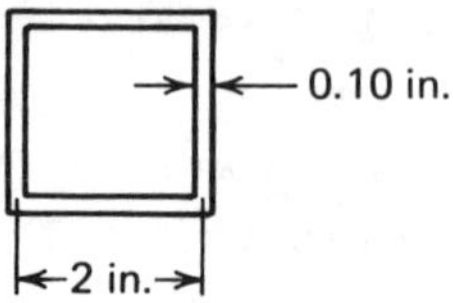

Problem 4. Calculate the maximum shear stress and the angle of twist for the sections shown. Each bar is a uniform straight bar, with torque T and length L as shown. Only the cross sections are shown. G = 4×10^6 psi for all problems.

Torque T = 2500 in-lb

L = 3'

104A

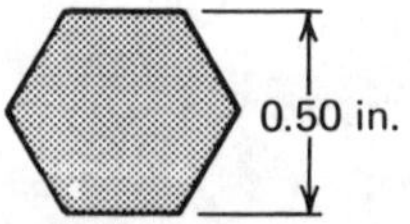

Problem 5. If the torque T = 3000 in-lb and L = 36", calculate the shear flow and the angle of twist for AlSl 1025 steel.

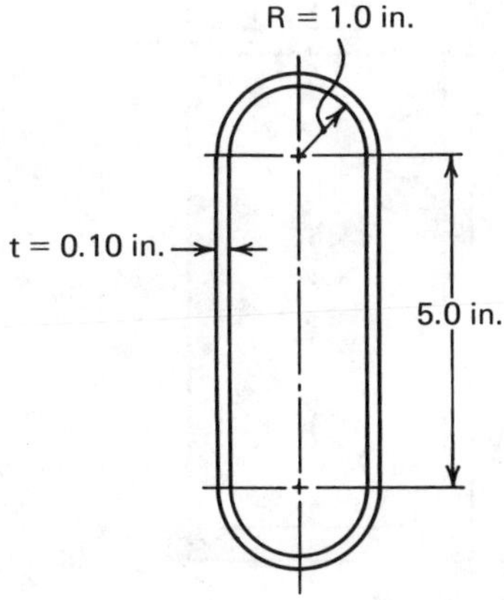

Problem 6. If the torque T = 600 in-lb and L = 60", calculate only the angle of twist for this section. $G = 5 \times 10^6$ psi.

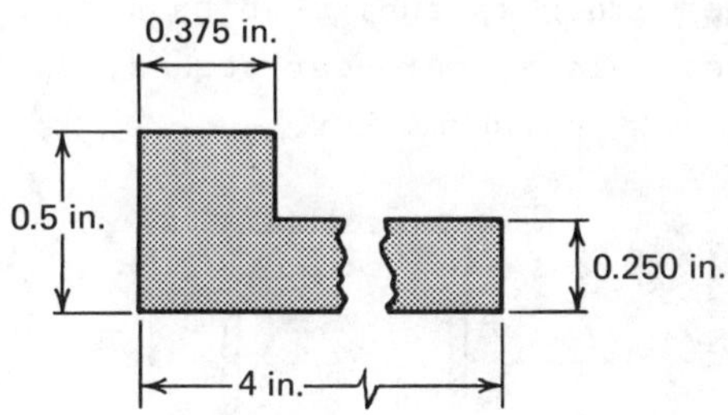

Problem 7. An equilateral triangular prism has a length of 12 inches and the triangle is one inch on each side. It is subjected to a moment of 2,000 in-lb. Find: (1) the torsional shear stress and (2) the angular deflection.

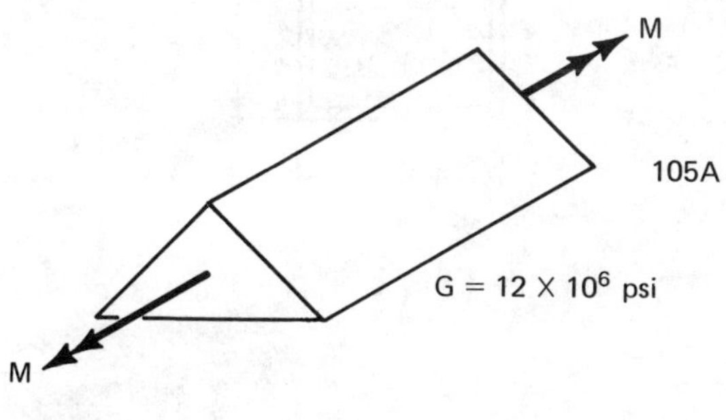

Problem 8. Calculate J for the section shown.

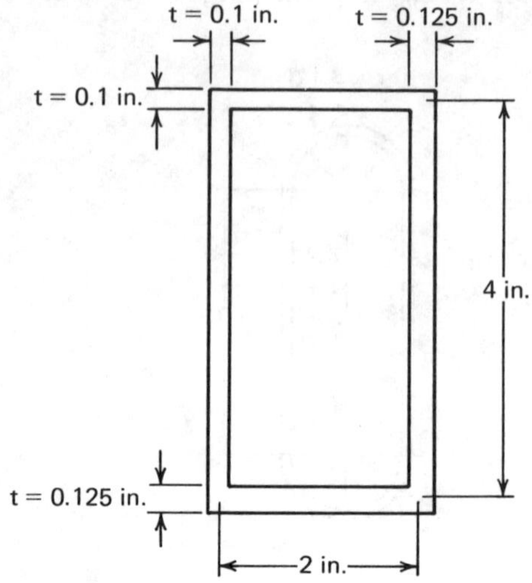

Problem 9. A member is made of 2024-T3 aluminum and has length of 12'. The cross section is as shown in the figure. If the member is subjected to a torque of 12,000 in-lb, calculate the shear stress, τ, and the angle of twist, θ, in the member.

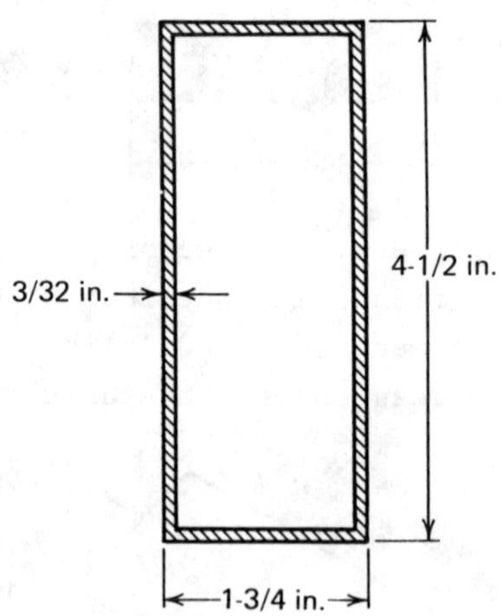

Problem 10. Compare the shear stress and angle of twist for the shaft of circular cross section with the shear stress and angle of twist of the same shaft with 1/2 in. diameter hole. T = 32,000 in-lb; L = 48 in.; G = 12 x 10^6 psi.

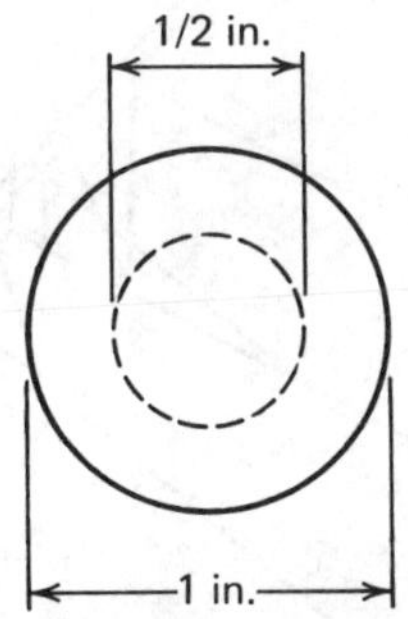

Problem 11. Calculate the angle of twist for the section shown.

T = 10,000 in-lb
L = 100 in.
G = 10^7 psi

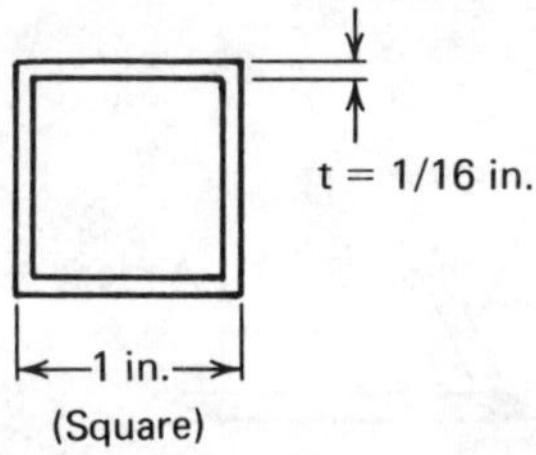

Problem 12. Calculate the moment, M, required to twist this bar through 1°. What is the max shear stress at this condition? Outside reference material will be needed.

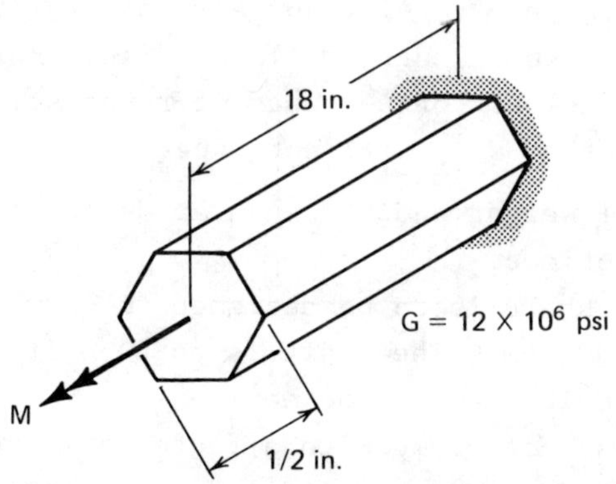

Problem 13.

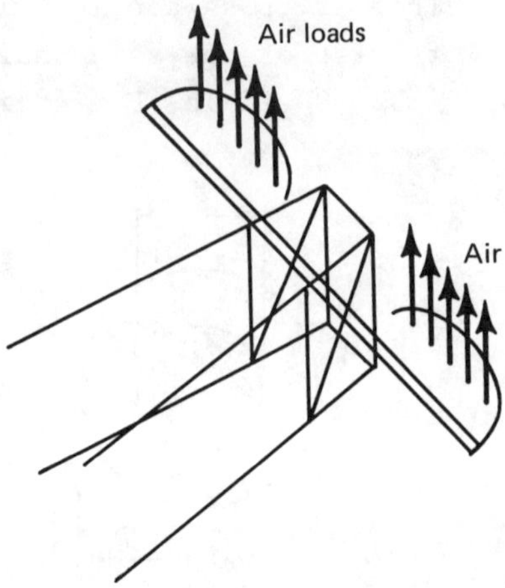

A small airplane is to have its elevators operate through a torque tube. Looking at 1/2 of the tube, a side view would be:

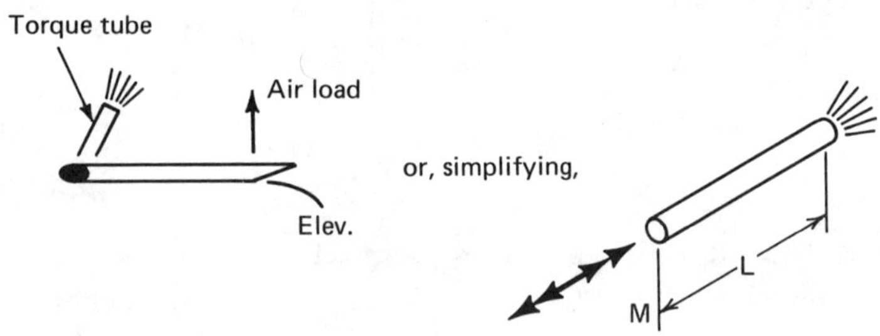

Consider a moment due to elevator air loads of 6,000 in-lb. The torque tube length will be 28 in. For material AlSl 4130 steel, select torque tubes of round-solid and round-tubular so that τ_{max} will not exceed 30,000 psi. Compare the total weight and the angular deflection for each of these sections.

Problem 14. A shaft is to be designed to carry 1,000 horsepower at 2500 rpm. The shaft is to be 4 ft. long.

The following criteria must be met:

A. The shaft must not yield with a factor of safety of 2.0.

B. The deflection must not exceed 1/3 of a degree.

1. Do the following:
   a. Size a solid, circular shaft to meet the criteria.
   b. Do this for at least three materials.
   c. State whether each shaft is stress or deflection governed.
2. Do at least one of the following:
   a. Take one of the shafts from part 1. Hollow out the inside and increase the O.D. in an attempt to minimize weight. The O.D. for the hollow shaft may not exceed 1.5 times the O.D. of the solid shaft. As the wall of the shaft becomes thin, it may buckle in torsion. See Chapter 10 for torsional buckling.
   b. Take one of the shafts from part 1. Try at least three "other" solid shapes of the same material and see if the circular cross section gives minimum weight.
   c. Take one of the shafts from part 1. Assume that this shaft is to have shoulders cut on the ends. The ratio of D/d must not exceed 1.25. For a solid circular shaft, size the D and d so that the shaft will not yield with a factor of safety of 2.0. (D will have to be larger than in part 1.) See Chapter 9 for stress concentrations in shafts like this.

Problem 15. Find the location and magnitude of the maximum shear stress in the circular torsion bar. If the material is AISC 36 steel, find the total twist of one end with respect to the other. Ignore stress concentration effects.

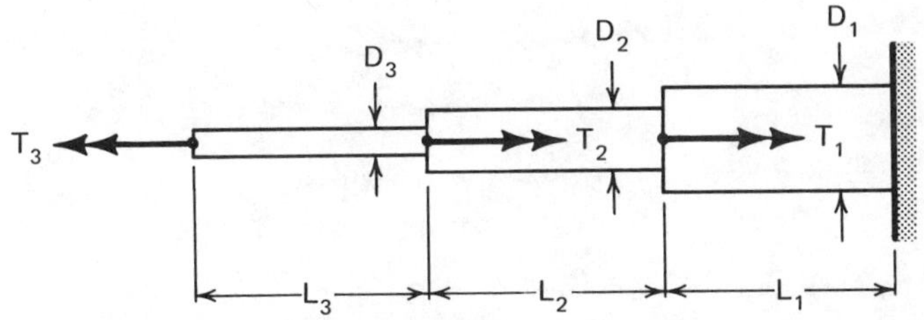

T_1 = 1000 in-lb D_1 = 1.0 in L_1 = 3 ft

T_2 = 200 in-lb D_2 = 0.75 in L_2 = 2 ft

T_3 = 100 in-lb D_3 = 0.5 in L_3 = 1 ft

Problem 16. Compare the maximum shear stress and the angle of twist, on a percentage basis, for the two sections. Let R = 1.5 in and t = 0.05 in.

References

1. "Formulas for Stress and Strain," Fourth Edition, Raymond J. Roark, McGraw-Hill, New York, 1965.
2. "Mechanics of Materials," F. R. Shanley, McGraw-Hill, New York, 1967.
3. "Aircraft Structures," David J. Peery, McGraw-Hill, 1950.
4. "Mechanics of Materials," Second Edition, E. P. Popov, Prentice-Hall, Englewood Cliffs, New Jersey, 1976.
5. "Analysis and Design of Flight Vehicle Structures," E. F. Bruhn, Tri-State Offset Co., 1973.

Chapter 4

THE BEAM ELEMENT IN BENDING AND SHEAR: STRESS AND DEFLECTION

4.1 INTRODUCTION

A beam is a relatively long, slender member which carries loads which are perpendicular to its longitudinal axis and which develops internal stresses due to its internal bending moment and shear force. A typical beam is depicted in Figure 4.1.1. The study of beams and their behavior has a long and fascinating history. Beams were first studied by Galileo (Reference 1). Figure 4.1.2 is a reproduction of Galileo's diagram for the bending of a cantilever beam.

Further developments in beam theory include the efforts of Jacob Bernoulli, Daniel Bernoulli and L. Euler (Reference 2). The beam is a very important structural element.

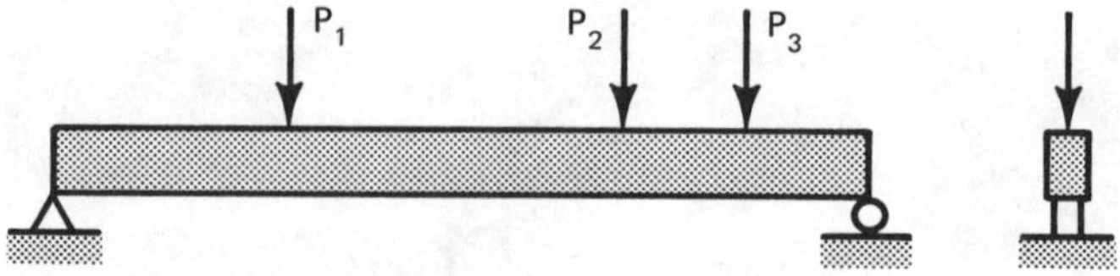

Figure 4.1.1

The first-approximation model of many structural elements is that they behave as a beam in bending. Some examples are: steel beams in a building, wooden floor joists in a home, freeways, freeway bridges and overpasses, railroad rails, power transmission shafting, airplane wings, fuselage and tail sections. An understanding of beam theory and behavior is essential to the understanding of many other kinds of structure.

Figure 4.1.2

Objectives for Beam Shear and Moment Diagrams

1. Be able to use the method of static equilibrium to calculate the reactions on the beam.
2. Be able to replace the graphical symbols for beam supports with the appropriate reaction(s).
3. Be able to draw a free-body diagram of the beam with the reactions in equilibrium with the applied loads.
4. Be able to state the definition of static indeterminancy. Be able to tell whether or not a beam is statically determinate.
5. Be able to slice a portion from the beam; to show the internal shear and bending moment in equilibrium with the external loads and reactions on this segment; and to use the equations of statics to calculate the internal shear and bending moment.
6. Be able to successively apply step number 5 and plot the shear and moment diagrams for the beam.
7. Be able to state and illustrate the definitions for positive and negative shear and moment.
8. Be able to replace distributed loadings with statically equivalent concentrated loads:
   A. On the entire beam to write external equilibrium equations and solve for the reactions.
   B. On an appropriate internal segment of the beam to write internal equilibrium equations and solve for the shear and moment.
9. Know from memory the relations:

$$q(x) = \frac{dV(x)}{dx} \tag{4.2.3}$$

and

$$V(x) = \frac{dM(x)}{dx} \tag{4.2.4}$$

10. Be able to integrate Equations (4.2.3) and (4.2.4). Be able to apply the boundary conditions to get the

constants of integration. Be able to determine and plot the relationships for V(x) and M(x) using this procedure.

11. Be able to use the five observations about the shear diagram and the seven observations about the moment diagram to sketch the proper shapes and slopes of V(x) and M(x) diagrams.
12. Be able to apply the method of superposition to obtain V(x) and M(x).

4.2 SHEAR AND MOMENT DIAGRAMS

It is expected that the student will have a background in shear and moment diagrams from his study of statics. A brief review of the essentials will be given. Consider the beam of Figure 4.2.1. (This particular example will be referred to several times.)

Let us first discuss the significance of the graphical symbols for the support of the beam. The beam of Figure 4.2.1 is said to be simply supported. The triangle support at the left end indicates a pinned connection. This symbol means that both a horizontal and vertical reaction can occur at this point. (See Figure 4.2.2a.) The second support will carry a vertical reaction only. (See Figure 4.2.2b.) This is often called a roller. (The pin and roller will keep the beam in position under up loading as well as down. The symbols are only graphical simplifications.) Figure 4.2.2c shows a cantilevered support which will carry both a vertical force and a moment. See Figure 4.1.2 again.

Returning now to Figure 4.2.1, the first step in the analysis of internal shear and moment is to SOLVE FOR THE REACTIONS. The reactions are assumed in accordance with the convention of Figure 4.2.2. It is not necessary to assume the correct direction. Then the principles of statics are used to write the external equilibrium equations in which the reactions will be in equilibrium with the applied loads. This is shown in Figure 4.2.1b from which

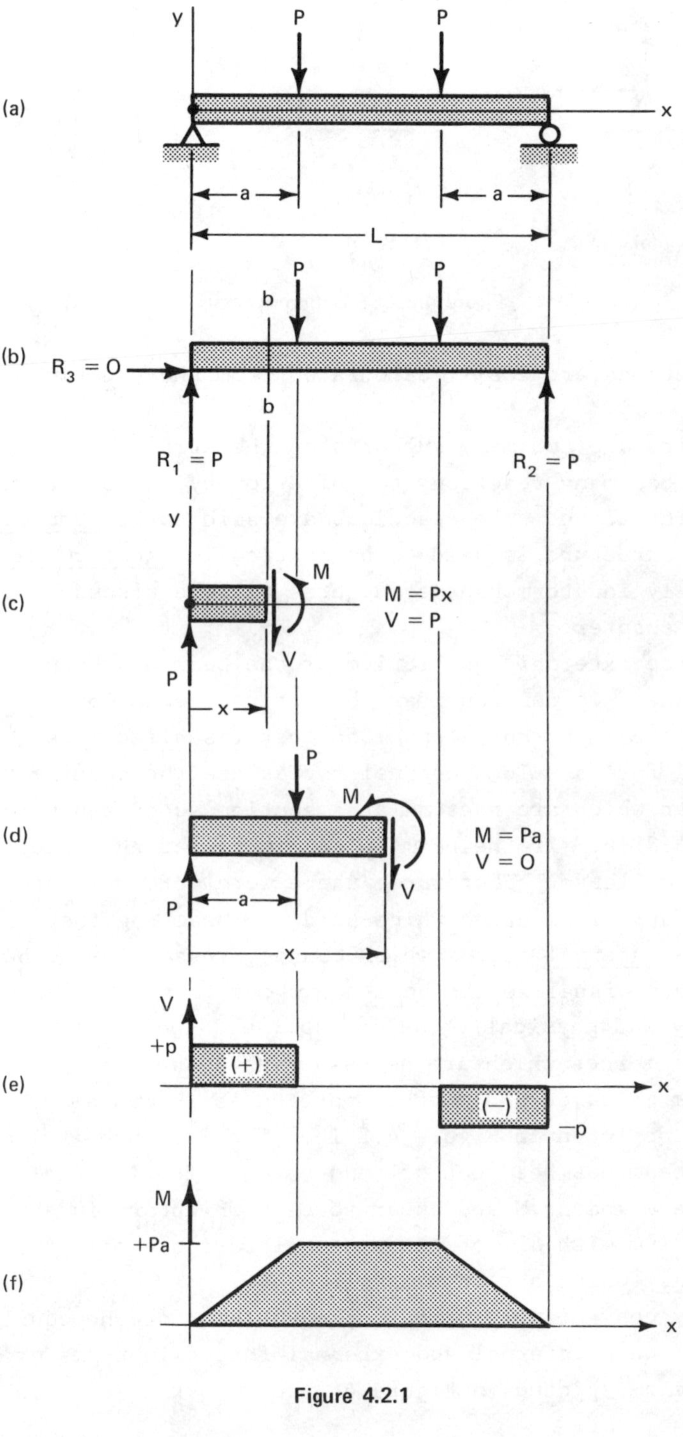
(a)
y
P
P
x
a
a
L
(b)
R3 = O
b
b
R1 = P
R2 = P
(c)
M
M = Px
V = P
V
P
x
(d)
P
M
M = Pa
V = O
V
a
x
(e)
V
+p
(+)
(–)
–p
x
(f)
M
+Pa
x

Figure 4.2.1

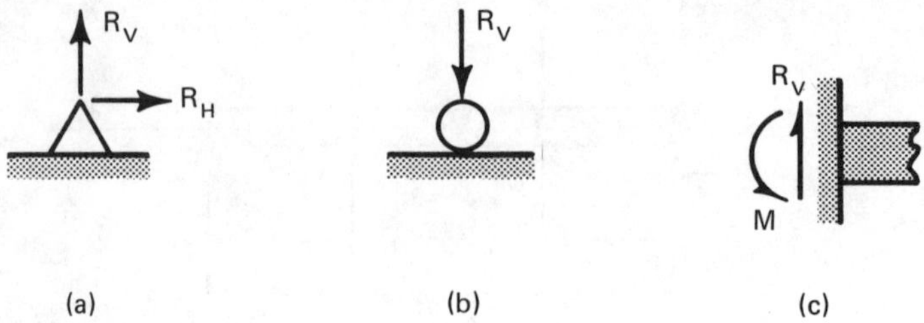

Figure 4.2.2 Reaction symbols

the reactions are easily calculated from the equations of statics.

If there are too many supports on the structure, there will be too many reactions to solve for by the equations of statics alone. The reactions are said to be redundant and the structure is said to be statically indeterminate. Statically indeterminant structures will be discussed in a later chapter.

Once the external equilibrium of the beam has been established, we can consider the internal equilibrium. This is the important step. One must visualize these internal forces. The internal forces are the bending moment and shear which are present at a section such as b-b in Figure 4.2.1b. If the beam is cut in two at this section, it will collapse. Therefore, there were internal forces present at this section which held the beam together. It is these internal forces which cause stresses in the beam. One should visualize the beam being cut; see it being cut mentally and physically and on paper and look at the internal forces which are necessary when one portion of the beam is saved and another portion is thrown away. This is depicted in Figure 4.2.1c. The left hand "slice" of the beam has been cut off and saved. The internal forces are moment M and shear force V which are in equilibrium with all of the external forces--including the reactions--which act on this slice.

We may conveniently write our equations for the equilibrium between internal and external forces in terms of the variable x depicted in Figure 4.2.1c. In this case, x may vary from x = 0.0 to x = a. If x is greater than a, then we will be on the other side of load P and our free body diagram will change.

The equations of static equilibrium now give

$$M = Px \tag{4.2.1}$$

$$V = P \tag{4.2.2}$$

When x is in the range a < x < (L-a), we must use the free body diagram of Figure 4.2.1d. We must change our equations each time we cross a concentrated load or moment.

The equations of equilibrium for the left hand segment of the beam shown in Figure 4.1.3d give

$$P - P - V = 0$$
$$V = 0 \tag{4.2.3}$$

$$Px - P(x - a) - M = 0 \tag{4.2.4}$$
$$M = Pa$$

To complete the shear and moment diagram for the entire beam, we could let x vary until x = L and redraw the free body diagrams each time we crossed a concentrated load or reaction. In this example, it is easier to take advantage of the <u>symmetry</u> of the problem and make the right half of the moment diagram look like the left. The shear will be <u>anti-symmetric</u> and the right half will have a negative sign. We shall discuss the sign conventions next. The shear and moment diagrams for the example are shown in Figures 4.2.1e and f.

A discussion of the <u>sign convention</u> for internal shear and bending moment is usually helpful. Students frequently have trouble writing the equations of statics and getting the shear and moment diagrams to come out right because they do not understand the sign conventions. Let us first look at the definition of positive bending moment in Figure 4.2.3a and b. It is extremely important to know

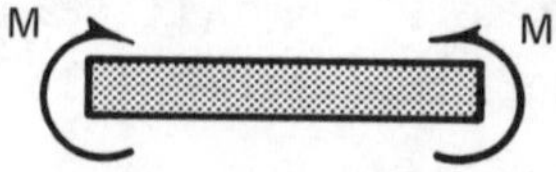

(a) Positive bending moment

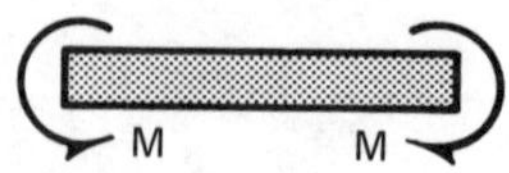

(b) Negative bending moment

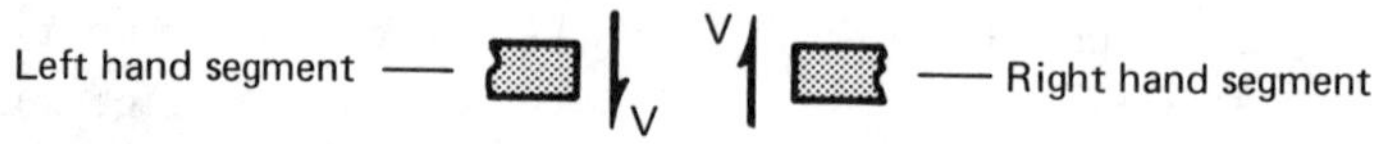

(c) Positive internal shear

(d) Negative internal shear

Figure 4.2.3

that we are looking at the internal moment (bending moment) and not any external moment.

The rules are simple and must be memorized. Positive bending moment puts the top of the beam in compression and the bottom in tension. (The beam bends concave up and holds water.) Negative bending moment puts the bottom of the beam in compression and the top in tension. Positive internal shear acts down on a left hand segment or up on a right hand segment. This is shown in Figure 4.2.3c and d. Negative shear is the opposite of positive shear. Note that up or down or clockwise or counterclockwise alone does not decide whether or not the shear or moment is positive or negative.

The kinds of loading which may be put on a beam are shown in Figure 4.2.4. This list is not exclusive, so that there may be various other kinds of loading as well. Note in particular in Figure 4.2.4b that there is an important distinction between an applied external moment,

M_e, and the internal bending moment in the beam. Any combination of the loadings shown may also be applied.

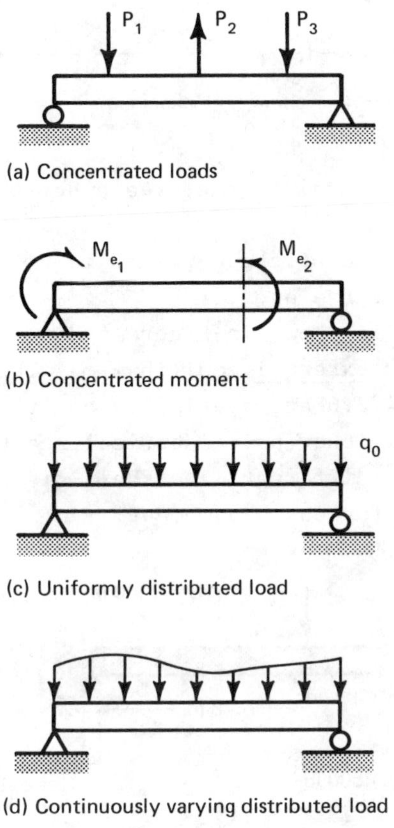

Figure 4.2.4 Types of beam loading

Many times beams are loaded by distributed loading. Examples of this are: occupant, equipment and structure loads in buildings, snow loads on roofs and air loads on aircraft. This situation is depicted in Figure 4.2.5, where the load intensity is given in terms of lb per running span of the beam.

Two principles can be used to replace the distributed load by its static equivalent for purposes of calculating the reactions:

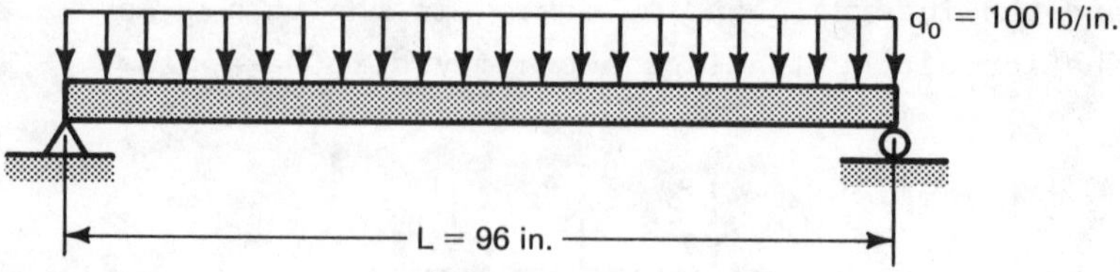

Figure 4.2.5

1. The net load is equal to the area under the loading curve.
2. The line of action of the net load is at the centroid of the loading curve area.

This replacement is shown in Figure 4.2.6. NOTE: This step works <u>only</u> for <u>external</u> equilibrium. The beams of Figures 4.2.5 and 4.2.6 are statically equivalent externally; that is, they have identical reactions. Their internal forces are different. We shall discuss how to treat the internal forces in a moment.

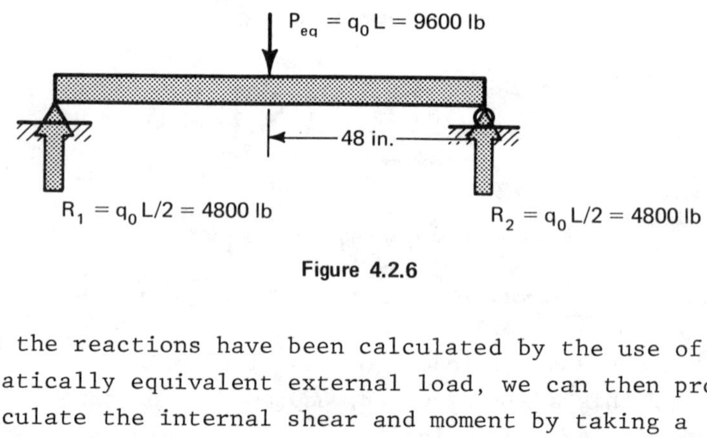

Figure 4.2.6

Once the reactions have been calculated by the use of the statically equivalent external load, we can then proceed to calculate the internal shear and moment by taking a slice of the beam and considering the real (distributed) load which acts on this portion. The distributed load <u>and</u> the previously calculated reactions will be in equilibrium with the internal shear and moment. It is convenient to assume that the shear and moment are positive. It is also convenient, when working with continuous functions, to work out the results for a general coordinate x. This is shown in Figure 4.2.7. Now we can replace the distributed loading acting on the appropriate <u>piece</u> of the beam with its static equivalent as shown.

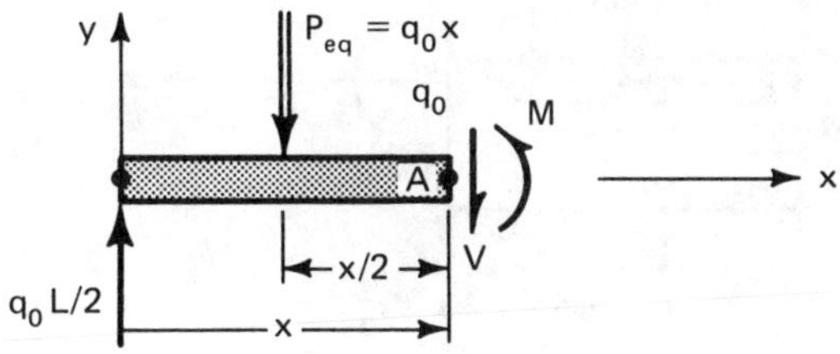

$$\Sigma F_V = 0$$

$$q_0 L/2 - q_0 x - V = 0$$

$$V = q_0 L/2 - q_0 x = 4800 - 100x$$

$$\Sigma M_A = 0$$

$$M + q_0 x(x/2) - (q_0 L/2)(x) = 0$$

$$M = (q_0/2)(Lx - x^2)$$

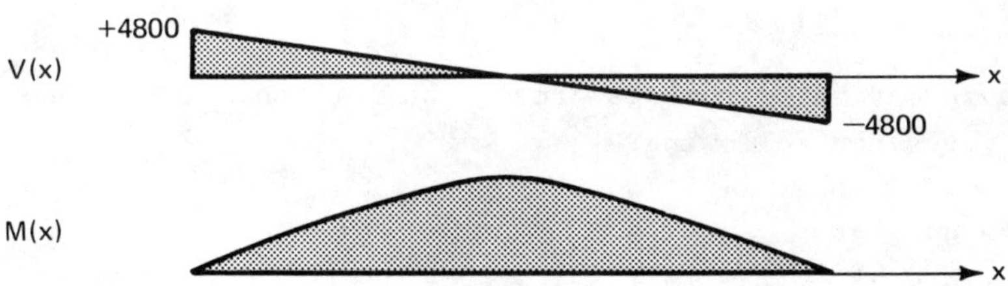

Figure 4.2.7

There is a convenient differential and integral calculus relationship which relates load, shear and bending moment. It is especially convenient to use these relationships for continuous loading functions. Consider Figure 4.2.8, where the loading function is shown, along with the shear and moment relations. All quantities are shown acting in their <u>positive</u> sense.

The load is shown as a varying function, but as $dx \to 0$ in the limit, the static equivalent force will be $q(x)dx$. This force acts at $dx/2$ as shown.

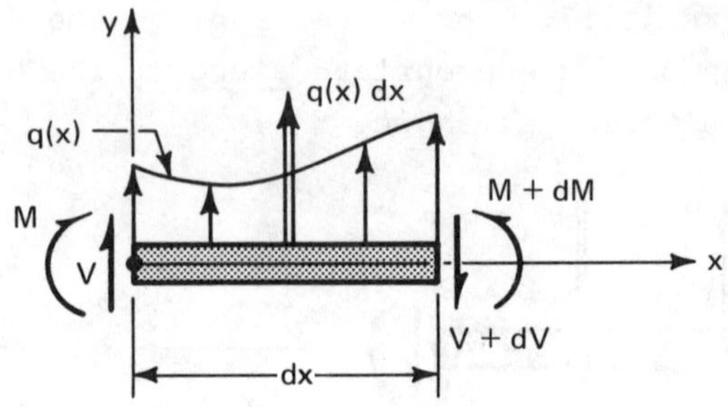

Figure 4.2.8

The equilibrium of forces in the vertical direction will give:

$$\sum F_v = 0$$

$$V + q(x)dx - (V + dV) = 0$$

$$q(x)dx - dV = 0$$

$$\boxed{q(x) = \frac{dV(x)}{dx}} \qquad (4.2.5)$$

The summation of moments around the left end of the segment gives the following:

$$\sum M = 0$$

$$-M + q(x)dx \cdot \frac{dx}{2} + (M + dM) - (V + dV)dx = 0$$

If we ignore higher order terms

$$dM - Vdx = 0$$

$$\boxed{V(x) = \frac{dM(x)}{dx}} \qquad (4.2.6)$$

Equations 4.2.5 and 4.2.6 will be illustrated with an example. Figure 4.2.9 shows a beam with continuous, linearly varying load.

The first step, as always, should be to find the reactions by statics. The statically equivalent external load is

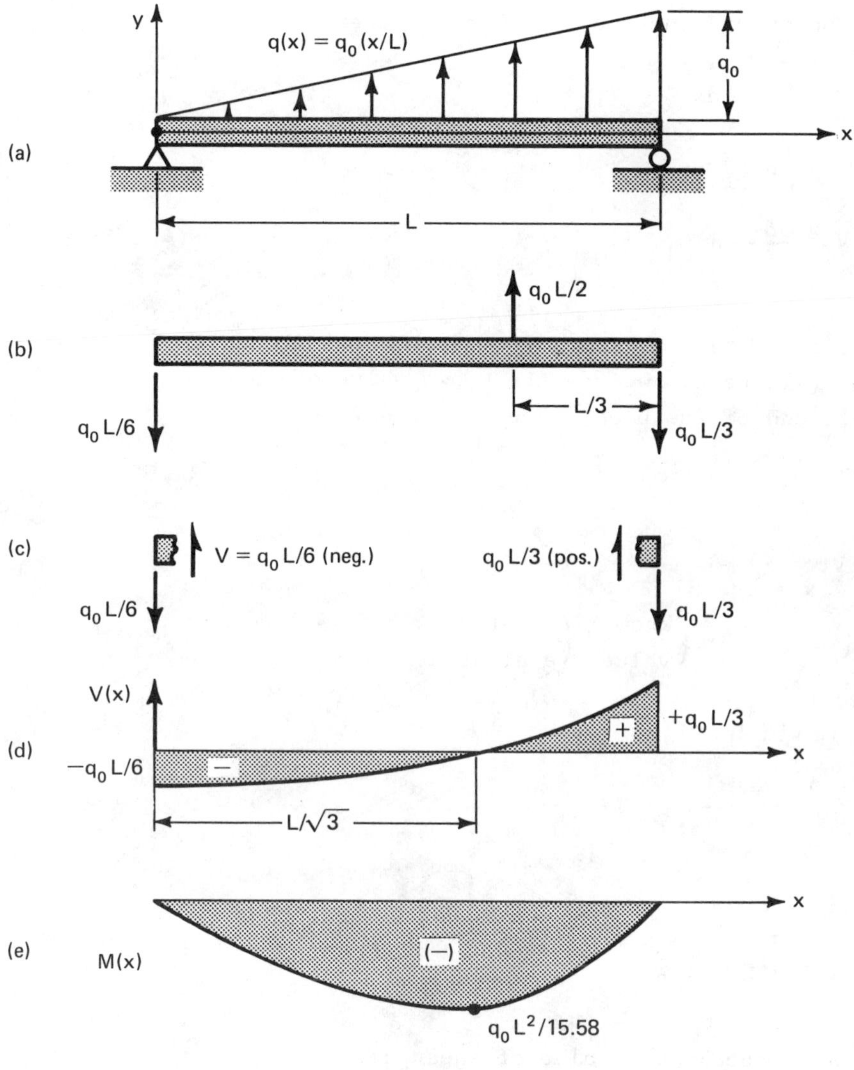

Figure 4.2.9

shown in Figure 4.2.9b. The equivalent load equals the area under the loading triangle, or $\frac{q_o L}{2}$. The line of action is through the centroid of the loading triangle. The reactions are easily calculated by the method of static equilibrium.

The shear function may then be determined by

$$V = \int q(x)\,dx$$

$$= \int q_o\left(\frac{x}{L}\right)dx$$

$$V = \frac{q_o x^2}{2L} + C_1 \qquad (4.2.7)$$

The constant C_1 comes from the <u>indefinite</u> <u>integral</u>. To evaluate C_1, we use the known value of shear at the left end of the beam. See Figure 4.2.9c.

$$V\Big|_{x=0} = -\frac{q_o L}{6} = \cancel{\frac{q_o x^2}{2L}}^{\,0} + C_1$$

$$\begin{pmatrix}\text{known}\\ \text{value}\end{pmatrix}\begin{pmatrix}\text{Equation 4.2.5}\\ \text{evaluated at } x = 0\end{pmatrix}$$

from which

$$C_1 = -\frac{q_o L}{6}$$

and

$$V = \frac{q_o x^2}{2L} - \frac{q_o L}{6} \qquad (4.2.8)$$

As a check, the value of shear at the right end of the beam should be

$$+\frac{q_o L}{3}$$

(See Figure 4.2.9c.) We use $x = L$ in Equation 4.2.8.

$$V\Big|_{x=L} = \frac{q_o L^2}{2L} - \frac{q_o L}{6}$$

$$V = \frac{q_o L}{3} \qquad \text{(checks)}$$

The shear function is plotted in Figure 4.2.9d. Note that the shear does the following:

1. Starts negative and ends positive.
2. Is second order in x. (Equation 4.2.8.)
3. Has a slope which is initially zero and then becomes increasingly positive. (Slope of shear curve equals magnitude of loading curve at any particular value of x.) (Equation 4.2.5.)
4. At any point x_A, the <u>change</u> in magnitude of the shear from another point x_B is called V_{A-B} and is equal to the area under the loading curve between the same two points. (Equation 4.2.5.)
5. V goes through zero at

$$V = 0 = \frac{q_o x^2}{2L} - \frac{q_o L}{6}$$

$$\frac{\not{q}_o x^2}{2L} = \frac{\not{q}_o L}{6}$$

$$x^2 = \frac{L^2}{3}$$

$$x = L/\sqrt{3}$$

This is also the point at which the area under the loading curve has magnitude $q_o L/6$.

The moment function is found from Equation (4.2.6).

$$M = \int V dx + C_2$$

$$M = \int \left(\frac{q_o x^2}{2L} - \frac{q_o L}{6} \right) dx + C_2$$

$$M = \frac{q_o x^3}{6L} - \frac{q_o L x}{6} + C_2$$

The constant C_2 is evaluated from the known value of bending moment at the left end of the beam. Because of the simple support, the beam cannot have a moment at the left end, therefore:

$$M(x)\Big|_{x=0} = 0 = \cancelto{0}{\frac{q_o x^3}{6L}} - \cancelto{0}{\frac{q_o Lx}{6}} + C_2$$

$$C_2 = 0$$

$$M(x) = \frac{q_o x^3}{6L} - \frac{q_o Lx}{6} \tag{4.2.9}$$

As a check, M(x) should also go to zero at the right end support.

$$M(x)\Big|_{x=L} = \frac{q_o L^2}{6} - \frac{q_o L^2}{6} = 0 \qquad \text{(checks)}$$

The moment function is plotted in Figure 4.2.9e. Some features of the moment curve are:

1. It is zero at $x = 0$ and $x = L$.
2. It is negative everywhere else which can be seen from the nature of the support, the direction of loading and the definition of negative bending.
3. It is third order in x.
4. Has an initial negative slope which decreases to zero and then goes positive. Equation (4.2.6).
5. Is a minimum at $x = L\sqrt{3}$. Equation (4.2.6) and Figure 4.2.9.
6. Has a <u>change</u> in magnitude between points A and B, called M_{A-B} which is equal to the area under the shear curve between the same two points.
7. Has a minimum of magnitude

$$M(x) = \frac{q_o L^2}{15.58}$$

The features of the shear and moment curves which have been discussed in this example can be used to sketch shear and moment curves and explore their significant features.

Problems

Problem 1. For the beam shown, draw a segment of the beam, draw the free-body diagram and calculate the values of V and M for each of the indicated segments.

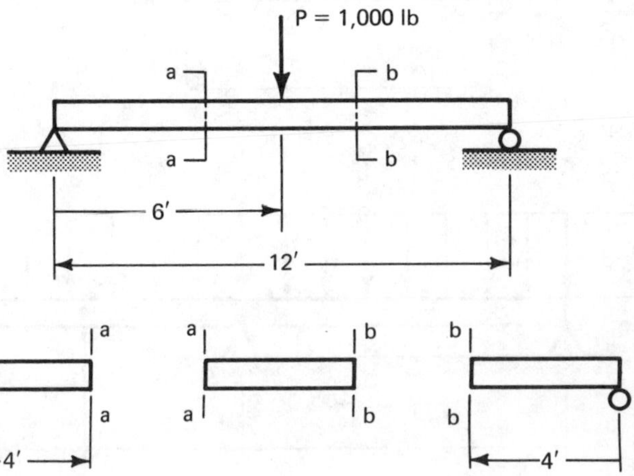

Problem 2. Rework the example of Figure 4.2.1 by starting at the right hand support and proceeding to the left.

Problem 3. Find the shear and moment diagrams for the beam shown by:

a) working with both loads at once

b) using superposition, as indicated, and adding the separate results

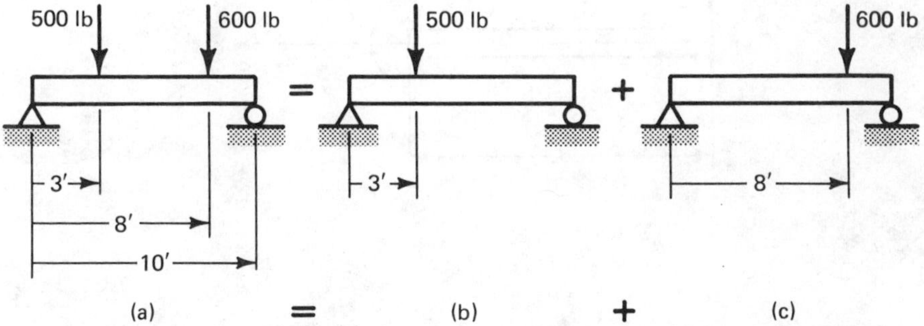

Problem 4. For the beam shown, find the maximum value of shear and moment and draw the shear and moment diagrams.

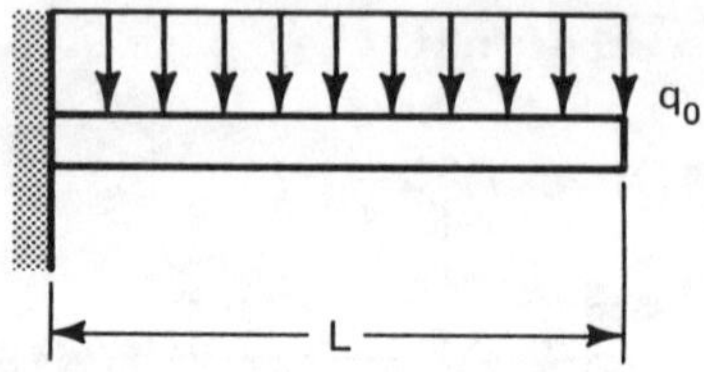

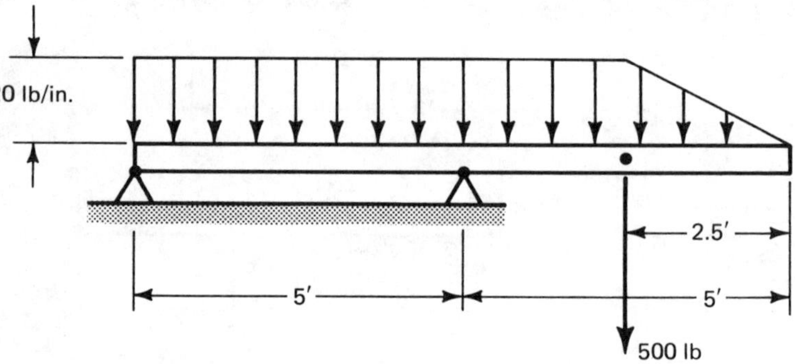

Problem 5. Draw this beam to scale: Find the shear and moment diagrams and draw them to scale. Label all the significant points. The three figures should fill one page.

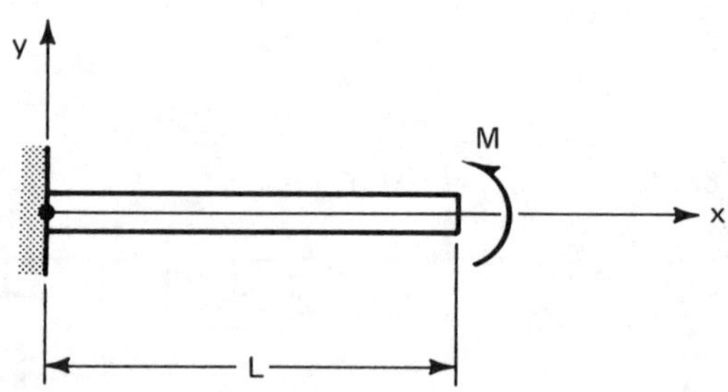

Problem 6. Find M(x) and V(x) for the beam shown.

Objectives for the Beam in Symmetric Bending

(Closed Book Unless Otherwise Noted)

1. Be able to define symmetric bending with words and illustrations.
2. Be able to distinguish between symmetric and unsymmetric bending.
3. Be able to state the assumptions and limitations of the bending stress equations.
4. Be able to define the neutral axis of a beam in words and illustrations.
5. Be able to state Equation (4.3.16) from memory and define the terms.
6. Be able to calculate the bending moment at any point on the span of a given beam. At this point on the span, be able to calculate the bending stress with respect to depth. Be able to sketch the stress distribution.

4.3 THE BEAM IN SYMMETRIC BENDING

At any point along the span of the beam the bending moment gives rise to internal stresses. Let us consider a cross section of the beam shown in Figure 4.3.1(a).

If we examine the end view of the cross section, we will view the moment vector and the beam cross section as shown in Figure 4.3.1(b). It is important at this point to visualize how the bending moment on the beam actually causes uniaxial tension and compression--the simple state of stress from Chapter 1--along the longitudinal directions of the beam. The beam may be visualized as a collection of closely, bundled longitudinal fibres--just like a handful of spaghetti. Some of the fibres are shown in end view in Figure 4.3.1(b). Figure 4.3.1(c) shows an isometric view. (Only a few of the "fibres" are shown.)

It can be seen that the bending moment causes the upper fibres to be compressed and the bottom fibres to be extend-

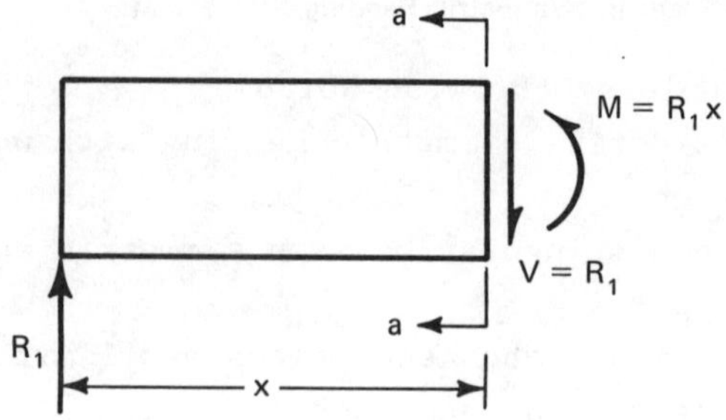

(a)

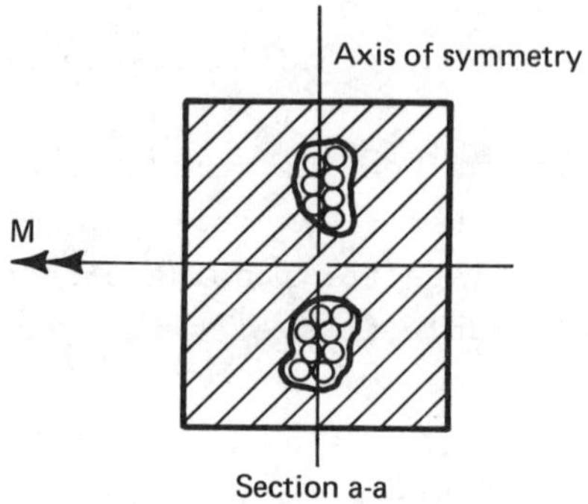

(b)

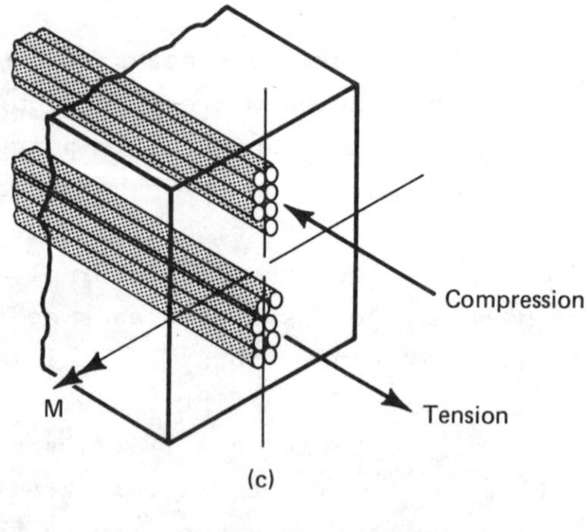

(c)

Figure 4.3.1

ed. Thus the bending moment causes beam tensile and compressive stresses which are uniaxial stresses along the longitudinal axis of the beam. These are not necessarily the only stresses which take place but these are the ones which we will examine in detail now.

We have considered the cross section to be a rectangle for the sake of simplicity. The remarks of this section apply equally well to sections meeting the requirements for symmetric bending. There are two requirements for symmetric bending:

1) The cross section must have an axis of symmetry.
2) The moment vector, as depicted in Figure 4.3.1(b) must be perpendicular to the axis of symmetry.

Some other examples of cross sections which meet the two requirements for symmetric bending are shown in Figure 4.3.2.

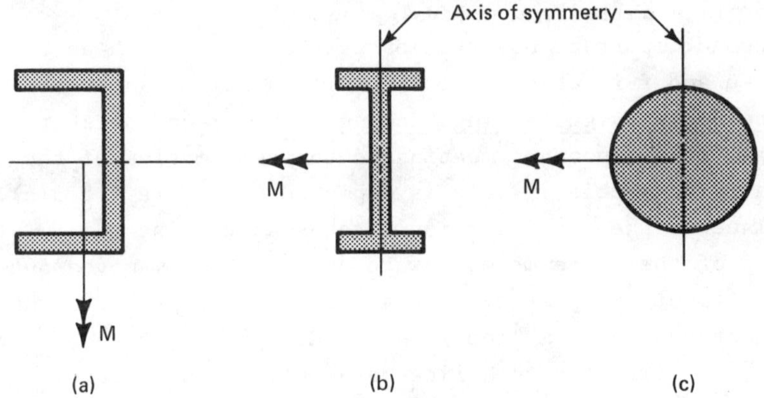

Figure 4.3.2

It is especially important to assure that the formulas which are developed in this cross section for symmetric bending are only used when appropriate. The methods for treating unsymmetric bending will be discussed in the next section.

We now turn our attention to the study of the geometry of bending--much as we did in the torsion derivation of Chapter 3. Let us consider the action of bending alone. For example, we can consider the mid-portion of the beam of Figure 4.2.1. In this region of the span, the bending moment is constant. The first observation that we make is that bending of this portion of the beam will cause the top to go in compression and the bottom to go in tension. Furthermore, the curvature of the beam will be circular. This is shown in Figure 4.3.3(a). (In this figure, the curvature is exaggerated for purpose of illustration.) The radius of curvature, R, is measured to the line which, in end view, passes through the centroid of the cross section. (Line a-a.) The line is called the neutral axis. All fibres above line a-a are in compression; all fibres below it are in tension. The line a-a and the line b-b (which is coincident with moment vector M) form a plane called the neutral plane.

A second important observation is that lines such as c-c, d-d, e-e and f-f, which were initially straight and vertical, remain straight during bending. The preceding statement has some important meaning to the deformation of the entire cross section. This is depicted in Figure 4.3.3(b). The plane of the cross section rotates about line b-b. The top of the plane moves inward and the bottom moves outward. All of the particles on a line such as g-g move the same distance to form line g'-g'. This is also true of any set of particles on a line parallel to b-b. All of the particles on such a line will move the same distance. (But not the same as the particles on line g-g.) This geometric behavior is customarily described by the statement that during bending plane sections remain plane.

Let us now return to Figure 4.3.3(a) and look at what happens with respect to the y axis (depth) as far as strain and deformation are concerned. Consider a portion of length dx as shown. At some distance y up from the neutral axis, the material is undergoing compression δ as shown. (We have previously defined positive moment as one which

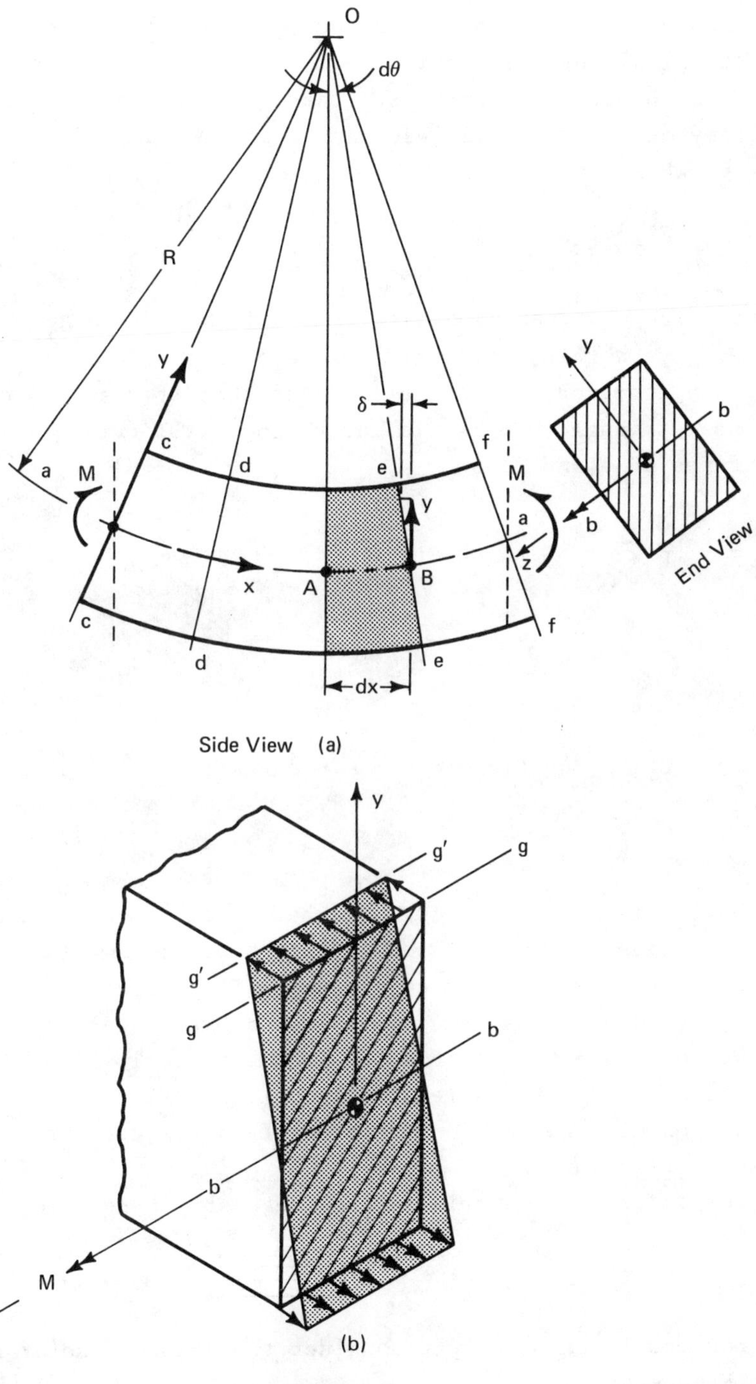
O
dθ
R
y
δ
c
d
e
f
a
M
y
a
x
A
B
z
c
d
e
f
dx
b
b
End View
Side View (a)
y
g′
g
g′
g
b
b
M
z
(b)

Figure 4.3.3

causes compression in the top of the beam.) Before bending, all of the longitudinal fibres in the shaded trapezoid were of length dx. Therefore, the axial strain of any one of these fibres, may be written as a function of y, which is

$$\varepsilon(y) = -\frac{\delta(y)}{dx} \tag{4.3.1}$$

(Positive y causes compressive strain which is negative.)

We have already discussed the fact that $\delta(y)$ is a linear function. To get $\delta(y)$, note that, for small deflections, triangle OAB is similar to the small triangle defined by y and δ, so

$$\frac{\delta(y)}{y} = \frac{dx}{R} \tag{4.3.2}$$

or

$$\delta(y) = \frac{y}{R}\,dx \tag{4.3.3}$$

Substituting Equation (4.3.3) into Equation (4.3.1) gives

$$\varepsilon(y) = -\frac{y}{R} \tag{4.3.4}$$

Now, this may be getting somewhat involved, so this may be the time to go back over the geometric arguments and the figure carefully--step-by-step. To finish up the derivation requires hanging in just a bit longer.

We have used the deformation--geometry tool which is available to us for stress analysis derivations. Our two remaining tools--for the present--are stress-strain and equilibrium.

For an elastic material, the stress-strain relation is

$$\sigma(y) = E\varepsilon(y) \tag{4.3.5}$$

For equilibrium we must consider two things. Refer now to Figure 4.3.4.

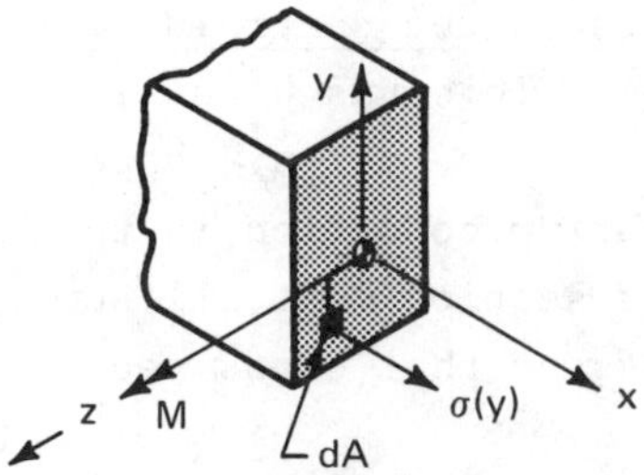

Figure 4.3.4

The first is that the sum of all the forces in the x direction must be zero. These are the forces down the span of the beam, which must be zero at every cross section, since there aren't any such forces present.

Thus

$$\int_A \sigma(y)\,dA = 0 \qquad (4.3.6)$$

Into Equation (4.3.6), substitute Equations (4.3.4) and (4.3.5)

$$\int_A \frac{-E\,y}{R}\,dA = 0$$

or

$$-\frac{E}{R}\int_A y\,dA = 0$$

In Equation (4.3.7), neither E or R can be zero. So

$$\int_A y\,dA = 0$$

In Equation (4.3.8) the $\int_A y\,dA$ is called the first moment of area. (Recall your work in statics on area properties.) If this integral is zero, then the reference axis for the integral is located at the centroid of the area. Since this integral must always be zero, (Equation (4.3.8) is general) then it can be said that the neutral axis of a beam always goes through the centroid (for homogeneous, isotropic materials). This has been illustrated for a rectangle but remains true for any shape.

The second equilibrium condition is that the internal moments on any cross section must all sum up to the internal moment, M, which acts at that cross section. Refer to Figure 4.3.4 again.

$$-\int_A \underbrace{\sigma(y) \quad dA}_{\text{force}}\ \underbrace{y}_{\text{arm}} = M \tag{4.3.9}$$

force x arm = applied moment

Since a negative y gives a positive moment as in Figure 4.3.4 use now Equation (4.3.5)

$$-\int_A E\ \varepsilon(y)\ dA\ y = M \tag{4.3.10}$$

Use now Equation (4.3.4)

$$\int_A \frac{E}{R}\ y^2\ dA = M \tag{4.3.11}$$

or

$$\frac{E}{R}\int_A y^2 dA = M \tag{4.3.12}$$

In Equation (4.3.12) the term $\int_A y^2 dA$ is the centroidal moment of inertia $\equiv$ I. Thus:

$$\frac{EI}{R} = M \tag{4.3.13}$$

Use again Equation (4.3.5)

$$\frac{\sigma(y)\ I}{\varepsilon(y)\ R} = M \tag{4.3.14}$$

and Equation (4.3.4)

$$\frac{\sigma(y)\ \cancel{R}I}{-y\ \cancel{R}} = M \tag{4.3.15}$$

$$\boxed{\sigma(y) = \frac{-My}{I}} \qquad (4.3.16)$$

This equation is limited to elastic behavior.

If we summarize the results of this section up to Equation (4.3.16), we could make the following observations.

1. The presence of a bending moment will compress one side of the beam and extend the other. This bends the beam to the shape of Figure 4.3.3(a). The stresses which are present are compressive stresses in the longitudinal fibres of one part of the beam and tensile stresses in the other.
2. There is a line called the neutral axis. It is a longitudinal axis passing through the centroids of all the cross sections. This line separates the tension side from the compression side of the beam.
3. Figure 4.3.3(b) shows the linear relation of displacements of the longitudinal fibres (shortening and extension) up and down from the neutral axis. The strains are linearly related to the displacements, Equation (4.3.1) and the stresses are linearly related to the displacements, Equation (4.3.1) and the stresses are linearly related to the strains, Equation (4.3.5). Therefore, the strains and stresses are also distributed as shown in Figure 4.3.3(b).
4. In Equation (4.3.16); $\sigma = \frac{-My}{I}$. The z axis is along the moment vector and the y axis is perpendicular to it as shown in Figure 4.3.4. Along any line defined by y = constant, the displacements, strains and stresses are equal. Positive y will result in compressive (negative) stresses.

<u>Example: Beam bending stress</u>

The beam of example problem of Figure 4.2.1 has the cross section shown below. Calculate the bending stress at the mid-span and at the cross section locations a-a, b-b and c-c.

Let P = 1,000 lb and a = 48 in.

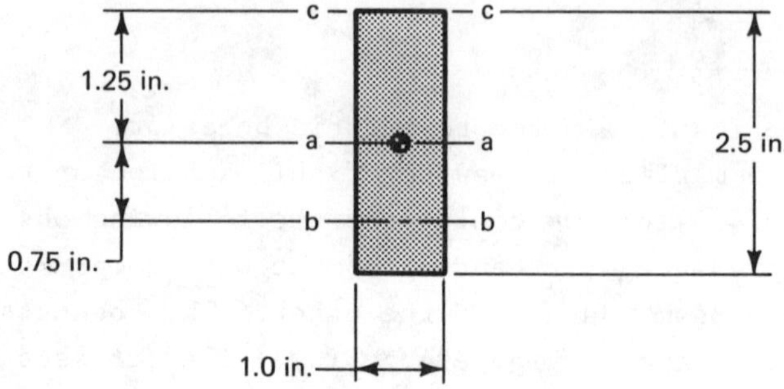

Solution:

$I = bh^3/12 = (1)(2.5)^3/12 = 1.302 \text{ in}^4$

The moment, from Figure 4.2.1 = Pa = 1,000(48) = 48,000 in · lb

$\sigma = My/I$

$\sigma_{a-a} = 0$ (since $y = 0$)

$\sigma_{b-b} = My/I = 48{,}000(0.75)/1.302 = 27{,}650$ psi (Tensile)

$\sigma_{c-c} = My/I = 48{,}000(1.25)/1.302 = 46{,}083$ psi (Compression)

Problems

Symmetric vs. Unsymmetric bending

Given the following cross sections and corresponding load vectors, identify the loading as symmetric or unsymmetric.

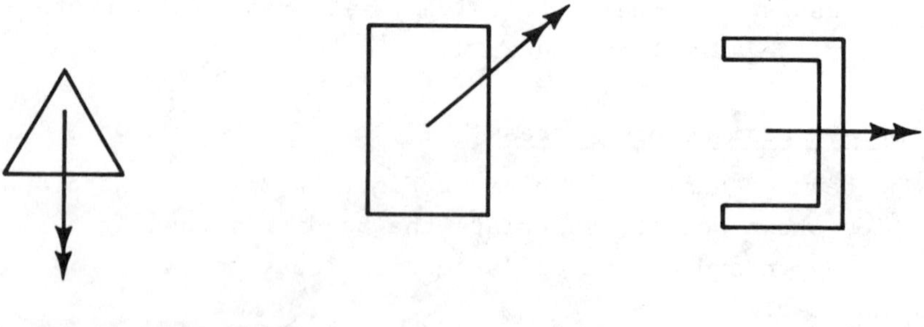

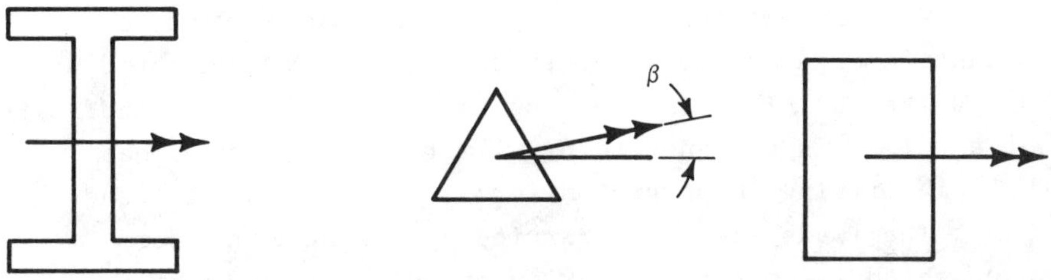

Problems: Symmetric Bending Stress

Problem 1.

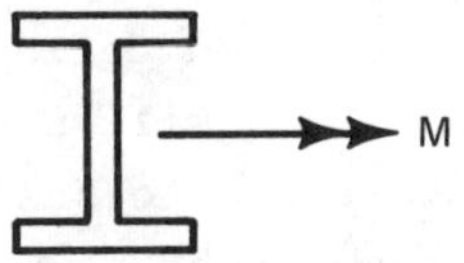

If the section shown is a 12 WF 190, what value of moment, applied as shown, will cause a bending stress of 46,000 psi?

Problem 2. Given the beam of triangular cross section as shown below, with the concentrated force acting as the load, calculate the bending stress incurred along the top of the beam at a point 5 inches from the wall. Label all steps CLEARLY.

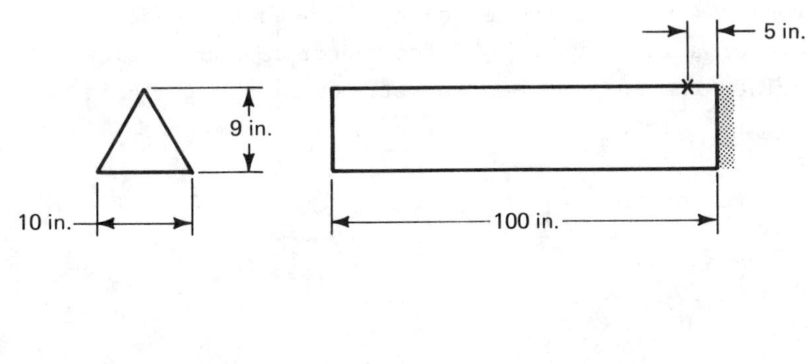

Problem 3. The wing of an airplane may be assumed to be cantilevered at the aircraft center line as indicated in the sketch. The wing loading is 10 lb/ft^2 at 1g. and the airplane is designed to 6g. limit load. Assume that the main lifting loads are uniformly distributed over the span. The loads are to be carried by a single, uniform spar, which is an I beam configuration. The depth of the beam is fixed at 7". Standard aluminum 7075-T6 sheet is to be formed and riveted as shown. Do the following:

a. Size the beam for limit load bending stress.
b. Calculate the associated deflection. Express this as a % of the span. (You will need an aluminum handbook.)
c. Draw your cross section to scale.
d. Calculate the beam weight.

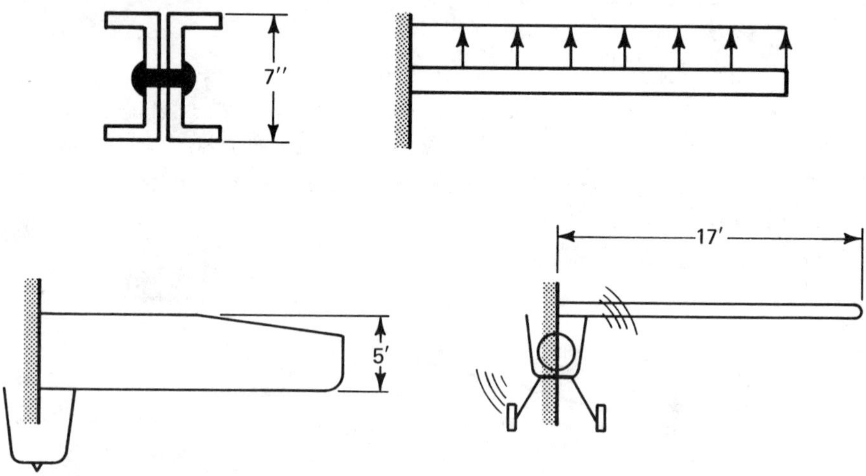

Problem 4. A person weighing 185 pounds stands in the middle of a 2 x 4 over an 8 foot span as shown below. Calculate the maximum bending stress in the 2 x 4 if

a. It is laid on its side 

b. It is upright 

Objectives for Centroid, Moment and Product of Inertia

(Closed Book Unless Otherwise Stated)

1. Be able to state Equations (4.4.1) and (4.4.2). Be able to use them to derive the centroids of "simple" areas.
2. Be able to state Equations (4.4.3) and (4.4.4). Be able to use them to calculate the moment of inertia of "simple" areas.
3. Be able to state Equation (4.4.5). Be able to use it to find the product of inertia of "simple" areas.
4. Be able to state Equations (4.4.8) and (4.4.9) and use them to calculate the centroids of composite areas.
5. Be able to state Equations (4.4.10), (4.4.11), and (4.4.12). Be able to use them to calculate the moments and products of inertia of composite areas.
6. Given a figure with the centroid, I_x, I_y and I_{xy} specified, be able to find I_{max}, I_{min}, $I_{xy\ max}$ and their orientations.

4.4 CENTROID, MOMENT OF INERTIA AND PRODUCT OF INERTIA

The important area properties discussed in this section are usually taught in statics courses. Because this may not be true for all the users of this text, a brief discussion of these properties and their calculation will be given. Figure 4.4.1 shows two areas and a set of reference axes for each area. The area of Figure 4.4.1(a) can be described as a "continuous" shape; the one of Figure 4.4.1(b) as a "composite" shape. The composite shape is made up of two or more standard shapes. Standard shapes include rectangles, triangles, circles or other shapes for which area properties are commonly tabulated.

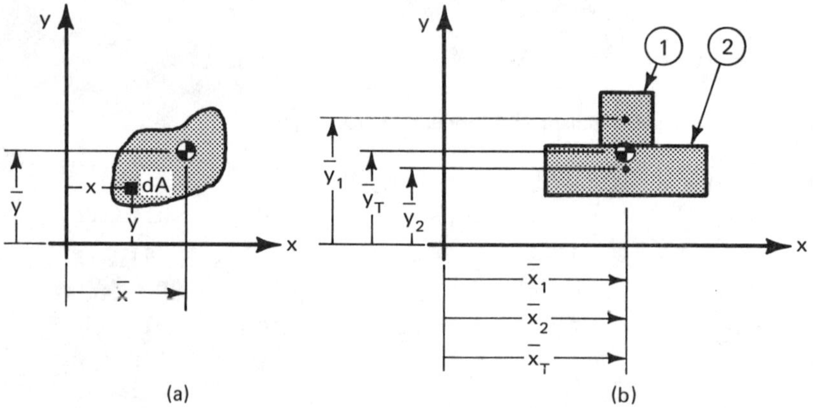

Figure 4.4.1

For the area of Figure 4.4.1(a) the centroid is defined as

$$\bar{x} = \frac{\int_A x dA}{\int_A dA} \tag{4.4.1}$$

$$\bar{y} = \frac{\int_A y dA}{\int_A dA} \tag{4.4.2}$$

These definitions are commonly used to find the centroids of the standard shapes. The <u>moment of inertia</u> of this area is defined to be

$$I_x = \int_A y^2 dA \qquad (4.4.3)$$

about the x axis and

$$I_y = \int_A x^2 dA \qquad (4.4.4)$$

about the y axis.

The <u>product of inertia is defined</u> as

$$I_{xy} = \int_A xy \, dA \qquad (4.4.5)$$

For the areas which are composites of standard shapes, such as the one shown in Figure 4.4.1(b), we assume that the centroid, moment of inertia and product of inertia of the standard shapes can be looked up in tables. Then the total properties are found in terms of component properties.

For the centroid we have

$$\bar{x}_T = \frac{A_1\bar{x}_1 + A_2\bar{x}_2}{A_1 + A_2} \qquad (4.4.6)$$

$$\bar{y}_T = \frac{A_1\bar{y}_1 + A_2\bar{y}_2}{A_1 + A_2} \qquad (4.4.7)$$

Equations (4.4.6) and (4.4.7) apply to Figure 4.4.1(b), and other figures with two component areas. In general, these two equations become

$$\bar{x}_T = \frac{\sum_{i=1}^{n} A_i\bar{x}_i}{\sum_{i=1}^{n} A_i} \qquad (4.4.8)$$

and

$$\bar{y}_T = \frac{\sum_{i=1}^{n} A_i \bar{y}_i}{\sum_{i=1}^{n} A_i} \qquad (4.4.9)$$

To calculate the moments and products of inertia of composite shapes, it is usually expedient to use the parallel axis, or transfer of axis formula. This formula relates the inertia properties of the areas about their own centroidal axis to some other axis which is parallel to the centroidal axis but is at some distance away. Refer to Figure 4.4.2.

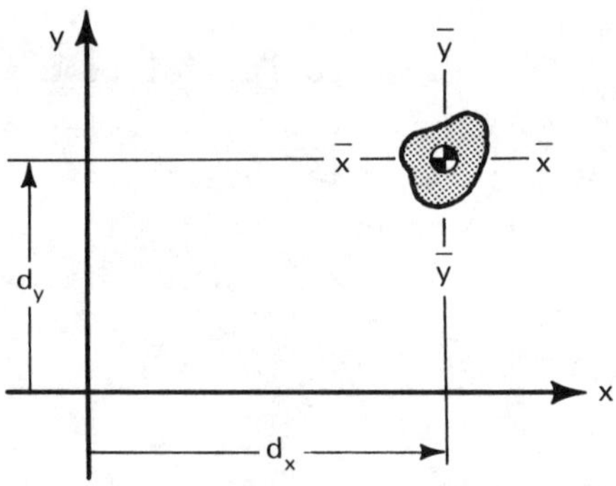

Figure 4.4.2

The formulas are:

$$I_x = I_{\bar{x}-\bar{x}} + Ad_y^2 \qquad (4.4.10)$$

(inertia any || axis = inertia own centroid + area x transfer distance squared.)

$$I_y = I_{\bar{y}-\bar{y}} + Ad_x^2 \qquad (4.4.11)$$

and

$$I_{xy} = I_{\overline{xy}} + A\, d_x d_y \qquad (4.4.12)$$

Note that in Equation (4.4.12) the signs of the terms d_x and d_y are significant. They are plus as shown in Figure 4.4.2.

Another important inertia relationship which is important in our next section on unsymmetric bending, is the effect of a rotation of axes on the moments and product of inertia. In this case we are considering the variation of inertias for a given body with given inertia properties and centroid; the only variable is the orientation of the axes with respect to an angle θ as shown in Figure 4.4.3.

It is emphasized that the angle θ is the only variable. This is the angle of rotation of the centroidal reference axes and expresses the change of reference axes from the x - y system to the x' - y' system.

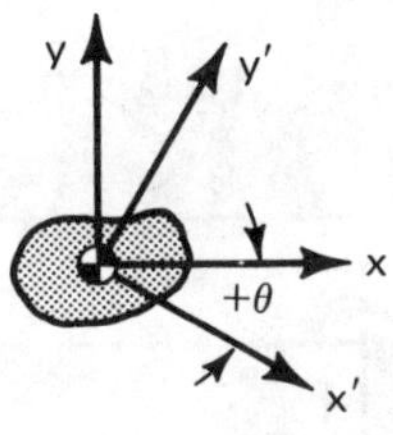

Figure 4.4.3

As the angle θ varies, we can make several observations:

1. There will be a value of θ for which $I_{x'y'}$ is zero. At this value of θ the principal axes of inertia are defined. At the principal axis orientation, the values of $I_{x'}$ and $I_{y'}$ will be maximum and minimum.
2. The value of θ at which the principal axes occur is given in terms of the known initial x - y axis properties by

$$\tan 2\theta = \frac{2I_{xy}}{I_x - I_y} \tag{4.4.13}$$

3. The principal moments of inertia are given by

$$I_{x'} = I_{max} = \frac{I_x + I_y}{2} + \sqrt{\left(\frac{I_x - I_y}{2}\right)^2 + I_{xy}^2} \tag{4.4.14}$$

$$I_{y'} = I_{min} = \frac{I_x + I_y}{2} - \sqrt{\left(\frac{I_x - I_y}{2}\right)^2 + I_{xy}{}^2} \qquad (4.4.15)$$

The following observations are frequently helpful:

1. An axis for which the product of inertia is zero is a principal axis.
2. If an area has an axis of symmetry, the symmetry axis will be a principal axis and an axis of zero product of inertia.

Example Problems: Centroid, moment and product of inertia.

Example 1. Find the centroid of the following cross section of a beam.

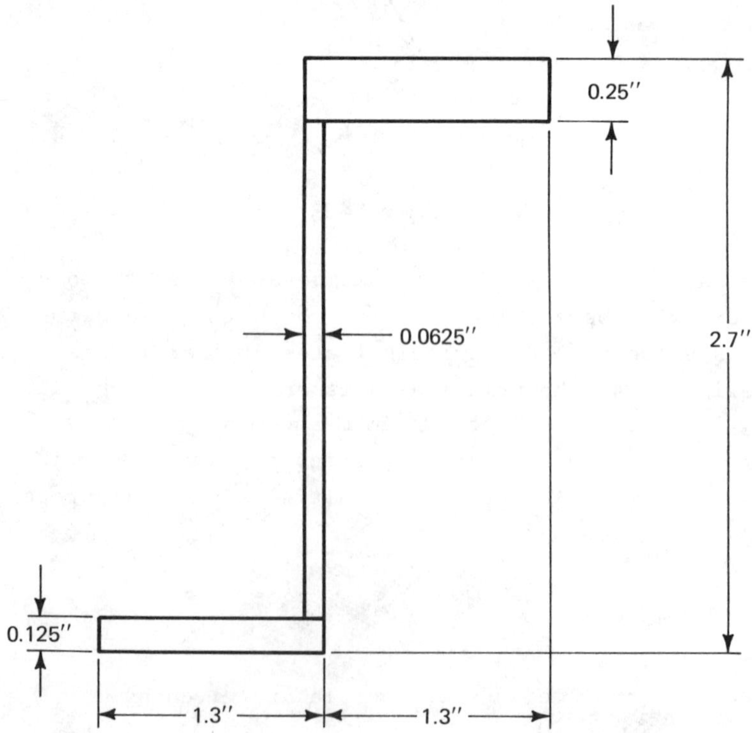

Example 2. Compute I_{yy}, I_{xx}, angle of inclination of principal axis, I_{max}, I_{min}.

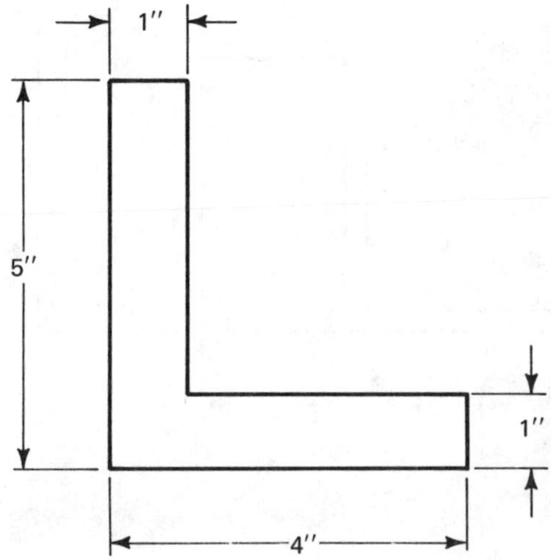

Example 3. Determine the product of inertia for the area shown.

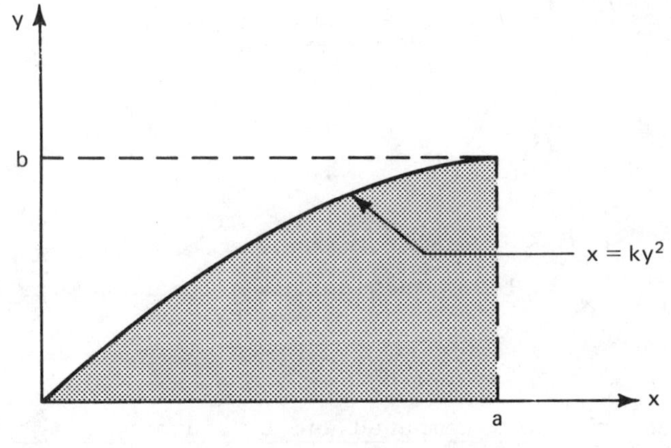

Example 1: Solution.

Find the centroid of the following cross section of a beam.

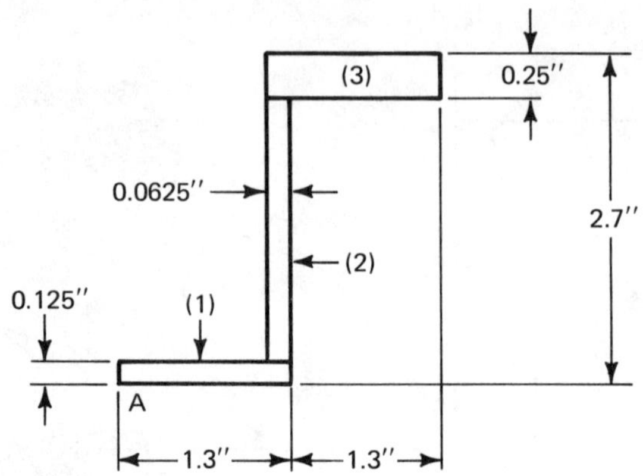

Solution

Start at A for origin. Use right hand coordinate system axis.

Member	$A(in^2)$	$\overline{X}$(in)	$\overline{Y}$(in)	$A \cdot \overline{X}$	$A\overline{Y}$
1	0.1625	0.65	0.0625	0.106	0.0102
2	0.145	1.269	1.288	0.184	0.187
3	0.341	1.919	2.575	0.654	0.877
Total	0.648			0.944	1.074

$$\overline{X} = \frac{\sum Ax}{\sum A} = \frac{0.913}{0.6325} = \overline{X} = 1.457 \text{ in.}$$

$$\overline{Y} = \frac{\sum Ay}{\sum A} = \frac{1.032}{0.6325} = \overline{Y} = 1.657 \text{ in.}$$

Example 2: Solution

Compute I_{yy}, I_{xx} angle of inclination of principal axis, I_{max}, I_{min}

Solution

I. The centroid was computed and is shown in diagram.

II. Using a right hand coordinate system axis at the centroid, one computes the following:

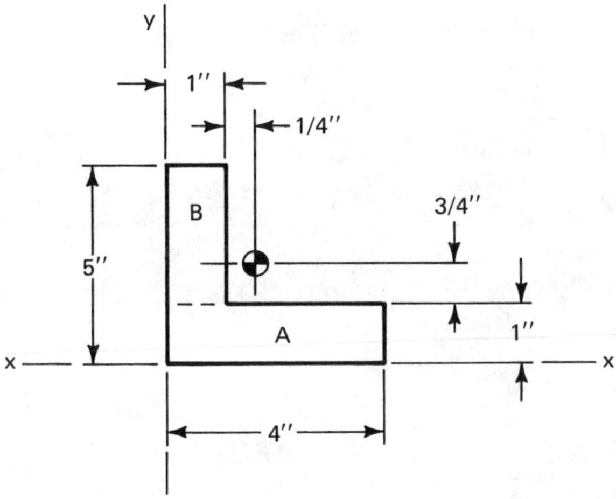

For Part A:

$$I_{xy_A} = \bar{I}_{xy} + dy\ dx\ A = 0 + (-5/4)(3/4)(4)$$

$$= -3.75\ \text{in}^4$$

For Part B:

$$I_{xy_B} = \bar{I}_{xy} + dy\ dx\ A = 0 + (5/4)(-3/4)(4)$$

$$= -3.75\ \text{in}^4$$

$$I_{xy_{TOTAL}} = I_{xy_A} + I_{xy_B} = -3.75 - 3.75 = \underline{\underline{I_{xy_{TOTAL}}}}$$

$$\underline{\underline{= -7.50\ \text{in}^4}}$$

III. $\underline{I = \bar{I} + A\ d^2}$

A. For Part A:

$I_x = (1/12)(4)(1^3) + (5/4)^2\ 4 = 6.583\ \text{in}^4$

$I_y = (1/12)(1)(4^3) + (3/4)^2\ 4 = 7.583\ \text{in}^4$

B. For Part B:

$I_x = 11.583\ \text{in}^4$

$I_y = 2.583\ \text{in}^4$

C. For the entire structure:

$$I_{xx} = 6.583 + 11.583 = \underline{\underline{I_{xx} = 18.67\ (in^4)}}$$

$$I_{yy} = 7.583 + 2.583 = \underline{\underline{I_{yy} = 10.167\ (in^4)}}$$

<u>This is with respect to centroidal axis</u>.

IV. <u>Angle of inclination</u>

$$\text{Tan } 2\theta = \frac{2I_{xy}}{I_x - I_y} = \frac{-2(7.50)}{10.167 - 18.167} = 1.875$$

$$2\theta = 61^{o}56'$$
$$\theta = 30^{o}58'$$

V. $\underline{I_{max}, I_{min}}$

$$I_{x'} = \frac{I_{xx} + I_{yy}}{2} + \left(\frac{I_{xx} - I_{yy}}{2}\right)\left(\cos 2\theta\right) + I_{xy} \sin 2\theta$$

$$I_{y'} = \frac{I_{xx} + I_{yy}}{2} - \left(\frac{I_{xx} - I_{yy}}{2}\right)\left(\cos 2\theta\right) - I_{xy} \sin 2\theta$$

$$I_{x'} = \frac{18.167 + 10.167}{2} + \frac{18.167 - 10.167}{2}(0.4705)$$

$$+ (7.5)(0.8824) = I_{x'} = 22.67\ in^4$$

$$I_{y'} = \frac{18.167 + 10.167}{2} - \frac{18.167 - 10.167}{2}(0.4705)$$

$$- (7.5)(0.8824) \quad I_{y'} = 5.67\ in^4$$

Because of definition of principal axis, one concludes that

$$\underline{\underline{I_{x'} = I_{max} = 22.67\ in^4}}$$

$$\underline{\underline{I_{y'} = I_{min} = 5.67\ in^4}}$$

One could also use the equivalent formulas to obtain:

$$I_{x'} = I_x \cos^2\theta + I_y \sin^2\theta - 2I_{xy} \sin\theta\cos\theta$$

$$I_{y'} = I_x \sin^2\theta + I_y \cos^2\theta + 2I_{xy} \sin\theta\cos\theta$$

Example 3: Solution

Determine the product of inertia for the following curve.

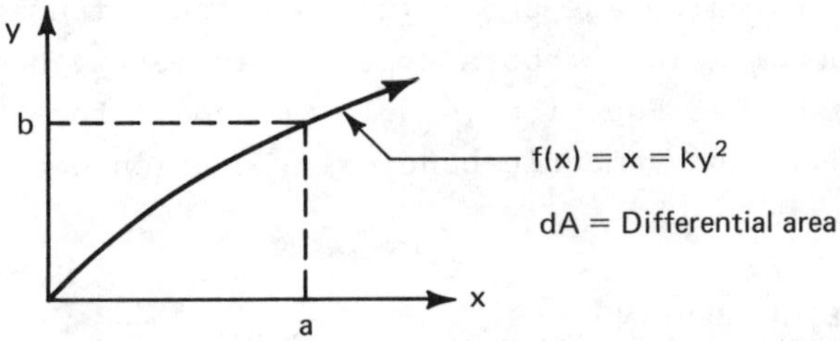

Solution

I.

$$I_{xy} = \int_0^b \int_{\frac{ay^2}{b^2}}^{a} xy \, dx \, dy = \int_0^b \frac{1}{2}\left(a^2 - \frac{a^2y^4}{b^4}\right) y \, dy$$

$$I_{xy} = \frac{1}{6}(a^2b^2)$$

II. Also

$$I_{xy} = \int_0^a \int_{y=0}^{\sqrt{\frac{x}{K}}} xy \, dy \, dx \qquad K = \frac{x}{y^2} = \frac{a}{b^2}$$

III. Or

$$dI_{xy} = xydA = xyxdy = xy \frac{a}{b^2}y^2dy = \frac{a}{b^2}xy^3dy = \frac{a^2}{b^4}y^5dy$$

$$\int_0^{I_{xy}} dI_{xy} = I_{xy} = \int_0^b \frac{a^2}{b^4} y^5 dy = \frac{a^2}{b^4}\frac{y^6}{6}\Big|_0^b = \frac{1}{6}a^2b^2 = I_{xy}$$

Objectives for Unsymmetric Bending

(Closed book unless otherwise noted.)

1. Be able to distinguish between two kinds of unsymmetric bending. (Figure 4.5.1).
2. Be able to state and sketch the positive sign conventions for unsymmetric bending.
3. Be able to use Equation (4.5.1) to solve for the stresses in unsymmetric bending. Equation need not be memorized.
4. Be able to calculate and illustrate the orientation of the neutral axis for both types of symmetric bending.
5. Be able to use Equation (4.5.15) to solve for the stresses in unsymmetric bending. Equation need not be memorized.

4.5 UNSYMMETRIC BENDING

We now consider the methods for treating unsymmetric bending. The question of symmetry is, as previously stated, really concerned with the relationship of the moment vector to any axis of symmetry that the cross section in bending may have. Let us distinguish two categories of unsymmetric bending:

1. The cross section has symmetry but the orientation of the moment vector is incorrect for symmetric bending. Figure 4.5.1(a).

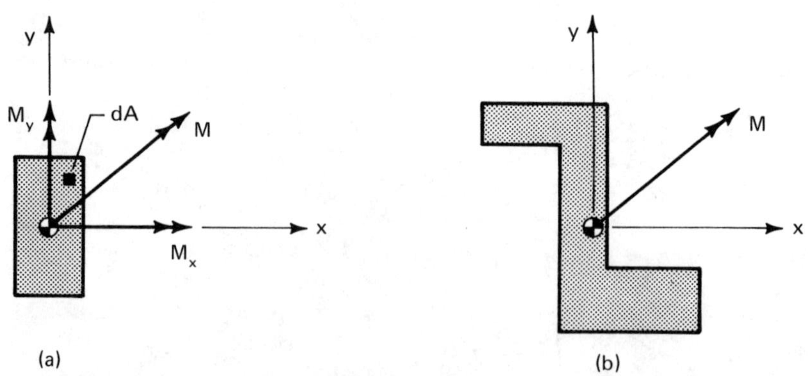

Figure 4.5.1

2. The cross section has no axis of symmetry. Figure 4.5.1(b).

In the first category, Figure 4.5.1(a), it is relatively easy to see from the figure that this case can be treated as a superposition of two symmetric bending cases. It is only necessary to define the sign conventions and keep them straight when we superimpose the stresses.

The sign conventions are:

1. Positive axes are as shown in Figure 4.5.1(a).
2. Moment components are positive when their direction is coincident with the positive x or y coordinate. (Figure 4.5.1(a).)
3. Tensile stresses are positive; compressive stresses are negative.

Then we can superimpose the stresses due to M_x and M_y by adding the two symmetric bending cases. (NOTE: In Figure 4.5.1(a) we have a cross section with double symmetry. We only need one axis of symmetry to use this method, however, because one axis of symmetry will be enough to give a zero product of inertia. This will be explained more thoroughly later in this section.) The formula is obtained by using Equation (4.3.16) twice and watching the signs.

$$\sigma = \frac{M_x y}{I_x} - \frac{M_y x}{I_y} \tag{4.5.1}$$

The small element dA in Figure 4.5.1(a) experiences positive values of M_x, M_y, x and y. M_x will cause tensile (plus) stress and M_y will cause compressive (negative) stress.

An important consideration in this kind of bending is the location of the neutral axis in the face of the cross section. It isn't where you would think it is. This can be better understood by contrasting the symmetric and unsymmetric cases of Figure 4.5.2(a) and (b).

In Figure 4.5.2(a), the line b-b forms the neutral plane and is coincident with the moment vector. This was mentioned previously in our discussion about symmetric bending. The deflection of the cross section will be perpendicular to

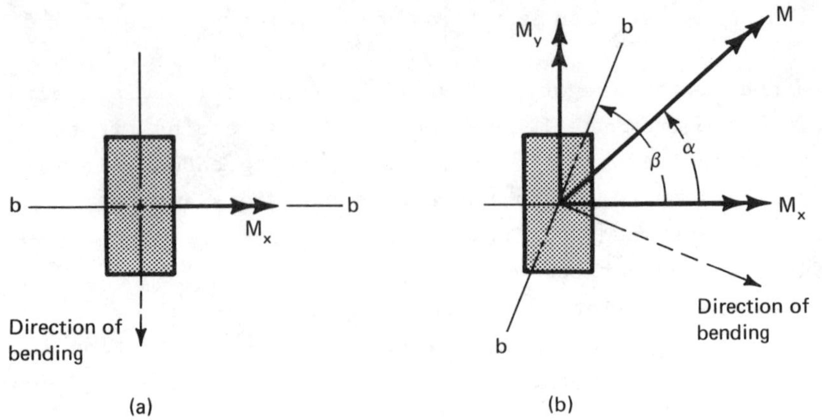

Figure 4.5.2

both the moment vector and line b-b as shown in Figure 4.5.2(a). In the case of unsymmetric bending, Figure 4.5.2(b), we can observe experimentally and mathematically that the line b-b, which forms the neutral plane, is no longer coincident with the moment vector M. Let us denote the orientation of line b-b with respect to the x-y coordinate system by the angle β, as shown in the figure. The angle β will, in general, be different than the angle α. Angle α is the orientation of the moment vector in the x-y coordinate system.

The orientation of line b-b can be determined mathematically by realizing that the neutral plane is the locus of all points in the cross section which have zero stress. So we can set Equation (4.5.1) equal to zero.

$$\frac{M_x y}{I_x} - \frac{M_y x}{I_y} = 0 \qquad (4.5.2)$$

Tan β in this coordinate system is defined to be y/x, so

$$\tan \beta = \frac{y}{x} = \frac{M_y I_x}{I_y M_x} \qquad (4.5.3)$$

This equation will permit us to solve for angle β. The direction of deflection of the beam will be as shown in Figure 4.5.2(b). This direction is perpendicular to axis b-b. It is interesting to note that the orientation of the neutral axis is now a function of the magnitude of loading, the direction of loading and the x- and y- inertia properties of the area.

The preceding method may be named the principal axis method, and this method is one of the ways to solve the problem of Figure 4.5.1(b). The problem is simply resolved into principal coordinates as shown in Figure 4.5.3.

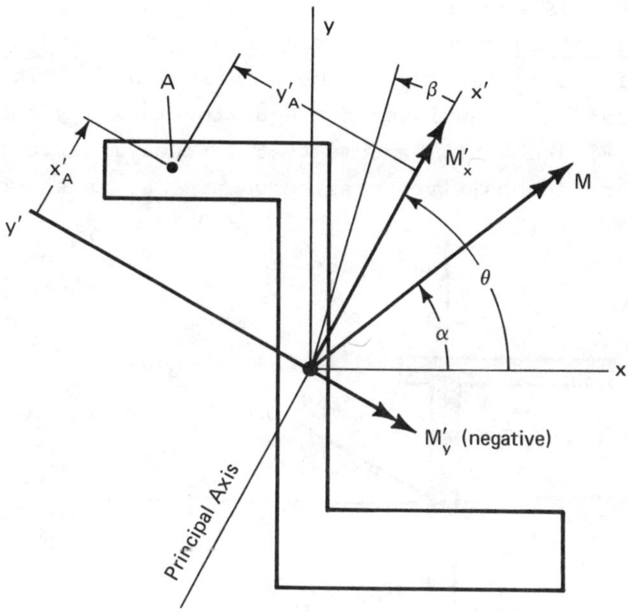

Figure 4.5.3

When the principal axes x' - y' are found, the moments $M_{x'}$ and $M_{y'}$ are resolved in this system. For a point such as A, Figure 4.5.3, the coordinates $x_{A'}$ and $y_{A'}$ are measured in the x' - y' system.

Equation (4.5.1) is then written as

$$\sigma = \frac{M_{x'} y'}{I_{x'}} - \frac{M_{y'} x'}{I_{y'}} \tag{4.5.4}$$

and Equation (4.5.3) as

$$\tan \beta = \frac{M_{y'} I_{x'}}{I_{y'} M_{x'}} \tag{4.5.5}$$

The angle β is now referenced to the x' - y' system as shown in Figure 4.5.3. Note that $I_{x'}$ and $I_{y'}$ are principal moments of inertia and that $I_{x'y'} = 0$ in the x' - y' system.

The general method for unsymmetric bending eliminates the requirement for transferring to the principal axes. This is best understood by deriving the equation in x - y coordinates. In Figure 4.5.4, assume that the unsymmetric cross section is bent in positive total curvature, R, as shown.

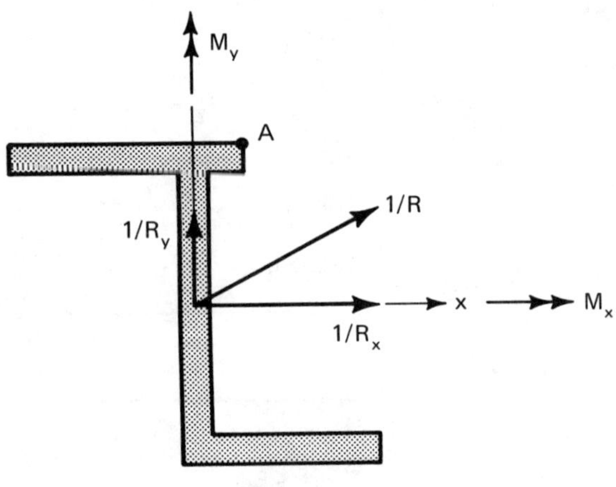

Figure 4.5.4

The strain at a typical point such as A is given as a superposition of the linear strains by

$$\epsilon = \left(\frac{1}{R_x}\right) y - \left(\frac{1}{R_y}\right) x \tag{4.5.6}$$

Also, the stress on a typical point is

$$\sigma = E\varepsilon = E\left[\left(\frac{1}{R_x}\right)y - \left(\frac{1}{R_y}\right)x\right] \tag{4.5.7}$$

The force at this point is

$$dP = \sigma dA = E\left[\left(\frac{1}{R_x}\right)y - \left(\frac{1}{R_y}\right)x\right]dA \tag{4.5.8}$$

The moment about the x and y axes are

$$dM_x = ydP = E\left[y^2\left(\frac{1}{R_x}\right) - xy\left(\frac{1}{R_y}\right)\right]dA \tag{4.5.9}$$

$$dM_y = -xdP = E\left[-xy\left(\frac{1}{R_x}\right) + x^2\left(\frac{1}{R_y}\right)\right]dA \tag{4.5.10}$$

Integration of Equations (4.5.9) and (4.5.10) will give the following. (We recall the mathematical definitions of the moments and products of inertia.)

$$M_x = E\left[I_x\left(\frac{1}{R_x}\right) - I_{xy}\left(\frac{1}{R_y}\right)\right] \tag{4.5.11}$$

$$M_y = E\left[-I_{xy}\left(\frac{1}{R_x}\right) + I_y\left(\frac{1}{R_y}\right)\right] \tag{4.5.12}$$

Equations (4.5.11) and (4.5.12) may be treated as two equations in two unknowns, with unknowns R_x and R_y. Solving the equation will give

$$\left(\frac{1}{R_x}\right) = \frac{1}{E}\,\frac{M_x I_y + M_y I_{xy}}{I_x I_y - I_{xy}^2} \tag{4.5.13}$$

$$\left(\frac{1}{R_y}\right) = \frac{1}{E}\,\frac{M_y I_x + M_x I_{xy}}{I_x I_y - I_{xy}^2} \tag{4.5.14}$$

Substituting equations (4.5.13) and (4.5.14) into Equation (4.5.7) gives

$$\sigma(x,y) = \left(\frac{M_x I_y + M_y I_{xy}}{I_x I_y - I_{xy}^2}\right) y - \left(\frac{M_y I_x + M_x I_{xy}}{I_x I_y - I_{xy}^2}\right) x \qquad (4.5.15)$$

This is the solution for stress in the x-y coordinate system.

There are several noteworthy features in Equation (4.5.15). First we can observe that if (say) the y axis is an axis of symmetry, then $I_{xy} = 0$ and Equation (4.5.15) reduces to Equation (4.5.2), then, if only M_x is present, Equation (4.5.15) reduces to Equation (4.3.16).

Consider also what happens to an unsymmetric section such as that of Figure 4.5.4 when M_x is present and $M_y \equiv 0$. Equation (4.5.15) becomes

$$\sigma(x,y) = \left(\frac{M_x I_y}{I_x I_y - I_{xy}^2}\right) y - \left(\frac{M_x I_{xy}}{I_x I_y - I_{xy}^2}\right) x \qquad (4.5.16)$$

This means that, unlike symmetric bending, the application of M_x alone on an unsymmetric cross section will induce a twisting about the y axis. This comes as a result of I_{xy} being non-zero for the unsymmetric section.

Worked Example

Unsymmetric Bending

For the beam cross section as shown and loaded below, calculate the bending stress at point A, and calculate the orientation of the neutral axis.

$I_x = 693.3$ in^4
$I_y = 173.3$ in^4
$I_{xy} = -240$ in^4

Solution:

The stress at any point on the beam is given by the unsymmetrical bending formula

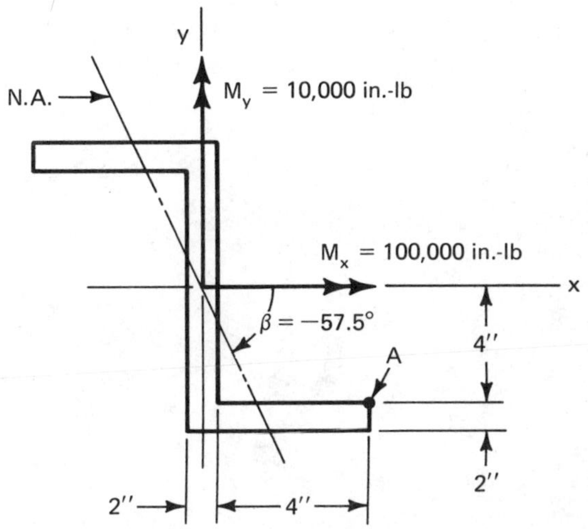

$$\sigma = \left(\frac{-M_x I_{xy} - M_y I_x}{I_x I_y - I_{xy}^2}\right)x + \left(\frac{M_y I_{xy} + M_x I_y}{I_x I_y - I_{xy}^2}\right)y$$

Substituting in the proper values we have

$$\sigma = \frac{-(100,000)(-240) - (10,000)(693.3)}{(693.3)(173.3) - (-240)^2}\, x$$

$$+ \frac{(10,000)(-240) + (100,000)(173.3)}{(693.3)(173.3) - (-240)^2}\, y$$

which yields the following equation for bending stress at any particular point on the cross section.

$$\sigma = 272.9x + 238.7y \qquad (1)$$

Point A is located by coordinates

$$x_A = 5 \text{ in} \qquad y_A = -4 \text{ in}$$

Hence:

$$\sigma_A = 272.9(5) + 238.7(-4)$$

$$\sigma_A = 409.7 \text{ psi tension}$$

The orientation of the neutral axis is obtained from equation (1) by substituting $\sigma = 0$. Namely,

$$0 = 272.9x + 238.7y$$

or

$$\frac{y}{x} = \tan\beta = -\frac{272.9}{238.7}$$

which leads to

$$\beta = -48.8^{\circ} \quad \text{(See Diagram)}$$

Problems

Problem 1. The beam shown below has the section properties shown in the Table. Calculate the bending stress at point A.

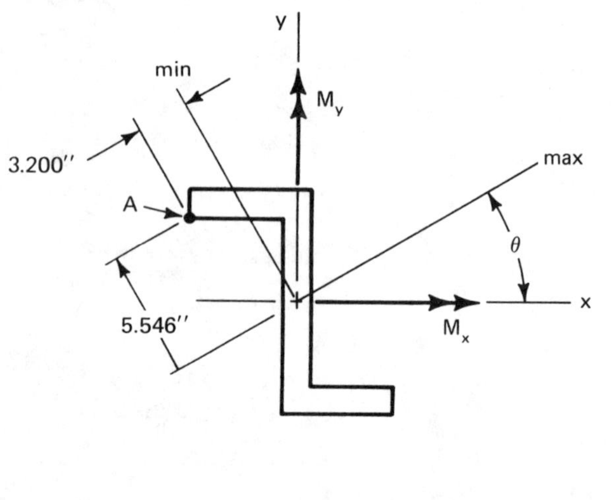

$I_x = 693.3$ in^4
$I_y = 173.3$ in^4
$I_{xy} = -240$ in^4
$I_{max} = 787.1$ in^4
$I_{min} = 79.5$ in^4
$\theta = 21.35^{\circ}$
$M_x = 10^5$ in-lb
$M_y = 10^3$ in-lb

Problem 2. For the beam cross section as shown and as loaded below, calculate the bending stress at point C, and calculate the orientation of the neutral axis. Show all work. SKETCH IN NEUTRAL AXIS.

$I_x = 500 \text{ in}^4$
$I_y = 100 \text{ in}^4$
$I_{xy} = 200 \text{ in}^4$

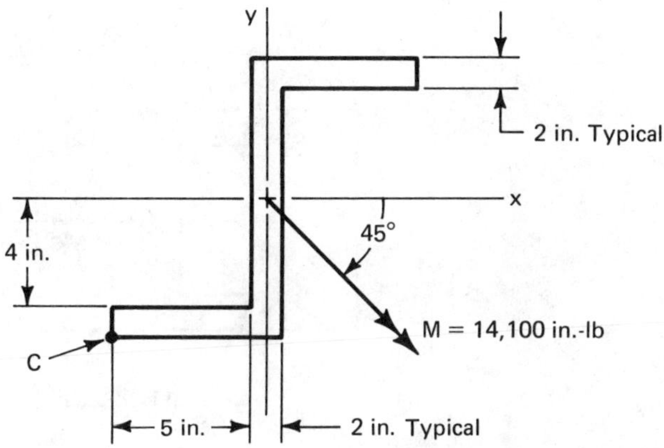

Problem 3. (a) An equal angle channel is 8 x 8 x 56.9 lb/ft. $I_{xx} = I_{yy} = 98.0 \text{ in}^4$. ρ_{min} is 1.56 in. Knowing I_{xx} and I_{yy}, and knowing that $\rho_{min} = I_{min}/A$, find I_{min} and I_{max} and I_{xy}.

(b) If $M_x = M_y = 1{,}000$ in-lb as shown below, calculate the unsymmetrical bending stress $\sigma(x,y)$ (stress as a function of x and y).

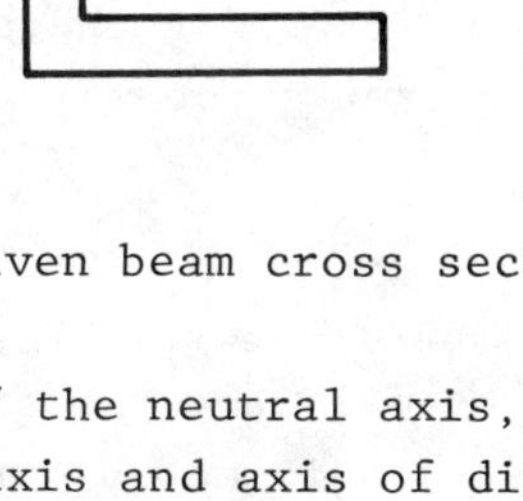

Problem 4. For the given beam cross section loaded as shown below,

a) find orientation of the neutral axis,

b) sketch in neutral axis and axis of displacement,

c) calculate the magnitude of the stress at point F.

(INDICATE TENSION OR COMPRESSION)

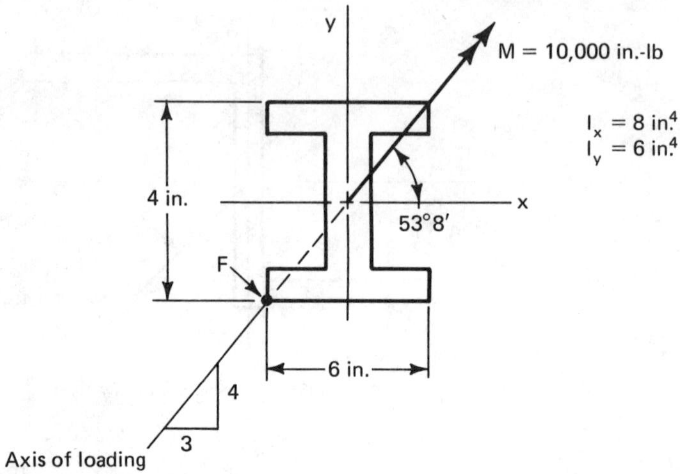

Objectives for Shear Stresses in Beams

(Closed Book Unless Otherwise Noted)

1. Be able to use Equation (4.6.9) to calculate shear stresses in beam cross sections.
2. Be able to identify the proper sub-area of the cross section and calculate $Q = A\bar{y}$.
3. Be able to calculate and sketch the shear stress distribution vs. depth for "simple" cross sections.
4. Be able to calculate the shear center for "simple" cross sections.

4.6 SHEAR STRESSES IN BEAMS

We know that at any point along the span of the beam where the moment is changing that a shear force exists since

$$V = \frac{dM}{dx} \tag{4.2.6}$$

The manner in which this shear force is carried over the cross section of the beam is the point of study of this section. The shear stress is not distributed across the cross section uniformly, so that $\tau \neq P/A$, as in bolt or rivet shear. Consider the segment of the beam shown in Figure 4.6.1. We will only consider the shear stresses in symmetric bending.

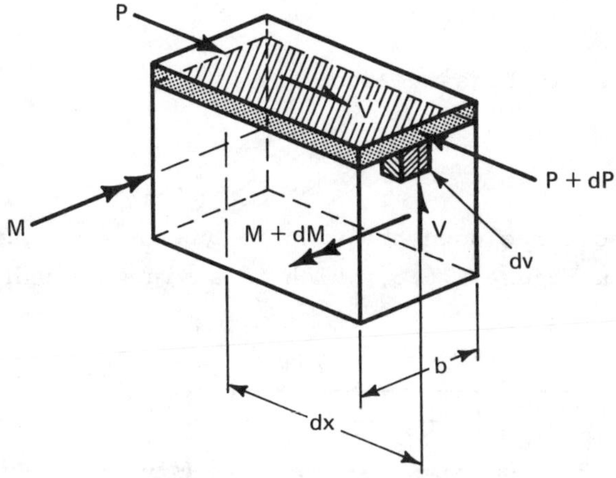

Figure 4.6.1

The change in moment along the portion of span dx will cause an unbalanced force along the span. The shaded area on the front face of the cross section will see the force P + dP while the force on the corresponding area of the back face is P. This requires a shear force V down the span as shown.

If we now take a close look at the side view of the volume dV which is indicated in Figure 4.6.1, we see the depiction of Figure 4.6.2.

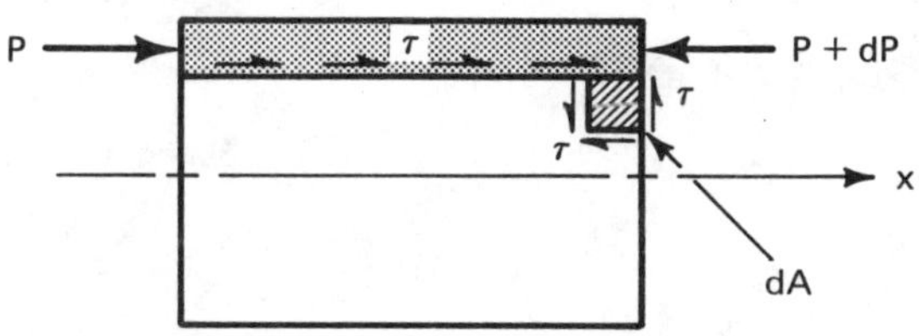

Figure 4.6.2 Side view of beam

The spanwise shear is accompanied by vertical and horizontal shear stresses as shown on area dA.

Returning to Figure 4.6.1, the force acting on the front face of the shaded area is given by integrating the quantity $\sigma \cdot dA$ over the area which is shaded in Figure 4.6.1.

$$P + dP = \int_A (\sigma + d\sigma)\, dA \tag{4.6.1}$$

On the back face of the same area

$$P = \int_A \sigma dA \tag{4.6.2}$$

The unbalanced force down the span is put in equilibrium. This is shown in Figure 4.6.3, which is a view through the top of the beam.

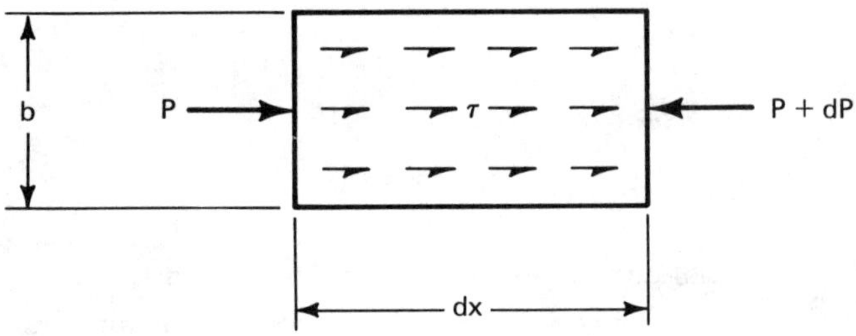

Figure 4.6.3 View of spanwise forces looking down through top of beam

Writing the equation of equilibrium down the span gives

$$P - (P + dP) + \tau b dx = 0 \tag{4.6.3}$$

or

$$\cancel{\int_A \sigma dA} - \cancel{\int_A \sigma dA} - \int_A d\sigma dA + \tau b dx = 0 \tag{4.6.4}$$

or

$$-\int_A d\sigma\, dA + \tau b dx = 0 \tag{4.6.5}$$

We know that

$$\sigma = \frac{My}{I} \tag{4.3.16}$$

So

$$d\sigma = \frac{dMy}{I} \tag{4.6.6}$$

Since M is not a function of y, we can combine Equations (4.6.5) and (4.6.6) in the following form.

$$- \frac{dM}{I} \int_A y dA + \tau b dx = 0 \qquad (4.6.7)$$

or

$$\tau = \frac{dM}{dx} \frac{\int_A y dA}{IB} \qquad (4.6.9)$$

The term dM/dx is equal to V. The $\int_A y dA$ of the <u>shaded</u> portion of the cross section is defined as the first moment of this <u>shaded portion</u> of area with respect to the neutral axis of the cross section. The first moment of area is usually given the symbol Q. Equation (4.6.8) becomes

$$\boxed{\tau = \frac{VQ}{Ib}} \qquad (4.6.9)$$

It is likely that reviewing each term of Equation (4.6.9) will be helpful. We are calculating the shear stress at a particular level, say y, on a cross section such as the one shown in Figure 4.6.4.

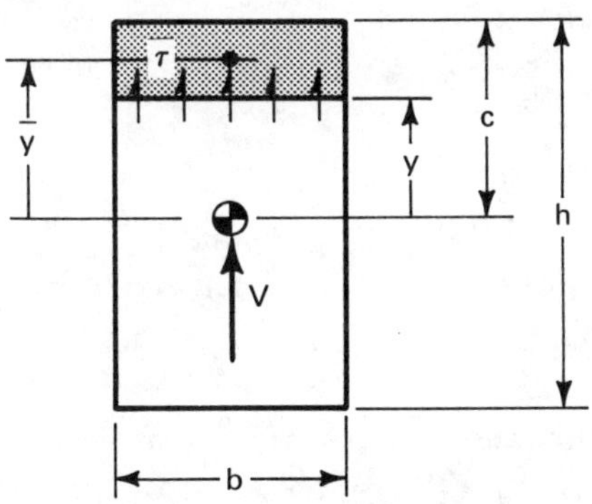

Figure 4.6.4 View of cross section of beam

The change in moment M down the span results in shear force V which is carried by the cross section as shown.

Shear force V causes shear stress τ which is non-uniformly distributed over the depth of the beam as a function of y. The width of the beam is given as b. The centroidal moment of inertia is I. Q is the first moment of the shaded area with respect to the neutral axis.

Q is given as $\int_A y\,dA$. For simple shapes, such as the shaded rectangle shown in Figure 4.6.4

$$Q = A\bar{y} \tag{4.6.10}$$

where A is the area of the shaded portion of the cross section and $\bar{y}$ is the distance from the centroid of the entire cross section to the centroid of the shaded area.

Example

Find the shear stress as a function of depth for the beam shown in Figure 4.6.4.

The centroid of the rectangular cross section will be at the center. At a general depth y, the moment of area of the shaded portion will be

$$Q = A\bar{y}$$
$$A = b(c - y)$$
$$\bar{y} = y + 1/2\ (c - y) = 1/2\ (c + y)$$
$$Q = \frac{b}{2}(c^2 - y^2)$$

Then from Equation (4.6.9)

$$\tau = \frac{VQ}{Ib} = \frac{V(c^2 - y^2)}{2I}$$

The distribution of shear stress with depth is parabolic and goes to zero at the top and bottom. The maximum value of shear stress is at the center of the beam. The maximum value is at the centerline. This is shown in Figure 4.6.5.

By use of the relationship $I = bh^3/12$, we can show that

$$\tau_{max} = \frac{3}{2}\frac{V}{A} \tag{4.6.11}$$

The maximum shear stress is 1.5 times the average shear stress obtained by just dividing the force by the area.

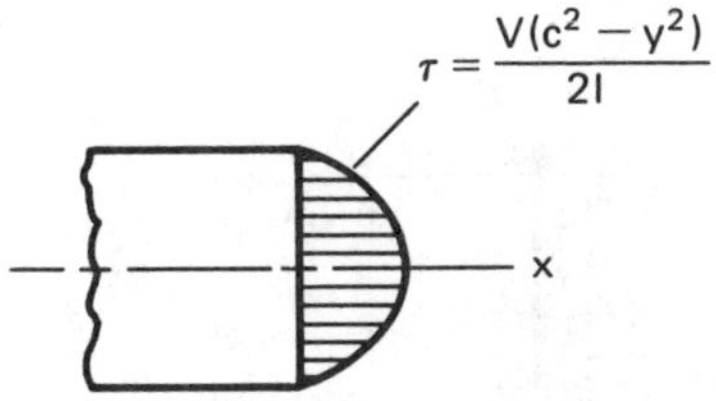

Figure 4.6.5 Side view of beam showing distribution of shear stress with depth

If Equation (4.6.11) is written as

$$\tau_{max} = K_s \frac{V}{A} \tag{4.6.12}$$

the term K_s can be defined as the shear factor. This factor is frequently tabulated for standard shapes so that the maximum shear stress may be obtained by a simple calculation.

A significant property of the beam cross section is the shear center. A beam which is loaded vertically through the centroid of the cross section as shown in Figure 4.6.6 will twist. This will happen if the load is applied at the centroid of the cross section, or at any other point except the shear center.

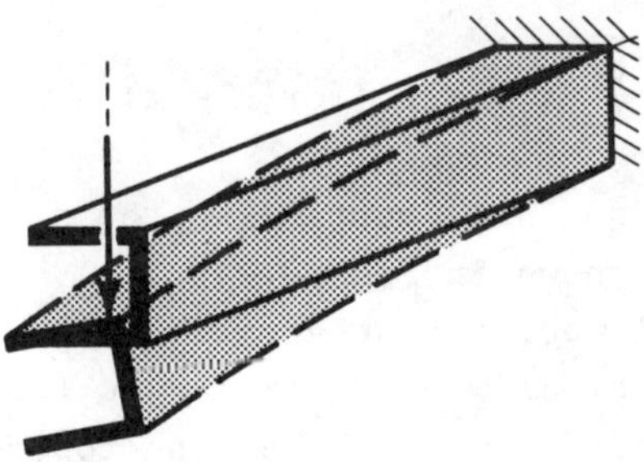

Figure 4.6.6 A beam loaded at any point other than the shear center will twist

Each cross section will have a shear center. This can be calculated by balancing out the shear stresses which "flow" in the cross section. This is shown in Figure 4.6.7 for the channel section.

The shear stress distribution along the top flange is obtained by considering the coordinate x, which defines a

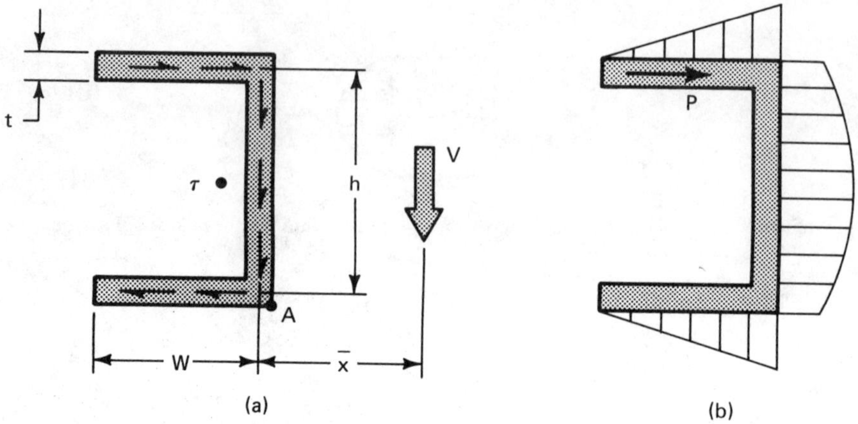

Figure 4.6.7

shaded area along the top leg of the channel section. The coordinate y will remain constant along this flange.

The first moment of area will be, (see Figure 4.6.8)

$$Q = A\bar{y} \qquad (4.6.13)$$

$$Q = tx_1c$$

$$\tau = \frac{VQ}{Ib} = \frac{Vc\not{t}x_1}{I\not{t}} = \frac{Vc}{I}x_1 \qquad \text{(for } b = t\text{)} \qquad (4.6.14)$$

The terms V, C and I are constant so the shear stress varies linearly as shown in Figure 4.6.7(b).

Down the vertical web, the shear stress variation is parabolic. The vertical web is similar to the rectangle of Figure 4.6.4; the difference is that the top flange adds to the Q term and thus the shear stresses do not go to zero at the top and bottom of this section.

For the shaded area of Figure 4.6.8(b), the shear stress down the web will go as

$$\tau = \frac{VQ}{Ib}$$

$$Q = A_1\bar{y}_1 + A_2\bar{y}_2$$

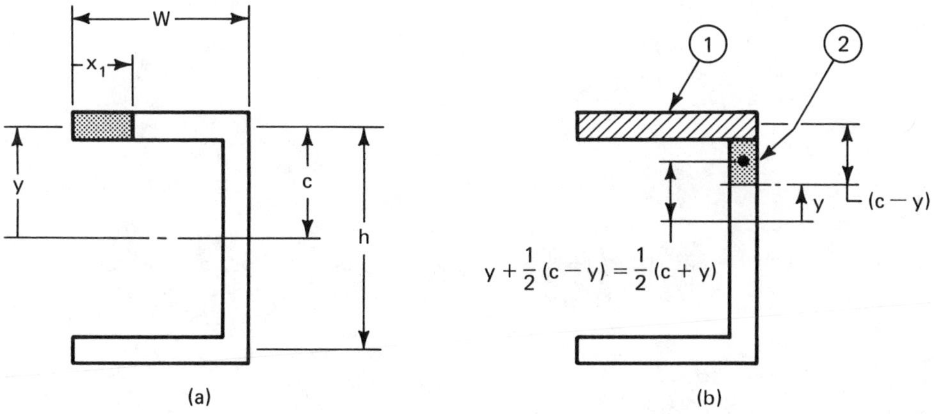

Figure 4.6.8

$$= wtc + 1/2t(c - y)(c + y)$$
$$= wtc + 1/2t(c^2 - y^2)$$

or

$$\tau = \frac{V[wtc + 1/2t(c^2 - y^2)]}{It} \tag{4.6.15}$$

The stress in the bottom flange is mathematically the same as the top flange, but the direction is as shown in Figure 4.6.7(a).

To calculate the shear center, we may take the equilibrium of inplane moments about the point A in Figure 4.6.7(a). When we take moments about point A, the force in the vertical web and the bottom flange do not appear in the equation because they have zero lever arm.

The <u>total force</u>, P, in the top flange is

$$P = \int_0^W \tau(x_1)\ t dx_1 \tag{4.6.16}$$

$$= \frac{Vct}{I}\int_0^W x_1 dx_1 = \frac{Vctw^2}{2I} \tag{4.6.17}$$

The equation of in-plane moment equilibrium is, from Figure 4.6.7(a),

$$V\bar{x} = \frac{Vctw^2h}{2I}$$

or

$$\bar{x} = \frac{tw^2h^2}{I} \tag{4.6.18}$$

By writing I as $2wtc^2 + \frac{th^3}{12}$ with $h = 2c$, it can be shown that

$$\bar{x} = \frac{w}{2 + h/3w} \tag{4.6.19}$$

Shear centers for other sections may be found by similar calculations. Reference 3 also gives extensive tabulations.

EXAMPLE: Plot the shear stress for the cross section shown in Figure 4.6.9.

Solution: It is sufficient to plot the distribution for only the top half of the cross section since the bottom half will be the same as the top. The appropriate equation is

$$\tau(y) = \frac{VQ(y)}{Ib(y)}$$

where both b and Q are functions of y.

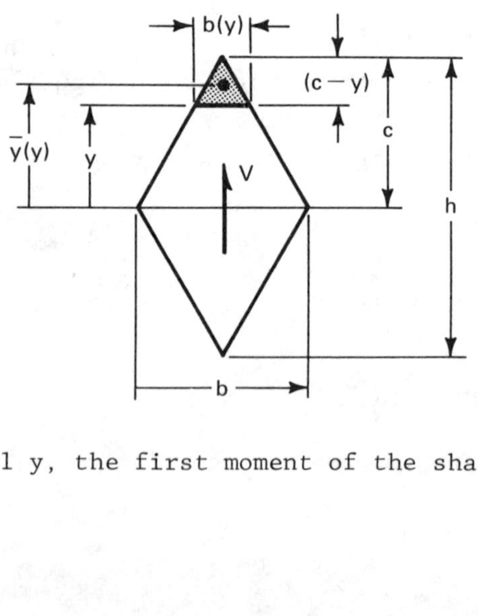

For a general y, the first moment of the shaded area is required.

$$b(y) = b\left(1 - \frac{y}{c}\right)$$

$$@\ y = 0;\ b(y) = b$$

$$@\ y = c;\ b(y) = 0$$

$$Q(y) = A(y)\ \overline{y}\ (y)$$

$$A(y) = \frac{1}{2}\ b(y)\ (c - y)$$

$$= \frac{1}{2}\ b\left(1 - \frac{y}{c}\right)(c - y)$$

$$= \frac{1}{2}\ b\left(c - 2y + \frac{y^2}{c}\right)$$

$$\overline{y}\ (y) = y + \frac{1}{3}(c - y) = \left(\frac{2}{3}y + \frac{1}{3}c\right)$$

$$Q(y) = \frac{1}{2b}\left(c - 2y + \frac{y^2}{c}\right)\left(\frac{2}{3}y + \frac{1}{3}c\right)$$

$$= \frac{1}{6b}\left(-y^2 + \frac{2}{3}\frac{y^3}{c} + \frac{1}{3}c^2\right)$$

Then

$$\tau = \frac{V}{I}\left[\frac{\frac{1}{6}\left(-y^2 + \frac{2}{3}\frac{y^3}{c} + \frac{1}{3}c^2\right)}{\left(1 - \frac{y}{c}\right)}\right]$$

It is interesting to plot $\tau(y)$ for this section.

Problems

Problem 1.

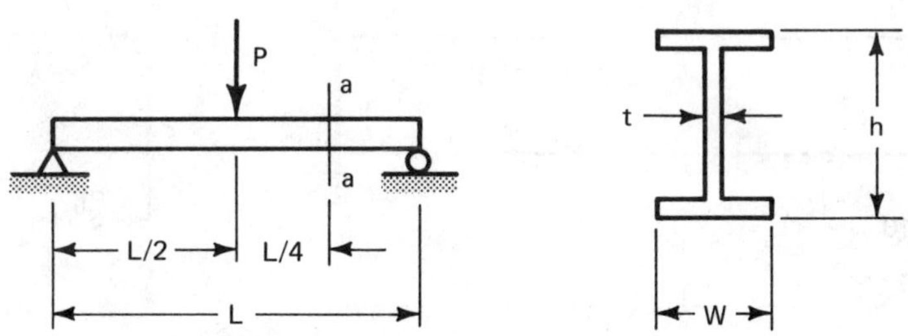

Plot the shear stress vs. depth for this section if: P = 400 lb, L = 60 in, h = 3 in, t = 0.100 in, w = 1.5 in.

Problem 2. Calculate the shear stress and shear flow at points A, B, and C. Sketch in the shear flow distribution around the entire cross section. Locate the shear center.

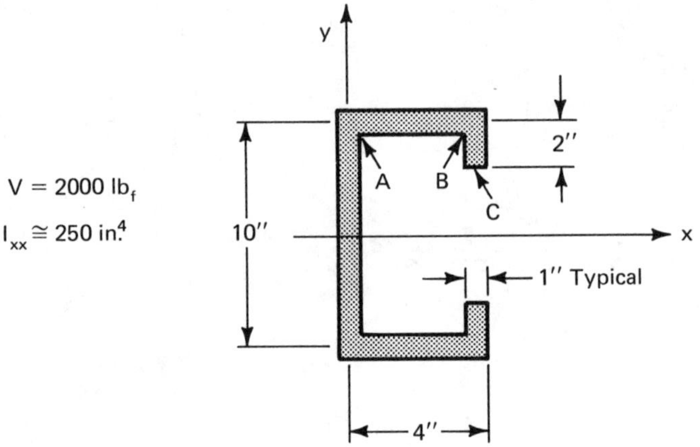

Problem 3. For the beam as shown and loaded below, find:

a. The location of maximum shear stress on the beam.
b. The moment of inertia for the uniform cross section.
c. The shear stress through a cross section at the wall on the neutral axis.
d. The shear stress on the bottom surface of the beam at a point 10 in. from the free end.

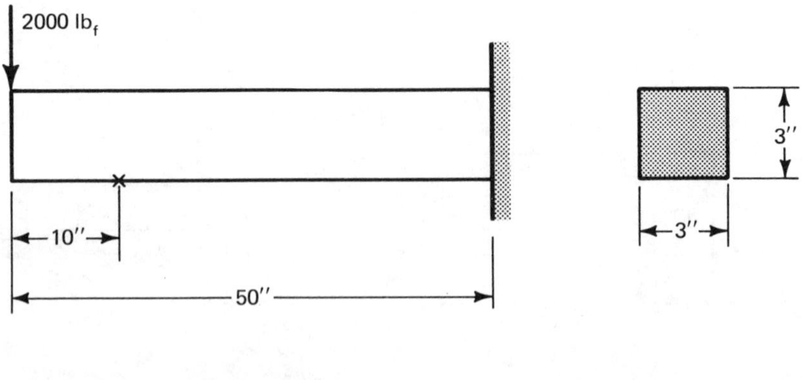

Problem 4. Calculate Q and sketch in b for each of the shaded portions of the cross sections shown below.

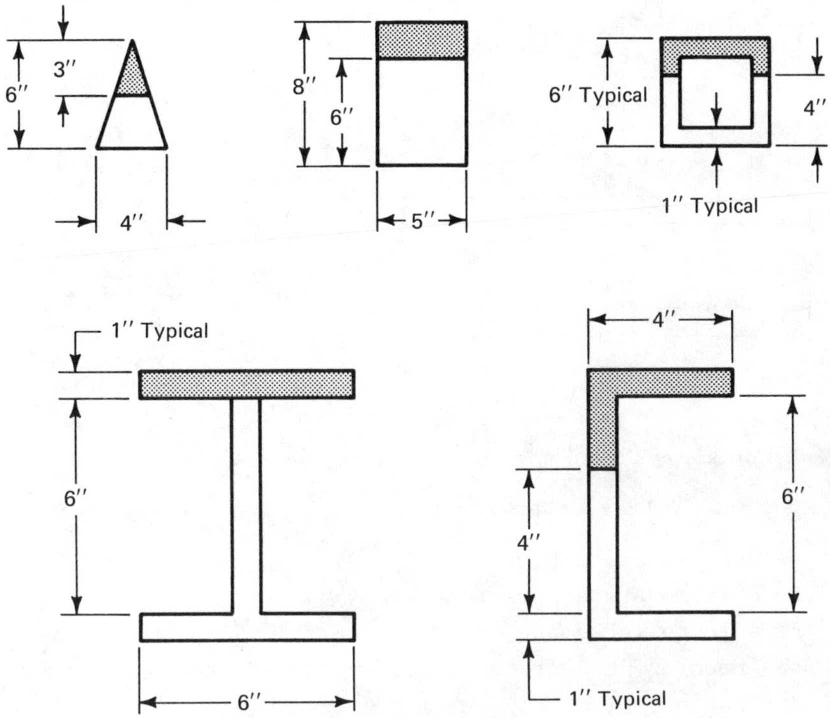

Problem 5. A uniform simply supported beam is loaded as shown. If W = 10 lb/in and L = 120", calculate the distribution of shear stress as a function of depth at the point of maximum shear.

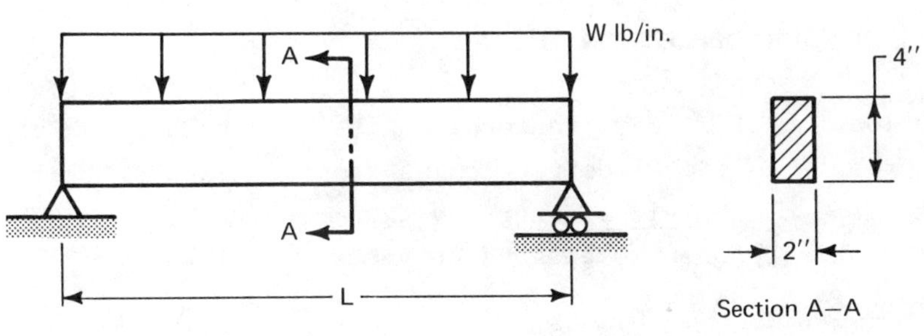

Problem 6. (a) Calculate the maximum shear force in lb., with respect to span. (b) At the point of maximum shear, calculate the shear stress at levels a, b, c and d on the cross section.

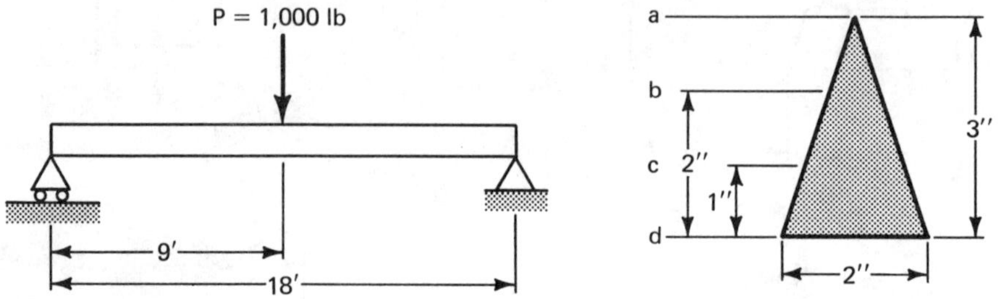

Cross Section
(enlarged view)

Objectives of Bending Deflection of Beams

(Closed Book Unless Otherwise Noted)

1. Be able to state Equation (4.7.3).
2. Be able to state Equation (4.7.4).
3. Be able to integrate Equation (4.7.3) twice and apply the boundary conditions in order to get the slope and deflection curves of a beam.
4. Be able to integrate Equation (4.7.4) four times and apply boundary conditions in order to get the shear, moment, slope and deflection curves for a beam.
5. Be able to use tables such as Table 4.7.1 and the method of superposition to obtain beam deflections.

4.7 THE BENDING DEFLECTION OF BEAMS

The loading of a beam tranverse to its longitudinal axis will produce a lateral deflection as shown in Figure 4.7.1.

The slope of the elastic curve y (x) is $dy/dx \equiv \phi(x)$. The relationship between moment and curvature is given by Equation (4.3.13).

$$\frac{1}{R} = \frac{M(x)}{EI} \tag{4.3.13}$$

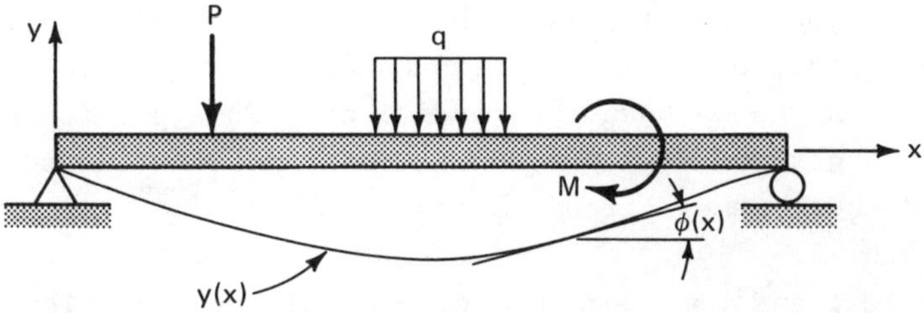

Figure 4.7.1

We also recall from calculus the definition of curvature

$$\frac{1}{R} = \frac{(d^2y/dx^2)}{\left[1 + \left(\frac{dy}{dx}\right)^2\right]^{3/2}} \qquad (4.7.1)$$

which, for small dy/dx becomes

$$\frac{1}{R} = \frac{d^2y}{dx^2} \qquad (4.7.2)$$

Combining Equations (4.3.13) and (4.7.2) gives

$$\frac{d^2y}{dx^2} = \frac{M(x)}{EI} \qquad (4.7.3)$$

Utilization of previous relationships

$$q = \frac{dV}{dx}$$

and

$$V = \frac{dM}{dx}$$

shows that

$$\frac{d^4y}{dx^4} = \frac{q(x)}{EI} \qquad (4.7.4)$$

The solution for $y(x)$ is found by integrating Equation (4.7.3) twice, or by integrating Equation (4.7.4) four times and applying known boundary conditions. This procedure is easy when $q(x)$ or $M(x)$ are continuous functions and somewhat awkward under other circumstances.

Example:

Find $y(x)$ and y_{max} for the uniformly loaded cantilever beam.

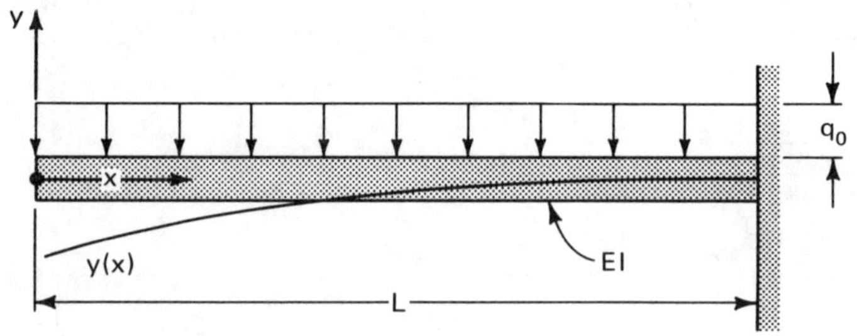

Solution:

Since $q(x) = q_o$ is a continuous function of x, we can use Equation (4.7.4).

$$\frac{d^4y}{dx^4} = \frac{q(x)}{EI} = \frac{-q_o}{EI} \qquad \text{(downward negative)}$$

or

$$EIy^{IV} = -q_o$$

Integrating once

$$EIy^{III} = -q_o\ x + C_1$$

When the loading is integrated once, we get the shear. Since the shear at the free end of the beam is zero, we can use the condition that $V = 0$ at $x = 0$ to get constant C_1.

$$EIy^{III} = V(x) = -q_o x + C_1$$

$$EIy^{III}\Big|_{x=0} = 0 = 0 + C_1$$

$$C_1 = 0$$

Then

$$EIy^{III} = -q_o x$$

Another integration gives EIy^{II} which is the moment, M(x).

$$EIy^{II} = M(x) = \frac{-q_o x^2}{2} + C_2$$

Since the moment at the free end of a beam is zero

$$EIy^{II}\Big|_{x=0} = 0 = 0 + C_2$$

$$C_2 = 0$$

and

$$EIy^{II} = \frac{-q_o x^2}{2}$$

a third integration will give y^I, which is the slope of the beam $\theta(x)$.

$$EIy^{I} = \theta(x) = \frac{-q_o x^3}{6} + C_3$$

The slope is zero at the built-in end, so

$$EIy^{I}\Big|_{x=L} = 0 = \frac{-q_o L^3}{6} + C_3$$

$$C_3 = \frac{q_o L^3}{6}$$

And

$$EIy^{I} = \frac{q_o}{6}(L^3 - x^3)$$

The fourth integration gives the deflection curve y(x).

$$EIy = \frac{q_o}{6}(L^3 x - \frac{x^4}{4}) + C_4$$

The deflection will be zero at x = L.

$$EIy\Big|_{x = L} = 0 = \frac{q_o}{6}(L^4 - 1/4L^4) + C_4$$

$$= \frac{q_o L^4}{8} + C_4$$

$$C_4 = \frac{-q_o L^4}{8}$$

So that

$$y = \frac{q_o}{24EI}(4L^3 x - x^4 - 3L^4)$$

The maximum deflection is at the tip where x = 0

$$y_{max} = \frac{-q_o L^4}{8EI}$$

Example:

Find the deflection curve for the cantilever beam shown.

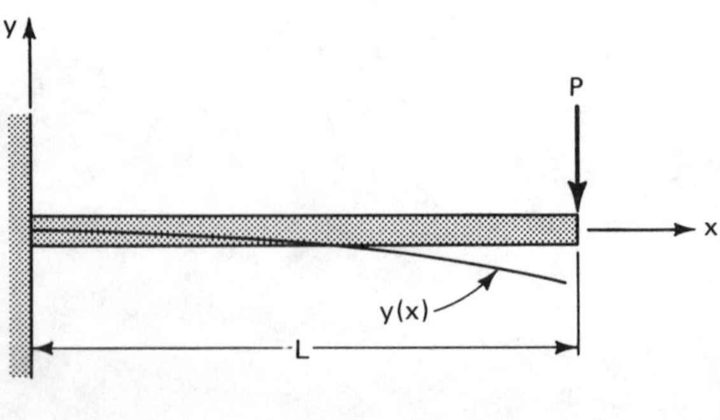

Solution:

In this example, P is not a continuous function of x. We usually cannot integrate the load function four times except by special methods known as singularity functions. We will not discuss singularity functions in this text. (See for example Reference 4.)

In this example it is easier to find the moment function and integrate it twice. M(x) is shown in the sketch. It is necessary to calculate the reactions.

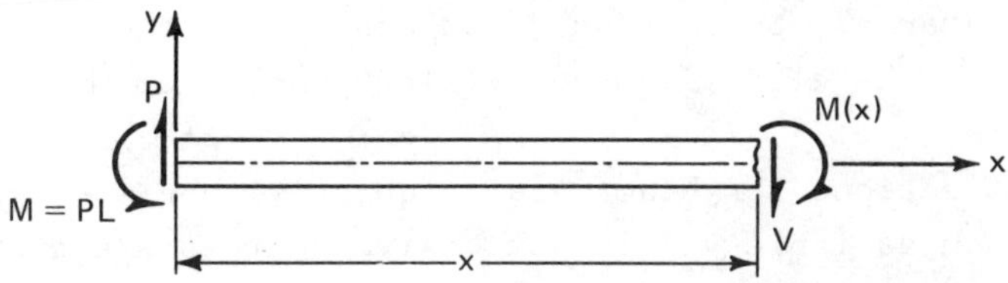

This is negative bending moment in accordance with Equation (4.7.3).

$$EIy^{II} = M(x) = P(x - L)$$

$$EIy^{I} = P\left(\frac{x^2}{2} - Lx\right) + C_1$$

The slope is zero at x = 0, therefore

$$C_1 = \theta_o = 0$$

and

$$EIy^{I} = \frac{P}{2}(x^2 - 2Lx)$$

Another integration gives

$$EIy = \frac{P}{2}\left(\frac{x^3}{3} - Lx^2\right) + C_2$$

The deflection is zero at x = 0, so

$$C_2 = 0$$

and

$$y = \frac{P}{6EI}(x^3 - 3Lx^2)$$

The maximum deflection will be at x = L,

$$y_{max} = \frac{P(L^3 - 3L^3)}{6EI} = \frac{-PL^3}{3EI}$$

It can be noted that for small deflections and a linear elastic material, the method of superposition can be used to superimpose the elastic curves y(x) or the maximum deflections, etc.

The deflections for beams are frequently tabulated in tables such as Table 4.7.1. Extensive sources are available; see for example Reference 3.

Table 4.7.1

Loading	Equation of Elastic Curve	
	Maximum Deflection	Slope at End
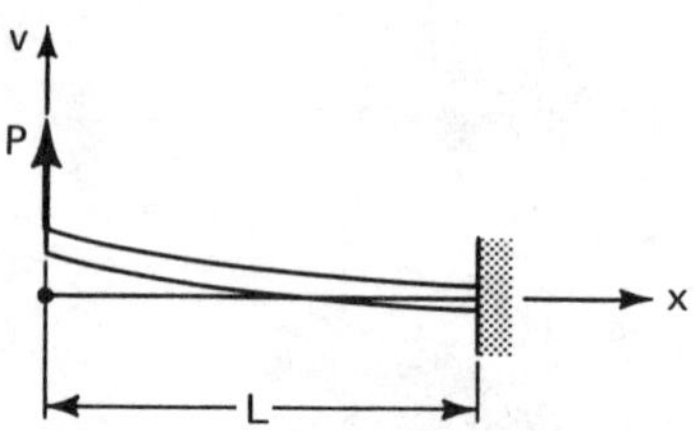	$v = \frac{P}{6EI}(2L^3 - 3L^2x - x^3)$ $v_{max} = v(0) = \frac{PL^3}{3EI}$	$\theta(0) = -\frac{PL^3}{2EI}$
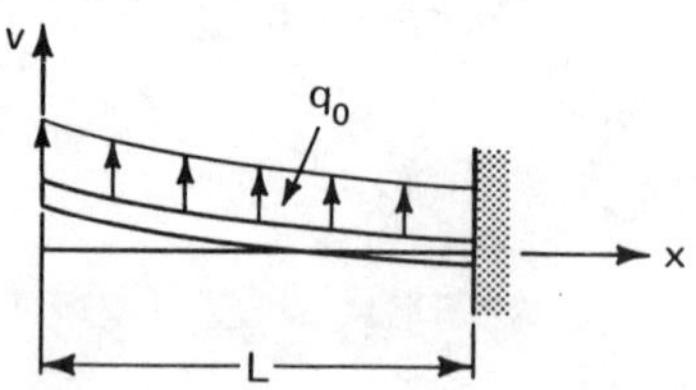	$v = \frac{q_o}{24EI}(x^4 - 4L^3x - 3L^4)$ $v_{max} = v(0) = \frac{q_oL_4}{8EI}$	$\theta(0) = -\frac{q_oL^3}{6EI}$

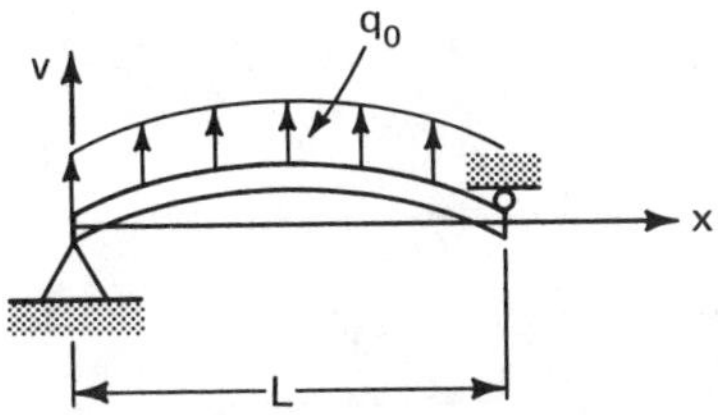

$$v = \frac{q_o x}{24EI}(L^3 - 2Lx^2 + x^3)$$

$$v_{max} = v(L/2) = \frac{5q_o L^4}{384EI} \qquad \theta(0) = -\theta(L) = \frac{q_o L^3}{24EI}$$

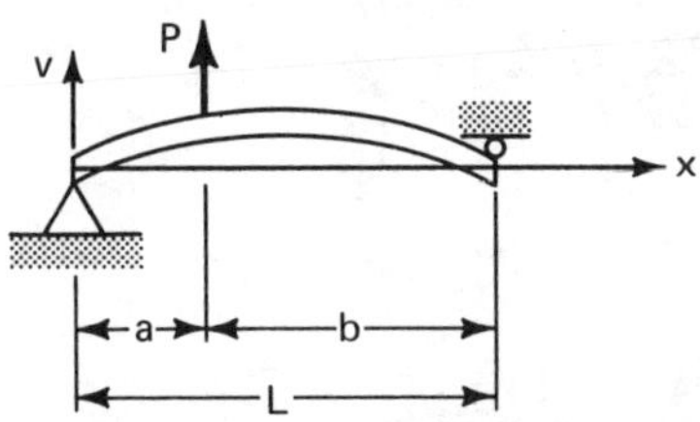

When $0 \leq x \leq a$, then

$$v = \frac{Pb}{6EIL}[(L^2 - b^2)x - x^3]$$

See Example 11-4.

When $a = b = \frac{L}{2}$, then

$$v = \frac{Px}{48EI}(3L^2 - 4x^2) \qquad (0 \leq x \leq \frac{L}{2})$$

$$v_{max} = v(L/2) = \frac{PL^3}{48EI} \qquad \theta(0) = -\theta(L) = \frac{PL^2}{16EI}$$

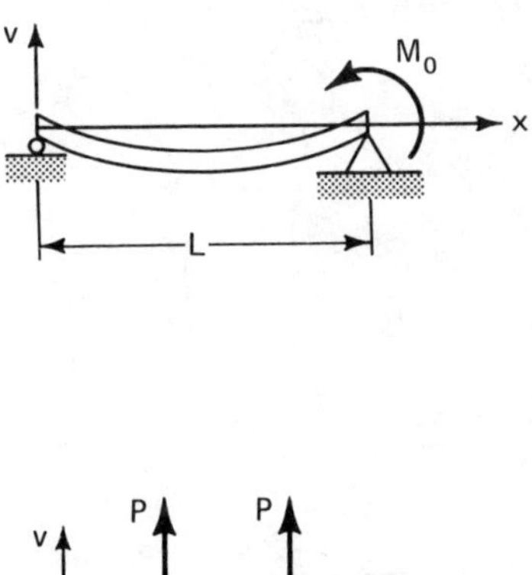

$$v = -\frac{M_o x}{6EIL}(L^2 - x^2)$$

$$v_{max} = v(L/\sqrt{3}) = -\frac{M_o L^2}{9\sqrt{3}\ EI} \qquad \theta(0) = -\frac{\theta(L)}{2} = -\frac{M_o L}{6EI}$$

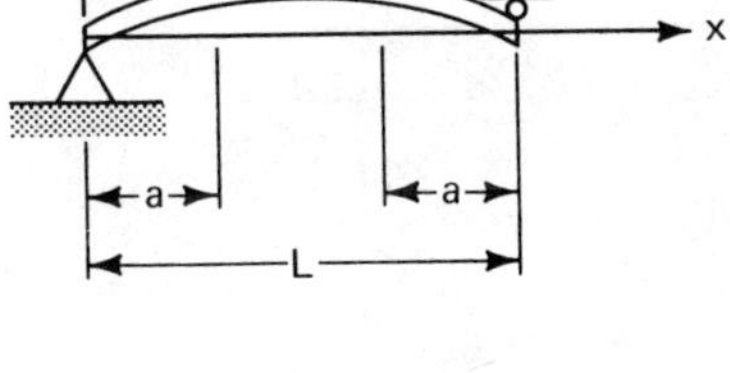

$$v_a = v(a) = \frac{Pa^2}{6EI}(3L - 4a)$$

$$v_{max} = v(L/2) = \frac{Pa}{24EI}(3L^2 - 4a^2)$$

$$\theta(0) = \frac{Pa}{2EI}(L - a)$$

Problems

The Bending Deflection of Beams

Problem 1. Find V(x), M(x), θ(x), y(x) and y_{max} for the beams shown. Assume that each beam has modulus of elasticity E and moment of inertia I.

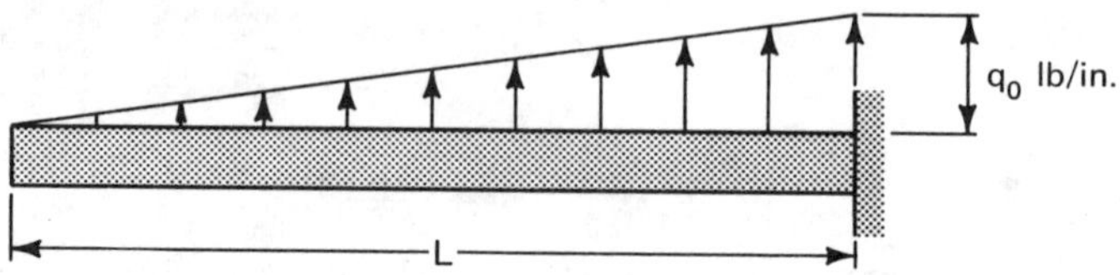

(a)

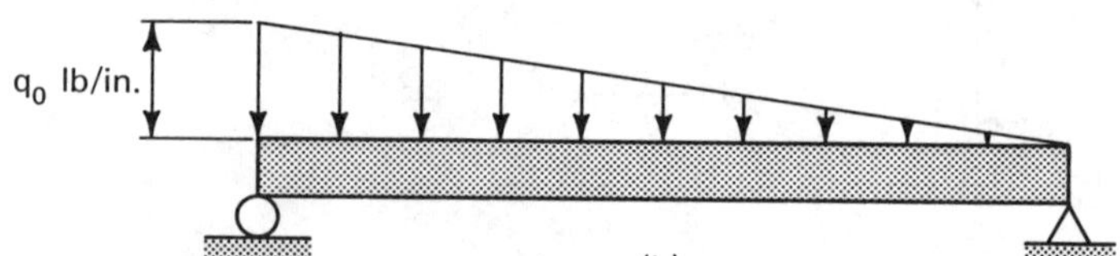

(b)

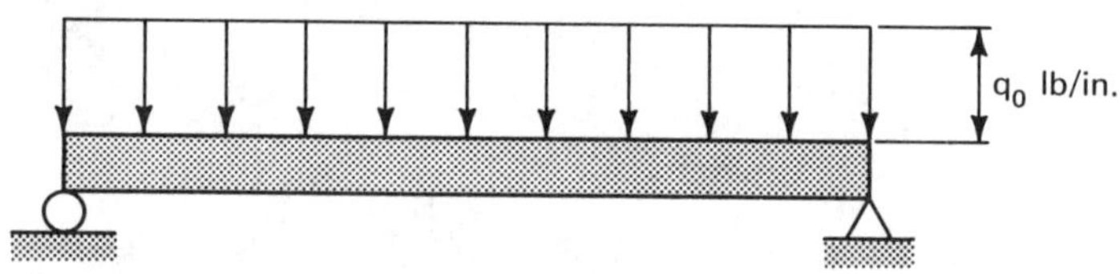

(c)

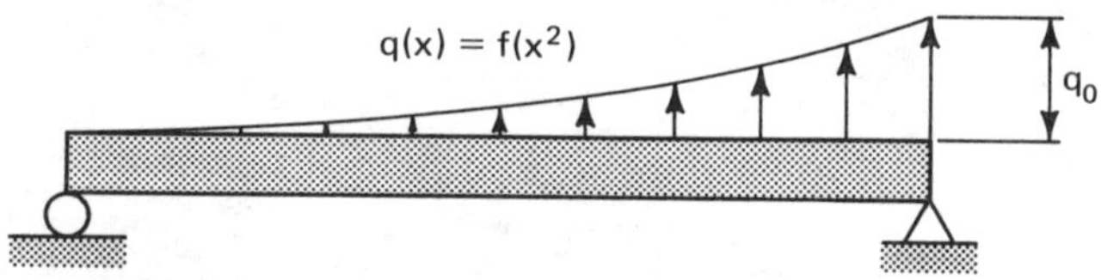

(d)

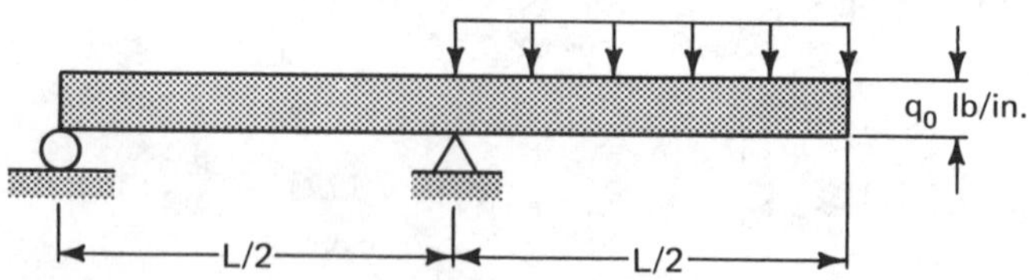

(e)

Problem 2. For the beams shown, find $\theta(x)$ and $y(x)$. Assume that each beam has modulus of elasticity E and moment of inertia I.

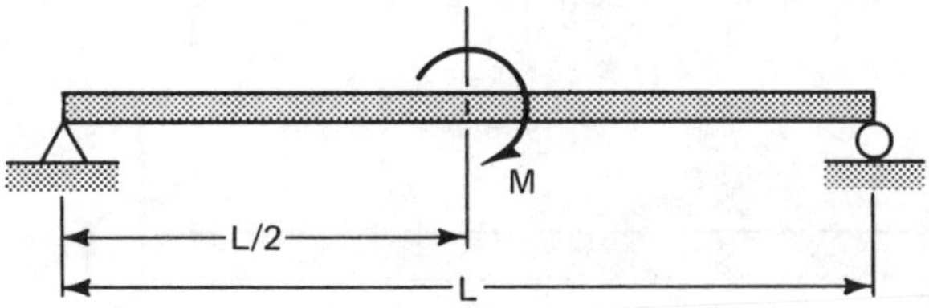

(a)

(b)

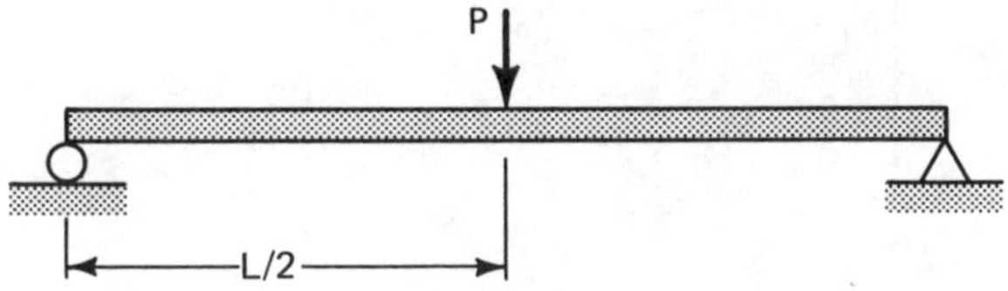

(c)

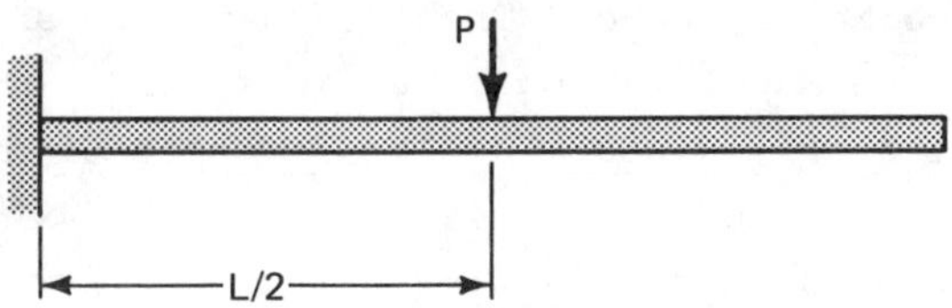

(d)

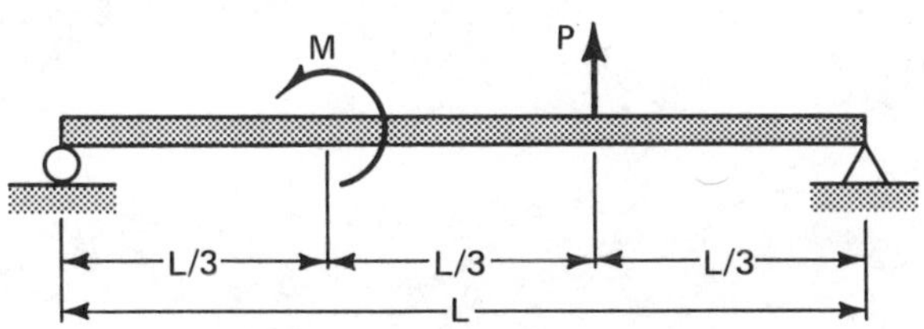

(e)

4.8 BEAM ELEMENT STIFFNESS MATRIX

The stiffness matrix for a beam element with forces and displacements as shown below is given without derivation.

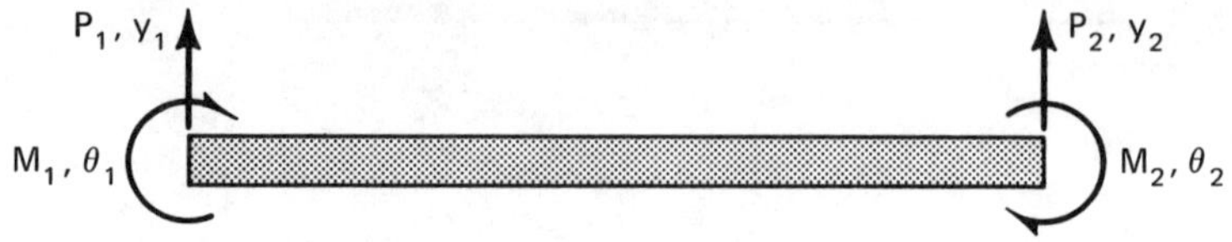

$$\begin{bmatrix} \frac{12}{L^3} & \frac{-6}{L^2} & \frac{-12}{L^3} & \frac{-6}{L^2} \\ \frac{-6}{L^2} & \frac{4}{L} & \frac{-6}{L^2} & \frac{2}{L} \\ \frac{-12}{L^3} & \frac{6}{L^2} & \frac{12}{L^3} & \frac{6}{L^2} \\ \frac{-6}{L^2} & \frac{2}{L} & \frac{6}{L^2} & \frac{4}{L} \end{bmatrix} \times EI$$

References

1. Galileo, Two New Sciences, Trans. Henry Crew and Alfonso de Salvio, The Macmillan Co., New York, 1933.
2. Timoshenko, S. P., History of the Strength of Materials, McGraw-Hill Book Co., Inc., 1953.
3. Roark, Raymond J., Formulas for Stress and Strain, 4th edition, McGraw-Hill, 1965.
4. Panlilio, F., Elementary Theory of Structural Strength, John Wiley and Sons, Inc., 1963.

Chapter 5

BIAXIAL STRESSES AND THE THEORY OF ELASTICITY

5.1 INTRODUCTION

Many real-life stress situations are biaxial. When this occurs it is very important to see how the stresses combine to affect the structure. Any structure or portion of structure in a general load situation such as shown in Figure 5.1.1 will experience σ_x, σ_y and τ_{xy} as functions of x and y which vary in a (usually) continuous fashion throughout the region.

Some examples of regions where these kinds of stress situations might occur are:

* the surface of a pressure vessel
* the skin of an aircraft or missile
* the floor of a building

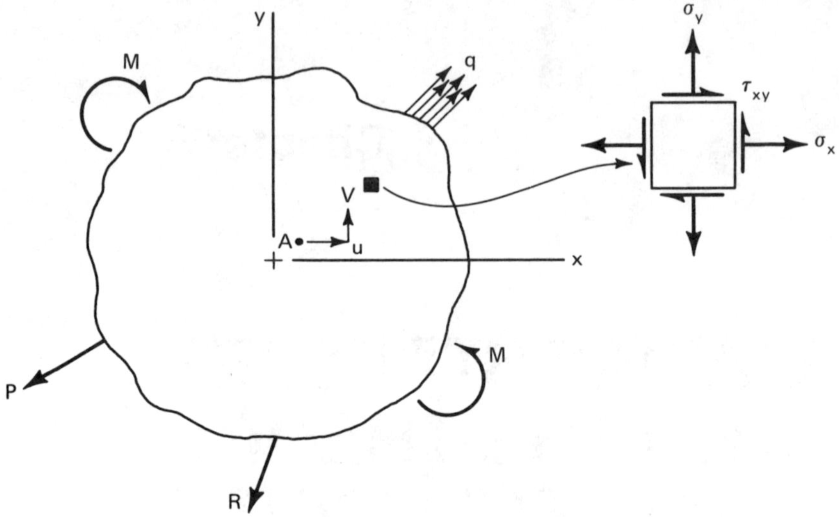

Figure 5.1.1

Many other examples can and will be cited throughout this text. It is very important to recognize biaxial stress situations and treat them properly.

In this chapter we are going to concern ourselves with two kinds of problems:

1. We know σ_x, σ_y and τ_{xy} at a particular point on the structure and we wish to investigate the effect of a rotation of x-y axis system at this point. All of the other variables are held constant.
2. We have a structure and a loading and we wish to determine

$$\sigma_x = f_1\ (x,y)$$

$$\sigma_y = f_2\ (x,y)$$

$$\tau_{xy} = f_3\ (x,y)$$

throughout the structure.

The first of the problems above will be treated in the sections on combined stresses, principal stresses and Mohr's circle. The second problem will be treated in the section on elementary thoery of elasticity.

Objectives

1. Be able to depict the state of stress variation with angle as shown in Figure 5.2.1 and 5.2.3.
2. Be able to convert stress to area and write equations of equilibrium such as the equations in the derivation of Equations (5.2.1) and (5.2.2).
3. Be able to state Equations (5.2.1), (5.2.2), (5.2.3) and (5.2.4) from memory. Be able to define the terms.
4. Be able to state Equations (5.3.1), (5.3.2) and (5.3.3) from memory. Be able to explain the use of the equations.
5. Be able to write and illustrate the positive sign conventions for biaxial stress.
6. Given any biaxial stress element, be able to:
   a) locate the principal axis (max axis) to within 45° by inspection.
   b) Calculate θ_p, the angle to the max axis.
   c) Calculate the principal stresses and show them on a properly oriented element.
   d) Calculate τ_{max} and the associated normal stresses σ_a and show them on a properly oriented element.

5.2 BIAXIAL STRESSES

In this section we consider the effect of orientation of axes on the results of our stress analysis. It is often surprising for students in stress analysis when it is first pointed out that the results are dependent upon the orientation of axes or in "how we choose to look at it."

This can be illustrated quite easily by considering a uniaxial rod which is in "pure" tension, Figure 5.2.1(a).

It is usual to consider the stress acting normal to the x axis as $\sigma_x = \frac{P}{A}$ and state that no shear stress act in this

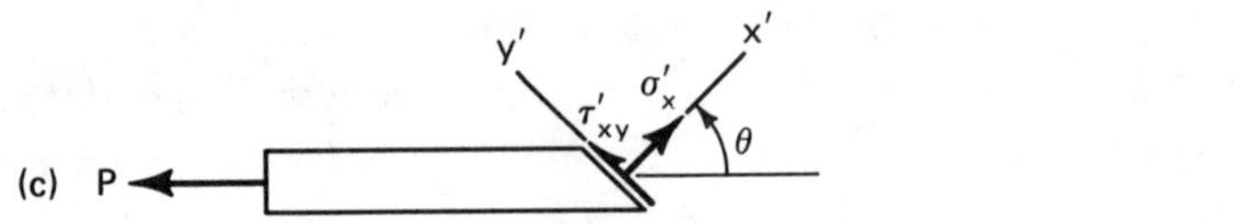

Figure 5.2.1

rod. (Figure 5.2.1(b).) However, if we rotate the axis system at the same point and without changing the load on the rod, to a new orientation of axes x'-y', defined by angle θ, Figure 5.2.1(c), then our results will be different.

This illustration for the uniaxial rod is part of a more general problem illustrated in Figure 5.2.2.

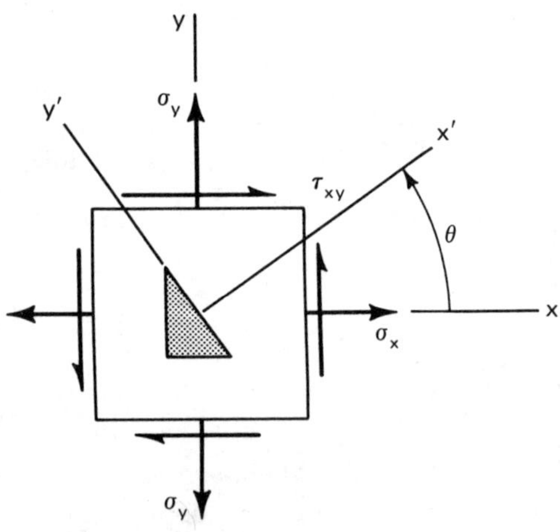

Figure 5.2.2

We wish to determine how the rotation of axes through angle θ effects the stress on the large element. If we isolate a free body of the shaded portion of the element we can derive the desired results with the aid of Figure 5.2.3.

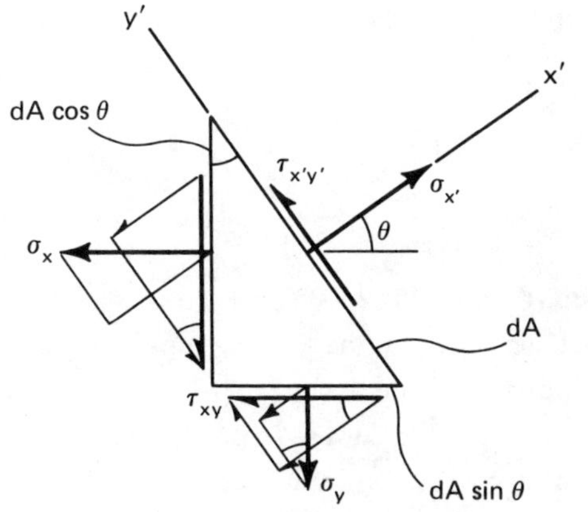

Figure 5.2.3

In Figure 5.2.3 we sum <u>forces</u> along the x'-y' directions. The forces are obtained by multiplying the stresses times the area the stresses are acting on. (Stresses themselves are not in equilibrium and are not vectors.)

$$\sum F_{x'} = \sigma_{x'} dA - \sigma_x dA\cos^2\theta - \sigma_y dA\sin^2\theta - 2\tau_{xy} dA\sin\theta\cos\theta = 0$$

or

$$\sigma_{x'} = \sigma_x\cos^2\theta + \sigma_y\sin^2\theta + 2\tau_{xy}\sin\theta\cos\theta \qquad (5.2.1)$$

Similarly,

$$\sum F_{y'} = 0$$

gives

$$\tau_{x'y'} = (\sigma_y - \sigma_x)\sin\theta\cos\theta + \tau_{xy}(\cos^2\theta - \sin^2\theta) \qquad (5.2.2)$$

Equations (5.2.1) and (5.2.2) can be combined with trigonometric identities to give

$$\sigma_{x'} = \frac{\sigma_x + \sigma_y}{2} + \frac{\sigma_x - \sigma_y}{2}\cos 2\theta + \tau_{xy}\sin 2\theta \qquad (5.2.3)$$

$$\tau_{x'y'} = -\frac{\sigma_x - \sigma_y}{2}\sin 2\theta + \tau_{xy}\cos 2\theta \qquad (5.2.4)$$

Equations (5.2.3) and (5.2.4) give the desired variation of $\sigma_{x'}$ and $\tau_{x'y'}$ with θ. The stress $\sigma_{y'}$ can be obtained from Equation (5.2.3) by setting $\theta = \theta + 90^o$.

5.3 PRINCIPAL STRESSES

As θ varies in Equations (5.2.3) and (5.2.4), $\sigma_{x'}$ and $\tau_{x'y'}$ vary as functions of θ. There are some special orientations of θ which are of special interest. The first θ of special interest is the θ for which $\sigma_{x'}$ is a maximum. This can, of course, be obtained by taking the derivative of Equation (5.2.3) with respect to θ and setting the result = 0. If this is done, we note that the derivative of Equation (5.2.3) is Equation (5.2.4) so we simply set Equation (5.2.4) = 0 and get the result.

$$\tan 2\theta_p = \frac{\tau_{xy}}{\frac{1}{2}(\sigma_x - \sigma_y)} \qquad (5.3.1)$$

Equation (5.3.1) gives the θ for which $\sigma_{x'}$ is maximum. Let us call this angle θ_p (for "theta principal"). The value of $\sigma_{x'}$ at θ_p is

$$\sigma_{max} = \frac{\sigma_x + \sigma_y}{2} + \sqrt{\left(\frac{\sigma_x - \sigma_y}{2}\right)^2 + \tau_{xy}^2} \qquad (5.3.2)$$

There is a corresponding σ_{min} given by

$$\sigma_{min} = \frac{\sigma_x + \sigma_y}{2} - \sqrt{\left(\frac{\sigma_x - \sigma_y}{2}\right)^2 + \tau_{xy}^2} \qquad (5.3.3)$$

The stresses σ_{max} and σ_{min} are called the <u>principal stresses</u>.

Another particular θ of interest is the one for which $\tau_{x'y'}$ is a maximum. This angle can be found by taking the derivative of Equation (5.2.4) with respect to θ and setting it equal to zero from which

$$\tan 2\theta_m = -\frac{1}{2}\frac{(\sigma_x - \sigma_y)}{\tau_{xy}} \tag{5.3.4}$$

We call this angle θ_m ("theta max-shear"). If we compare Equations (5.3.1) and (5.3.4) we see that θ_p and θ_m are 45° away from one another.

The magnitude of the maximum shear stress is

$$\tau_{xy\ max} = \sqrt{\left(\frac{\sigma_x - \sigma_y}{2}\right)^2 + \tau_{xy}^{\ 2}} \tag{5.3.5}$$

Equation (5.3.5) is the same as the radical terms of Equations (5.3.2) and (5.3.3).

We are now able to find the principal axes, the principal stress and the maximum shear stress. These values are of special interest as we will see shortly.

Occasionally there may be difficulty in determining which of the solutions of Equations (5.3.1) gives the angle θ_p. (The other root will be the θ to σ_{min}.) It is convenient to estimate the direction of the σ_{max} axis by the following device. Figure 5.3.1 shows an element with normal and shear stresses acting. We can consider the element to be a superposition of pure tension + pure shear. We can get the direction of σ_{max} from the pure tension portion by inspection. (σ_{max} is the algebraically larger of the two principal stresses.) Then we can get the direction of the σ_{max} axis on the pure shear portion by realizing that the σ_{max} axis lies on a line through the shear arrowhead. The original element, which is the sum of the two portions will have σ_{max} located within the defined 45° sector.

Another useful observation is that when we know the orientation of θ_p, then the maximum shear element is oriented at 45° to θ_p with the shear arrows pointing toward the σ_{max} axis. (Figure 5.3.2.)

On the τ_{max} element, we have normal stresses σ_a which are given by

$$\sigma_a = \frac{(\sigma_x + \sigma_y)}{2} \tag{5.3.6}$$

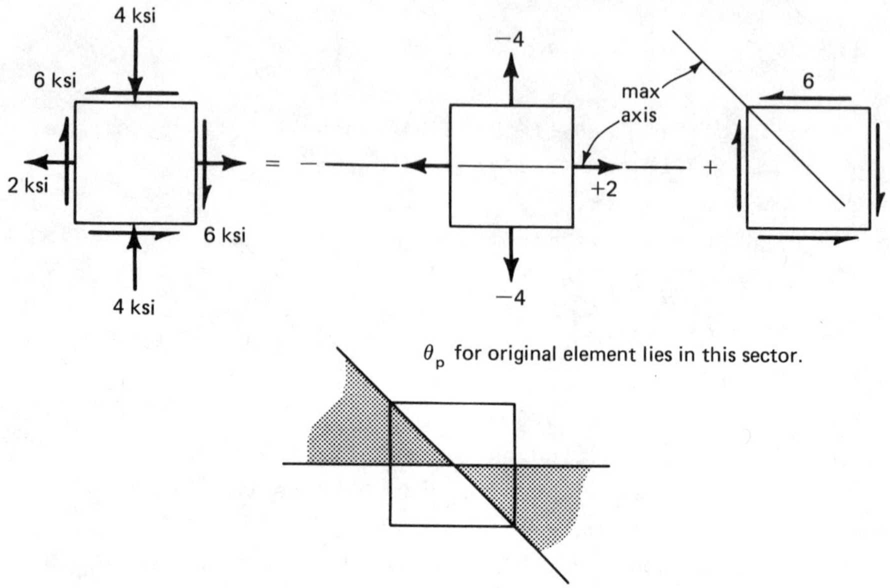

Figure 5.3.1

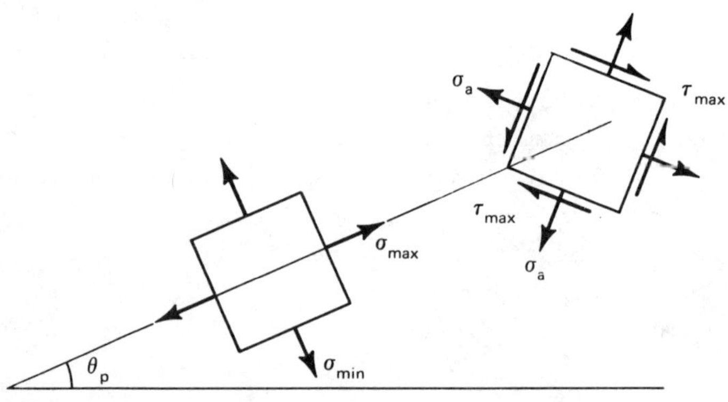

Figure 5.3.2

Summary of Findings for Principal Stress and Maximum Shear Stress

Let us list the significant findings for this section. These findings should be memorized.

1. Given a general element, we can find the normal and shear stresses as functions of rotation of axes as defined by θ.

2. There is a θ, θ_p, for which the normal stress will be a maximum. This angle defines the principal axis. The corresponding normal stress on the other face of the element will be a minimum. At the principal stress orientation, the shear stress will be zero (always!).
3. There will be an angle θ, θ_m, for which the shear stress is maximum. This angle is 45° away from θ_p. The maximum shear stress angle does not usually have zero normal stresses on it but has normal stresses σ_a. These normal stresses are equal in magnitude and direction.
4. If we examine Equation (5.3.2) we see that all the information is contained in it.

   a.

$$\sigma_{max} = \left(\frac{\sigma_x + y}{2}\right) + \sqrt{\left(\frac{\sigma_x - \sigma_y}{2}\right)^2 + \tau_{xy}{}^2}$$

   b. The minimum stress is the same except we have a minus sign.

$$\sigma_{min} = \left(\frac{\sigma_x + \sigma_y}{2}\right) - \sqrt{\left(\frac{\sigma_x - \sigma_y}{2}\right)^2 + \tau_{xy}{}^2}$$

   c. The maximum shear stress is the radical term

$$\tau_{x - y\ max} = \sqrt{\left(\frac{\sigma_x - \sigma_y}{2}\right)^2 + \tau_{xy}{}^2}$$

   d. The normal stresses associated with $\tau_{xy\ max}$ are given by the leading term

$$\sigma_a = \left(\frac{\sigma_x + \sigma_y}{2}\right)$$

<u>Example Problem</u>:

A biaxially stressed element has stresses as shown. (The units of stress are omitted.) Do the following:

a. Show the location of the principal stress axes to the nearest 45° by inspection.

b. Calculate the principal angle, θ_p, and show an element oriented in this direction.

c. Calculate the principal stresses σ_{max} and θ_{min} and show them on the principal element.

d. Calculate the maximum shear stress, $\tau_{xy\ max}$ and the associated normal stresses σ_a and show them on a properly oriented element.

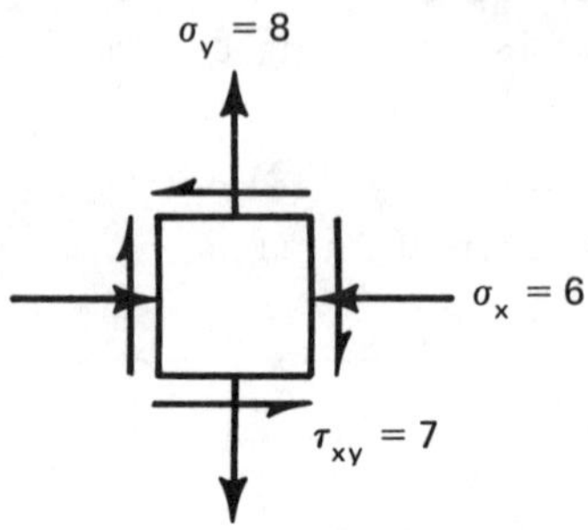

Solution:

The input stresses with correct sign are

$\sigma_x = -6$

$\sigma_y = +8$

$\tau_{xy} = -7$ (minus!)

The principal axis (max axis) to nearest 45° is shown below:

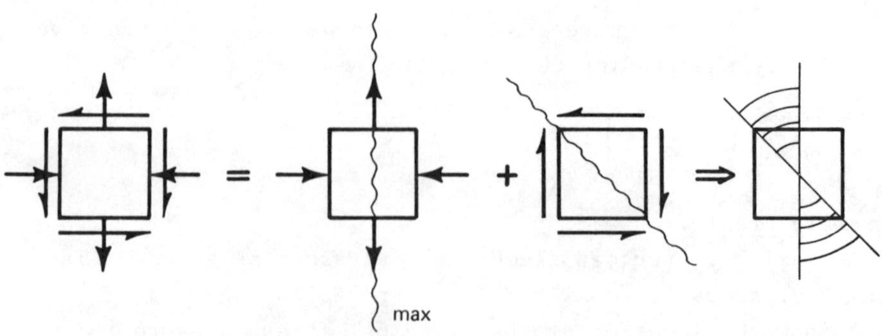

The principal angle θ_p is given by:

$$\tan 2\theta_p = \frac{\tau_{xy}}{\frac{1}{2}(\sigma_x - \sigma_y)} = \frac{-7}{\frac{1}{2}(-6 - 8)} = \frac{-7}{-7} = 1$$

$2\theta = 45^o, 225^o$
$\theta = 22.5^o$, $\underline{112.5^o}$ (from 2, must be this one.)

The principal stresses are

$$\sigma_{max,min} = \frac{\sigma_x + \sigma_y}{2} \pm \sqrt{\left(\frac{\sigma_x - \sigma_y}{2}\right)^2 + \tau_{xy}^{\ 2}}$$

$$= \frac{-6 + 8}{2} \pm \sqrt{\left(\frac{-6 - 8}{2}\right)^2 + (-7)^2}$$

$$= 1 \pm 9.899$$

$\sigma_{max} = \underline{\underline{10.899}}$ $\sigma_{min} = \underline{\underline{-8.899}}$

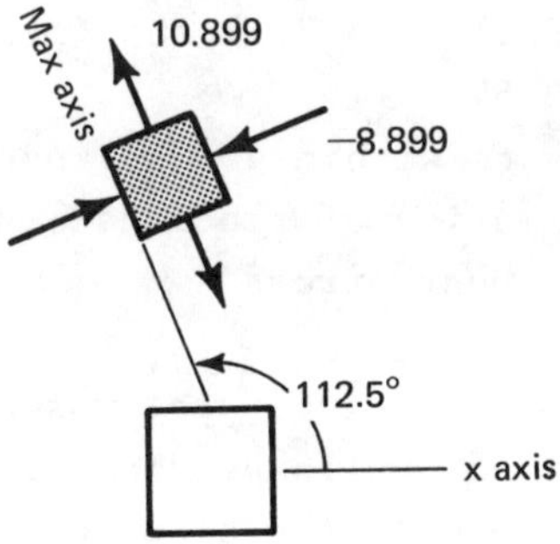

From Part (c), we can identify τ_{max} as

$$\tau_{max} = \sqrt{\left(\frac{x - y}{2}\right)^2 + \tau_{xy}^{\ 2}} = \underline{\underline{9.899}}$$

$$\sigma_a = \frac{\sigma_x + \sigma_y}{2} = \underline{\underline{1}}$$

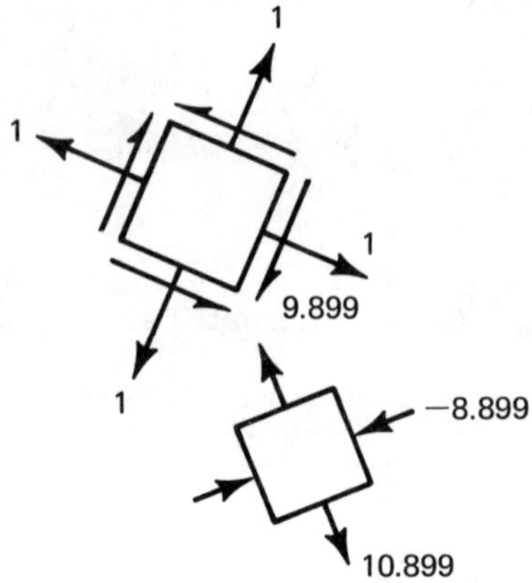

Problems

Closed Book:

You should be able to do these problems without reference to any aids.

1. Derive Equations (5.2.1) and (5.2.2).
2. State Equations (5.2.3) and (5.2.4).
3. State Equations (5.3.1), (5.3.2) and (5.3.3).
4. State Equation (5.3.5).
5. Locate the max axis to within 45° by inspection.
6. State the four "summaries of findings" for principal stress and maximum shear stress.

Open Book:

Problem 1. For one or more input columns below, find the max axis to the nearest 45° by inspection; find the inclination of θ_p; find σ_{max} and σ_{min} and show them on a properly oriented element; find τ_{max} and σ_a and show them on a properly oriented element.

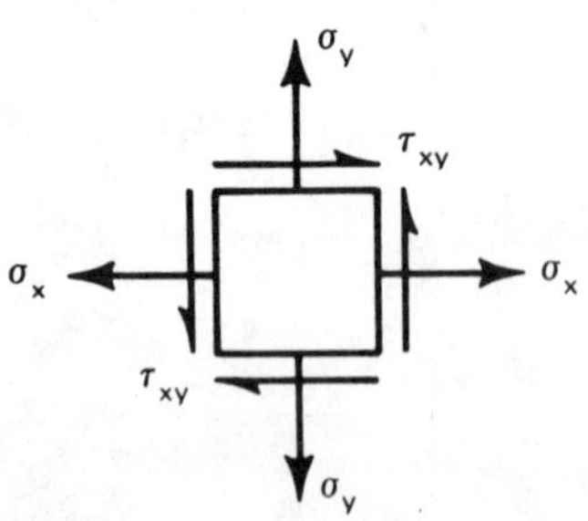

	(a)	(b)	(c)	(d)	(e)
σ_x	5	−6	−2.7	4	9
σ_y	2	−3	7	4	−5
τ_{xy}	−3	−6	2	9	3

Units are ksi

Problem 2.

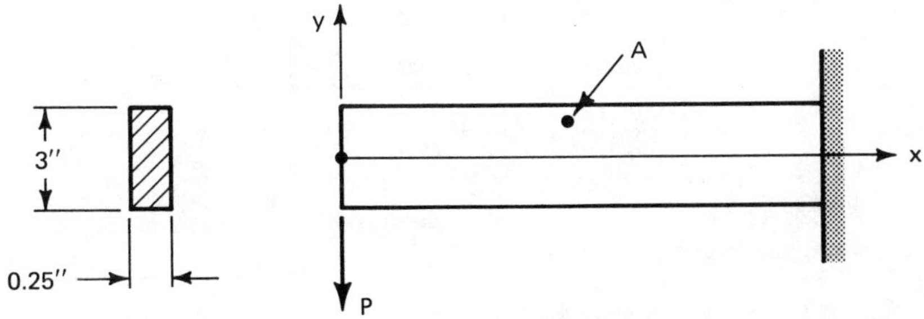

For one or more columns in the table below: (1) find the state of stress at point a, which is specified by coordinates x and y. (2) Find the principal stresses and their orientation for this element. (3) Find the max shear stresses, the associated normal stresses and the max shear element orientation.

	(a)	(b)	(c)	(d)	(e)
x (in.)	30.	30.	20.	30	0.
y (in.)	0.	1.0	−1.5	1.5	−1.5
p (lb)	900	1000	1200	1200	900

Problem 3.

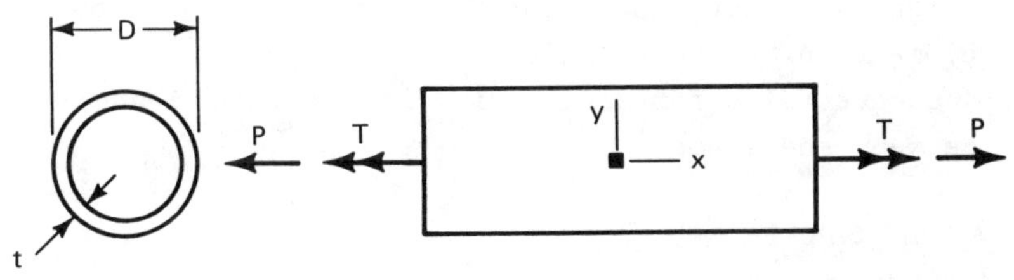

For one or more columns in the table, find the x-y plane stresses. Assume the curvature is negligible. Find the principal stresses and their orientation. Find the max shear stress and the associated normal stresses and the max shear element orientation.

	(a)	(b)	(c)	(d)
P (lb)	1200	900	1000	1000
T (in.-lb)	900	1200	1000	500
D (in.)	10	12	9	8
t (in.)	0.1	0.125	0.1	0.125

Objectives for Mohr's Circle

1. Be able to draw and illustrate the sign conventions for Mohr's circle.
2. Be able to state the formula for the center of the circle. (From memory.)
3. Given any general biaxial stress element, be able to plot Mohr's circle and find:
   a) The principal angle θ_p
   b) σ_{max} and σ_{min}
   c) τ_{max} and σ_a
   d) The orientation of all elements with the stresses on them properly shown
   e) The state of stress for an element oriented at any angle. Be able to show the element and the stresses.
4. Be able to answer (correctly) true-false questions on the eight observations about Mohr's circle.
5. Be able to label the identification of any axis on the element and on the circle.
6. Know the rules for angle (direction and magnitude) between the element and the circle.

5.4 MOHR'S CIRCLE

In 1882 Professor Otto Mohr published his findings on a graphical method for the solution of stress at a point. (Reference 1.) Mohr's circle is a convenient way of obtaining numerical solutions. It is more important as a way of developing visual powers for "seeing" what happens to the stresses on the element as θ varies. Let us consider Figure 5.4.1 in our discussions.

The element in Figure 5.4.1 states the problem and defines the sign conventions. Theta is positive measured counter-clockwise from the x axis. The normal stresses are positive if tensile; negative if compressive. The plus x face is the face which intersects the plus x axis. A positive shear stress acts on the plus x face in the plus y direction. Its companion shear arrow is negative.

On the circle part of the figure, the coordinates of the "circle space" are σ and τ. Any state of stress of the element for any θ can be plotted on the σ-τ coordinate system. The plot of all such σ and τ for any element will trace out a circle. The center of the circle will be at $(\sigma_x + \sigma_y)/2$. To plot the circle, do the following:

1. Find the center at $\sigma = \left(\frac{\sigma_x + \sigma_y}{2}\right)$; $\tau = 0$.
2. Plot the plus x face; $\sigma = \sigma_x$; $\tau = \tau_{xy}$.
3. Plot the plus y face; $\sigma = \sigma_y$; $\tau = -\tau_{xy}$.
4. Draw the diameter and draw the circle. See Figure 5.4.1.

Now observe the following:

1. The intersections of the circle with the σ axis gives σ_{max} and σ_{min}.
2. The x, y, max and min axes can be labeled on the circle.
3. The principal angle, $2\theta_p$, <u>from the x axis to the max axis</u> can be found on the circle. For the <u>element</u>, θ_p, from the x axis to the max axis is half as much and <u>goes the other direction</u>. A similar angle exists from the y axis to the min axis on either element or circle.
4. The maximum shear stress is at the top and bottom of the circle and has associated normal stress $\sigma_a = \left(\frac{\sigma_x + \sigma_y}{2}\right)$, the coordinate of the center of the circle.
5. The angle from the x axis to the max shear axis is 2β on the circle, will be β on the element and go the other way.
6. The angle between the max (or min) axis and the max shear axis will always be 90°.

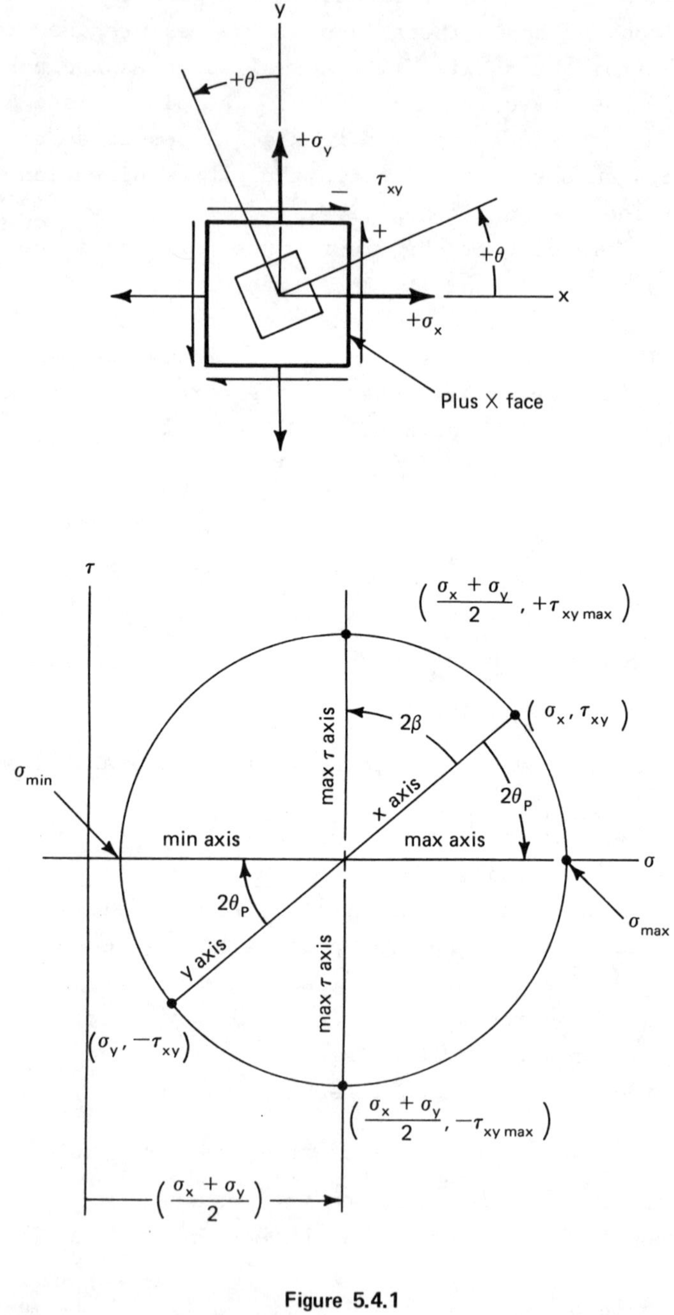
y
+θ
$+\sigma_y$
τ_{xy}
+
+θ
x
$+\sigma_x$
Plus X face
τ
$\left(\frac{\sigma_x + \sigma_y}{2}, +\tau_{xy\,max}\right)$
(σ_x, τ_{xy})
2β
max τ axis
x axis
$2\theta_P$
σ_{min}
min axis
max axis
σ
σ_{max}
$2\theta_P$
y axis
max τ axis
$(\sigma_y, -\tau_{xy})$
$\left(\frac{\sigma_x + \sigma_y}{2}, -\tau_{xy\,max}\right)$
$\left(\frac{\sigma_x + \sigma_y}{2}\right)$

Figure 5.4.1

7. The σ_{max} and σ_{min} axes are always orientations of zero shear stress.
8. The angles between axes on the circle and axes on the element are always twice as much on the circle as on the element and the directions are always opposite. This statement can be used to find the stresses acting on an element at any desired θ.

Example Problem:

(a) Solve the last example problem (all parts) by Mohr's circle.

(b) In addition, find the stresses on an element oriented at $\theta = -60^{\circ}$.

Solution:

The plus x and plus y faces are labeled and the stresses are given signs.

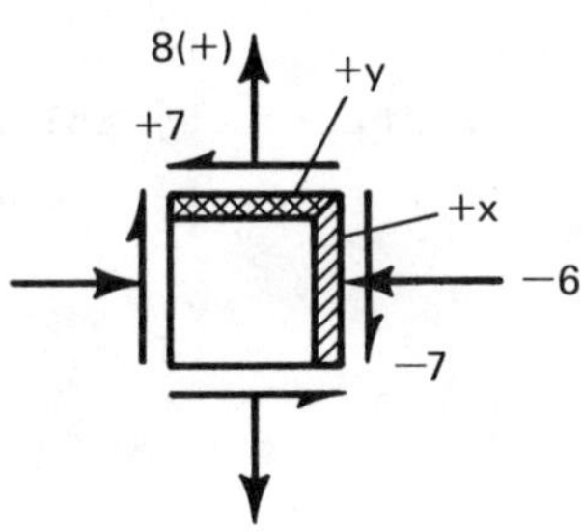

The center of the circle is at $\frac{(\sigma_x + \sigma_y)}{2} = \frac{-6 + 8}{2} = 1.$

The center, the plus x and plus y stresses are plotted and the circle is drawn.

From the circle, 2θ is measured to be 45° C. W. from the y axis to the max axis.

The following are also measured off of the circle:

$\sigma_{max} = 10.9$

$\sigma_{min} = -8.9$

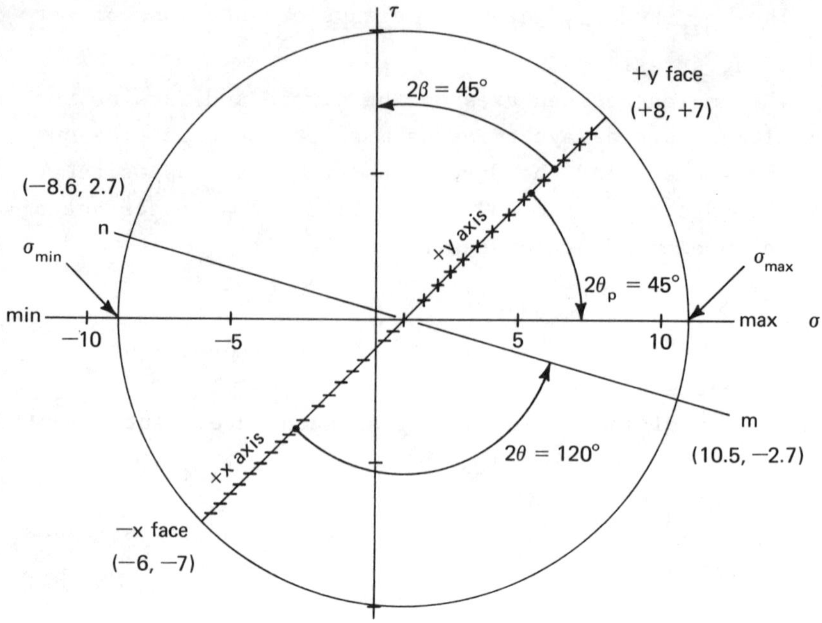

The principal element is located by going the opposite direction and one-half the magnitude of the angle on the circle from y to max.

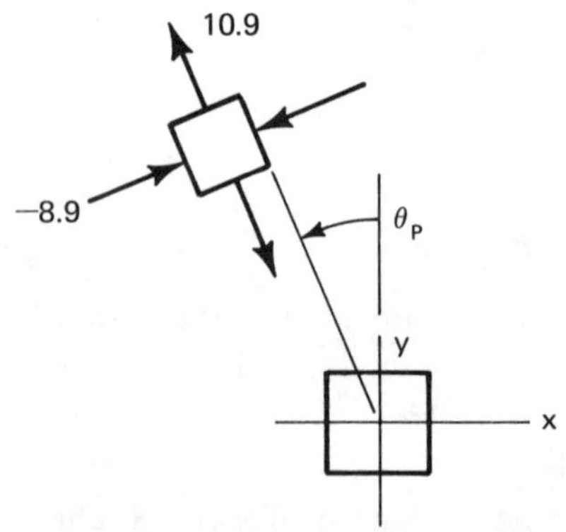

The max shear stress is scaled from the circle.

$\tau_{max} = 9.9$

The associated normal stress is the distance to the center of the circle.

$\sigma_a = 1$

The maximum shear element is shown below. (Note the angle β between the y axis and the τ_{max} axis for the element and the circle.)

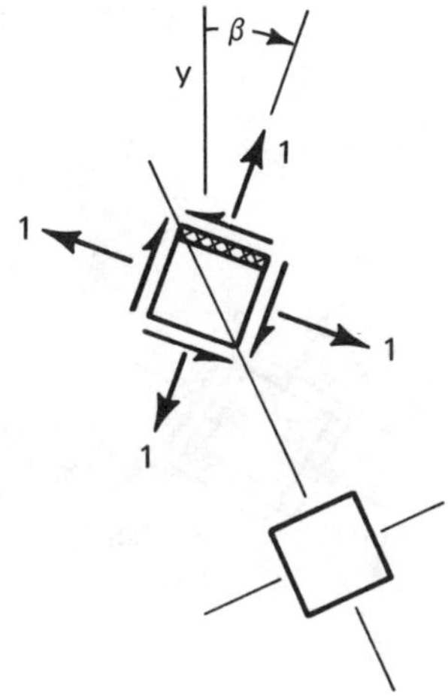

To obtain the state of stress for the element at $\theta = -60^\circ$, draw the element at proper orientation. (The angle is 60° CW from the x axis to the m axis.)

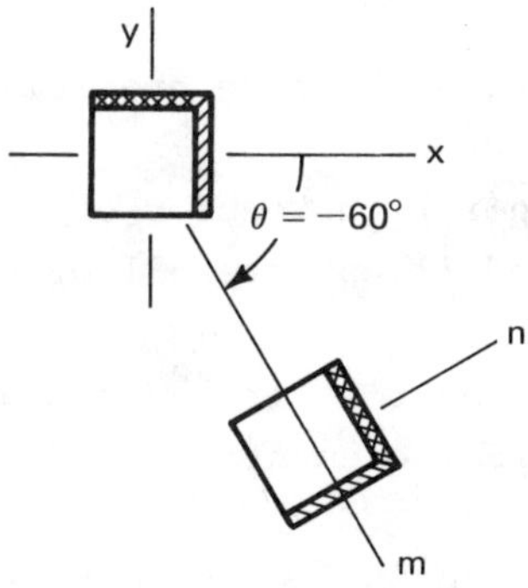

Now go back to the circle and go 120° CCW from the x axis to the m axis.

The m and n axes are identified on the circle and the stresses scaled off.

For the m axis:

$$\sigma_m = 10.5$$

$$\tau = -2.7$$

For the n axis:

$$\sigma_n = -8.6$$

$$\tau = +2.7$$

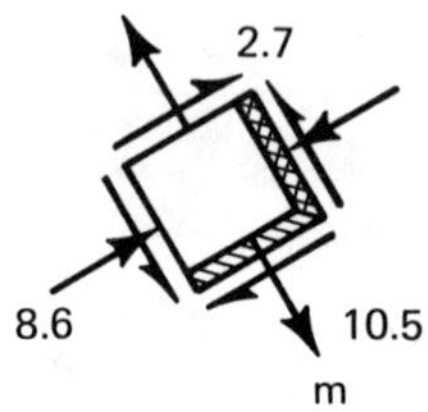

The proper directions for the shear arrows can be obtained by keeping track of the "plus x" and "plus y" faces and following the usual sign convention.

Problems

<u>Closed Book</u>:

You should be able to work these problems without any aids.

1. State the equation for the center of Mohr's circle.
2. Draw an element and illustrate all of the sign conventions.
3. Plot the center of Mohr's circle and the plus x and y faces for any typical element.

<u>Open Book</u>:

1. Work problem 1 of Section 5.3 by Mohr's circle.
2. Work problem 2 of Section 5.3 by Mohr's circle.

3. Work problem 3 of Section 5.3 by Mohr's circle.
4. For one or more columns in the table shown, use Mohr's circle to find:
   a) θ_p
   b) σ_{max} and σ_{min}
   c) τ_{max} and σ_a
   d) The state of stress on an element oriented at angle α.

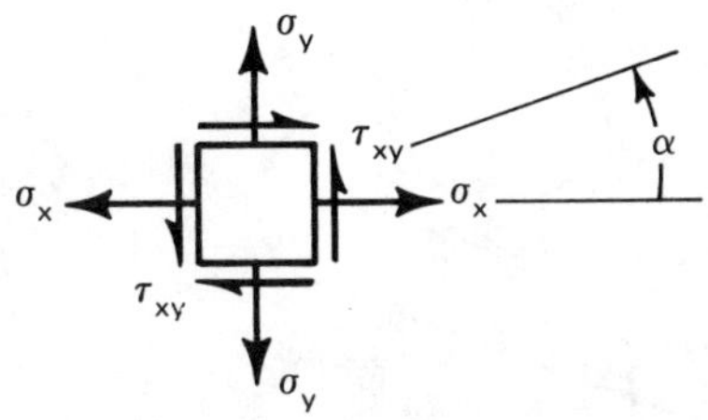

Show all elements with proper orientation and with all stresses depicted.

	(a)	(b)	(c)	(d)
σ_x (ksi)	5	4	3	−2
σ_y (ksi)	−7	3	4	−9
τ_{xy} (ksi	2	2	−5	−3
α (deg)	−21	31	−71	71

Objectives

1. Be able to derive Equations (5.5.1) and (5.5.2).
2. Be able to draw the three Mohr's circles for any pressure vessel and find $\tau_{max\text{-}max}$.
3. Be able to state, explain and correctly use the two theories of failure.
4. Be able to correctly distinguish between ductile and brittle failure of materials.

5.5 THE PRESSURE VESSEL AND $\tau_{MAX-MAX}$

One quite important, but subtle, observation can be made about the third dimension, or z direction even though we will not consider three dimensional stresses in general. The fact that $\sigma_3 = 0$ may not always be trivial, as we can see by considering the example of the pressure vessel. Figure 5.5.1(a) shows a vessel with internal pressure. The walls are considered to be "thin;" t << R. (Figure 5.5.1(b).)

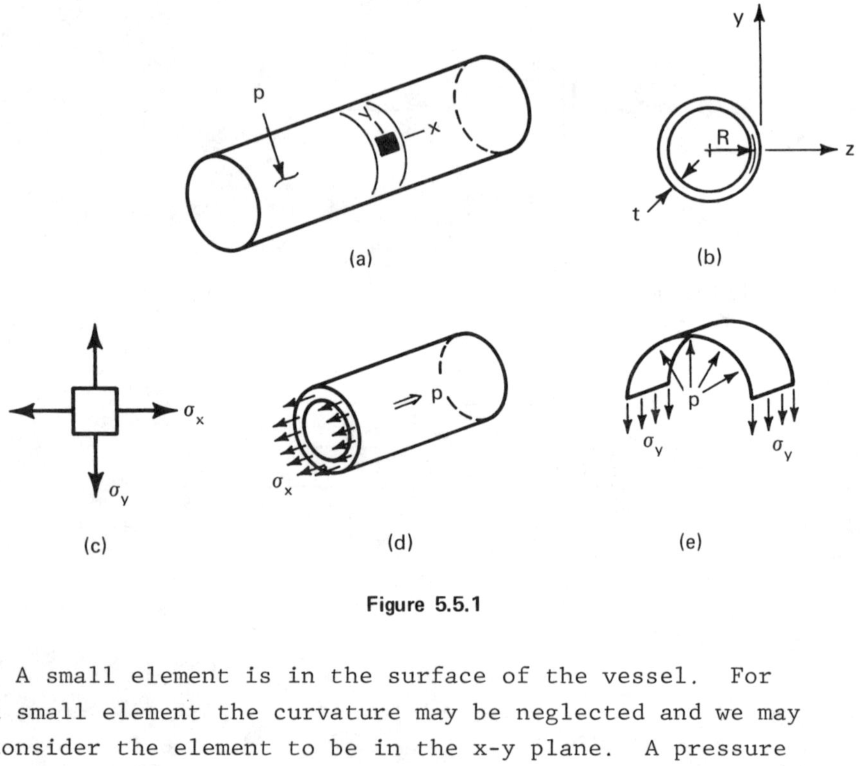

Figure 5.5.1

A small element is in the surface of the vessel. For a small element the curvature may be neglected and we may consider the element to be in the x-y plane. A pressure vessel develops normal stresses in two dimensions (Figure 5.5.1(c)). The stress σ_x develops because the pressure tries to blow the ends apart. In Figure 5.5.1(d) if we equate forces along the x-axis,

$$\sigma_x \ (2\pi \ R \ t) = p(\pi R^2)$$

$$\boxed{\sigma_x = \frac{pR}{2t}} \qquad (5.5.1)$$

Similarly, the pressure will try to blow the hoops apart in the y-direction. From Figure 5.5.1(e),

$$p(2R\ \Delta x) = 2\sigma y\ \Delta x \cdot t$$

$$\boxed{\sigma_y = \frac{pR}{t}} \qquad (5.5.2)$$

Equations (5.5.1) and (5.5.2) give σ_x and σ_y for a pressure vessel in terms of pressure p, radius (mean) R and thickness t. There are no shear stresses in the surface planes of a thin-wall pressure vessel. Therefore the x and y (longitudinal and hoop) directions are principal directions. Figure 5.5.2 shows the principal element for a pressure vessel.

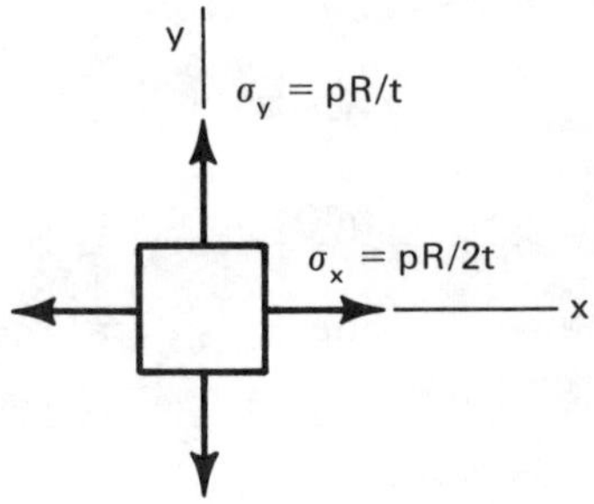

Figure 5.5.2

Let us examine Mohr's circle for this pressure vessel, Figure 5.5.3. $\sigma_{max} = \frac{pR}{t}$ and $\sigma_{min} = \frac{pR}{2t}$ in the x-y plane. $\tau_{xy\ max}$ is shown and equals $\frac{pR}{4t}$. <u>However</u>, if we consider the z direction, there are two more circles we can draw. $\sigma_z = \sigma_3 = 0$ by inspection, see Figure 5.5.1(b).

The x-z circle joins $\sigma_3 = 0 = \sigma_z$ and $\sigma_y = \sigma_{min} = \frac{pR}{2t}$. A third circle may be drawn for the y-z plane which joins

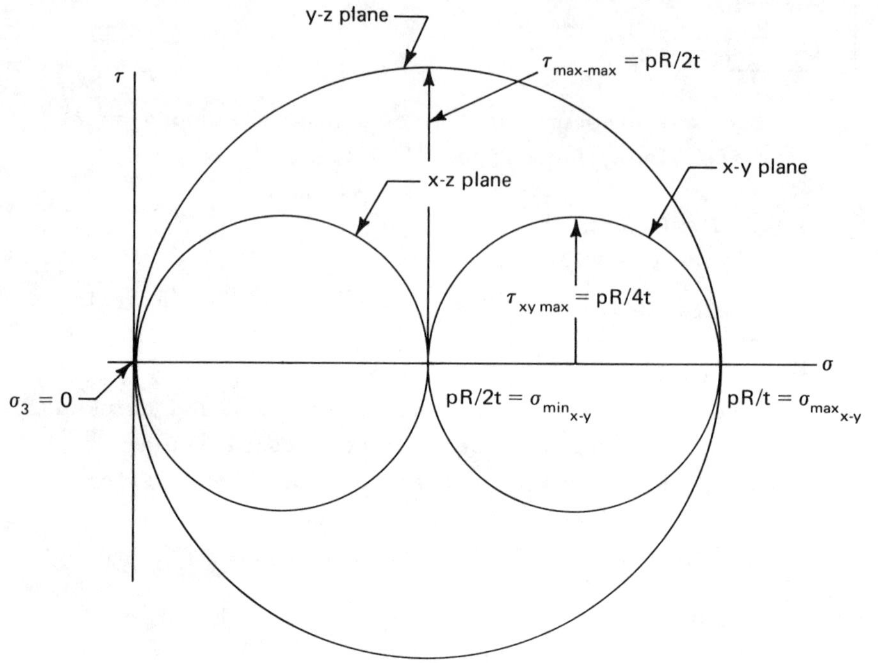

Figure 5.5.3

$\sigma_3 = 0 = \sigma_z$ and $\sigma_x = \sigma_{max} = \frac{pR}{t}$. The maximum of the τ_{max} in each plane is the y-z plane τ_{max}, called $\tau_{max\text{-}max}$. Its direction will be 45° through the wall of the vessel as shown in Figure 5.5.4.

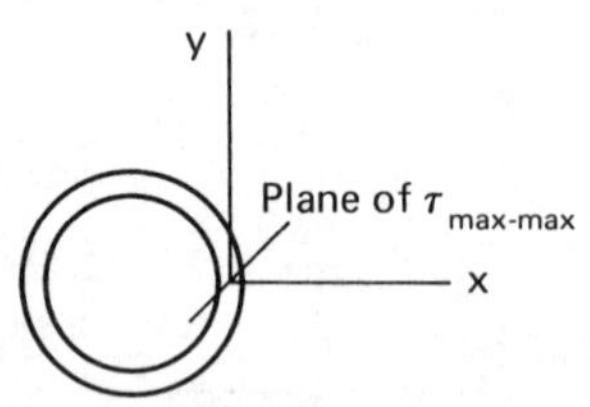

Figure 5.5.4

This is the maximum shear stress which is significant for the pressure vessel, as we shall see in the next section.

5.6 THEORIES OF FAILURE

We are now prepared to state two vitally important theories of failure for materials:

1. A <u>ductile</u> material will fail when the shear stress $\tau_{max-max}$ equals the shear stress τ_{max} which occured at yield for the same material in a uniaxial tension test. (Reference (2) contains a more comprehensive discussion of theories of failure.)
2. A <u>brittle</u> material will fail when the maximum normal stress, σ_{max} equals σ_{max} at fracture for the same material in a uniaxial tension test. (F_{tu}.)

These two theories have been shown to correlate quite well with experimentally observed failures.

<u>Example</u>:

A pressure vessel with the dimensions shown is made of AISI 1025 steel. This is a ductile steel with F_{ty} of 36,000 psi. At what pressure will the vessel fail by yielding? In what direction will the failure occur?

<u>Solution</u>:

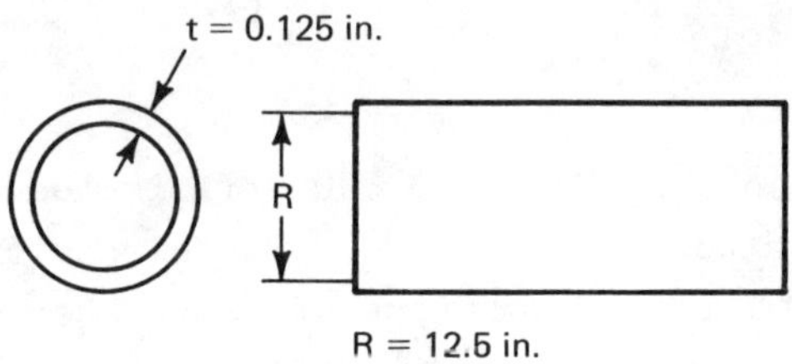

1. <u>Calculate</u> $\tau_{max-max}$ --

   The solution for all pressure vessels is the same in general terms and is shown in Figure 5.5.3 where

$$\tau_{max-max} = \frac{pR}{2t}$$

for our vessel

$$\tau_{max-max} = \frac{pR}{2t} = \frac{p\ 12.5}{2(.125)} = \underline{\underline{50\ p\ psi}}$$

2. Calculate τ_{max} at yield in uniaxial tension.
The Mohr's circle for uniaxial tension at the yield point of the material is shown below, where $\sigma_{max} = \sigma_{yield}$ and $\sigma_y = 0$.

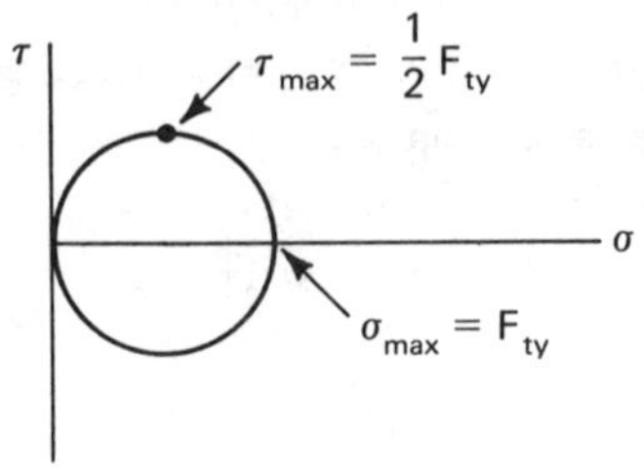

The ductile material is said to fail when $\tau_{max\text{-}max} = \tau_{max}$ at yield in uniaxial tension.

For our pressure vessel

$$50\ p = 1/2\ F_{ty}$$
$$50\ p = 1/2\ (36{,}000)$$
$$p = \frac{18{,}000}{50} = \underline{360 \text{ psi (answer)}}$$

The direction of failure is shown in Figure 5.5.4 (answer)

Example:

A brittle cast iron bar (F_{tu} = 20,000 psi) is loaded in combined tension and torsion as shown. The dimensions of the bar and the axial load and torque are such that σ_x = 12,000 psi and τ_{xy} = 8,000 psi. Will the bar fail? What is the factor of safety? What is the direction of the critical stress?

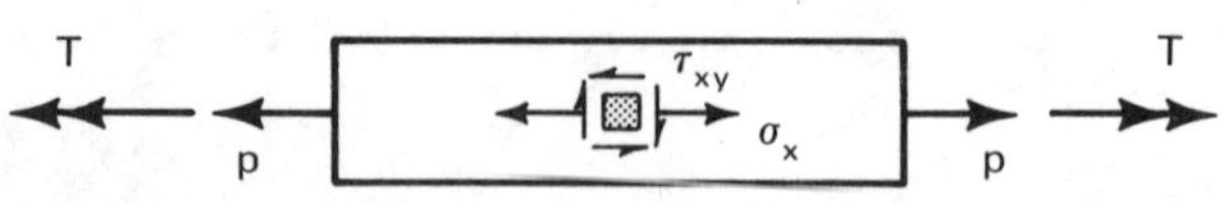

Solution:

The principal stress σ_{max} is most easily found by using Mohr's circle. This is shown below. The center of the circle is at

$$\frac{\sigma_x + \sigma_y}{2} = \frac{12,000}{2} = 6,000 \text{ psi}$$

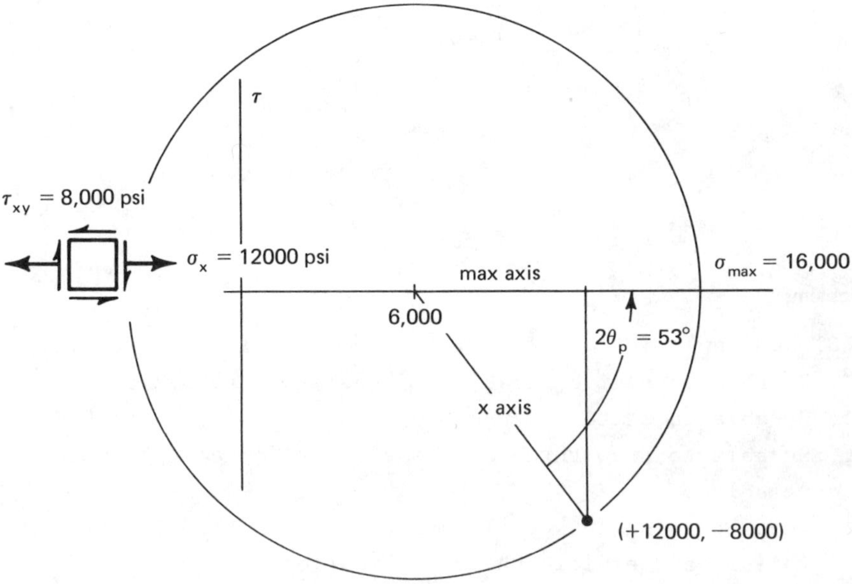

From the Mohr's circle

σ_{max} = 16,000 psi

We know that

F_{tu} = 20,000 psi

Since 16,000 < 20,000 or $\sigma_{max} < F_{tu}$ the brittle material will not fail (answer).

The factor of safety is

$$\text{F. S.} = \frac{20,000}{16,000} = 1.25 \text{ (answer)}$$

The critical direction is the direction of the σ_{max} axis. Using the Mohr's circle, we see that 2θ from the x axis to the max axis is 53° as shown. The critical direction is 26.5° the other way and is shown below. Note: The crack direction in this part, if it were to fail in brittle tension is shown by the wavy line.

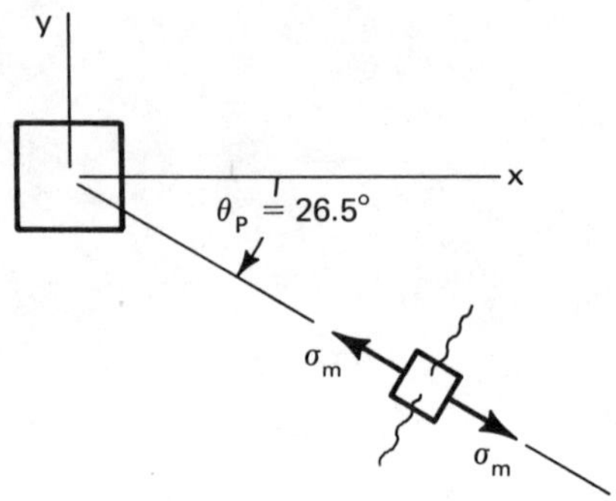

Problems

Closed Book:

1. Be able to derive Equations (5.5.1) and (5.5.2).
2. Be able to state and use Equations (5.5.1) and (5.5.2).
3. Be able to draw three Mohr's circles for any pressure vessel.
4. Be able to state and illustrate the two theories of failure and explain their application.

Open Book: For each of the problems below:

A. State what yield stress that the material from which the part is made would have to be a 1.0 F. S. based on ductile failure theory.

B. State what ultimate stress that the material from which the part is made would have to be for a 1.0 F. S. based on brittle failure theory.

1. Problem 1 of Section 5.3.
2. Problem 2 of Section 5.3.
3. Problem 3 of Section 5.3.
4. One or more of the columns below.

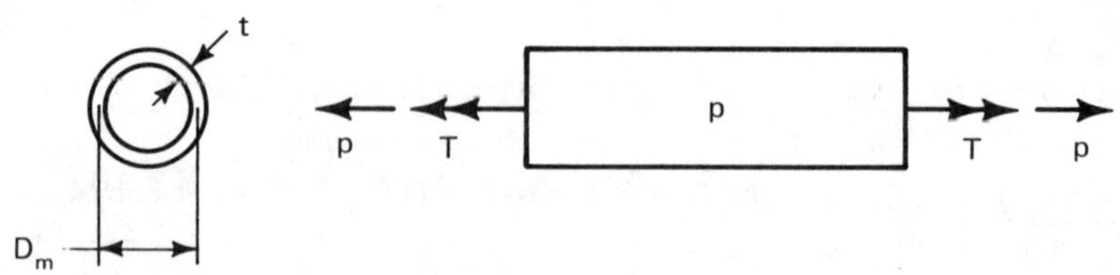

	(a)	(b)	(c)	(d)	(e)
P (lb)	30000	−30000	40000	50000	60000
T (lb-in.)	10000	10000	20000	50000	60000
Dm (in.)	5	5	4	5	5
t (in.)	0.1	0.1	0.125	0.1	0.1
p (psi)	200	200	500	500	500

Objectives

1. Be able to derive, closed book, the following equations: equilibrium equations, strain-displacement equations, stress-strain equations, compatibility equations, biharmonic equation.
2. Be able to solve problems similar to Problem 1 at the end of this section (open book).

5.7 ELEMENTARY THEORY OF ELASTICITY IN PLANE STRESS

The mathematical theory of elasticity is a considerably extensive subject. References (3), (4), and (5) offer excellent introductory to intermediate treatments. The portion we can treat in this text can only be of the most introductory nature. Nevertheless, it is worth doing because it introduces one of the most <u>fundamental</u> areas of stress analysis; because it increases our understanding of how stresses behave in structures and structural components; because it introduces the concepts of advanced mathematics, i.e. partial differential equations, to the study of stress analysis.

We shall restrict ourselves to <u>plane stress</u>. By this we understand that the stresses such as those in Figure 5.1.1 vary with x and y, but that all z-direction stresses are zero. We can then proceed to derive relationships for the x-y plane stresses and displacements.

A typical two dimensional stress element is shown in Figure 5.7.1.

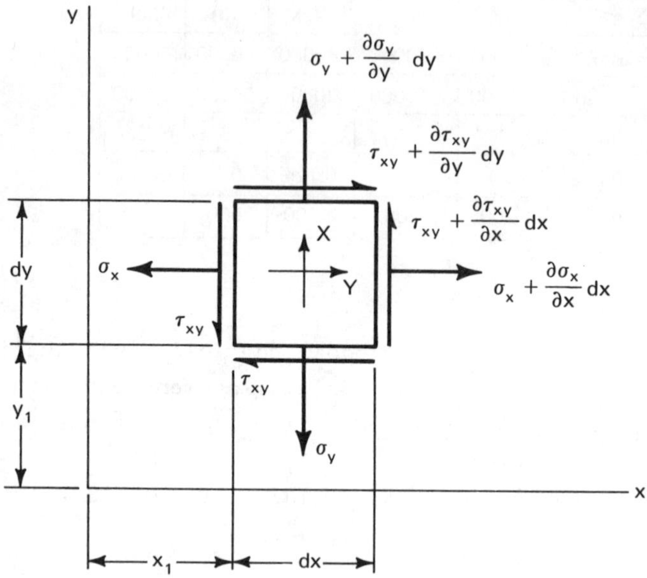

Figure 5.7.1

Forces X and Y are body forces (gravity, electrical or inertial fields, etc.).

In the x-direction, we can write $\sum F_x = 0$. We consider the z-direction thickness to be unity.

$$-\sigma_x dy + \left(\sigma_x + \frac{\partial \sigma_x}{\partial x} dx\right) dy$$

$$+ \left(\tau_{xy} + \frac{\partial \tau_{xy}}{\partial y} dy\right) dx - \tau_{xy} dx + X dx dy = 0$$

$$\frac{\partial \sigma_x}{\partial x} + \frac{\partial \tau_{xy}}{\partial y} + X = 0 \qquad (5.7.1)$$

Similarly in the y direction

$$\frac{\partial \sigma_y}{\partial y} + \frac{\partial \tau_{xy}}{\partial x} + Y = 0 \qquad (5.7.2)$$

These are the two-dimensional, plane-stress equations of equilibrium.

Let us now consider the displacements of particles in a two dimensional stress field such as that illustrated in Figure 5.7.2. Particle A will actually move when the structure is loaded. We may think of its motion, or displacement, as being composed of "horizontal" displacement u and "vertical" displacement v. If we consider ABC and D to form the corners of our element shown in Figure 5.7.1, then we can illustrate the deformation of our element in Figure 5.7.2.

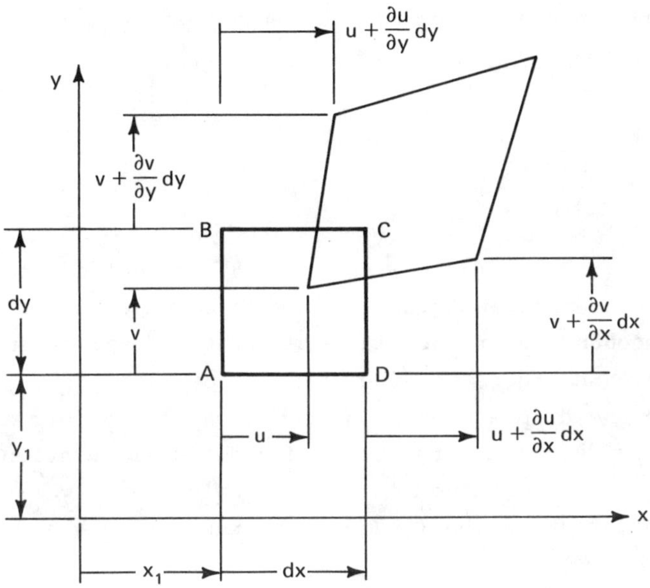

Figure 5.7.2

If we restrict ourselves to small displacements, then the tangent of an angle equals the angle and the displacements are as shown in the Figure.

We now write the relation between strain and displacement. Recall that normal strain is elongation per unit length and shear strain is change in angle from original right angle. Then

$$\varepsilon_x = \frac{\text{change in length}}{\text{original length}}$$

$$\varepsilon_x = \frac{u + \frac{\partial u}{\partial x}\,dx - u}{dx}$$

$$\varepsilon_x = \frac{\partial u}{\partial x} \tag{5.7.3}$$

Similarly

$$\varepsilon_y = \frac{\partial v}{\partial y} \tag{5.7.4}$$

The change in angle from original right angle is

$$\gamma_{xy} = \frac{u + \frac{\partial u}{\partial y}\,dy - u}{dy} + \frac{v + \frac{\partial v}{\partial x}\,dx - v}{dx}$$

$$\gamma_{xy} = \frac{\partial u}{\partial y} + \frac{\partial v}{\partial x} \tag{5.7.5}$$

Equations (5.7.3), (5.7.4) and (5.7.5) are called the strain-displacement equations. In this case we have three strain components expressed in terms of two displacement components. In order that the strains are ultimately related to the displacements in such a way as to form a continuously deformed structure, an additional relationship is needed. It is obtained in the following way.

Differentiate Equation (5.7.5) with respect to x and y

$$\frac{\partial^2 \gamma_{xy}}{\partial x \partial y} = \frac{\partial^2}{\partial x \partial y} \frac{\partial u}{\partial y} + \frac{\partial^2}{\partial x \partial y} \frac{\partial v}{\partial x} \tag{5.7.6}$$

We postulate that u and v are continuous functions of x and y. Therefore the order of differentiation is interchangeable and Equation (5.7.6) can be written,

$$\frac{\partial^2 \gamma_{xy}}{\partial x \partial y} = \frac{\partial^2}{\partial y^2} \frac{\partial u}{\partial x} + \frac{\partial^2}{\partial x^2} \frac{\partial v}{\partial y} \tag{5.7.6}$$

Using Equations (5.7.3) and (5.7.4) in Equation (5.7.6) gives,

$$\frac{\partial^2 \varepsilon_x}{\partial y^2} + \frac{\partial^2 \varepsilon_y}{\partial x^2} = \frac{\partial^2 \gamma_{xy}}{\partial x \partial y} \qquad (5.7.7)$$

Equation (5.7.7) is called the equation of compatibility.

The two dimensional stress-strain relationships for plane stress may be written as

$$\varepsilon_x = \frac{1}{E}(\sigma_x - \nu\sigma_y) \qquad (5.7.8)$$

$$\varepsilon_y = \frac{1}{E}(\sigma_y - \nu\sigma_x) \qquad (5.7.9)$$

$$\gamma_{xy} = \frac{2(1 + \nu)}{E}\tau_{xy} \qquad (5.7.10)$$

Everything is now getting rather complicated and scattered out. To tie it all together we introduce the concept of a stress function, $\psi(x,y)$ such that

$$\sigma_x = \frac{\partial^2 \psi}{\partial y^2} \qquad (5.7.11)$$

$$\sigma_y = \frac{\partial^2 \psi}{\partial x^2} \qquad (5.7.12)$$

$$\tau_{xy} = \frac{-\partial^2 \psi}{\partial x \partial y} \quad \text{(if body forces are omitted)} \qquad (5.7.13)$$

That such a stress function $\psi(x,y)$ does exist will be shown shortly.

If we substitute Equations (5.7.8, 9 and 10) in Equation (5.7.7) we get

$$\frac{1}{E}\frac{\partial^2}{\partial y^2}(\sigma_x - \nu\sigma_y) + \frac{1}{E}\frac{\partial^2}{\partial x^2}(\sigma_y - \nu\sigma_x)$$

$$= \frac{2(1 + \nu)}{E}\frac{\partial^2 \tau_{xy}}{\partial x \partial y} \qquad (5.7.14)$$

Now we use Equations (5.7.11, 12 and 13)

$$\frac{1}{E}\frac{\partial^2}{\partial y^2}\left(\frac{\partial^2\psi}{\partial y^2} - \nu\frac{\partial^2\psi}{\partial x^2}\right) + \frac{1}{E}\frac{\partial^2}{\partial x^2}\left(\frac{\partial^2\psi}{\partial x^2} - \nu\frac{\partial^2\psi}{\partial y^2}\right)$$

$$= \frac{2(1+\nu)}{E}\frac{\partial^2}{\partial x\partial y}\left(\frac{-\partial^2\psi}{\partial x\partial y}\right) \qquad (5.7.15)$$

Or

$$\frac{1}{E}\frac{\partial^4\psi}{\partial y^4} - \frac{\nu}{E}\frac{\partial^4\psi}{\partial x^2\partial y^2} + \frac{1}{E}\frac{\partial^4\psi}{\partial x^4} - \frac{\nu}{E}\frac{\partial^4\psi}{\partial x^2\partial y^2} \qquad (5.7.16)$$

Cancelling the common terms,

$$\frac{\partial^4\psi}{\partial x^4} + 2\frac{\partial^4\psi}{\partial x^2\partial y^2} + \frac{\partial^4\psi}{\partial y^4} = 0 \qquad (5.7.17)$$

The stress function ψ which satisfies Equation (5.7.17) will satisfy equilibrium, compatibility and stress-strain. Equation (5.7.17) is called the biharmonic equation.

As an illustration let us consider the beam of Figure 5.7.3. This beam has been treated in Chapter 4 by elementary beam theory. Now we shall use the results of the two-dimensional theory of elasticity.

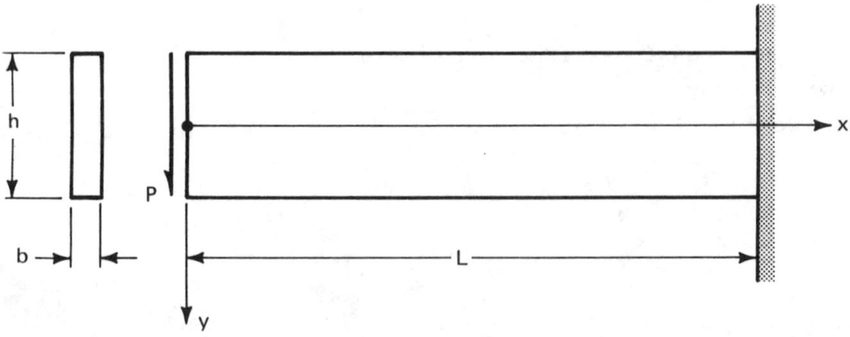

Figure 5.7.3

The knowledge we have of beam theory will help us select a stress function ψ. We know that the bending moment, and therefore the stress, increases linearly with x. At any station x, the stress is a linear function of y.

Therefore it is reasonable to assume,

$$\frac{\partial^2\psi}{\partial y^2} = \sigma_x = C_1 xy \tag{5.7.18}$$

We integrate Equation (5.7.18) twice. Because of the partial differentiation, we get functions of x when we integrate instead of just constants. Thus we have

$$\psi = \frac{C_1}{6} xy^3 + yf_1(x) + f_2(x) \tag{5.7.19}$$

Substitute ψ of Equation (5.7.19) in (5.7.17)

$$\nabla^4 \psi = 0$$

Yields

$$\frac{y d f_1{}^4(x)}{dx^4} + \frac{d^4 f_2(x)}{dx^4} = 0 \tag{5.7.20}$$

The only way Equation (5.7.20) can be satisfied for all x and y is

$$\frac{d f_1{}^4(x)}{dx^4} = 0 \tag{5.7.21}$$

and

$$\frac{d^4 f_2(x)}{dx^4} = 0 \tag{5.7.22}$$

These equations are easily solved for $f_1(x)$ and $f_2(x)$.

$$f_1(x) = c_2x^3 + c_3x^2 + c_4x + c_5 \tag{5.7.23}$$

and

$$f_2(x) = c_6x^3 + c_7x^2 + c_8x + c_9 \tag{5.7.24}$$

Recall Equation (5.7.19); then ψ becomes

$$\psi = \frac{c_1}{6}xy^3 + y(c_2x^3 + c_3x^2 + c_4x + c_5) + c_6x^3 + c_7x^2 + c_8x + c_9 \tag{5.7.25}$$

We will ignore the weight of the beam and the body forces are zero.

$$\sigma_y = \frac{\partial^2\psi}{\partial x^2} = 6c_2yx + 6c_6x + 2c_3y + 2c_7 \tag{5.7.26}$$

$$\tau_{xy} = \frac{-\partial^2\psi}{\partial x\partial y} = \frac{-c_1}{2}y^2 - 3c_2x^2 - 2c_3x - c_4 \tag{5.7.27}$$

We now apply the known <u>boundary conditions</u>, at $y = \pm h/2$, $\sigma_y = 0$ because these surfaces are not loaded. Therefore

$$6(c_2\frac{h}{2} + c_6)x + 2(c_3\frac{h}{2} + c_7) = 0$$

$$6(-c_2\frac{h}{2} + c_6)x + 2(-c_3\frac{h}{2} + c_7) = 0 \tag{5.7.28}$$

In order that Equations (5.7.28) be zero for all values of x

$$c_2\frac{h}{2} + c_6 = 0; \quad c_3\frac{h}{2} + c_7 = 0; \quad -c_2\frac{h}{2} + c_6 = 0;$$

$$-c_3\frac{h}{2} + c_7 = 0$$

From these equations

$$c_2 = c_3 = c_6 = c_7 = 0.$$

We now have, from Equation (5.7.27)

$$\tau_{xy} = \frac{-c_1y^2}{2} - c_4 \tag{5.7.29}$$

The surfaces are free of shear stress for all $y = \pm \frac{h}{2}$, thus,

$$\frac{-c_1}{8} h^2 - c_4 = 0$$

or

$$c_4 = \frac{-c_1 h^2}{8}$$

The integration of the shear stress times area over the end of the beam must add up to P.

$$-\int_{-\frac{h}{2}}^{\frac{h}{2}} \tau_{xy}\ b\ dy = \int_{-\frac{h}{2}}^{\frac{h}{2}} \frac{c_1}{8} b\ (4y^2 - h^2)\ dy = P \qquad (5.7.30)$$

or

$$c_1 = \frac{-12P}{h^3 b} \qquad (5.7.31)$$

Since $I = \frac{bh^3}{12}$; we finally obtain,

$$\sigma_x = \frac{-Pxy}{I}$$

$$\sigma_y = 0$$

$$\tau_{xy} = \frac{-P}{2I}\left(\frac{h^2}{4} - y^2\right) \qquad (5.7.32)$$

Equations (5.7.32) correspond exactly with elementary beam theory.

To obtain the displacements, we integrate Equations (5.7.3, 4 and 5). We use the stress-strain relations also, with

$$\varepsilon_x = \frac{\partial u}{\partial x} = \frac{\sigma_x}{E} = \frac{-Pxy}{EI} \qquad (5.7.33)$$

$$\varepsilon_y = \frac{\partial v}{\partial y} = \frac{-\nu\sigma_x}{EI} = \frac{\nu Pxy}{EI} \tag{5.7.34}$$

$$\gamma_{xy} = \frac{\partial u}{\partial y} + \frac{\partial v}{\partial x} = \frac{2(1 + \nu)}{E}\tau_{xy} = \frac{-(1 + \nu)P}{EI}\left(\frac{h^2}{4} - y^2\right) \tag{5.7.35}$$

Integrate Equation (5.7.33)

$$u = \frac{-P}{2EI} x^2 y + g_1(y) \tag{5.7.36}$$

Integrate Equation (5.7.34)

$$v = \frac{\nu P}{2EI} xy^2 + g_2(x) \tag{5.7.37}$$

Substitute Equations (5.7.36) and (5.7.37) into Equation (5.7.35)

$$\frac{dg_1}{dy} - \frac{P}{EI}\left(1 + \frac{\nu}{2}\right) y^2 = \frac{-dg_2}{dx} + \frac{P}{2EI} x^2 - \frac{(1 + \nu)P}{4EI} h^2 \tag{5.7.38}$$

The left hand terms of Equation (5.7.3) are functions of y alone and the terms on the right hand side are functions of x alone. This can only happen if both functions are equal to the same constant. Let us use a_1 for this constant.

Setting the left and right sides of Equation (5.7.38) equal to a_1 and integrating, we have

$$g_1(y) = \frac{P}{3EI}\left(1 + \frac{\nu}{2}\right) y^3 + a_1 y + a_2 \tag{5.7.39}$$

$$g_2(x) = \frac{P}{6EI} x^3 - \frac{(1 + \nu)P}{4EI} h^2 x - a_1 x + a_3 \tag{5.7.40}$$

Substituting back into Equation (5.7.36) and (5.7.37)

$$u = \frac{-P}{2EI} x^2 y + \frac{P}{3EI}\left(1 + \frac{\nu}{2}\right) y^3 + a_1 y + a_2 \tag{5.7.41}$$

$$v = \frac{\nu P}{2EI} xy^2 + \frac{P}{6EI} x^3 - \frac{(1 + \nu)P}{4EI} h^2x - a_1 x + a_3 \tag{5.7.42}$$

The boundary conditions can be written

$$u, v \Big|_{\substack{x = L \\ y = 0}} = 0$$

$$\frac{\partial v}{\partial x} \Big|_{\substack{x = L \\ y = 0}} = 0$$

Substituting the boundary conditions results in

$$u = \frac{-P}{2EI} x^2y + \frac{P}{3EI} (1 + \frac{\nu}{2}) y^3 + \frac{P}{2EI} \left[L^2 - (1 + \nu)\frac{h^2}{2} \right] y \tag{5.7.43}$$

$$v = \frac{\nu P}{2EI} xy^2 + \frac{P}{6EI} x^3 - \frac{PL^2}{2EI} x + \frac{PL^3}{3EI} \tag{5.7.44}$$

The equation of the elastic axis is v evaluated at $y = 0$.

$$v = \frac{Px^3}{6EI} - \frac{PL^2x}{2EI} + \frac{PL^3}{3EI} \tag{5.7.45}$$

Equation (5.7.45) checks with elementary beam theory.

Problems

Problem 1. Solve for the stress and deflection in the cantilever beam with applied moment on the end.

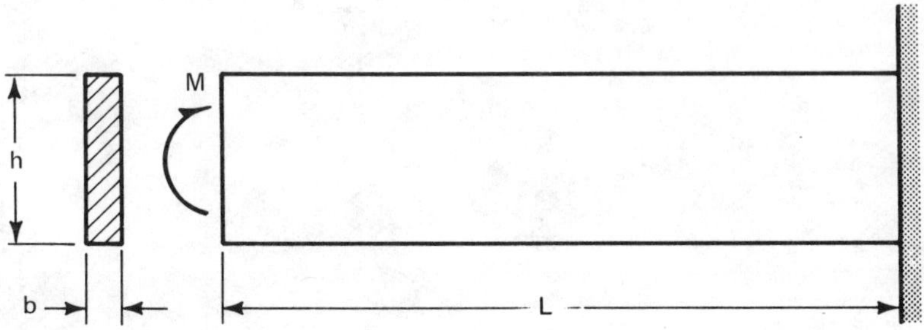

References

1. History of the Strength of Materials. S. P. Timoshenko, McGraw-Hill, New York, 1953.
2. Mechanical Engineering Design, Joseph E. Shigley, McGraw-Hill, New York, 1963.
3. Analysis of Stress and Strain, A. G. Durelli, E. A. Phillips, C. H. Tsao, McGraw-Hill, New York, 1958.
4. Theory of Elasticity, S. P. Timoshenko and J. N. Goodier, McGraw-Hill, New York, 1951.
5. Applied Elasticity, C. T. Wang, McGraw-Hill, New York, 1953.

Chapter 6

INTRODUCTION TO THE FINITE ELEMENT METHOD

6.1 AN OVERVIEW OF THE FINITE ELEMENT METHOD

Occasionally a revolutionary method will come along in a certain engineering field. The finite element method has revolutionized stress analysis and is currently still having its impact felt, not only in stress analysis, but in areas such as aerodynamics and heat transfer. We shall deal only with the structural applications. The finite element method brings together three areas: 1) matrix algebra (those not familiar with matrix algebra should read Appendix B.); 2) the digital computer and accompanying numerical analysis; and 3) stress analysis.

The finite element method began in the aerospace industry in the 1950's and gained real impetus in the early 1960's.

Today it is both a vital, developed and useful method, while at the same time it is an area of ongoing research and development. Reference (2) gives a good historical development of this method.

There are several reasons for the revolutionary effect of the finite element method:

1. It utilizes a digital computer and thus handles large amounts of processing in short time. This includes automation of input and graphic display of output.
2. It has the feature of permitting the user to assemble the properties of a complex, large structure from a number of elements with simple properties.
3. It is capable of handling static, dynamic, stability and non-linear problems for a very broad range of structures.

Let us consider how item number two above would work on a simple structure. As we do this, we will also introduce some terminology. Look at Figure 6.1.1, where a simple example structure (a truss) is shown. The example shown will be two-dimensional but extending the finite element method to three dimensions presents no big problems.

The truss is assembled as a collection of axial (two-force) members. Each of the truss members is referred to as an element. Since the elements are finite in size, as opposed to the differentially small element in the "classical" method, we get the name finite element. The usual name for the axial force element is the rod element. Other types of elements will be discussed near the end of this chapter.

A point where two elements intersect, or a point where the structure is supported is called a node. In Figure 1 the nodes are numbered one through six. The elements are also numbered: $\boxed{1}$, $\boxed{2}$, etc.

We are discussing a particular sub-topic of the finite element method known as the stiffness method. It is also commonly called the displacement method. In this method, the displacements of the nodes are the unknowns and the general solution of the structural equations centers on finding the nodal displacements. Other information, such as the element forces and stresses, can be calculated once the structural displacements are known. In the truss of Figure 6.1.1, the significant displacements are the vertical and horizontal displacements of each of the nodes. The displacements of node six (u_{11} and u_{12}) are shown for illustration. A node can also be loaded and R_3, R_4, R_9, R_{10} are shown in Figure 6.1.1 as examples.

The displacements and loads are positive if they are in the plus x and y directions of the user-defined coordinate axes as shown; otherwise they are negative. The location of each node is specified by giving its x and y coordinates in the system shown.

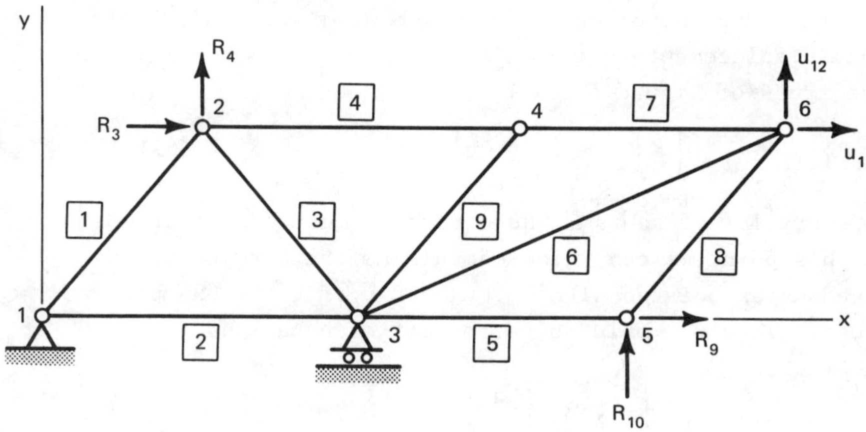

Figure 6.1.1 Two-dimensional example truss

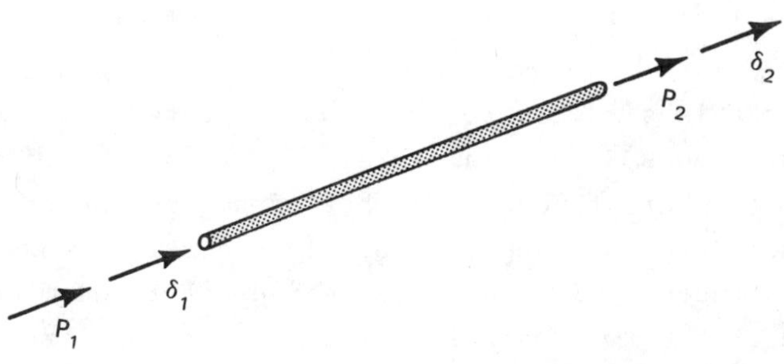

Figure 6.1.2 Element nodes, displacements and freedoms

A very important point is that the structure is mathematically modeled by examining the load-deflection behavior of its nodes. This is true of this truss and it is true of all structures, in general, which are analyzed by the finite element method. This means that all the loading and displacement occurs at the nodes and that all the distributed stiffness properties of the structure must be "lumped" into the nodes.

Lumping the structural stiffness properties at the nodes is accomplished by assembling the structural stiffness matrix. Recall that the general entry of a stiffness matrix K_{ij} is, by definition, the force at the ith node which is present when a unit displacement is given to the jth node and the other nodes are held fixed. Mathematically, the load-displacement relationship is given in matrix form for the truss of Figure 6.1.1 by

$$\left\{R\right\}_{12 \times 1} = \left[K\right]_{12 \times 12} \left\{u\right\}_{12 \times 1} \qquad (6.1.1)$$

Generally, of course, the structure must be supported. At this point we can think of this as the zeroing of the freedoms associated with u_1, u_2 and u_6 for the example truss. Thus we would solve the set of equations

$$\left\{R\right\}_{9 \times 1} = \left[K\right]_{9 \times 9} \left\{u\right\}_{9 \times 1} \qquad (6.1.2)$$

The solution is: (The loads are known)

$$\left\{u\right\} = \left[K\right]^{-1} \left\{R\right\} \qquad (6.1.3)$$

We see that this simple structure has generated a system of nine simultaneous, linear algebraic equations. This is equal to the (number of nodes x number of degrees of freedoms per node)--freedoms suppressed. In general, a three-dimensional structure will have six degrees of freedom per node instead of the two degrees of freedom for this truss example. For large structures, a lot of equations in a lot of unknowns will be generated. (Up to several thousand unknowns.) Thus the efficient use of a computer is essential.

The over-all structural stiffness matrix is actually affected by several factors:

1. The location of the nodes.
2. The size of the elements and the element material properties.
3. The way the elements are connected (to which nodes).

It is extremely fortunate that all of the factors can be dealt with by considering the elements one-by-one, and that the relatively complex job of assembling the structural stiffness matrix consists of taking each element stiffness matrix, transforming the stiffness matrix to structural coordinates, and adding the result to the proper location of the structural stiffness matrix. This will be discussed in more detail in subsequent sections.

The elements are also characterized by their load-displacement relationships. Figure 6.1.2 shows a rod element. Such elements make up the example truss of Figure 6.1.1. The element extremities are called nodes. The nodal loads P_1 and P_2 must be related to the nodal displacements δ_1 and δ_2. This is done by writing the element stiffness matrix. The element stiffness matrix is: (It will be derived later. See also Section 1.4.)

$$\left\{P\right\}_{2 \times 1} = \left[k\right]_{2 \times 2} \left\{\delta\right\}_{2 \times 1} \qquad (6.1.4)$$

or

$$\begin{Bmatrix} P_1 \\ P_2 \end{Bmatrix} = \frac{AE}{L}\begin{bmatrix} 1 & -1 \\ -1 & 1 \end{bmatrix}\begin{Bmatrix} \delta_1 \\ \delta_2 \end{Bmatrix} \qquad (6.1.5)$$

The only difficult part is to transform the element stiffness matrix (Equation 6.1.5), which is written in element coordinates, to the x-y structural coordinate system. However, once we understand this transformation, the computer will do it for us quite handily. It is the element-by-element assembly process which gives the finite element method its power. If we have a library of element types available to us, we can assemble more complicated structures --much like building it up from a set of tinker toys. The elements are each described completely by their (relatively) simple stiffness matrices. Of course, we must have other elements in addition to rod elements if this is going to work. We shall now discuss some of the more common types of elements.

Most large-scale programs will have the following element types: (All are generally three-dimensional.)

1. Beam element (Bar). Takes bending in two planes, shear in two planes and torsion. (Figure 6.1.3.)
2. Rod element. Takes axial forces. May take torsion. (Figure 6.1.4.)
3. Shear panels. In-plane shear. (Figure 6.1.5.)
4. Membrane elements. In-plane shear and normal stresses. (Also Figure 6.1.5.)
5. Plate elements. In-plane shear and normal stress plus bending stress and plate shear. (Figure 6.1.6.)
6. Axi-symmetric elements. Represent the cross section of axi-symmetric structures such as shells. (Figure 6.1.7.)
7. Solid polyhedral elements. For building up "bulky" structures which call for three-dimensional elasticity solutions. (Figure 6.1.8.)

It is not our intention in this book to go into the finite element method in depth. We will consider the fundamentals. The fundamental ideas of the finite element method are easily illustrated and understood by using the rod element and the two-dimensional truss. Once the fundamentals are understood, the extension of the finite element method to more advanced elements will be easily understandable. We shall therefore proceed to develop the method based on the two-dimensional truss.

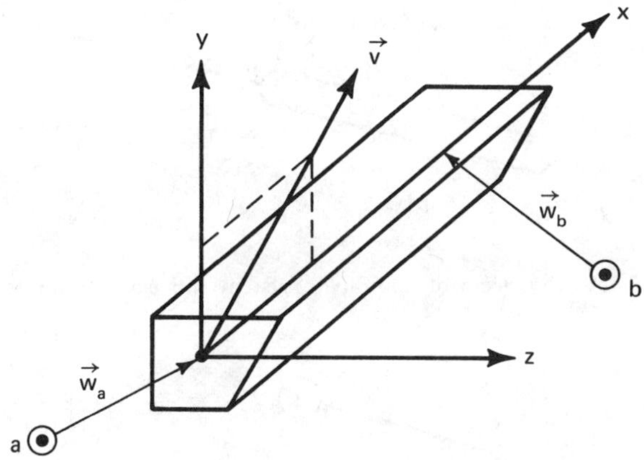

(a) Element coordinate system

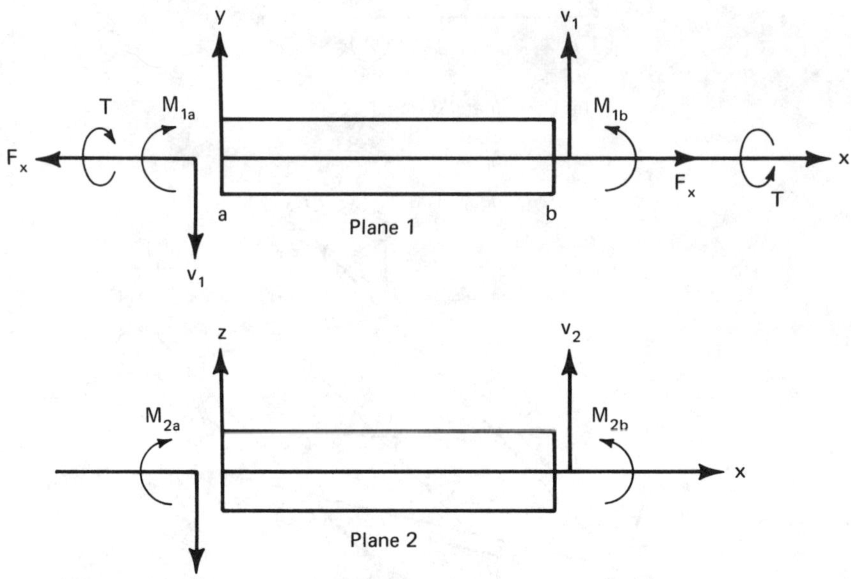

(b) Element forces

Figure 6.1.3 Bar element coordinate system and element forces

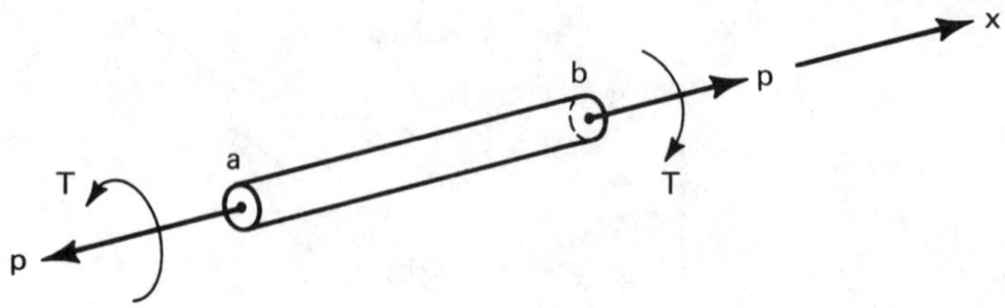

Figure 6.1.4 Rod element coordinate system and element forces

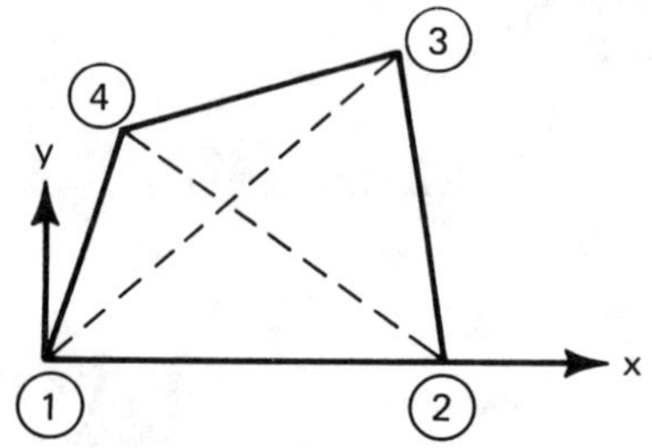

(a) Coordinate System

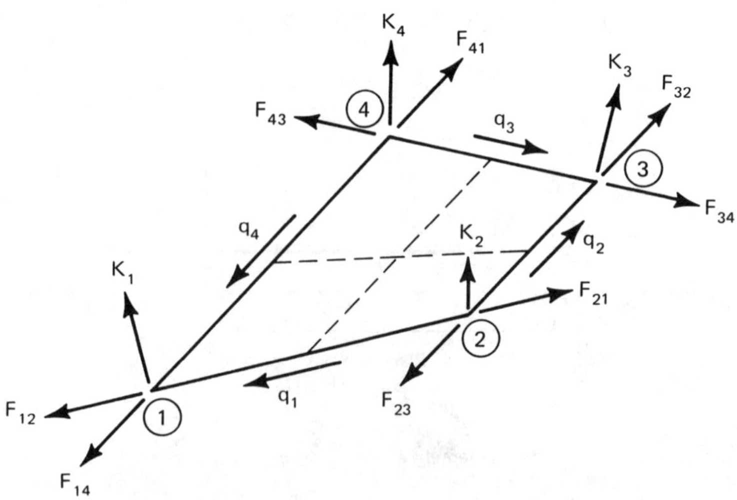

(b) Corner forces and shear flows

Figure 6.1.5 Coordinate system and element forces for shear panel elements

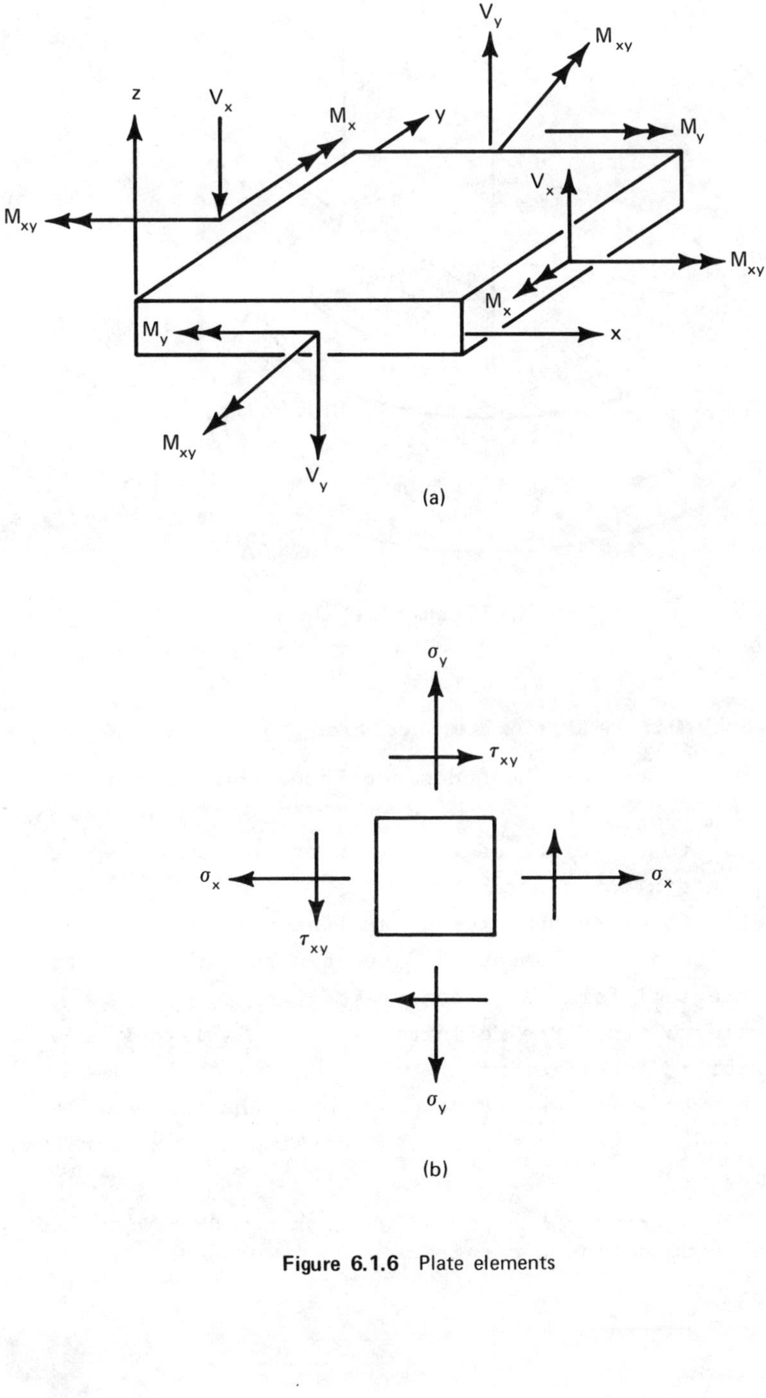

Figure 6.1.6 Plate elements

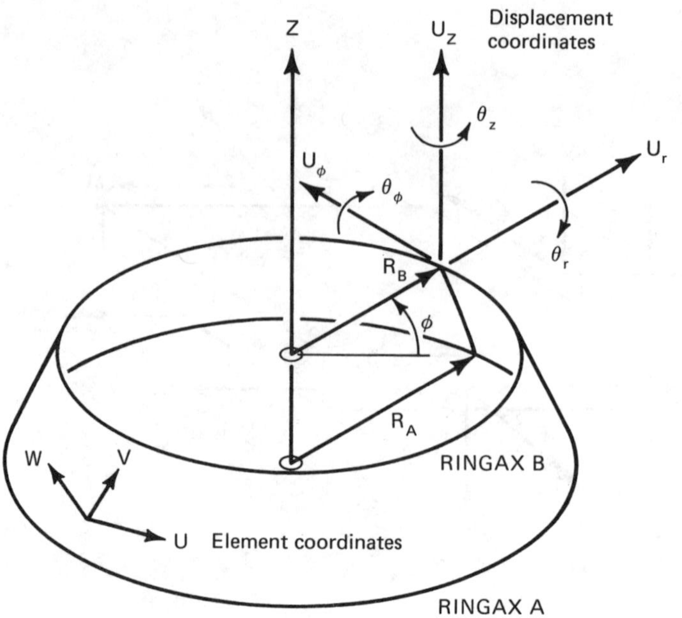

Figure 6.1.7 Conical shell elements

Objectives

(Closed Book Unless Otherwise Specified)

1. Be able to number the nodes and label the freedoms for a 2-d truss.
2. Be able to write the general form of the load-displacement matrix for a structure.
3. Be able to write the element stiffness matrix in general terms for a rod element, and be able to get this matrix in numerical form for a specific element.
4. Be able to state the definitions of the stiffness and flexibility coefficients.
5. Be able to write out the expansion of the force-deflection and deflection-force relationships and explain the terms.
6. Be able to derive the rod element stiffness matrix.
7. Be able to define contragradience in words and with equations.

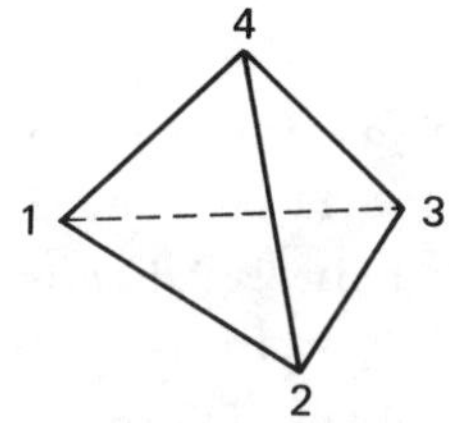

(a) Tetrahedron

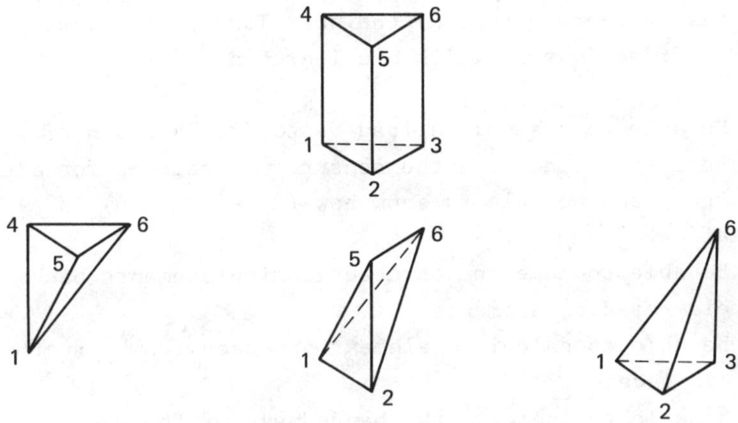

(b) Wedge and One of its Six Decompositions

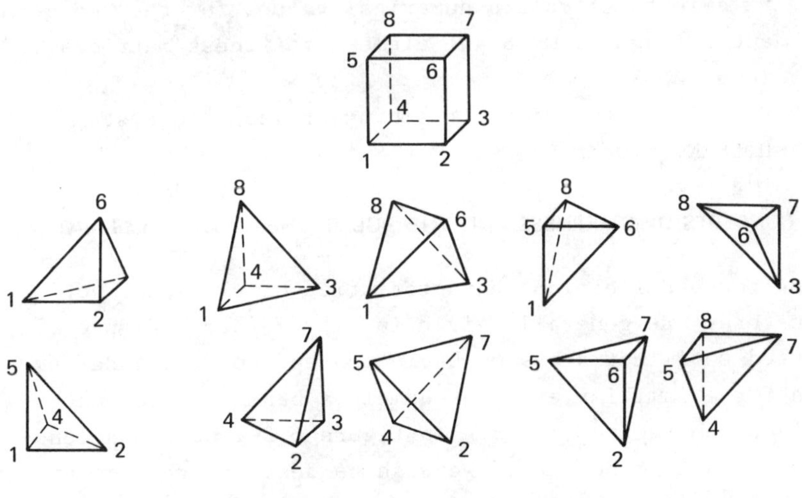

(c) Hexahedron and its Two Decompositions

Figure 6.1.8 "Solid" or three-dimensional finite elements

8. Be able to derive Equation (6.3.14).
9. Be able to derive Equation (6.3.17).
10. Be able to number the nodes and label the freedoms for a structure composed of 2-d beams or triangular membranes.
11. Be able to calculate the following for a specified rod element in a particular truss: a) element stiffness in structural coordinates, b) element stiffness in master coordinates, c) the positions of the master stiffness matrix which are incremented by the results of (b).
12. Be able to assemble a load vector for a given truss.
13. Be able to identify the constraint freedoms for a given truss and be able to show how $(R) = K_M \; (u)$ is affected.
14. Be able to take the structural displacements back to element displacements.
15. Be able to calculate element stresses from structure displacements.
16. Be able to indicate the bandedness of the master stiffness matrix for any structure.
17. Be able to calculate numerical values for the 2-d beam and triangular membrane element stiffness matrices. (Open Book.)
18. Be able to run simple training programs on system HILSEN. (Open Book.)

6.2 CONCEPTS OF STIFFNESS AND THE ROD ELEMENT STIFFNESS MATRIX

The force-deflection characteristics of elements and structures are generally given in matrix form. Figure 6.2.1 is used to illustrate some basic concepts of force-deflection for a familiar example--namely a beam. These concepts will be generalized later. In Figure 6.2.1 we see a long, slender uniform beam. If we wish to describe the position of this beam in a discrete, or finite way, we would pick n nodes: 1, 2, 3, . . ., n as shown. We may choose an arbitrary number of nodes. The deflected position of the beam is then approximated by giving the n displacements δ_1, δ_2, δ_3, . . . δ_n. The loads are applied only at the nodes (the loads must be lumped there) and are given by P_3, P_4, . . ., P_n. Of course, a node may or may not be loaded, depending on the load case being considered.

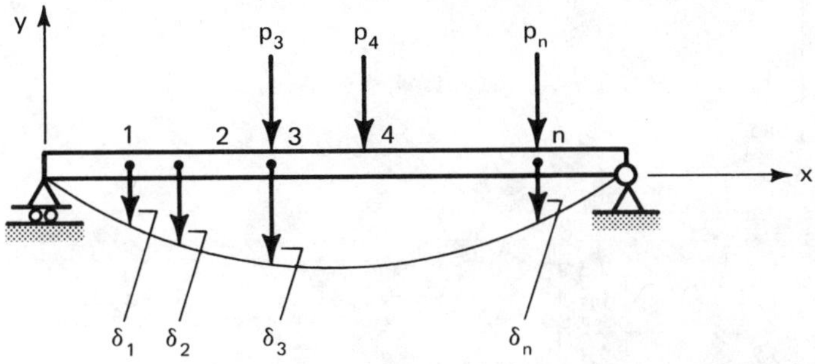

Figure 6.2.1

The load-deflection relationship for this beam can then be written as:

$$\{P\} = [K]\ \{\delta\} \tag{6.2.1}$$

where

$$\begin{bmatrix} P_1 \\ P_2 \\ P_3 \\ . \\ . \\ . \\ P_n \end{bmatrix}$$ is the load matrix

and

$$\begin{bmatrix} \delta_1 \\ \delta_2 \\ \delta_3 \\ \cdot \\ \cdot \\ \cdot \\ \delta_n \end{bmatrix}$$ is the displacement matrix

and

$$\begin{bmatrix} k_{11} & k_{12} & - - - & k_{1n} \\ k_{21} & k_{22} & - - - & k_{2n} \\ \cdot & & & \cdot \\ \cdot & & & \cdot \\ \cdot & & & \cdot \\ \cdot & & & \cdot \\ k_{n1} & - - - & - - - & k_{nn} \end{bmatrix}$$ is the stiffness matrix (a symmetric matrix)

The definition of the ij^{th} term of the stiffness matrix is given as follows:

k_{ij} = the force necessary at the ith node to cause the structure to have a unit displacement at the jth node while the other nodes are held at zero displacement.

The stiffness matrix relationships are usually a little difficult to see clearly. Visualization is clearer when the flexibility matrix is written. If we look at the inverse relationship of Equation (6.2.1) we see

$$\left\{P\right\} = \left[k\right]\left\{\delta\right\}$$

$$\left[k\right]^{-1}\left\{P\right\} = \left[k\right]^{-1}\left[k\right]\left\{\delta\right\}$$

$$\left[F\right]\left\{P\right\} = \left\{\delta\right\}$$

$$\left\{\delta\right\} = \left[F\right]\left\{P\right\} \qquad (6.2.2)$$

where $\left[F\right] = \left[k\right]^{-1}$ is the flexibility matrix of the structure. A symmetric matrix.

We note that for the supported beam shown, $\left[k\right]$ will not be singular. The definition of the ij^{th} coefficient of the flexibility matrix is given as follows:

f_{ij} = the displacement of the ith node of the structure due to a unit load at the jth node.

It is enlightening to look at any particular row of Equation (6.2.2) written out. Row three, for example, says

$$\delta_3 = f_{31} P_1 + f_{32} P_2 + f_{33} P_3 + \ . \ . \ . + f_{3n} P_n \qquad (6.2.3)$$

This shows clearly that the total deflection, δ_3, of the structure is a linear combination of all the loads. The f_{3j} gives the relative contribution of each of the P_j to the total displacement.

Based on the preceding definitions, we can now derive the element stiffness matrix for the rod element. To do this, we must relate the forces to the displacements within the element. This is relatively easy for the rod element, but may be difficult for elements such as plates and shells, etc.

From this definition of stiffness and a little reflection, we can observe a general truth. If we give a unit displacement to the jth node of an element (or structure) and hold other displacements equal zero, we can write down the jth column of the stiffness matrix as the nodal forces which are required to do this. For the rod, we can do as shown below. See Figure 6.2.2.

set $\delta_1 = 1$ hold $\delta_2 = 0$

$$\begin{Bmatrix} P_1 \\ P_2 \end{Bmatrix} = \begin{bmatrix} K_{11} & / \\ K_{21} & / \end{bmatrix} \begin{Bmatrix} \delta_1 \\ \delta_2 \end{Bmatrix}$$

↑ write down the first column of the stiffness matrix as the forces required to do this.

Mathematically, in matrix form, we could have Equation (6.2.1) with $\delta_1 = 1$, $\delta_2 = 0$

$$\begin{Bmatrix} P_1 \\ P_2 \end{Bmatrix} = \begin{bmatrix} K_{11} & K_{12} \\ K_{21} & K_{22} \end{bmatrix} \begin{Bmatrix} \delta_1 \\ \delta_2 \end{Bmatrix} = \begin{bmatrix} K_{11} & K_{12} \\ K_{21} & K_{22} \end{bmatrix} \begin{Bmatrix} 1 \\ 0 \end{Bmatrix} \tag{6.2.4}$$

from which

$$P_1 = K_{11} \tag{6.2.5}$$

and

$$P_2 = K_{21} \tag{6.2.6}$$

for this particular set of displacements.

Proceeding with this method for the rod element, we would have the configuration depicted in Figure 6.2.2 with $\delta_1 = 1$ and $\delta_2 = 0$. Since we know that

$$\delta = \frac{PL}{AE} \tag{1.3.9}$$

for a rod, the situation in Figure 6.2.2 gives the first column of the stiffness matrix as

$$[K] = \begin{bmatrix} +\frac{AE}{L} & / \\ -\frac{AE}{L} & / \end{bmatrix} \tag{6.2.7}$$

Repeating this process with $\delta_2 = 1$ and $\delta_1 = 0$ gives the second column of the rod element stiffness matrix as

$$[K] = \begin{bmatrix} / & -\frac{AE}{L} \\ / & +\frac{AE}{L} \end{bmatrix} \tag{6.2.8}$$

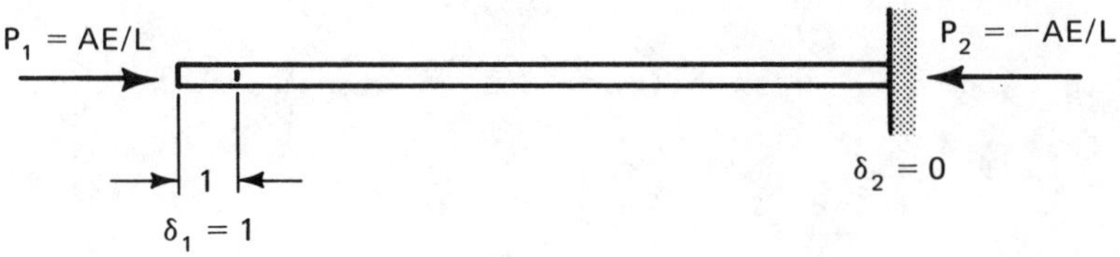

Figure 6.2.2

From which we can write the rod element stiffness matrix as

$$[K] = \frac{AE}{L} \begin{bmatrix} 1 & -1 \\ -1 & 1 \end{bmatrix} \tag{6.2.9}$$

We will use this stiffness matrix in our subsequent derivations.

6.3 MATRIX COORDINATE TRANSFORMATIONS AND THE ROD ELEMENT TRANSFORMATION MATRIX

If we consider the forces and displacements expressed in two coordinate systems

System 1	System 2	
$\{F\}$	$\{Q\}$	Forces
$\{X\}$	$\{q\}$	Displacements

we can generally find a scheme for transferring the displacements from one coordinate system to another, say

$$\{q\} = [T]\{X\} \tag{6.3.1}$$

Then the work expressed in the two coordinate systems is an invariant and this work can be expressed by

$$\{X\}^T \{F\} = \{q\}^T \{Q\}$$

or

$$\{X\}^T \{F\} = \{X\}^T [T]^T \{Q\}$$

and, since the $\{X\}$ are arbitrary displacements, we have

$$\{F\} = [T]^T \{Q\} \qquad (6.3.2)$$

and we see that the force transformation from $\{Q\}$ to $\{F\}$ is given by Equation (6.3.2). This type of transformation is called a <u>CONTRAGRADIENT</u> transformation.

For example, consider the transformation of coordinates as the translation of axes shown in Figure 6.3.1. One coordinate system has forces $\{A\}$ and displacements $\{a\}$. The second has forces $\{B\}$ and displacements $\{b\}$.

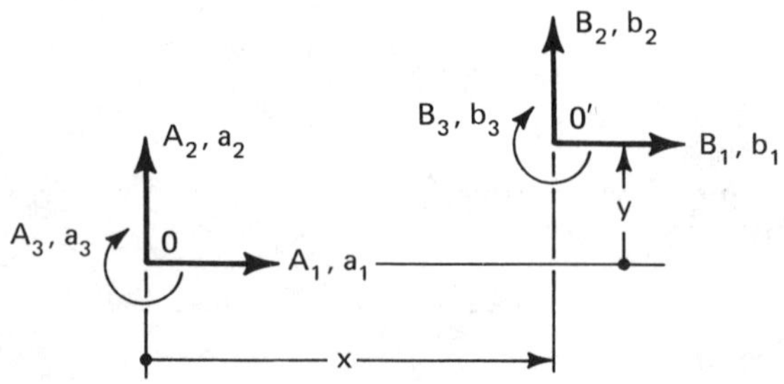

Figure 6.3.1

The relationship between forces in the two systems is

$$B_1 = A_1$$
$$B_2 = A_2$$
$$B_3 = A_3 - A_1 y = A_2 X$$

, or

$$\begin{Bmatrix} B_1 \\ B_2 \\ B_3 \end{Bmatrix} = \begin{bmatrix} 1 & 0 & 0 \\ 0 & 1 & 0 \\ -y & X & 1 \end{bmatrix} \begin{Bmatrix} A_1 \\ A_2 \\ A_3 \end{Bmatrix} \qquad (6.3.3)$$

For small displacements, we have the displacement relationship

$$b_1 = a_1 + a_3 y \qquad b_2 = a_2 - a_3 x \qquad b_3 = a_3 \quad , \text{ or } \quad \begin{Bmatrix} b_1 \\ b_2 \\ b_3 \end{Bmatrix} = \begin{bmatrix} 1 & 0 & y \\ 0 & 1 & -X \\ 0 & 0 & 1 \end{bmatrix} \begin{Bmatrix} a_1 \\ a_2 \\ a_3 \end{Bmatrix} \tag{6.3.4}$$

Observing the above relationships, the following can be seen by inspection, or by multiplying out

$$\{B\} = [T_1]\{A\} \tag{6.3.3}$$

$$\{b\} = [T_2]\{a\} \tag{6.3.4}$$

$$[T_1][T_2]^T = [I] \quad \text{Multiply and check} \tag{6.3.5}$$

$$[T_2]^T[T_1] = [I] \quad \text{Multiply and check} \tag{6.3.6}$$

From the above $\{a\} = [T_1]^T\{b\}$ (6.3.7)

Thus the transformation is contragradient.

Let us consider how the stiffness matrix and load-deflection relationship of Equation (6.2.9) will transform under a general transformation of coordinates. We will assume that the transformation is contragradient.

We wish to have the transformation from the element coordinate system where

$$\{P\} = [K_e]\{\delta\} \tag{6.3.8}$$

where $[K_e]$ is the element stiffness matrix to the structure coordinate system where

$$\{R\} = [K_{M_e}]\{u\} \tag{6.3.9}$$

where $\left[K_{M_e}\right]$ is the element stiffness matrix transformed to master, or structure, coordinates. We assume that we can find a displacement transformation

$$\left\{\delta\right\} = \left[T\right] \left\{u\right\} \tag{6.3.10}$$

From Equation (6.3.2) the principle of contragradience says

$$\left\{R\right\} = \left[T\right]^T \left\{P\right\} \tag{6.3.11}$$

Substitute Equation (6.3.8) in Equation (6.3.11)

$$\left\{R\right\} = \left[T\right]^T \left[K_e\right] \left\{\delta\right\} \tag{6.3.12}$$

Substitute Equation (6.3.10) in Equation (6.3.12)

$$\left\{R\right\} = \left[T\right]^T \left[K_e\right] \left[T\right] \left\{u\right\} \tag{6.3.13}$$

Compare Equation (6.3.13) with

$$\left\{R\right\} = \left[K_{M_e}\right] \left\{u\right\}$$

we see that

$$\left[K_{M_e}\right] = \left[T\right]^T \left[K_e\right] \left[T\right] \tag{6.3.14}$$

Equation (6.3.14) will provide the transformation of the element stiffness matrix in element coordinates to the element stiffness matrix in master coordinates provided that we can write the displacement transformation $\left[T\right]$. We shall illustrate this procedure for the rod element.

We wish to derive the relation

$$\left\{\delta\right\} = \left[T\right] \left\{u\right\} \tag{6.3.10}$$

for the rod element. This means that we will transform from element displacements δ_1 and δ_2 to structural displacements u_1, u_2, u_3 and u_4 as shown in Figure 6.2.3.

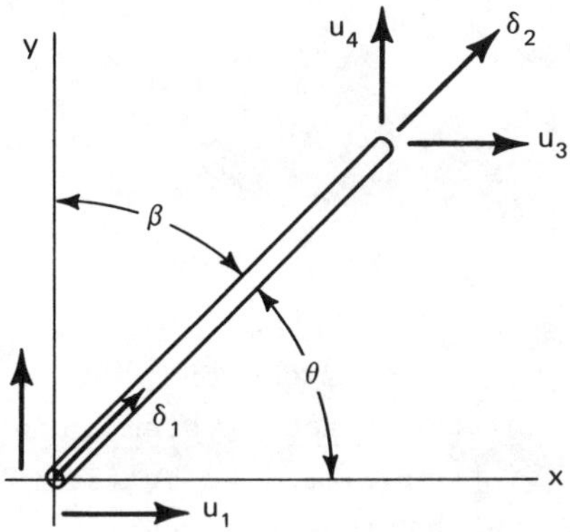

Figure 6.3.2

The element is oriented in the x-y coordinate system at angles θ and β as shown in Figure 6.2.3. The direction cosines of the angles may be expressed as

$$\ell = \cos\theta$$
$$m = \cos\beta$$

We may then write

$$\delta_1 = \ell u_1 + m u_2 \tag{6.3.15}$$

$$\delta_2 = \ell u_3 + m u_4 \tag{6.3.16}$$

Writing Equations (6.3.15) or (6.3.16) in matrix form gives,

$$\begin{Bmatrix} \delta_1 \\ \delta_2 \end{Bmatrix}_{2 \times 1} = \begin{bmatrix} \ell & m & o & o \\ o & o & \ell & m \end{bmatrix}_{2 \times 4} \begin{Bmatrix} u_1 \\ u_2 \\ u_3 \\ u_4 \end{Bmatrix}_{4 \times 1} \tag{6.3.17}$$

Now using Equation (6.3.14), we have

$$\left[K_{e_M}\right]_{4 \times 4} = \begin{bmatrix} \ell & o \\ m & o \\ o & \ell \\ o & m \end{bmatrix}_{4 \times 2} \times \frac{AE}{L} \begin{bmatrix} 1 & -1 \\ -1 & 1 \end{bmatrix}_{2 \times 2}$$

$$\begin{bmatrix} \ell & m & o & o \\ o & o & \ell & m \end{bmatrix}_{2 \times 4} \qquad (6.3.18)$$

This expression shows how the rod element, which "stiffens" coordinates δ_1 and δ_2 in the <u>element</u> system in accordance with Equation (6.2.9) is transformed to "stiffen" coordinates u_1, u_2, u_3 and u_4 in the <u>structure</u> system in accordance with Equation (6.3.18).

At this point, we can begin to visualize the assembly of the structure. Each rod element is attached to two structure nodes. The element stiffness matrix $\left[K_e\right]$ is transformed to element stiffness in master coordinates $\left[K_{ME}\right]$. In this way the element stiffness <u>adds</u> <u>to</u> the structure stiffness for the four structure freedoms associated with the two nodes to which the element is attached. This will be demonstrated in detail in a subsequent section.

If Equation (6.3.18) is multiplied out, we will have

$$\left[K_{M_e}\right] = \frac{AE}{L} \begin{bmatrix} \ell^2 & \ell m & -\ell^2 & -\ell m \\ \ell m & m^2 & -\ell m & -m^2 \\ -\ell^2 & -\ell m & \ell^2 & \ell m \\ -\ell m & -m^2 & \ell m & m^2 \end{bmatrix}_{4 \times 4} \qquad (6.3.19)$$

This is the final result for the rod element.

6.4 STRUCTURAL MODELING, NODE NUMBERING AND FREEDOMS, LOADING, ASSEMBLY OF MASTER STIFFNESS, SOLUTION FOR STRUCTURAL DISPLACEMENTS, SOLUTION FOR ELEMENT FORCES AND STRESSES

Once we have studied a number of element types and have derived the element stiffness matrices, we can catalog these different elements and use them for the purpose of assembling complex structures. In practice this is done by writing a computer subroutine which will compute the known element stiffness matrix from the element properties and assemble it into the structural stiffness matrix for the complete structure. This is done element by element until the entire structure is assembled.

To the question of how to model the structure there is no definitive answer. It is at this point where the engineer and his knowledge of structures plays an important role. He must examine the structure and its components and make a decision as to what role each component is playing in the structural assembly. Naturally structural components which behave as rods are modeled as rods, those which behave as beams are modeled as beams, etc. There is some room for imagination and even artistic decision in this process.

A similar situation exists when it comes to the question of how many elements to use, or how fine a mesh to make for the nodal points. There are no clear-cut answers to this question either, although studies have been made, References (1, 2, 3, 4, 5) which give quantitative information which applies to specific studies and can also serve as guidelines on other problems.

Fortunately, this is a situation in which we find that an intuitive approach is ordinarily satisfactory and "eye-balling" the structure will lead to good results.

Let us begin our study of this subject by following the development of structural modeling, node point numbering and freedoms, . . ., down to the solution for structural stresses, for the structure shown in Figure 6.4.1.

We will use a truss as a simple example and will later discuss the principles as applied to more complicated structures.

In Figure 6.4.1, a set of axes for the structure is defined. This is the x-y system, as shown. The structural coordinate system adopted is the one which most naturally fits the structure. This is usually referred to as the structural, global, or master system. We will most frequently use the term "master."

The nodes are then numbered as shown in Figure 6.4.1. The way (order) in which the nodes are numbered is significant. The nodes should be numbered in such a way that the numerical difference between node numbers at which an element is connected is minimized. In Figure 6.4.1, for example, the typical maximum numerical difference (the worst case for this structure) is in the diagonal members. Here, the difference is, say

$$12 - 9 = 3$$

The truss should not be numbered as shown in Figure 6.4.2, where we have, typically

$$11 - 4 = 7$$

a higher numerical difference. The importance of this step will be seen soon.

Once the nodes are numbered, the geometric position of the nodes is given by specifying their coordinates in the structural system. This is shown in Figure 6.4.1.

The next step is to number the elements. The numbering of the elements is insignificant, unless some particular methods of solving the structural equations are being used (Reference 9). We will just number the elements in any order, such as is shown in Figure 6.4.1, numbered (1) - (25).

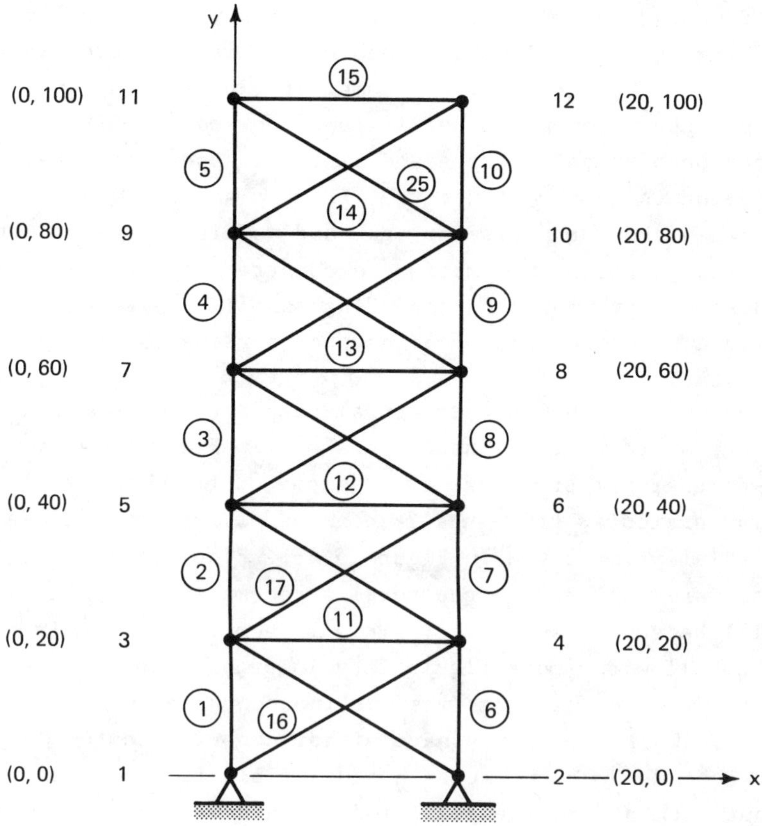

Figure 6.4.1

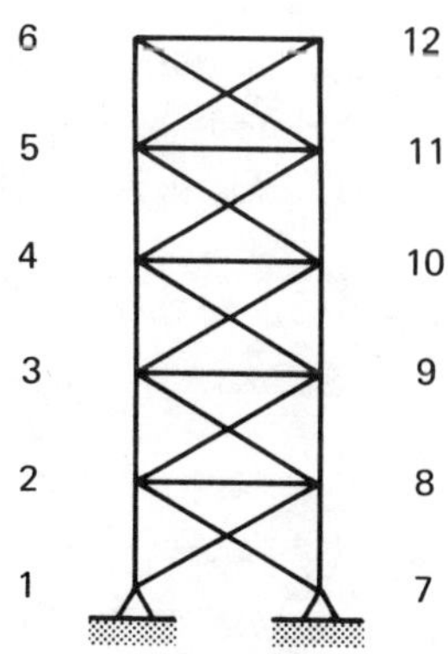

Figure 6.4.2

The fact is that we have now created a way of logically specifying where an element is and how it is connected in the structure. If we tell the computer (in its own language) that element number 17 is a rod and that it connects between nodes three and six, this specifies where the element is in the structure. If we also supply the information about the element area and modulus of elasticity, there exists enough information to construct the element stiffness matrix in element coordinates. (The element length can be calculated by using the coordinates of the end points.)

Let us now examine the structural displacements and loads which are appropriate. These are shown in Figure 6.4.3. A two-dimensional truss has two degrees of freedom per node, so this particular truss has 24 degrees of freedom. These are partially indicated in Figure 6.4.3. The support freedoms are included in the total freedoms at this stage and will be suppressed later. This is generally done in order to allow for more flexibility in programming.

The loading on the truss must be in accordance with the freedoms. That is, we may have either a horizontal or vertical load, or both, at each node. This is consistent with our element and structure models, and with our physical conception of a truss. A number of loads, R, are shown acting on the truss in Figure 6.4.3. In general, of course, any node may or may not have a load.

The numbering of nodal displacements and nodal loading follows a pattern. Given the node number, we can determine the x- and y-direction displacements by the following formula.

$$u_x = (2 \times \text{Node Number}) - 1 \qquad (6.4.1)$$

$$u_y = (2 \times \text{Node Number})$$

The structural freedoms for the truss run from one to twenty-four. The load numbers, or subscripts, are calculated the same way.

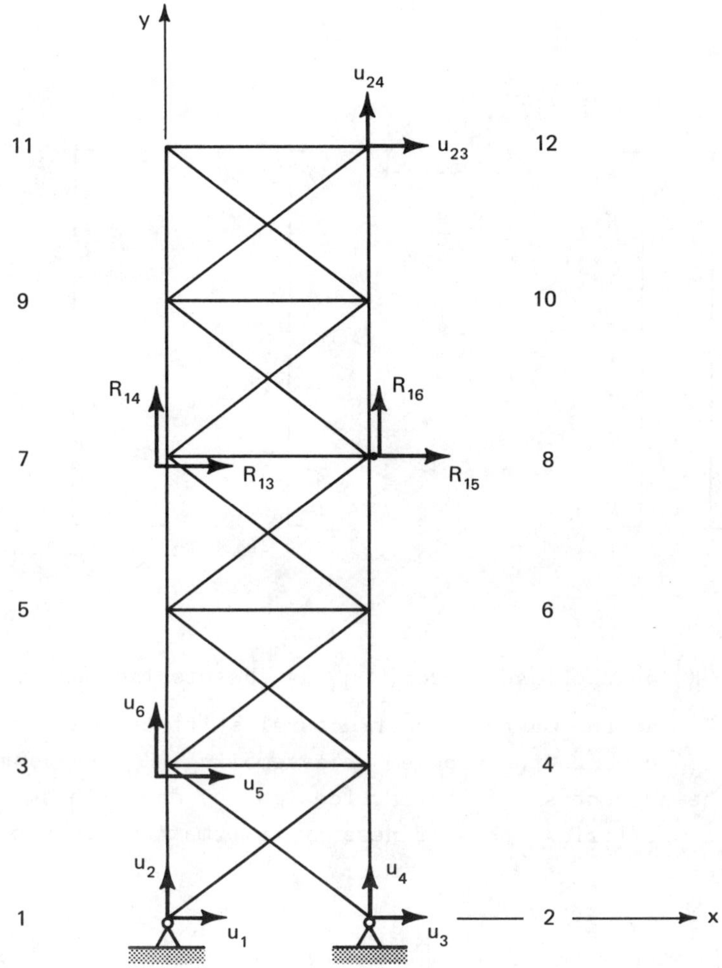

Figure 6.4.3 Not all freedoms have been labeled

$$R_x = (2 \text{ x Node Number}) - 1$$

$$R_y = (2 \text{ x Node Number}) \tag{6.4.2}$$

Note that the positive sense of the x-y axis also defines the positive sense of load and displacement.

The structural load-displacement relationship is now given by

$$\{R\} = [K_M]\{u\} \qquad (6.4.3)$$

or, in this case, by

$$\begin{Bmatrix} R_1 \\ R_2 \\ R_3 \\ \cdot \\ \cdot \\ \cdot \\ \cdot \\ \cdot \\ R_{24} \end{Bmatrix} = \begin{bmatrix} K_{11} & \text{----------}K_{1\ 24} \\ K_{21} & \\ K_{31} & \\ \cdot & \\ \cdot & \\ \cdot & \\ \cdot & \\ \cdot & \\ K_{24} & K_{24\ 24} \end{bmatrix}_{24 \times 24} \begin{Bmatrix} u_1 \\ u_2 \\ u_3 \\ \cdot \\ \cdot \\ \cdot \\ \cdot \\ \cdot \\ u_{24} \end{Bmatrix} \qquad (6.4.4)$$

with $\{R\}$ as the load vector, $\{u\}$ as the displacement vector, and $[K_M]$ as the master, or structural stiffness matrix.

We now come to the problem of assembling $[K_M]$ by considering the element stiffness matrices one by one. This is done by utilizing the stiffness transformation derived earlier

$$[K_{Me}]_{\text{each element}} = [T]^T\,[K_e]\,[T] \qquad (6.4.5)$$

Figure 6.4.4 shows a typical element from the truss of Figure 6.4.1. (Or any truss for that matter.) For purposes of specific example in the ensuing discussion, let us take it to be element number 17. Let us now derive a slightly different transformation matrix. This form is in frequent use as is the form of Equation (6.3.17).

We recall that the element of Figure 6.4.2 has element stiffness matrix in element coordinates of

$$\begin{Bmatrix} P_1 \\ \\ P_2 \end{Bmatrix} = \frac{AE}{L} \begin{bmatrix} 1 & -1 \\ \\ -1 & 1 \end{bmatrix} \begin{Bmatrix} \delta_1 \\ \\ \delta_2 \end{Bmatrix} \qquad (6.4.6)$$

The basic size here is 2 x 2. However, when we transform to structural coordinates, we must have the form

$$\begin{Bmatrix} R_{x_1} \\ R_{y_1} \\ R_{x_2} \\ R_{y_2} \end{Bmatrix} = \begin{bmatrix} K_{M_e} \end{bmatrix}_{4 \times 4} \begin{Bmatrix} u_{x_1} \\ u_{y_1} \\ u_{x_2} \\ u_{y_2} \end{Bmatrix} \qquad (6.4.7)$$

which is of basic size 4 x 4. We must express Equation (6.4.6) in a 4 x 4 system before we transform it. This is easily done by supplying (zero magnitude) forces and displacements normal to the element in element coordinates and adding zeroes into the stiffness matrix. Equation (6.4.6) then becomes

$$\begin{Bmatrix} P_1 \\ P_{n_1} \\ P_2 \\ P_{n_2} \end{Bmatrix} = \frac{AE}{L} \begin{bmatrix} 1 & 0 & -1 & 0 \\ 0 & 0 & 0 & 0 \\ -1 & 0 & 1 & 0 \\ 0 & 0 & 0 & 0 \end{bmatrix} \begin{Bmatrix} \delta_1 \\ \delta_{n_1} \\ \delta_2 \\ \delta_{n_2} \end{Bmatrix} \qquad (6.4.8)$$

Equation (6.4.8) is now ready to be transformed to master coordinates. In Figure 6.4.4, we can see the form that the transformation of forces will take as

$$P_1 = R_{x_1} \cos\theta - R_{y_1} \sin\theta$$

$$P_{1_n} = -R_{x_1} \sin\theta + R_{y_1} \cos\theta = 0$$

$$P_2 = R_{x_2} \cos\theta + R_{y_2} \sin\theta$$

$$P_{2_n} = -R_{x_2} \sin\theta + R_{y_2} \cos\theta = 0 \qquad (6.4.9)$$

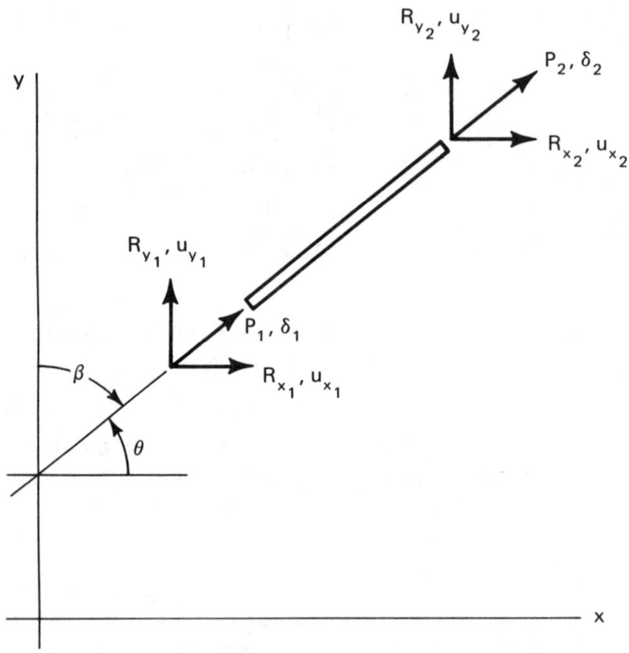

Figure 6.4.4

If we let $\lambda = \cos\theta$ and $\mu = \sin\theta$, we can write Equation (6.4.9) as

$$\begin{bmatrix} P_1 \\ P_{1_n} \\ P_2 \\ P_{2_n} \end{bmatrix} = \begin{bmatrix} \lambda & \mu & 0 & 0 \\ -\mu & \lambda & 0 & 0 \\ 0 & 0 & \lambda & \mu \\ 0 & 0 & -\mu & \lambda \end{bmatrix} \begin{bmatrix} R_{x_1} \\ R_{y_1} \\ R_{x_2} \\ R_{y_2} \end{bmatrix} \qquad (6.4.10)$$

Or,

$$[P] = [T]\,[R] \tag{6.4.11}$$

Now, contragradience tells us that

$$\{u\} = [T]^T\,\{\delta\} \tag{6.4.12}$$

Multiplying the product $\underline{T}\ \underline{T}^T$ will show $\underline{T}$ to be orthogonal, i.e., $\underline{T}^T = \underline{T}^{-1}$. Then we have

$$\{\delta\} = [T]\,\{u\} \tag{6.4.13}$$

and we see that the forces and the displacements transform in the same way. We now know that

$$\left[K_{M_E}\right] = [T]^T\left[K_e\right][T] \tag{6.4.14}$$

That is, the stiffness element for this element, transformed to structural coordinates, is given by Equation (6.4.14).

It is more common to express the $[T]$ matrix as

$$[T] = \begin{bmatrix} \ell & m & 0 & 0 \\ \ell & m & 0 & 0 \\ 0 & 0 & \ell & m \\ 0 & 0 & \ell & m \end{bmatrix} \tag{6.4.15}$$

where $\ell = \cos\theta$ and $m = \cos\beta$.

This process will be illustrated by using numerical values for the member of Figure 6.4.4, which is member number 17 from the truss of Figure 6.4.1. Let us take $AE = 10^7$ psi.

The member length, L, is easily calculated from the coordinates of the nodal points to which member 17 is connected, that is

$$L = \sqrt{(X_6 - X_3)^2 + (Y_6 - Y_3)^2} \tag{6.4.16}$$

The direction cosines ℓ and m may be calculated from

$$\ell = \frac{X_6 - X_3}{L} \; ; \qquad m = \frac{Y_6 - Y_3}{L} \tag{6.4.17}$$

For this element, we have

$$L_{17} = \sqrt{(20 - 0)^2 + (40 - 20)^2}$$

$$L_{17} = 28.284 \text{ in.}$$

Once we have $\left[K_{M_E}\right]$ which is the element stiffness transformed to master coordinates, we must now enter the elements from this $\left[K_{M_E}\right]$ matrix into the proper position in the $\left[K_M\right]$ (24 x 24) master matrix. This is done by incrementing, or adding in, the stiffness terms from $\left[K_{M_E}\right]$ into the appropriate place in $\left[K_M\right]$. The structure is literally assembled numerically by following this procedure for all elements. The stiffness coefficients of all elements coming into a given node may be added directly. The elements essentially act in parallel to stiffen a node.

The entry of $\left[K_M\right]$ into which a given term of $\left[K_{M_E}\right]$ is added is determined from the nodes to which the element is connected. We know that element 17, for example, connects nodes 3 and 6. The structural freedoms (which follow the nodal numbering scheme) that are affected by element 17 are therefore:

(2* Node No.) - 1 = (2 x 3) - 1 = 5
(2* Node No.) = (2 x 3) = 6
(2* Node No.) - 1 = (2 x 6) - 1 = 11
(2* Node No.) = (2 x 6) = 12

The freedoms associated with $\left[K_{M_E}\right]$ for element number 17 are shown in Figure 6.4.5. These are added one-by-one into the appropriate entries of $\left[K_M\right]$ as illustrated in the Figure.

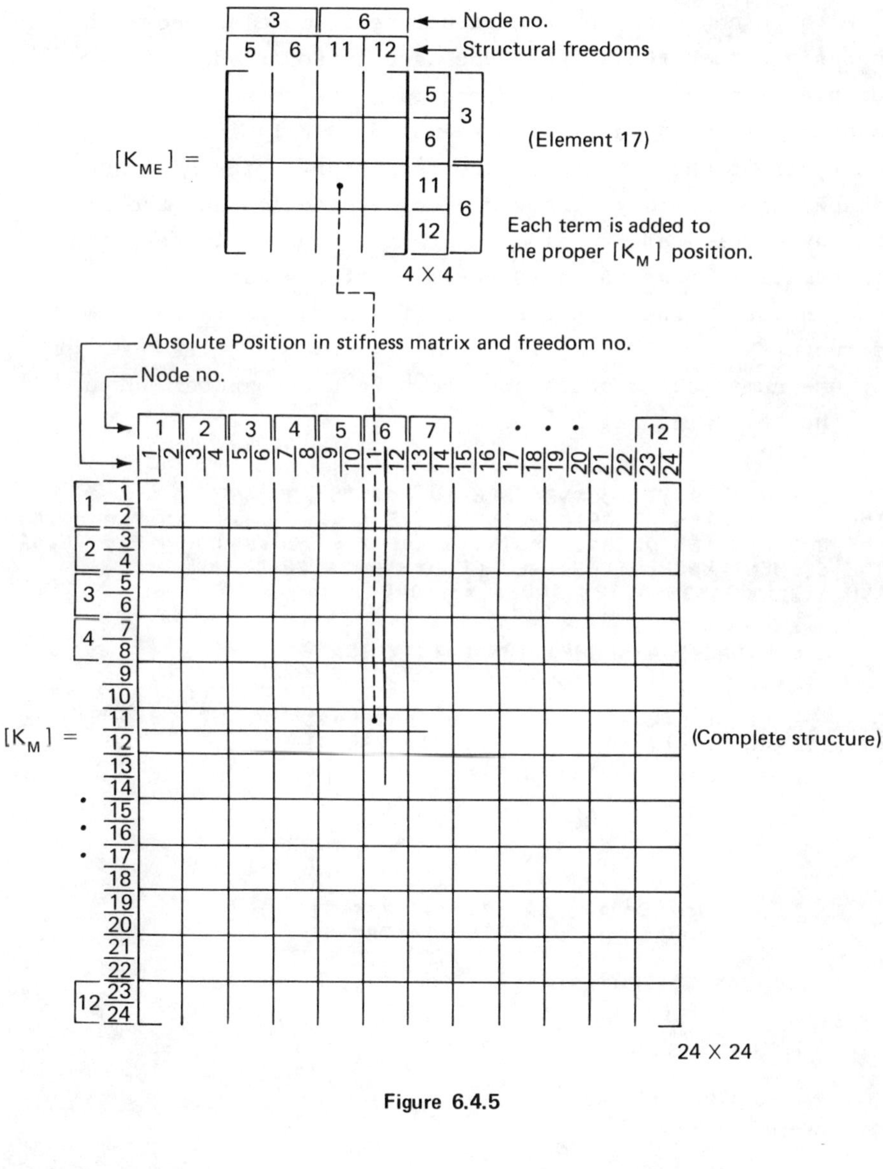

Figure 6.4.5

Completing this process for all of the elements will assemble the structural stiffness matrix of size 24 x 24 which relates the twenty-four nodal loads to the twenty-four nodal displacements. Essentially $\left[K_M\right]$ in Equation (6.4.4) is thus assembled.

We have now seen how the structure is modeled, how the nodes are numbered and how the associated freedoms for displacement and load are numbered. Now we see how the master stiffness matrix is assembled. It is interesting that all of the features of the structure: its size and shape, the topology of the element connection and the size and properties of the element are all completely described mathematically by the structural stiffness matrix. To the computer, the structural stiffness matrix <u>is</u> the structure.

Some numerical results are shown in the computer output on the following pages.

```
ELEMENT NUMBER 17  CONNECTS GRID POINTS  3  AND  6
THE COORDINATES OF GRID POINT 3 ARE X =          0 AND Y = 20.000
THE COORDINATES OF GRID POINT 6 ARE X =  20.000 AND Y = 40.000
THE ELEMENT PROPERTIES ARE  LENGTH = 28.284
AND AE/L = 353553.391 AND A = 1.000

THE ELEMENTAL TRANSFORMATION MATRIX IS

               1                        2
1       7.0710678E-01           7.0710678E-01
2      -7.0710678E-01           7.0710678E-01
3                   0                       0
4                   0                       0

               3                        4
                    0                       0
                    0                       0
        7.0710678E-01           7.0710678E-01
       -7.0710678E-01           7.0710678E-01

 THE ELEMENT STIFFNESS MATRIX FOR THIS ELEMENT IS

               1                        2
1       3.5355339E 05                       0
2                   0                       0
3      -3.5355339E 05                       0
4                   0                       0
```

3	4
-3.5353339E 05	0
0	0
3.5353339E 05	0
0	0

THE TRANSPOSE OF THE TRANSFORMATION MATRIX IS

	1	2
1	7.0710678E-01	-7.0710678E-01
2	7.0710678E-01	7.0710678E-01
3	0	0
4	0	0

3	4
0	0
0	0
7.0710678E-01	-7.0710678E-01
7.0710678E-01	7.0710678E-01

THE ELEMENT STIFFNESS MATRIX IN STRUCTURAL COORDINATES IS

	1	2
1	1.7677670E 05	1.7677670E 05
2	1.7677670E 05	1.7677670E 05
3	-1.7677670E 05	-1.7677670E 05
4	-1.7677670E 05	-1.7677670E 05

3	4
-1.7677670E 05	-1.7677670E 05
-1.7677670E 05	-1.7677670E 05
1.7677670E 05	1.7677670E 05
1.7677670E 05	1.7677670E 05

THE MASTER STIFFNESS MATRIX INCLUDING THIS ELEMENT IS

	1		3
1	1.7677670E 05	1.7677670E 05	0
2	1.7677670E 05	6.7677670E 05	0
3	0	0	0
4	0	0	0
5	0	0	0
6	0	-5.0000000E 05	0
7	-1.7677670E 05	-1.7677670E 05	0
8	-1.7677670E 05	-1.7677670E 05	0
9	0	0	0
10	0	0	0
11	0	0	0
12	0	0	0
13	0	0	0
14	0	0	0
15	0	0	0
16	0	0	0

	4	5	6
1	0	0	0
2	0	0	-5.0000000E 05
3	0	0	0
4	5.0000000E 05	0	0
5	0	6.7677670E 05	1.7677670E 05
6	0	1.7677670E 05	1.1767767E 06
7	0	-5.0000000E 05	0
8	-5.0000000E 05	0	0
9	0	0	0
10	0	0	-5.0000000E 05
11	0	-1.7677670E 05	-1.7677670E 05
12	0	-1.7677670E 05	-1.7677670E 05
13	0	0	0
14	0	0	0
15	0	0	0
16	0	0	0

	7	8	9
1	-1.7677670E 05	-1.7677670E 05	0
2	-1.7677670E 05	-1.7677670E 05	0
3	0	0	0
4	0	-5.0000000E 05	0
5	-5.0000000E 05	0	0
6	0	0	0
7	6.7677670E 05	1.7677670E 05	0
8	1.7677670E 05	1.1767767E 06	0
9	0	0	5.0000000E 05
10	0	0	0
11	0	0	-5.0000000E 05
12	0	-5.0000000E 05	0
13	0	0	0
14	0	0	0
15	0	0	0
16	0	0	0

	10	11	12
1	0	0	0
2	0	0	0
3	0	0	0
4	0	0	0
5	0	-1.7677670E 05	-1.7677670E 05
6	-5.0000000E 05	-1.7677670E 05	-1.7677670E 05
7	0	0	0
8	0	0	-5.0000000E 05
9	0	-5.0000000E 05	0
10	1.0000000E 06	0	0
11	0	6.7677670E 05	1.7677670E 05
12	0	1.7677670E 05	1.1767767E 06
13	0	0	0
14	-5.0000000E 05	0	0
15	0	0	0
16	0	0	-5.0000000E 05

```
              13                         14                        15

 1                     0                          0                          0
 2                     0                          0                          0
 3                     0                          0                          0
 4                     0                          0                          0
 5                     0                          0                          0
 6                     0                          0                          0
 7                     0                          0                          0
 8                     0                          0                          0
 9                     0                          0                          0
10                     0            -5.0000000E 05                          0
11                     0                          0                          0
12                     0                          0                          0
13         5.0000000E 05                          0            -5.0000000E 05
14                     0             5.0000000E 05                          0
15        -5.0000000E 05                          0             5.0000000E 05
16                     0                          0                          0

              16

 1                     0
 2                     0
 3                     0
 4                     0
 5                     0
 6                     0
 7                     0
 8                     0
 9                     0
10                     0
11                     0
12        -5.0000000E 05
13                     0
14                     0
15                     0
16         5.0000000E 05

THE IMAGE OF THE MASTER STIFFNESS MATRIX IS

     1         6         11          16
 1   1 1 * * * * 1 1 * * * * * * * *
     1 1 * * * 1 1 1 * * * * * * * *
     * * * * * * * * * * * * * * * *
     * * * 1 * * * 1 * * * * * * * *
     * * * * 1 1 1 * * * 1 1 * * * *
 6   * 1 * * 1 1 * * * 1 1 1 * * * *
     1 1 * * 1 * 1 1 * * * * * * * *
     1 1 * 1 * * 1 1 * * * 1 * * * *
     * * * * * * * * 1 * 1 * * * * *
     * * * * * 1 * * * 1 * * * 1 * *
11   * * * * 1 1 * * 1 * 1 1 * * * *
     * * * * 1 1 * 1 * * 1 1 * * * 1
     * * * * * * * * * * * * 1 * 1 *
     * * * * * * * * * 1 * * * 1 * *
     * * * * * * * * * * * * 1 * 1 *
16   * * * * * * * * * * * 1 * * * 1
```

Once the master structural matrix, K_M, has been assembled from the structural configuration, we must then specify the load vector. This is done by using the structural freedom numbers as calculated from Equation (6.3.1). That is, a load in the X-(structural) direction at node 8 would be R_{15} and one in the Y direction at node 8 would be R_{16}. If the structure is loaded as shown in Figure 6.4.6, the load vector $\{R\}$ would be

$$\{R\} = \begin{bmatrix} 0. \\ 0. \\ 0. \\ 0. \\ 0. \\ 0. \\ 0. \\ 0. \\ 250. \\ 0. \\ 0. \\ 0. \\ 250. \\ 0. \\ 0. \\ 0. \\ 250. \\ 0. \\ 0. \\ 0. \\ 250. \\ 0. \\ 0. \\ 250. \end{bmatrix} \begin{matrix} 1 \\ 2 \\ 3 \\ 4 \\ 5 \\ 6 \\ 7 \\ 8 \\ 9 \\ 10 \\ 11 \\ 12 \\ 13 \\ 14 \\ 15 \\ 16 \\ 17 \\ 18 \\ 19 \\ 20 \\ 21 \\ 22 \\ 23 \\ 24 \end{matrix}$$

The structural stiffness matrix of Figure 6.4.6 which is assembled from all of the elements in the structure, may be thought of as the unsupported structure stiffness matrix, since we have not taken the pins at nodes one and two into account. This is the usual procedure--to assemble the structural stiffness matrix and then constrain the supports.

The structural freedom directions which are affected by the supports are calculated from Equation (6.4.2). In our example truss, we are talking of freedoms 1, 2, 3 and 4. There are several ways of modifying

$$\{R\} = [K_M]\{u\} \tag{6.4.3}$$

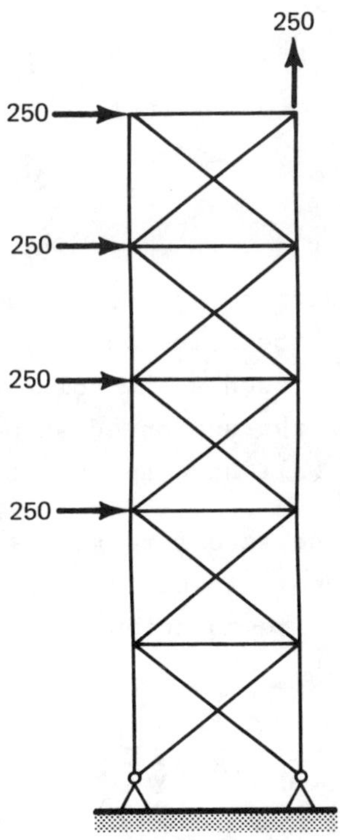

Figure 6.4.6

to mathematically indicate that the nodal freedoms at the support have been suppressed. One convenient way is as follows:

1. Knowing the structural freedom number, F, which is to be suppressed, zero out row F and column F in the

master stiffness matrix; EXCEPT, put a 1 on the diagonal.
2. Zero out row F of the load matrix.
3. Zero out row F of the displacement matrix.

This has the effect of partitioning Equation (6.3.3) in the following way: (subscript s indicates suppressed freedoms; subscript a indicates active freedoms).

$$\begin{Bmatrix} \underline{R}_s = \underline{0} \\ \hline \underline{R}_a \end{Bmatrix} \left[\begin{array}{c|c} \underline{I} & \underline{0} \\ \hline \underline{0} & \underline{K}_{M_a} \end{array} \right] \begin{Bmatrix} \underline{u}_s = \underline{0} \\ \hline \underline{u}_a \end{Bmatrix} \qquad (6.4.18)$$

From this it can be shown that the solution for the active structural freedoms is

$$\left\{u_a\right\} = \left[K_{M_a}\right]^{-1} \left\{R\right\}_a \qquad (6.4.19)$$

although in actual practice, the zeroes are carried right along with the rest of the solution.

Equation (6.4.19) represents the solution for the structural displacements of the supported structure. For a stable structure, $\underline{K}_{M_a}$ will have an inverse. The vector $\left\{u_a\right\}$ then represents the actual x- and y- component displacements of each node, according to the structural freedom numbers of each node. The solution vector for our example follows:

X 1=	0	Y 1=	0
X 2=	0	Y 2=	0
X 3=	8.61632E-03	Y 3=	5.94460E-03
X 4=	8.54054E-03	Y 4=	-5.55540E-03
X 5=	2.67910E-02	Y 5=	9.92422E-03
X 6=	2.65227E-02	Y 6=	-9.07578E-03
X 7=	4.99022E-02	Y 7=	1.21764E-02
X 8=	4.96541E-02	Y 8=	-1.08235E-02
X 9=	7.50675E-02	Y 9=	1.31761E-02
X10=	7.48173E-02	Y10=	-1.13239E-02
X11=	1.00274E-01	Y11=	1.34262E-02
X12=	1.00025E-01	Y12=	-1.10735E-02

Note that the zeroes for the suppressed nodes are included as part of the solution.

An interesting observation is the $\underline{K}_M$ for the unsupported structure, Equation (6.4.3), is singular and does not have an inverse. This is reflected in the physical interpretation that the unsupported structure is not statically stable. It has rigid-body degrees of freedom.

Inspection of Equation (6.4.19) shows that we may easily consider multiple load cases. The constrained structural stiffness matrix, $\underline{K}_{M_a}$ is a function of structure and geometry and constraint, but not of loading. Therefore, $\underline{K}_{M_a}^{-1}$ will be the same for all load cases. Multiple load cases may be solved for by placing multiple columns on the right hand side of Equation (6.4.19) as follows:

$$\left[\left\{ u_I \right\} \left\{ u_{II} \right\} \right]_{N \times n} = \left[\underline{K}_{M_a} \right]^{-1}_{N \times N} \times \left[\left\{ R_I \right\} \left\{ R_{II} \right\} \right]_{N \times n}$$

where N = the number of active structural freedoms and n = the number of load cases. $\left\{ u_I \right\}$ will be the solution for load case $\left\{ R_I \right\}$, etc.

We can now observe the importance of the order of node numbering. If we minimize the numerical difference between adjacent nodes, we are creating a "banded" stiffness matrix. That is, the non-zero elements are grouped near the diagonal. This is because the entry in K_M is calculated from the node number at each end of the element. Bandedness of the stiffness matrix has at least two advantages:

1. Better "conditioning" of the stiffness matrix and thus, better numerical accuracy of the solution.
2. More efficient storage, which can be accomplished by writing routines which do not store the zeroes outside the band. (It should be noted that we usually

do not actually find the inverse of $\underline{K}_{M_a}$, but often just solve Equation (6.4.19) by Gauss elimination or a similar technique. Programs which take advantage of bandedness do not invert $\underline{K}_{M_a}$.)

The stiffness matrix banded pattern for our example truss may be seen in Figure 6.4.7.

After we obtain the solution for structural displacements, the $\{u\}$ from the solution of Equation (6.4.19), we are then concerned with the element forces and stresses. The element stiffness matrix and element transformation matrix may be used to calculate element displacements and forces.

Consider element 17 once more. We can transform the structural displacements $\{r\}$ back to element displacements $\{\delta\}$ in the following ways.

$$\underbrace{\begin{bmatrix} \delta_1 \\ 0 \\ \delta_2 \\ 0 \end{bmatrix}}_{\text{Element 17}} = [T]_{\text{Element 17}} \times \underbrace{\begin{bmatrix} u_5 \\ u_6 \\ u_{11} \\ u_{12} \end{bmatrix}}_{\substack{\text{structural displacements} \\ \text{for nodes of element 17}}} \tag{6.4.20}$$

(Recall Equation 6.4.13.)

We now know the $\{\delta\}$ and may use $\underline{K}_e$, Equation (6.4.8), to find element forces P_1 and P_2.

$$\begin{bmatrix} P_1 \\ 0 \\ P_2 \\ 0 \end{bmatrix} = [K_e]_{\substack{\text{Element} \\ 17}} \times \{\delta\}_{\substack{\text{Element} \\ 17}} \tag{6.4.21}$$

For the rod element, of course, the stress in the element is just

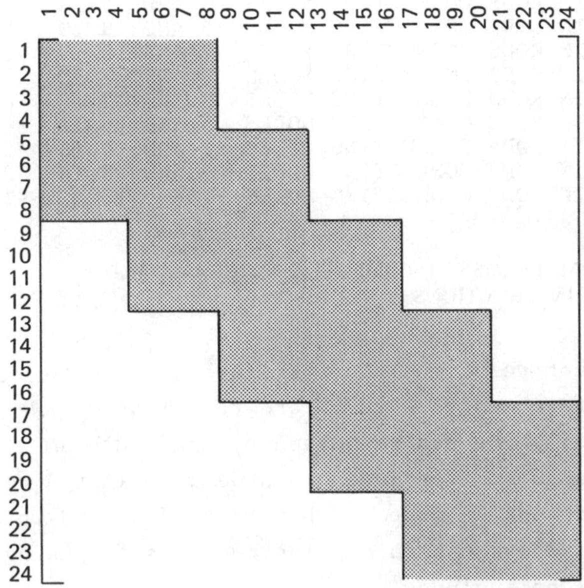

Figure 6.4.7 Illustrates bandedness of stiffness matrix for truss of Figure 6.4.1. Non-shaded entries are zero.

$$\sigma = \frac{P_2}{A} = -\frac{P_1}{A} \tag{6.4.22}$$

(See Figure 6.4.4.)

The results for element 17 of our truss are shown below:

```
THE STRESS IN RODS CONNECTING NODES  1 AND  3 IS  2972.2978
THE STRESS IN RODS CONNECTING NODES  3 AND  5 IS  1989.8141
THE STRESS IN RODS CONNECTING NODES  5 AND  7 IS  1126.0663
THE STRESS IN RODS CONNECTING NODES  7 AND  9 IS   499.8884
THE STRESS IN RODS CONNECTING NODES  9 AND 11 IS   125.0116
THE STRESS IN RODS CONNECTING NODES  2 AND  4 IS -2777.7022
THE STRESS IN RODS CONNECTING NODES  4 AND  6 IS -1760.1859
THE STRESS IN RODS CONNECTING NODES  6 AND  8 IS  -873.9337
THE STRESS IN RODS CONNECTING NODES  8 AND 10 IS  -250.1116
THE STRESS IN RODS CONNECTING NODES 10 AND 12 IS   125.0116
THE STRESS IN RODS CONNECTING NODES  3 AND  4 IS   -37.8882
THE STRESS IN RODS CONNECTING NODES  5 AND  6 IS  -134.1196
THE STRESS IN RODS CONNECTING NODES  7 AND  8 IS  -124.0453
THE STRESS IN RODS CONNECTING NODES  9 AND 10 IS  -125.1001
THE STRESS IN RODS CONNECTING NODES 11 AND 12 IS  -124.9884
THE STRESS IN RODS CONNECTING NODES  1 AND  4 IS   746.2836
```

```
THE STRESS IN RODS CONNECTING NODES  3 AND  6 IS   721.5119
THE STRESS IN RODS CONNECTING NODES  5 AND  8 IS   528.8220
THE STRESS IN RODS CONNECTING NODES  7 AND 10 IS   353.7112
THE STRESS IN RODS CONNECTING NODES  9 AND 12 IS   176.7603
THE STRESS IN RODS CONNECTING NODES  3 AND  2 IS  -667.9299
THE STRESS IN RODS CONNECTING NODES  5 AND  4 IS  -692.7017
THE STRESS IN RODS CONNECTING NODES  7 AND  6 IS  -531.8381
THE STRESS IN RODS CONNECTING NODES  9 AND  8 IS  -353.3955
THE STRESS IN RODS CONNECTING NODES 11 AND 10 IS  -176.7930
```

6.5 ADDITIONAL ELEMENTS, COMPUTER PROGRAMS AND PROGRAM INSTRUCTIONS

We shall not delve much further into the finite element method. References (1 thru 5) are recommended. We will mention briefly the 3-d rod element, the 2-d beam element and the triangular membrane. Then we will list a computer program which can be prepared by the student and which utilizes the above elements. Instructions for using the program will be included.

The 3-d rod has the same element stiffness matrix as the 2-d rod. To use the rod element in 3-d space, we need only to consider the rod element to be a 3-d line and find direction cosines l, m, and n. Then the transformation matrix $\left[T\right]$ for this element is,

$$\left[T\right]_{2 \times 6} = \begin{bmatrix} l & m & n & 0 & 0 & 0 \\ 0 & 0 & 0 & l & m & n \end{bmatrix} \qquad (6.5.1)$$

The remainder of the derivations proceed in the same fashion as those of the 2-d rod.

The 2-d beam element is given in Equation (4.8.1). If we use the direction conventions of Figure 6.5.1, the following transformation matrix is obtained. (The freedoms are relabeled in accordance with the notation of this chapter.)

$$\left[T\right]_{4 \times 6} = \begin{bmatrix} -m & l & 0 & 0 & 0 & 0 \\ 0 & 0 & 1 & 0 & 0 & 0 \\ 0 & 0 & 0 & -m & l & 0 \\ 0 & 0 & 0 & 0 & 0 & 1 \end{bmatrix} \qquad (6.5.2)$$

where $l = \cos\theta$; $m = \cos\beta$; $l = 1.0$ in Figure 6.5.1.

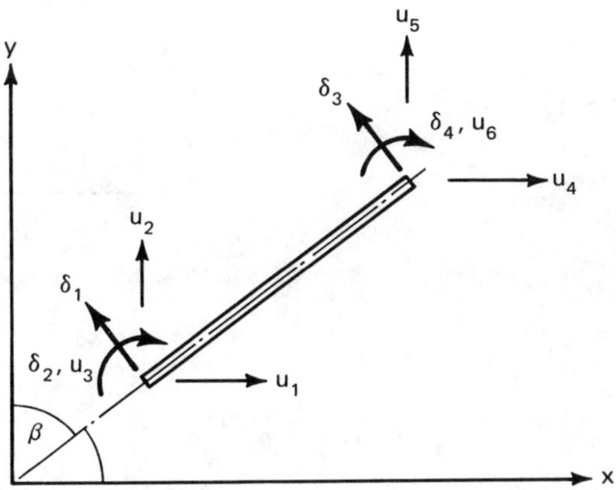

Figure 6.5.1

The triangular membrane element is used to model two dimensional elasticity problems and allows the in-plane stresses σ_x, σ_y and τ_{xy}. The basic geometry is shown in Figure 6.5.2.

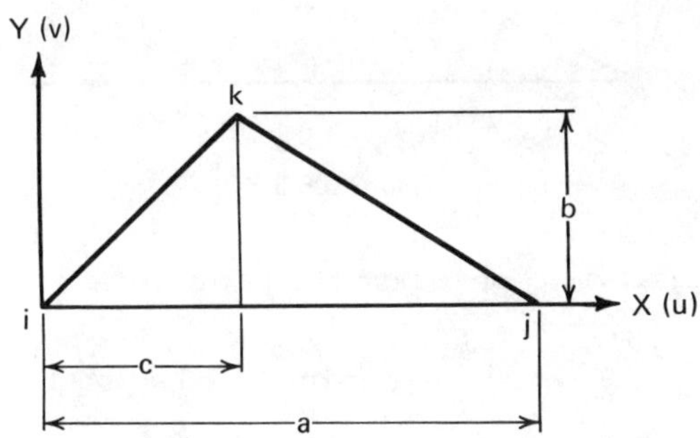

Figure 6.5.2

The thickness of the triangle is t and the definitions of dimensions a, b, and c are given in Figure 6.5.2. Then the element stiffness matrix is

$$K_e = \frac{Et}{2(1 - \nu^2)ab} \times$$

$$
\begin{array}{l}
1) \\ 2) \\ 3) \\ 4) \\ 5) \\ 6)
\end{array}
\left[\begin{array}{llllll}
b^2 + \lambda_1(c - a)^2 & & & & & \\
-b^2 - \lambda_1 c(c - a) & b^2 + \lambda_1 c^2 & & \text{SYM.} & & \\
\lambda_1 a(c - a) & -\lambda_1 ac & \lambda_1 a^2 & & & \\
-\lambda_2 b(c - a) & \lambda_1 cb + vb(c - a) & -\lambda_1 ab & \lambda_1 b^2 + (c - a)^2 & & \\
\lambda_1 b(c - a) + vcb & -\lambda_2 bc & \lambda_1 ab & -\lambda_1 b^2 - c(c - a) & \lambda_1 b^2 + c^2 & \\
-vab & vab & 0 & a(c - a) & -ac & a^2
\end{array}\right]
$$

where: $\lambda_1 = \frac{1}{2}(1 - \nu)$; $\lambda_2 = \frac{1}{2}(1 + \nu)$

If θ is defined according to Figure 6.5.3:

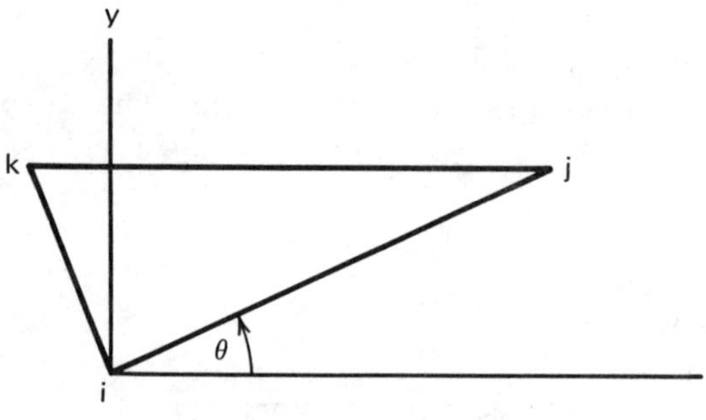

Figure 6.5.3

Then the transformation matrix $[T]$ for this element is:

$$
[T] = \begin{bmatrix}
\cos\theta & 0 & 0 & -\sin\theta & 0 & 0 \\
0 & \cos\theta & 0 & 0 & -\sin\theta & 0 \\
0 & 0 & \cos\theta & 0 & 0 & -\sin\theta \\
\sin\theta & 0 & 0 & \cos\theta & 0 & 0 \\
0 & \sin\theta & 0 & 0 & \cos\theta & 0 \\
0 & 0 & \sin\theta & 0 & 0 & \cos\theta
\end{bmatrix}_{6 \times 6}
$$

The following FORTRAN IV computer program comprises system HILSEN. (HILSEN was written by former Cal Poly,

Pomona student Donald Hilsen. Mr. Hilsen is now employed at WED Enterprises, Glendale, California.) This program can be punched and run for student training in the finite element method. The expanded output of this program shows the finite element method step-by-step. An abbreviated instruction manual concludes this chapter.

1.0 Introduction

HILSEN is a general purpose finite element system. It can handle several elements, one at a time and prints out at each step of the program. It is limited to core storage only.

2.0 Elements

The following elements are available:

a. 2-D beam
b. 2-D truss
c. 3-D truss
d. 2-D triangle plate

These elements are shown in Figures 2.1, 2.2, 2.3, 2.4.

3.0 Units of Measure

All dimensions must be kept in consistent units. The type does not matter. For example if the location, dimension and property of the elements are entered in inches, the output will be in inches.

4.0 Limitations

The limitations are:

a. 2-D beam: 20 nodes
b. 2-D truss: 30 nodes
c. 3-D truss: 20 nodes
d. 2-D triangle plate: 30 nodes

Only 1 element can be used in the structure.

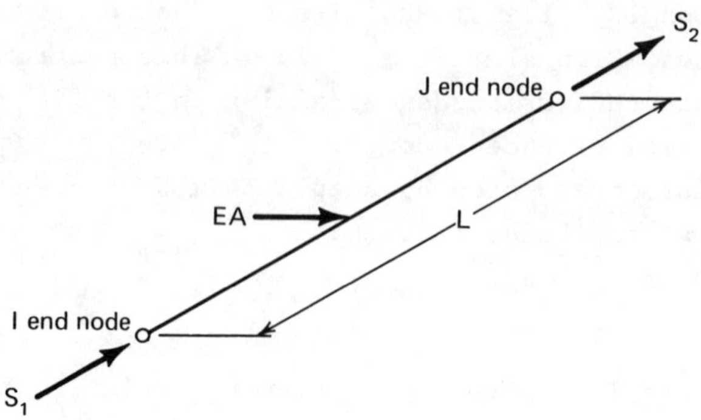

Figure 2.1 Truss element

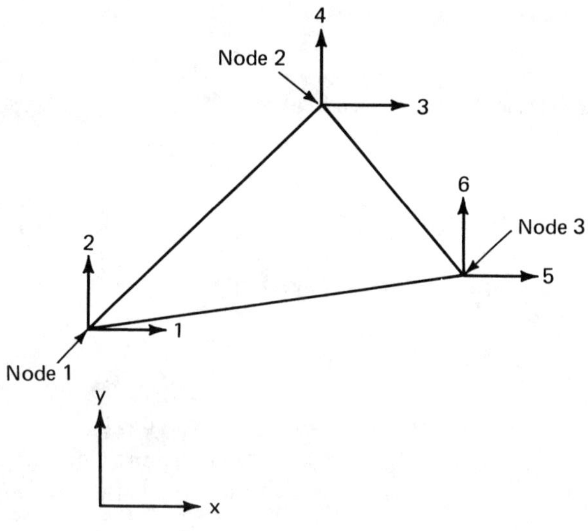

Figure 2.2 2-D plate element

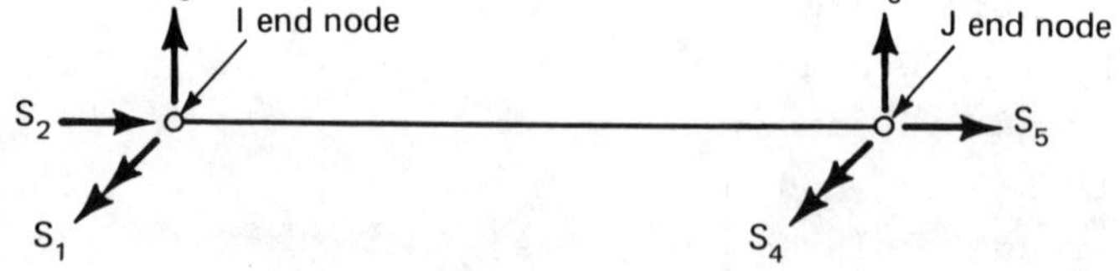

Figure 2.3 2-D frame element

5.0 Error Messages

None

6.0 Input Cards

The following pages describe the data input to the program. To set up a data deck simply set each card up as described by the TYPE and PURPOSE OR VALUE column and in the card field shown by the FIELDS column. The number of card is shown by the NUMBER OF CARDS column. If a 1 appears in the NUMBER OF CARDS column only one card is needed for the program for that entry. The data matrixes are entered with 1 card/node or 1 card/element.

There are 4 sets of data cards, one for each element and only 1 element can be used per structure. Each element has a common method of constraining the nodes and loading the nodes. This is the final set of cards for the data deck and is in the section "final common data section." A typical data deck would include the input deck for a particular element plus this final section.

NUMBER OF CARDS	TYPE	FIELDS	PURPOSE OR VALUE
INPUT DECK FOR BEAMS			
1	I1	1	Put a 1
1	I2	1-2	Number of nodes
	I2	3-4	Number of elements
	I2	5-6	Number of constrained freedoms
1	F20.3	1-20	Modulus E of beam
1	F10.3	1-10	Area A of beam
Next set of cards is the element data matrix, with 1 card/element equal to the number of beams			
	I4	1-4	Element number
	I3	5-7	I end node
	I3	8-10	J end node
	F10.3	11-20	Moment of inertia

NUMBER OF CARDS	TYPE	FIELDS	PURPOSE OR VALUE
	F10.3	21-30	X coordinate of I end node
	F10.3	31-40	X coordinate of J end node
	F10.3	41-50	Y coordinate of I end node
	F10.3	51-60	Y coordinate of J end node

NUMBER OF CARDS	TYPE	FIELDS	PURPOSE OR VALUE
INPUT DECK FOR PLATE			
1	I1	1	Put a 4
1	I2	1-2	Number of nodes
	I2	3-4	Number of elements
	I2	5-6	Number of constrained freedoms
1	F10.3	1-10	Plate thickness
	F10.3	11-20	Poissons ratio
	F10.3	21-30	Modulus E of plate

Next set of cards is the element data matrix, entered with 1 card/element equal to the number of elements.

NUMBER OF CARDS	TYPE	FIELDS	PURPOSE OR VALUE
	I5	1-5	Element number
	I5	6-10	Node 1
	I5	11-15	Node 2
	I5	16-20	Node 3
	F10.3	21-30	X coordinate of node 1
	F10.3	31-40	Y coordinate of node 1
	F10.3	41-50	X coordinate of node 2
	F10.3	51-60	Y coordinate of node 2
	F10.3	61-70	X coordinate of node 3
	F10.3	71-80	Y coordinate of node 3

NUMBER OF CARDS	TYPE	FIELDS	PURPOSE OR VALUE
INPUT DECK FOR 2-D TRUSS			
1	I1	1	Put a 2
1	I2	1-2	Number of nodes
	I2	3-4	Number of elements
	I2	5-6	Number of constrained freedoms
1	F20.3	1-20	Modulus E of elements

Next set of cards is the element data matrix, entered with 1 card/element equal to the number of elements.

NUMBER OF CARDS	TYPE	FIELDS	PURPOSE OR VALUE
	I4	1-4	Element number
	I3	5-7	I end node
	I3	8-10	J end node
	F10.3	11-20	Area
	F10.3	21-30	X coordinate of I end node
	F10.3	31-40	X coordinate of J end node
	F10.3	41-50	Y coordinate of I end node
	F10.3	51-60	Y coordinate of J end node

INPUT DECK FOR 3-D TRUSS

1	I1	1	Put a 3
1	I2	1-2	Number of nodes
	I2	3-4	Number of elements
	I2	5-6	Number of constrained freedoms
1	F20.3	1-20	Modulus of elasticity

Next set of cards is the element data matrix, entered with 1 card/element equal to the number of elements.

	I4	1-4	Element number
	I3	5-7	I end node
	I3	8-10	J end node
	F10.3	11-20	Area of truss
	F10.3	21-30	X coordinate of I end node
	F10.3	31-40	X coordinate of J end node
	F10.3	41-50	Y coordinate of I end node
	F10.3	51-60	Y coordinate of J end node

F10.3	61-70	Z coordinate of I end node
F10.3	71-80	Z coordinate of J end node

FINAL COMMON DATA SECTION

The next set of cards, in ascending numerical order, are the structural degrees of freedom that are constrained. For example, if freedom 3 is to be constrained, put a three on the data card. 12 freedoms can be constrained per card.

The next set of cards are the node point loads on the structure. Each degree of freedom must be represented in this set of cards as either having a load or not. Eight degrees of freedom can be read in on one card, with an F10.3 format.

7.0 Results

The print out is self-explanatory.

```
      PROGRAM STIFFNES
      COMMON TEE(6,6),ESM(6,6),TEETRAN(6,6),PRODUCT(6,6),
     1BIGK(48,48),MM(48),DEF(48),R(48),M(48),PROP(48,10),LL(48)
     1,ZEE(48)
      INTEGER ORD,FRED
      DO 60 KKK=1,11
      PRINT 1234
 1234 FORMAT(1H1)
      ORD = 48
      FRED = 0.
      DO 18 I=1,ORD
      DO 18 J=1,ORD
   18 BIGK(I,J) = 0.
      PRINT 8000
 8000 FORMAT(/////4H      )
      READ 1720,KONST
 1720 FORMAT(I1)
      READ 100,NNODES,NELMS,NSUP
  100 FORMAT(3I2)
      IF(KONST.EQ.4)5000,5001
 5001 IF(KONST.EQ.3)1710,2010
 2010 IF(KONST.EQ.2)1800,2011
 5000 CALL TRIANGLE(NNODES,NELMS,FRED)
      GO TO 10
 1710 CALL TRUSS3D(NNODES,NELMS,FRED)
      GO TO 10
 1800 CALL TRUSS2D(FRED,NELMS,NNODES)
      GO TO 10
 2011 CALL BEAM(NNODES,NELMS,FRED)
   10 CONTINUE
      CALL SUPPORT(NSUP,FRED)
      PRINT 1
    1 FORMAT(10X,30HTHE NODE POINT COORDINATES ARE,/////)
      DO 5 JD = 1,NNODES
      IF(KONST.EQ.3)6,2
    6 PRINT 3,JD,DEF(JD),R(JD),ZEE(JD)
```

```
    3 FORMAT(10X,4HNODE,I3,5X,4HX = ,F10.3,5X,4HY = ,F10.3,5X,4HZ = ,F10
     1.3)
      GO TO 5
    2 PRINT 44,JD,DEF(JD),R(JD)
   44 FORMAT(10X,4HNODE,I3,5X,4HX = ,F10.3,5X,4HY = ,F10.3)
    5 CONTINUE
      DO 604 I=1,ORD
  604 R(I) = 0.
      READ 602,(R(I), I=1,FRED)
  602 FORMAT(8F10.3)
      PRINT 20
   20 FORMAT(/////4H    )
      J = FRED+1
      DO 605 I=J,ORD
  605 BIGK(I,I) = 1.
      CALL MINV(BIGK,ORD,DET,LL,MM)
      CALL MMULTB(FRED)
      IF(KONST.EQ.4)5002,5003
 5003 IF(KONST.EQ.1)50,51
 5002 CALL TRISTRS(NELMS,NNODES)
      GO TO 60
   51 CALL TRUSSTRS(KONST,NNODES,NELMS)
      GO TO 60
   50 CALL FRMDEL(NNODES,NELMS)
   60 CONTINUE
      RETURN
      END

          FORTRAN DIAGNOSTIC RESULTS  FOR     STIFFNES
```

```
      SUBROUTINE BEAM (NNODES,NELMS,FRED)
      COMMON TEE(6,6),ESM(6,6),TEETRAN(6,6),PRODUCT(6,6),
     1BIGK(48,48),MM(48),DEF(48),R(48),M(48),PROP(48,10),LL(48)
     1,ZEE(48)
      INTEGER FRED,ELEN,GP1,GP2
      REAL LENGTH,LAMBDA,MU,INERT
      FRED = 3.*NNODES
      PRINT 5000
 5000 FORMAT(10X,40HTHE FOLLOWING ANALYSIS IS FOR A 2D FRAME    )
      PRINT 50,NELMS,NNODES,FRED
   50 FORMAT(10X,37HTHE TOTAL NUMBER OF ELEMENTS EQUALS   ,I3,2X,23HTHE N
     1UMBER OF NODES IS ,I3,2X,28HAND THE NUMBER OF UNKNOWNS =,I3,/////)
      KLM = 1
      READ 2,E
    2 FORMAT(4F20.3)
      READ 1,A
    1 FORMAT(F10.3)
 2121 READ 1201,ELEN,GP1,GP2,INERT,XZ,X1,YZ,Y1
 1201 FORMAT(I4,2I3,5F10.3)
      LENGTH = SQRTF((X1-XZ)*(X1-XZ)+(Y1-YZ)*(Y1-YZ))
      LAMBDA =(X1-XZ)/LENGTH
      MU = (Y1-YZ)/LENGTH
      PROP(ELEN,1) = GP1
      PROP(ELEN,2) = GP2
      PROP(ELEN,3) = E
      PROP(ELEN,4) = INERT
      PROP(ELEN,5) = A
      PROP(ELEN,6) = LENGTH
      PROP(ELEN,7) = MU
      PROP(ELEN,8) = LAMBDA
      R(GP2) = Y1
      R(GP1) = YZ
      DEF(GP2) = X1
      DEF(GP1) = XZ
      PRINT 706,ELEN,GP1,GP2
  706 FORMAT(10X,14HELEMENT NUMBER,I3,2X,20HCONNECTS GRID POINTS,I3,2X,3H
     1HAND,I3 )
```

```
      PRINT 707,GP1,XZ,YZ
      PRINT 707,GP2,X1,Y1
  707 FORMAT(10X,29HTHE COORDINATES OF GRID POINT,I3,2X,8HARE  X =,F10.3
     1,5X,8HAND  Y =,F10.3)
      PRINT 2708,LENGTH,INERT,A
 2708 FORMAT(10X,37HTHE ELEMENT PROPERTIES ARE  LENGTH = ,F10.3,8X,20HMO
     1MENT OF INERTIA = ,F10.3,3X,10HAND AREA =,F15.3)
      PRINT 2709,E
 2709 FORMAT(10X,28HTHE MODULUS OF ELASTICITY = ,E13.5,////)
      DO 2015 I=1,6
      DO 2015 J=1,6
 2015 TEE(I,J) = 0.
      TEE(1,1) = 1.
      TEE(2,2) = LAMBDA
      TEE(2,3) = MU
      TEE(3,2) = -MU
      TEE(3,3) = LAMBDA
      TEE(4,4) = 1.
      TEE(5,5) = LAMBDA
      TEE(5,6) = MU
      TEE(6,5) = -MU
      TEE(6,6) = LAMBDA
      IF(NELMS-10.)800,800,801
  800 PRINT 701
  701 FORMAT(10X,39HTHE ELEMENTAL TRANSFORMATION MATRIX IS )
      CALL PRIM(TEE,6,6)
  801 DO 2017 I =1,6
      DO 2017 J =1,6
 2017 ESM(I,J) = 0.
      EI = E*INERT
      ESM(2,2)  = (A*E)/LENGTH
      ESM(5,2)  = -(A*E)/LENGTH
      ESM(5,5)  = (A*E)/LENGTH
      ESM(2,5)  = -(A*E)/LENGTH
      ESM(1,1) =(4.*EI)/LENGTH
      ESM(1,3) =-(6.*EI)/(LENGTH*LENGTH)
      ESM(1,4) = (2.*EI)/LENGTH
```

```
      ESM(1,6) =  (6.*EI)/(LENGTH*LENGTH)
      ESM(3,1) =-(6.*EI)/(LENGTH*LENGTH)
      ESM(3,3) = (12.*EI)/(LENGTH*LENGTH*LENGTH)
      ESM(3,4) =-(6.*EI)/(LENGTH*LENGTH)
      ESM(3,6) =-(12.*EI)/(LENGTH*LENGTH*LENGTH)
      ESM(4,1) = (2.*EI)/LENGTH
      ESM(4,3) =-(6.*EI)/(LENGTH*LENGTH)
      ESM(4,4) =(4.*EI)/LENGTH
      ESM(4,6) =  (6.*EI)/(LENGTH*LENGTH)
      ESM(6,1) =  (6.*EI)/(LENGTH*LENGTH)
      ESM(6,3) =-(12.*EI)/(LENGTH*LENGTH*LENGTH)
      ESM(6,4) =  (6.*EI)/(LENGTH*LENGTH)
      ESM(6,6) = (12.*EI)/(LENGTH*LENGTH*LENGTH)
      IF(NELMS-10.)802,802,803
  802 PRINT 700
  700 FORMAT(10X,49HTHE ELEMENT STIFFNESS MATRIX FOR THIS ELEMENT IS   )
      CALL PRIM(ESM,6,6)
  803 DO 2020 I=1,6
      DO 2020 J=1,6
 2020 TEETRAN(J,I) = TEE(I,J)
      IF(NELMS-10.)804,804,805
  804 PRINT 703
  703 FORMAT(10X,46HTHE TRANSPOSE OF THE TRANSFORMATION MATRIX IS )
      CALL PRIM(TEETRAN,6,6)
  805 DO 2074 I=1,6
      DO 2074 J=1,6
 2074 PRODUCT(I,J) = 0.
      CALL MMULT(TEETRAN,ESM,PRODUCT,6,6,6)
      CALL MMULT(PRODUCT,TEE,ESM,6,6,6)
      IF(NELMS-10.)806,806,807
  806 PRINT 702
  702 FORMAT(10X,58HTHE ELEMENT STIFFNESS MATRIX IN STRUCTURAL COORDINAT
     1ES IS )
      CALL PRIM(ESM,6,6)
  807 CALL BIGK3D(GP1,GP2,0.)
      IF(NELMS-10.)4721,4721,5128
 4721 PRINT 704
```

```
 704 FORMAT(10X,54HTHE MASTER STIFFNESS MATRIX INCLUDING THIS ELEMENT I
    1S )
     CALL PRIMB(FRED)
     CALL IMAGES(FRED)
5128 IF(KLM-NELMS)3128,19,19
3128 KLM = KLM + 1
     GO TO 2121
  19 CONTINUE
     RETURN
     END
```

FORTRAN DIAGNOSTIC RESULTS FOR BEAM

```
      SUBROUTINE BIGK2D (GP1,GP2)
      COMMON TEE(6,6),ESM(6,6),TEETRAN(6,6),PRODUCT(6,6),
     1BIGK(48,48),MM(48),DEF(48),R(48),M(48),PROP(48,10),LL(48)
     1,ZEE(48)
      INTEGER XONE,XTWO,YONE,YTWO,GP1,GP2
      XONE = (2.*GP1)-1.
      YONE = (2.*GP1)
      XTWO = (2.*GP2)-1.
      YTWO = 2.*GP2
      BIGK(XONE,XONE) = BIGK(XONE,XONE)+ESM(1,1)
      BIGK(XONE,YONE) = BIGK(XONE,YONE)+ESM(1,2)
      BIGK(XONE,XTWO) = BIGK(XONE,XTWO) + ESM(1,3)
      BIGK(XONE,YTWO) = BIGK(XONE,YTWO) + ESM(1,4)
      BIGK(YONE,XONE) = BIGK(YONE,XONE) + ESM(2,1)
      BIGK(YONE,YONE) = BIGK(YONE,YONE)+ ESM(2,2)
      BIGK(YONE,XTWO) = BIGK(YONE,XTWO) + ESM(2,3)
      BIGK(YONE,YTWO) = BIGK(YONE,YTWO) + ESM(2,4)
      BIGK(XTWO,XONE) = BIGK(XTWO,XONE) + ESM(3,1)
      BIGK(XTWO,XTWO) = BIGK(XTWO,XTWO) + ESM(3,3)
      BIGK(XTWO,YTWO) = BIGK(XTWO,YTWO) + ESM(3,4)
      BIGK(XTWO,YONE) = BIGK(XTWO,YONE) + ESM(3,2)
```

```
      BIGK(YTWO,XONE) = BIGK(YTWO,XONE)  + ESM(4,1)
      BIGK(YTWO,YONE) = BIGK(YTWO,YONE)  + ESM(4,2)
      BIGK(YTWO,XTWO) = BIGK(YTWO,XTWO)  + ESM(4,3)
      BIGK(YTWO,YTWO) = BIGK(YTWO,YTWO)  + ESM(4,4)
      RETURN
      END

          FORTRAN DIAGNOSTIC RESULTS   FOR    BIGK2D


      SUBROUTINE BIGK3D (GP1,GP2,GP3)
      COMMON TEE(6,6),ESM(6,6),TEETRAN(6,6),PRODUCT(6,6),
     1BIGK(48,48),MM(48),DEF(48),R(48),M(48),PROP(48,10),LL(48)
     1,ZEE(48)
      INTEGER XONE,XTWO,YONE,YTWO,ZONE,ZTWO,GP1,GP2,GP3
      IF(GP3.EQ.0)2,1
    1 XONE = (2.*GP1)-1.
      YONE = 2.*GP1
      ZONE = (2.*GP2)-1.
      XTWO = 2.*GP2
      YTWO = (2.*GP3)-1.
      ZTWO = 2.*GP3
      GO TO 3
    2 XONE = (3.*GP1)-2.
      YONE = (3.*GP1)-1.
      ZONE = 3.*GP1
      XTWO = (3.*GP2)-2.
      YTWO = (3.*GP2)-1.
      ZTWO = (3.*GP2)
    3 BIGK(XONE,XONE) = BIGK(XONE,XONE)+ESM(1,1)
      BIGK(XONE,YONE) = BIGK(XONE,YONE)  + ESM(1,2)
      BIGK(XONE,ZONE) = BIGK(XONE,ZONE)  + ESM(1,3)
      BIGK(XONE,XTWO) = BIGK(XONE,XTWO)  + ESM(1,4)
      BIGK(XONE,YTWO) = BIGK(XONE,YTWO)  + ESM(1,5)
      BIGK(XONE,ZTWO) = BIGK(XONE,ZTWO)  + ESM(1,6)
```

```
      BIGK(YONE,XONE) = BIGK(YONE,XONE) + ESM(2,1)
      BIGK(YONE,YONE) = BIGK(YONE,YONE) + ESM(2,2)
      BIGK(YONE,ZONE) = BIGK(YONE,ZONE) + ESM(2,3)
      BIGK(YONE,XTWO) = BIGK(YONE,XTWO) + ESM(2,4)
      BIGK(YONE,YTWO) = BIGK(YONE,YTWO) + ESM(2,5)
      BIGK(YONE,ZTWO) = BIGK(YONE,ZTWO) + ESM(2,6)
      BIGK(ZONE,XONE) = BIGK(ZONE,XONE) + ESM(3,1)
      BIGK(ZONE,YONE) = BIGK(ZONE,YONE) + ESM(3,2)
      BIGK(ZONE,ZONE) = BIGK(ZONE,ZONE) + ESM(3,3)
      BIGK(ZONE,XTWO) = BIGK(ZONE,XTWO) + ESM(3,4)
      BIGK(ZONE,YTWO) = BIGK(ZONE,YTWO) + ESM(3,5)
      BIGK(ZONE,ZTWO) = BIGK(ZONE,ZTWO) + ESM(3,6)
      BIGK(XTWO,XONE) = BIGK(XTWO,XONE) + ESM(4,1)
      BIGK(XTWO,YONE) = BIGK(XTWO,YONE) + ESM(4,2)
      BIGK(XTWO,ZONE) = BIGK(XTWO,ZONE) + ESM(4,3)
      BIGK(XTWO,XTWO) = BIGK(XTWO,XTWO) + ESM(4,4)
      BIGK(XTWO,YTWO) = BIGK(XTWO,YTWO) + ESM(4,5)
      BIGK(XTWO,ZTWO) = BIGK(XTWO,ZTWO) + ESM(4,6)
      BIGK(YTWO,XONE) = BIGK(YTWO,XONE) + ESM(5,1)
      BIGK(YTWO,YONE) = BIGK(YTWO,YONE) + ESM(5,2)
      BIGK(YTWO,ZONE) = BIGK(YTWO,ZONE) + ESM(5,3)
      BIGK(YTWO,XTWO) = BIGK(YTWO,XTWO) + ESM(5,4)
      BIGK(YTWO,YTWO) = BIGK(YTWO,YTWO) + ESM(5,5)
      BIGK(YTWO,ZTWO) = BIGK(YTWO,ZTWO) + ESM(5,6)
      BIGK(ZTWO,XONE) = BIGK(ZTWO,XONE) + ESM(6,1)
      BIGK(ZTWO,YONE) = BIGK(ZTWO,YONE) + ESM(6,2)
      BIGK(ZTWO,ZONE) = BIGK(ZTWO,ZONE) + ESM(6,3)
      BIGK(ZTWO,XTWO) = BIGK(ZTWO,XTWO) + ESM(6,4)
      BIGK(ZTWO,YTWO) = BIGK(ZTWO,YTWO) + ESM(6,5)
      BIGK(ZTWO,ZTWO) = BIGK(ZTWO,ZTWO) + ESM(6,6)
      RETURN
      END

        FORTRAN DIAGNOSTIC RESULTS  FOR    BIGK3D
```

```
      SUBROUTINE SUPPORT (NSUP,FRED)
      COMMON TEE(6,6),ESM(6,6),TEETRAN(6,6),PRODUCT(6,6),
     1BIGK(48,48),MM(48),DEF(48),R(48),M(48),PROP(48,10),LL(48)
     1,ZEE(48)
      INTEGER FRED
      READ 30,(M(I), I=1,NSUP)
   30 FORMAT(12I2)
      PRINT 901
  901 FORMAT(10X,45HTHE DEGREES OF FREEDOM THAT ARE SUPPORTED ARE,/)
      DO 306 I=1,NSUP
  306 PRINT 307, M(I)
  307 FORMAT(10X,6I5)
      DO 303 I=1,NSUP
      DO 303 J=1,FRED
      L = M(I)
  303 BIGK(L,J) = 0.
      DO 304 J=1,NSUP
      DO 304 I=1,FRED
      K = M(J)
  304 BIGK(I,K) = 0.
      DO 403 I=1,NSUP
      KK = M(I)
  403 BIGK(KK,KK)= 1.
      PRINT 720
  720 FORMAT(10X,62HTHE MASTER STIFFNESS MATRIX FOR THE CONSTRAINED STRU
     1CTURE IS  )
      CALL PRIMB(FRED)
      CALL IMAGES(FRED)
      RETURN
      END

          FORTRAN DIAGNOSTIC RESULTS  FOR    SUPPORT
```

```
      SUBROUTINE FRMDEL(NNODES,NELMS)
      COMMON TEE(6,6),ESM(6,6),TEETRAN(6,6),PRODUCT(6,6),
     1BIGK(48,48),MM(48),DEF(48),R(48),M(48),PROP(48,10),LL(48)
     1,ZEE(48)
      INTEGER NNODES,NELMS,GP1,GP2,XONE,XTWO,YONE,YTWO,ZONE,ZTWO,ELEN
      REAL LENGTH,LAMBDA,MU,INERT
      PRINT 95
   95 FORMAT(10X,24HTHE NODE POINT LOADS ARE,/////)
      DO 4 JZ =1,NNODES
      JA = 3.*JZ-2.
      JB = 3.*JZ-1.
      JC = 3.*JZ
    4 PRINT 5,JZ,R(JA),R(JB),R(JC)
    5 FORMAT(10X,4HNODE,I3,5X,9HMOMENT = ,F10.3,5X,4HX = ,F10.3,5X,4HY =
     1 ,F10.3)
      PRINT 96
   96 FORMAT(10X,/////,30HTHE NODE POINT DEFLECTIONS ARE,/////)
      DO 2003 I=1,NNODES
      K = (3*I)-2
      L = (3*I)-1
      JK = (3*I)
      PRINT 2004,I,DEF(K),I,DEF(L),I,DEF(JK)
 2004 FORMAT(10X,5HTHETA,I2,1H=,E20.5,8X,1HX,I2,1H=,E20.5,8X,1HY,I2,1H=,
     1E20.5)
 2003 CONTINUE
      PRINT 29
   29 FORMAT(/////4H    )
      PRINT 30
      PRINT 29
   30 FORMAT(10X,23HTHE ELEMENT FORCES ARE  )
      PRINT 27
      DO 10 ELEN =1,NELMS
      GP1 = PROP(ELEN,1)
      GP2 = PROP(ELEN,2)
      E = PROP(ELEN,3)
      INERT = PROP(ELEN,4)
      A = PROP(ELEN,5)
```

```
      LENGTH = PROP(ELEN,6)
      MU = PROP(ELEN,7)
      LAMBDA = PROP(ELEN,8)
      XONE = (3.*GP1)-2.
      YONE = (3.*GP1)-1.
      ZONE = 3.*GP1
      XTWO = (3.*GP2)-2.
      YTWO = (3.*GP2)-1.
      ZTWO = 3.*GP2
      R(1) = DEF(XONE)
      R(2) = DEF(YONE)
      R(3) = DEF(ZONE)
      R(4) = DEF(XTWO)
      R(5) = DEF(YTWO)
      R(6) = DEF(ZTWO)
      EI = E*INERT
      DO 20 I=1,6
      DO 20 J=1,6
      ESM(I,J) = 0.
   20 TEE(I,J) = 0.
      TEE(5,6) = MU
      TEE(6,5) = -MU
      TEE(2,2) = LAMBDA
      TEE(5,5) = LAMBDA
      TEE(3,3) = LAMBDA
      TEE(1,1) = 1.
      TEE(6,6) = LAMBDA
      TEE(4,4) = 1.
      TEE(3,2) = -MU
      TEE(2,3) = MU
      ESM(1,1) =(4.*EI)/LENGTH
      ESM(1,3) =-(6.*EI)/(LENGTH*LENGTH)
      ESM(1,4) = (2.*EI)/LENGTH
      ESM(1,6) =   (6.*EI)/(LENGTH*LENGTH)
      ESM(3,1) =-(6.*EI)/(LENGTH*LENGTH)
      ESM(3,3) = (12.*EI)/(LENGTH*LENGTH*LENGTH)
      ESM(3,4) =-(6.*EI)/(LENGTH*LENGTH)
```

```
      ESM(3,6) =-(12.*EI)/(LENGTH*LENGTH*LENGTH)
      ESM(4,1) = (2.*EI)/LENGTH
      ESM(4,3) =-(6.*EI)/(LENGTH*LENGTH)
      ESM(4,4) =(4.*EI)/LENGTH
      ESM(4,6) =  (6.*EI)/(LENGTH*LENGTH)
      ESM(6,1) =  (6.*EI)/(LENGTH*LENGTH)
      ESM(6,3) =-(12.*EI)/(LENGTH*LENGTH*LENGTH)
      ESM(6,4) =  (6.*EI)/(LENGTH*LENGTH)
      ESM(6,6) = (12.*EI)/(LENGTH*LENGTH*LENGTH)
      ESM(2,2) =(A*E)/LENGTH
      ESM(2,5) = -(A*E)/LENGTH
      ESM(5,2) = -(A*E)/LENGTH
      ESM(5,5) = (A*E)/LENGTH
      CALL MMULT(TEE,R,PRODUCT,6,6,1)
      CALL MMULT(ESM,PRODUCT,TEE,6,6,1)
   27 FORMAT(10X,7HELEMENT,5X,4HNODE,20X,14HBENDING MOMENT,21X,11HAXIAL
     1FORCE,21X,11HSHEAR FORCE)
      PRINT 28,ELEN,GP1,TEE(1,1),TEE(2,1),TEE(3,1)
      PRINT 28,ELEN,GP2,TEE(4,1), TEE(5,1), TEE(6,1)
   28 FORMAT(13X,I2,8X,I2,14X,E20.5,13X,E20.5,12X,E20.5)
   10 CONTINUE
      RETURN
      END

          FORTRAN DIAGNOSTIC RESULTS  FOR    FRMDEL



      SUBROUTINE TRIANGLE(NNODES,NELMS,FRED)
      COMMON TEE(6,6),ESM(6,6),TEETRAN(6,6),PRODUCT(6,6),
     1BIGK(48,48),MM(48),DEF(48),R(48),M(48),PROP(48,10),LL(48)
     1,ZEE(48)
      INTEGER FRED,GP1,GP2,GP3,ELEN
      FRED = 2.*NNODES
      PRINT 5000
 5000 FORMAT(10X,71HTHE FOLLOWING ANALYSIS IS FOR AN ASSEMBLAGE OF TRIAN
```

```
      1GULAR FLAT PLATES      )
       PRINT 50,NELMS,NNODES,FRED
    50 FORMAT(10X,37HTHE TOTAL NUMBER OF ELEMENTS EQUALS  ,I3,2X,23HTHE N
      1UMBER OF NODES IS ,I3,2X,28HAND THE NUMBER OF UNKNOWNS =,I3,/////)
       KLM = 1
       READ 8,THICK,PR,E
     8 FORMAT(2F10.3,F20.3)
    13 READ 10,ELEN,GP1,GP2,GP3,X1,Y1,X2,Y2,X3,Y3
    10 FORMAT(4I5,6F10.3)
       PRINT 51,ELEN,GP1,GP2,GP3
    51 FORMAT(10X,14HELEMENT NUMBER,I3,2X,20HCONNECTS GRID POINTS,I3,2X,I
      13,2X,3HAND,I3)
       PRINT 40,GP1,X1,Y1
       PRINT 40,GP2,X2,Y2
       PRINT 40,GP3,X3,Y3
    40 FORMAT(10X,29HTHE COORDINATES OF GRID POINT,I3,2X,8HARE  X =,F10.3
      1,5X,8HAND  Y =,F10.3)
       TEMP = 0.
       IF(X1.EQ.X2)1,2
     1 TEMP = X1
       X1 = X3
       X3 = TEMP
       TEMP = Y1
       Y3 = TEMP
       TEMP = 1.
     2 SLOPE = (Y1-Y2)/(X1-X2)
       HEIGHT = (-SLOPE*X3+Y3+SLOPE*X2-Y2)/(SQRTF(1.+SLOPE*SLOPE))
       HEIGHT = SQRTF(HEIGHT*HEIGHT)
       BASE = SQRTF((X1-X2)*(X1-X2)+(Y1-Y2)*(Y1-Y2))
       AREA = .5*BASE*HEIGHT
       PRINT 42,AREA,THICK
    42 FORMAT(10X,34HTHE ELEMENT PROPERTIES ARE AREA = ,F10.3,8X,15HAND T
      1HICKNESS =,F10.3,///)
       IF(TEMP.EQ.1)3,4
     3 TEMP = X1
       X1 = X3
       X3 = TEMP
```

```
      TEMP = Y1
      Y1 = Y3
      Y3 = TEMP
    4 PROP(ELEN,1) = X1
      PROP(ELEN,2) = Y1
      PROP(ELEN,3) = X2
      PROP(ELEN,4) = Y2
      PROP(ELEN,5) = X3
      PROP(ELEN,6) = Y3
      PROP(ELEN,7) = AREA
      PROP(ELEN,8) = GP1
      PROP(ELEN,9) = GP2
      PROP(ELEN,10) = GP3
      DEF(GP1) = X1
      DEF(GP2) = X2
      DEF(GP3) = X3
      R(GP1) = Y1
      R(GP2) = Y2
      R(GP3) = Y3
      DO 9 I=1,6
      DO 9 J=1,6
    9 ESM(I,J) = 0.
      ESM(1,1) = E/(1.-(PR*PR))
      ESM(2,2) = ESM(1,1)
      ESM(1,2) = (E*PR)/(1.-(PR*PR))
      ESM(2,1) = ESM(1,2)
      ESM(3,3) = (E*(1.-PR))/(2.*(1.-(PR*PR)))
      IF(NELMS-10.)800,800,801
  800 PRINT 41,PR,E
   41 FORMAT(10X,45HTHE ELASTICITY MATRIX FOR A POISSONS RATIO OF,2X,F6.
     12,2X,30HAND A MODULUS OF ELASTICITY OF,E15.5,2X,2HIS )
      CALL PRIM(ESM,3,3)
  801 DO 5 I=1,6
      DO 5 J=1,6
      PRODUCT(I,J) = 0.
      TEE(I,J) = 0.
    5 TEETRAN(I,J) = 0.
```

```
      TEE(1,1) = (Y2-Y3)/(2.*AREA)
      TEE(1,3) = (Y3-Y1)/(2.*AREA)
      TEE(1,5) =(Y1-Y2)/(2.*AREA)
      TEE(2,2) = (X3-X2)/(2.*AREA)
      TEE(2,4) = (X1-X3)/(2.*AREA)
      TEE(2,6) = (X2-X1)/(2.*AREA)
      TEE(3,1) = TEE(2,2)
      TEE(3,2) = TEE(1,1)
      TEE(3,3) = TEE(2,4)
      TEE(3,4) = TEE(1,3)
      TEE(3,5) = TEE(2,6)
      TEE(3,6) = TEE(1,5)
      IF(NELMS-10.)802,802,803
  802 PRINT 44
   44 FORMAT(10X,23HTHE STRAIN MATRIX B IS   )
      CALL PRIM(TEE,3,6)
  803 DO 6 I=1,6
      DO 6 J=1,6
    6 TEETRAN(J,I) = TEE(I,J)
      IF(NELMS-10.)804,804,805
  804 PRINT 45
   45 FORMAT(10X,43HTHE TRANSPOSE OF THE STRAIN OR B MATRIX IS   )
      CALL PRIM(TEETRAN,6,3)
  805 CALL MMULT(TEETRAN,ESM,PRODUCT,6,3,3)
      CALL MMULT(PRODUCT,TEE,ESM,6,3,6)
      TAREA = THICK*AREA
      DO 7 I=1,6
      DO 7 J=1,6
    7 ESM(I,J) = TAREA*ESM(I,J)
      IF(NELMS-10.)806,806,807
  806 PRINT 46
   46 FORMAT(10X,32HTHE ELEMENT STIFFNESS MATRIX IS    )
      CALL PRIM(ESM,6,6)
  807 CALL BIGK3D(GP1,GP2,GP3)
      IF(NELMS-10.)808,808,809
  808 PRINT 704
  704 FORMAT(10X,54HTHE MASTER STIFFNESS MATRIX INCLUDING THIS ELEMENT I
```

```
   1S )
    CALL PRIMB(FRED)
    CALL IMAGES(FRED)
809 IF(KLM-NELMS)11,12,12
 11 KLM = KLM+1
    GO TO 13
 12 CONTINUE
    DO 15 I=1,6
    DO 15 J=1,6
 15 ESM(I,J) = 0.
    ESM(1,1) = E/(1.-(PR*PR))
    ESM(2,2) = ESM(1,1)
    ESM(1,2) = (E*PR)/(1.-(PR*PR))
    ESM(2,1) = ESM(1,2)
    ESM(3,3) = (E*(1.-PR))/(2.*(1.-(PR*PR)))
    RETURN
    END
```

FORTRAN DIAGNOSTIC RESULTS FOR TRIANGLE

```
    SUBROUTINE TRISTRS(NELMS,NNODES)
    COMMON TEE(6,6),ESM(6,6),TEETRAN(6,6),PRODUCT(6,6),
   1BIGK(48,48),MM(48),DEF(48),R(48),M(48),PROP(48,10),LL(48)
   1,ZEE(48)
    REAL LENGTH,LAMBDA
    INTEGER XONE,XTWO,YONE,YTWO,ZONE,ZTWO,GP1,GP2,GP3
    PRINT 95
 95 FORMAT(10X,24HTHE NODE POINT LOADS ARE,/////)
    DO 1 JZ = 1,NNODES
    JA = 2.*JZ-1.
    JB = 2.*JZ
  1 PRINT 2,JZ,R(JA),R(JB)
  2 FORMAT(10X,4HNODE,I3,5X,4HX = ,F10.3,5X,4HY = ,F10.3)
    PRINT 96
 96 FORMAT(10X,/////,30HTHE NODE POINT DEFLECTIONS ARE,/////)
```

```
      DO 2005 I=1,NNODES
      K = (2.*I)-1.
      L = 2.*I
 2005 PRINT 2006,I,DEF(K),I,DEF(L)
 2006 FORMAT(10X,1HX,I2,1H=,E20.5,8X,1HY,I2,1H=,E20.5)
      PRINT 20
   20 FORMAT(/////4H    )
      DO 50 KK=1,NELMS
      X1 = PROP(KK,1)
      Y1 = PROP(KK,2)
      X2 = PROP(KK,3)
      Y2 = PROP(KK,4)
      X3 = PROP(KK,5)
      Y3 = PROP(KK,6)
      AREA = PROP(KK,7)
      GP1 = PROP(KK,8)
      GP2 = PROP(KK,9)
      GP3 = PROP(KK,10)
      DO 11 I=1,6
      DO 11 J=1,6
   11 TEE(I,J) = 0.
      TEE(1,1) = (Y2-Y3)/(2.*AREA)
      TEE(1,3) = (Y3-Y1)/(2.*AREA)
      TEE(1,5) =(Y1-Y2)/(2.*AREA)
      TEE(2,2) = (X3-X2)/(2.*AREA)
      TEE(2,4) = (X1-X3)/(2.*AREA)
      TEE(2,6) = (X2-X1)/(2.*AREA)
      TEE(3,1) = TEE(2,2)
      TEE(3,2) = TEE(1,1)
      TEE(3,3) = TEE(2,4)
      TEE(3,4) = TEE(1,3)
      TEE(3,5) = TEE(2,6)
      TEE(3,6) = TEE(1,5)
      XONE = (2.*GP1)-1.
      YONE = 2.*GP1
      ZONE = (2.*GP2)-1.
      XTWO = 2.*GP2
```

```
      YTWO = (2.*GP3)-1
      ZTWO = 2.*GP3
      R(1) = DEF(XONE)
      R(2) = DEF(YONE)
      R(3) = DEF(ZONE)
      R(4) = DEF(XTWO)
      R(5) = DEF(YTWO)
      R(6) = DEF(ZTWO)
      CALL MMULT(ESM,TEE,PRODUCT,3,3,6)
      CALL MMULT(PRODUCT,R,TEE,3,6,1)
      PRINT 100,KK,TEE(1,1),TEE(2,1),TEE(3,1)
  100 FORMAT(10X,39HTHE STATE OF STRESS IN TRIANGLE NUMBER ,I2,2X,2HIS,3
     1X,9HSIGMA X = ,E15.5,3X,9HSIGMA Y = ,E15.5,3X,9HTAU XY = ,E15.5)
      AEOVERL = (((TEE(1,1)-TEE(2,1))*(TEE(1,1)-TEE(2,1)))/2.)+(TEE(3,1)
     1*TEE(3,1))
      LENGTH =((TEE(1,1)+TEE(2,1))/2.)+SQRTF(AEOVERL)
      EOVERL =((TEE(1,1)+TEE(2,1))/2.)-SQRTF(AEOVERL)
      LAMBDA = -(2.*TEE(3,1))/(TEE(1,1)-TEE(2,1))
      LAMBDA = ATANF(LAMBDA)
      LAMBDA = LAMBDA/2.
      LAMBDA = (360.*LAMBDA)/6.28
   50 PRINT 101,KK,LENGTH,EOVERL,LAMBDA
  101 FORMAT(10X,45HTHE PRINCIPLE STRESS INFORMATION FOR TRIANGLE,I2,2X,
     12HIS,2X,9HSIGMA 1 = ,E11.4,2X,9HSIGMA 2 = ,E11.4,2X,7HTHETA = ,E13
     1.5,//)
      RETURN
      END

          FORTRAN DIAGNOSTIC RESULTS  FOR    TRISTRS


      SUBROUTINE TRUSS2D (FRED,NELMS,NNODES)
      COMMON TEE(6,6),ESM(6,6),TEETRAN(6,6),PRODUCT(6,6),
     1BIGK(48,48),MM(48),DEF(48),R(48),M(48),PROP(48,10),LL(48)
     1,ZEE(48)
      REAL LENGTH,LAMBDA,MU
```

```
      INTEGER FRED,ELEN,GP1,GP2
      FRED = 2.*NNODES
      PRINT 5000
 5000 FORMAT(10X,40HTHE FOLLOWING ANALYSIS IS FOR A 2D TRUSS   )
      PRINT 50,NELMS,NNODES,FRED
   50 FORMAT(10X,37HTHE TOTAL NUMBER OF ELEMENTS EQUALS   ,I3,2X,23HTHE N
     1UMBER OF NODES IS ,I3,2X,28HAND THE NUMBER OF UNKNOWNS =,I3,/////)
      KLM = 1
      READ 2,E
    2 FORMAT(4F20.3)
   21 READ 1201,ELEN,GP1,GP2,A,XZ,X1,YZ,Y1
 1201 FORMAT(I4,2I3,5F10.3)
      LENGTH = SQRTF((X1-XZ)*(X1-XZ)+(Y1-YZ)*(Y1-YZ))
      LAMBDA =(X1-XZ)/LENGTH
      MU = (Y1-YZ)/LENGTH
      AEOVERL = (A*E)/LENGTH
      EOVERL =E/LENGTH
      PROP(ELEN,1) = GP1
      PROP(ELEN,2) = GP2
      PROP(ELEN,3) = EOVERL
      PROP(ELEN,4) = LAMBDA
      PROP(ELEN,5) = MU
      DEF(GP1) = XZ
      DEF(GP2) = X1
      R(GP1) = YZ
      R(GP2) = Y1
      PRINT 706,ELEN,GP1,GP2
  706 FORMAT(10X,14HELEMENT NUMBER,I3,2X,20HCONNECTS GRID POINTS,I3,2X,3H
     1HAND,I3 )
      PRINT 707,GP1,XZ,YZ
      PRINT 707,GP2,X1,Y1
  707 FORMAT(10X,29HTHE COORDINATES OF GRID POINT,I3,2X,8HARE  X =,F10.3H
     1,5X,8HAND  Y =,F10.3)
      PRINT 708,LENGTH,AEOVERL,A
  708 FORMAT(10X,37HTHE ELEMENT PROPERTIES ARE  LENGTH = ,F10.3,8X,11HAN
     1D AE/L = ,F20.3,8X,8HAND A = ,F10.3,/////)
      DO 15 I=1,4
```

```
      DO 15 J=1,4
   15 TEE(I,J) = 0.
      DO 16 I=1,4
   16 TEE(I,I) = LAMBDA
      TEE(1,2) = MU
      TEE(2,1) = -MU
      TEE(3,4) = MU
      TEE(4,3) = -MU
      IF(NELMS-15.)723,723,724
  723 PRINT 701
  701 FORMAT(10X,39HTHE ELEMENTAL TRANSFORMATION MATRIX IS )
      CALL PRIM(TEE,4,4)
  724 DO 17 I=1,4
      DO 17 J=1,4
   17 ESM(I,J) = 0.
      ESM(1,1)= AEOVERL
      ESM(1,3)= -AEOVERL
      ESM(3,1) = -AEOVERL
      ESM(3,3) = AEOVERL
      IF(NELMS-15.)725,725,726
  725 PRINT 700
  700 FORMAT(10X,49HTHE ELEMENT STIFFNESS MATRIX FOR THIS ELEMENT IS  )
      CALL PRIM(ESM,4,4)
  726 DO 20 I=1,4
      DO 20 J=1,4
   20 TEETRAN(J,I) = TEE(I,J)
      IF(NELMS-15.)727,727,728
  727 PRINT 703
  703 FORMAT(10X,46HTHE TRANSPOSE OF THE TRANSFORMATION MATRIX IS )
      CALL PRIM(TEETRAN,4,4)
  728 DO 74 J=1,4
      DO 74 I=1,4
   74 PRODUCT(I,J) = 0.0
      CALL MMULT(TEETRAN,ESM,PRODUCT,4,4,4)
      CALL MMULT(PRODUCT,TEE,ESM,4,4,4)
      IF(NELMS-15.)729,729,730
  729 PRINT 702
```

```
702 FORMAT(10X,58HTHE ELEMENT STIFFNESS MATRIX IN STRUCTURAL COORDINAT
   1ES IS )
    CALL PRIM(ESM,4,4)
730 CALL BIGK2D(GP1,GP2)
    IF(NELMS-15.)721,721,722
721 PRINT 704
704 FORMAT(10X,54HTHE MASTER STIFFNESS MATRIX INCLUDING THIS ELEMENT I
   1S )
    CALL PRIMB(FRED)
    CALL IMAGES(FRED)
722 IF(KLM-NELMS)28,19,19
 28 KLM = KLM + 1
    GO TO 21
 19 CONTINUE
    RETURN
    END
```

FORTRAN DIAGNOSTIC RESULTS FOR TRUSS2D

```
     SUBROUTINE TRUSS3D (NNODES,NELMS,FRED)
     COMMON TEE(6,6),ESM(6,6),TEETRAN(6,6),PRODUCT(6,6),
    1BIGK(48,48),MM(48),DEF(48),R(48),M(48),PROP(48,10),LL(48)
    1,ZEE(48)
     REAL LENGTH,LAMBDA,MU,NU
     INTEGER FRED,ELEN,GP1,GP2
     FRED = 3.*NNODES
     PRINT 5000
5000 FORMAT(10X,≠ THE FOLLOWING ANALYSIS IS FOR A 3D TRUSS≠)
     PRINT 50,NELMS,NNODES,FRED
  50 FORMAT(1X,≠THE TOTAL NUMBER OF ELEMENTS EQUALS≠,I3,2X,≠
    1NUMBER OF NODES IS≠,I3,2X,≠AND THE NUMBER OF UNKNOWNS≠,I3//////)
     KLM=1
     READ 2,E
   2 FORMAT(4F20.3)
1121 READ 201,ELEN,GP1,GP2,A,XZ,X1,YZ,Y1,ZZ,Z1
```

```
201 FORMAT(I4,2I3,7F10.3)
    LENGTH=
   1  SQRTF((X1-XZ)*(X1-XZ)+(Y1-YZ)*(Y1-YZ)+(Z1-ZZ)*(Z1-ZZ))
    LAMBDA =(X1-XZ)/LENGTH
    MU=(Y1-YZ)/LENGTH
    NU = (Z1-ZZ)/LENGTH
    AEOVERL = (A*E)/LENGTH
    EOVERL =E/LENGTH
    PROP(ELEN,1) = GP1
    PROP(ELEN,2) = GP2
    PROP(ELEN,3) = EOVERL
    PROP(ELEN,4) = LAMBDA
    PROP(ELEN,5) = MU
    PROP(ELEN,6) = NU
    DEF(GP1) = XZ
    DEF(GP2) = X1
    R(GP1) = YZ
    R(GP2) = Y1
    ZEE(GP1) = ZZ
    ZEE(GP2) = Z1
    PRINT 706,ELEN,GP1,GP2
706 FORMAT(10X,14HELEMENT NUMBER,I3,2X,20HCONNECTS GRID POINTS,I3,2X,3H
   1HAND,I3 )
    PRINT 1707,GP1,XZ,YZ,ZZ
    PRINT 1707,GP2,X1,Y1,Z1
1707 FORMAT(10X,29HTHE COORDINATES OF GRID POINT,I3,2X,8HARE  X =,F10.3
   1,5X,8HAND  Y =,F10.3,5X,8HAND  Z =,F10.3)
    PRINT 708,LENGTH,AEOVERL,A
708 FORMAT(10X,37HTHE ELEMENT PROPERTIES ARE  LENGTH = ,F10.3,8X,11HAN
   1D AE/L = ,F20.3,8X,8HAND A = ,F10.3,/////)
    DO 1015 I=1,6
    DO 1015 J=1,6
1015 TEE(I,J) = 0.
    TEE(1,1) = LAMBDA
    TEE(1,2) = MU
    TEE(1,3) = NU
    TEE(4,4) = LAMBDA
    TEE(4,5) = MU
```

```
      TEE(4,6) = NU
      IF(NELMS-10.)1723,1723,1724
 1723 PRINT 701
  701 FORMAT(10X,39HTHE ELEMENTAL TRANSFORMATION MATRIX IS )
      CALL PRIM(TEE,6,6)
 1724 DO 1017 I=1,6
      DO 1017 J=1,6
 1017 ESM(I,J) = 0.
      ESM(1,1) = AEOVERL
      ESM(1,4) = -AEOVERL
      ESM(4,1) = -AEOVERL
      ESM(4,4) = AEOVERL
      IF(NELMS-10.)1725,1725,1726
 1725 PRINT 700
  700 FORMAT(10X,49HTHE ELEMENT STIFFNESS MATRIX FOR THIS ELEMENT IS  )
      CALL PRIM(ESM,6,6)
 1726 DO 1020 I=1,6
      DO 1020 J=1,6
 1020 TEETRAN(J,I) = TEE(I,J)
      IF(NELMS-10.)1727,1727,1728
 1727 PRINT 703
  703 FORMAT(10X,46HTHE TRANSPOSE OF THE TRANSFORMATION MATRIX IS )
      CALL PRIM(TEETRAN,6,6)
 1728 DO 1074 I=1,6
      DO 1074 J=1,6
 1074 PRODUCT(I,J) = 0.
      CALL MMULT(TEETRAN,ESM,PRODUCT,6,6,6)
      CALL MMULT(PRODUCT,TEE,ESM,6,6,6)
      IF(NELMS-10.)1729,1729,1730
 1729 PRINT 702
  702 FORMAT(10X,58HTHE ELEMENT STIFFNESS MATRIX IN STRUCTURAL COORDINAT
     1ES IS )
      CALL PRIM(ESM,6,6)
 1730 CALL BIGK3D(GP1,GP2,0.)
      IF(NELMS-10.)1721,1721,1722
 1721 PRINT 704
  704 FORMAT(10X,54HTHE MASTER STIFFNESS MATRIX INCLUDING THIS ELEMENT I
```

```
     1S )
      CALL PRIMB(FRED)
      CALL IMAGES(FRED)
 1722 IF(KLM-NELMS)1128,19,19
 1128 KLM = KLM + 1
      GO TO 1121
   19 CONTINUE
      RETURN
      END

              FORTRAN DIAGNOSTIC RESULTS  FOR    TRUSS3D


      SUBROUTINE TRUSSTRS (KONST,NNODES,NELMS)
      COMMON TEE(6,6),ESM(6,6),TEETRAN(6,6),PRODUCT(6,6),
     1BIGK(48,48),MM(48),DEF(48),R(48),M(48),PROP(48,10),LL(48)
     1,ZEE(48)
      REAL LAMBDA,MU,NU
      INTEGER GP1,GP2,XONE,XTWO,YONE,YTWO,ZONE,ZTWO
      PRINT 95
   95 FORMAT(10X,24HTHE NODE POINT LOADS ARE,/////)
      IF(KONST.EQ.3)2001,2002
 2001 DO 10 JR=1,NNODES
      JS = 3.*JR-2.
      JT = 3.*JR-1.
      JU = 3.*JR
   10 PRINT 11,JR,R(JS),R(JT),R(JU)
   11 FORMAT(10X,4HNODE,I3,5X,4HX = ,F10.3,5X,4HY = ,F10.3,5X,4HZ = ,F10
     1.3)
      PRINT 96
   96 FORMAT(10X,/////,30HTHE NODE POINT DEFLECTIONS ARE,/////)
      DO 2003 I=1,NNODES
      K = (3*I)-2
      L = (3*I)-1
      JK = 3.*I
      PRINT 2004,I,DEF(K),I,DEF(L),I,DEF(JK)
```

```
2004 FORMAT(10X,1HX,I2,1H=,E20.5,8X,1HY,I2,1H=,E20.5,8X,1HZ,I2,1H=,E20.
    15)
2003 CONTINUE
     PRINT 20
  20 FORMAT(/////4H     )
     DO 1900 I=1,NELMS
     GP1 = PROP(I,1)
     GP2 = PROP(I,2)
     EOVERL = PROP(I,3)
     LAMBDA = PROP(I,4)
     MU = PROP(I,5)
     NU = PROP(I,6)
     XONE = (3.*GP1)-2.
     YONE = (3.*GP1)-1.
     ZONE = 3.*GP1
     XTWO = (3.*GP2)-2.
     YTWO = (3.*GP2)-1.
     ZTWO = (3.*GP2)
     XXONE = DEF(XONE)
     XXTWO = DEF(XTWO)
     YYONE = DEF(YONE)
     YYTWO = DEF(YTWO)
     ZZONE = DEF(ZONE)
     ZZTWO = DEF(ZTWO)
     STRESS = EOVERL*((XXTWO-XXONE)*LAMBDA+(YYTWO-YYONE)*MU+(ZZTWO-ZZON
    1E)*NU)
     PRINT 800,GP1,GP2,STRESS
1900 CONTINUE
     GO TO 2007
2002 DO 1 JZ = 1,NNODES
     JA = 2.*JZ-1.
     JB = 2.*JZ
   1 PRINT 2,JZ,R(JA),R(JB)
   2 FORMAT(10X,4HNODE,I3,5X,4HX = ,F10.3,5X,4HY = ,F10.3)
     PRINT 96
     DO 2005 I=1,NNODES
     K = (2.*I)-1.
     L = (2.*I)
```

```
2006 FORMAT(10X,1HX,I2,1H=,E20.5,8X,1HY,I2,1H=,E20.5)                    HL 00877
2005 PRINT 2006,I,DEF(K),I,DEF(L)                                        HL 00878
     PRINT 20                                                            HL 00879
     DO 900 I=1,NELMS                                                    HL 00880
     GP1 = PROP(I,1)                                                     HL 00881
     GP2 = PROP(I,2)                                                     HL 00882
     EOVERL = PROP(I,3)                                                  HL 00883
     LAMBDA = PROP(I,4)                                                  HL 00884
     MU = PROP(I,5)                                                      HL 00885
     XONE = (2.*GP1)-1.                                                  HL 00886
     XTWO = (2.*GP2)-1.                                                  HL 00887
     YONE = 2.*GP1                                                       HL 00888
     YTWO = 2.*GP2                                                       HL 00889
     YYTWO = DEF(YTWO)                                                   HL 00890
     YYONE = DEF(YONE)                                                   HL 00891
     XXTWO = DEF(XTWO)                                                   HL 00892
     XXONE = DEF(XONE)                                                   HL 00893
     STRESS = EOVERL*((XXTWO-XXONE)*LAMBDA+(YYTWO-YYONE)*MU)             HL 00894
     PRINT 800,GP1,GP2,STRESS                                            HL 00895
 800 FORMAT(10X,36HTHE STRESS IN RODS CONNECTING NODES ,I3,2X,3HAND,I3,HL 00896
    12X,2HIS,F15.4)                                                      HL 00897
 900 CONTINUE                                                            HL 00898
2007 CONTINUE                                                            HL 00899
     RETURN                                                              HL 00900
     END                                                                 HL 00901

             FORTRAN DIAGNOSTIC RESULTS  FOR    TRUSSTRS
```

```
      SUBROUTINE MMULT(A,B,R,M,N,L)                                     HL 00903
      DIMENSION A(6,6), B(6,6), R(6,6)                                  HL 00904
      DO 23 I=1,M                                                       HL 00905
      DO 23 J=1,L                                                       HL 00906
      R(I,J) = 0.                                                       HL 00907
      DO 23 K =1,N                                                      HL 00908
   23 R(I,J) = R(I,J) +A(I,K)*B(K,J)                                    HL 00909
      RETURN                                                            HL 00910
      END                                                               HL 00911

             FORTRAN DIAGNOSTIC RESULTS  FOR    MMULT


      SUBROUTINE MMULTB(FRED)                                           HL 00913
      COMMON TEE(6,6),ESM(6,6),TEETRAN(6,6),PRODUCT(6,6),               HL 00914
     1BIGK(48,48),MM(48),DEF(48),R(48),M(48),PROP(48,10),LL(48)         HL 00915
     1,ZEE(48)                                                          HL 00916
      INTEGER FRED                                                      HL 00917
      DO 23 I=1,FRED                                                    HL 00918
      DO 23 J=1,1                                                       HL 00919
      DEF(I) = 0.                                                       HL 00920
      DO 23 K=1,FRED                                                    HL 00921
   23 DEF(I) = DEF(I) + BIGK(I,K) * R(K)                                HL 00922
      RETURN                                                            HL 00923
      END                                                               HL 00924

             FORTRAN DIAGNOSTIC RESULTS  FOR    MMULTB
```

```
      SUBROUTINE PRIMB(FRED)                                            HL 00925
      COMMON TEE(6,6),ESM(6,6),TEETRAN(6,6),PRODUCT(6,6),               HL 00926
     1BIGK(48,48)                                                       HL 00927
     1,ZEE(48)                                                          HL 00928
      DIMENSION JP(6)                                                   HL 00929
      INTEGER FRED                                                      HL 00930
      LM=20                                                             HL 00931
      LN = 6                                                            HL 00932
      J1=1                                                              HL 00933
      N = FRED                                                          HL 00934
C   NEW VERTICAL BLOCK                                                  HL 00935
   11 I1=1                                                              HL 00936
      M = FRED                                                          HL 00937
C   SET N LIMITS                                                        HL 00938
      IF(N-LN) 13,13,14                                                 HL 00939
   13 NP=N                                                              HL 00940
      GO TO 15                                                          HL 00941
   14 NP=LN                                                             HL 00942
C   SET M LIMITS                                                        HL 00943
   15 IF (M-LM) 16,16,17                                                HL 00944
   16 MP=M                                                              HL 00945
      GO TO 18                                                          HL 00946
   17 MP=LM                                                             HL 00947
C   SET HEADING                                                         HL 00948
   18 DO 19 L=1,NP                                                      HL 00949
   19 JP(L)=J1+L-1                                                      HL 00950
      PRINT 20,(JP(L), L=1,NP)                                          HL 00951
   20 FORMAT (///4H     ,6I20/)                                         HL 00952
C   PRINT BLOCK                                                         HL 00953
      I2=I1+MP-1                                                        HL 00954
      J2=J1+NP-1                                                        HL 00955
      DO 21 L=I1,I2                                                     HL 00956
   21 PRINT 22,L,(BIGK(L,KJ), KJ=J1,J2)                                 HL 00957
   22 FORMAT(I10,6E20.7)                                                HL 00958
      I1=I1+MP                                                          HL 00959
      M=M-MP                                                            HL 00960
      IF (M) 24,24,15                                                   HL 00961
```

```
 24 N=N-NP
    IF (N) 25,25,26
 26 J1=J1+NP
    GO TO 11
 25 PRINT 60
 60 FORMAT(/////4H     )
    RETURN
    END
```

FORTRAN DIAGNOSTIC RESULTS FOR PRIMB

```
      SUBROUTINE MINV(A,N,D,L,M)
      REAL BIGA,HOLD
      INTEGER I,IJ,IK,IZ,J,JI,JK,JP,JQ,JR,K,KI,KJ,KK,NK
      DIMENSION A(1),L(1),M(1)
C        SEARCH FOR LARGEST ELEMENT
      D=1.0
      NK=-N
      DO 80 K=1,N
      NK=NK+N
      L(K)=K
      M(K)=K
      KK=NK+K
      BIGA=A(KK)
      DO 20 J=K,N
      IZ=N*(J-1)
      DO 20 I=K,N
      IJ=IZ+I
      IF(ABS(BIGA)-ABS(A(IJ)))15,20,20
   15 BIGA=A(IJ)
      L(K)=I
      M(K)=J
   20 CONTINUE
C        INTERCHANGE ROWS
      J=L(K)
```

```
      IF(J-K) 35,35,25                                                   HL 00995
   25 KI=K-N                                                             HL 00996
      DO 30 I=1,N                                                        HL 00997
      KI=KI+N                                                            HL 00998
      HOLD=-A(KI)                                                        HL 00999
      JI=KI-K+J                                                          HL 01000
      A(KI)=A(JI)                                                        HL 01001
   30 A(JI) =HOLD                                                        HL 01002
C        INTERCHANGE COLUMNS                                             HL 01003
   35 I=M(K)                                                             HL 01004
      IF(I-K) 45,45,38                                                   HL 01005
   38 JP=N*(I-1)                                                         HL 01006
      DO 40 J=1,N                                                        HL 01007
      JK=NK+J                                                            HL 01008
      JI=JP+J                                                            HL 01009
      HOLD=-A(JK)                                                        HL 01010
      A(JK)=A(JI)                                                        HL 01011
   40 A(JI) =HOLD                                                        HL 01012
C        DIVIDE COLUMN BY MINUS PIVOT (VALUE OF PIVOT ELEMENT IS         HL 01013
C        CONTAINED IN BIGA)                                              HL 01014
   45 IF (ABS(BIGA)-1.E-20)46,46,48                                      HL 01015
   46 D=0.0                                                              HL 01016
      RETURN                                                             HL 01017
   48 DO 55 I=1,N                                                        HL 01018
      IF(I-K) 50,55,50                                                   HL 01019
   50 IK=NK+I                                                            HL 01020
      A(IK)=A(IK)/(-BIGA)                                                HL 01021
   55 CONTINUE                                                           HL 01022
C        REDUCE MATRIX                                                   HL 01023
      DO 65 I=1,N                                                        HL 01024
      IK=NK+I                                                            HL 01025
      HOLD=A(IK)                                                         HL 01026
      IJ=I-N                                                             HL 01027
      DO 65 J=1,N                                                        HL 01028
      IJ=IJ+N                                                            HL 01029
      IF(I-K) 60,65,60                                                   HL 01030
   60 IF(J-K) 62,65,62                                                   HL 01031
```

```
   62 KJ=IJ-I+K
      A(IJ)=HOLD*A(KJ)+A(IJ)
   65 CONTINUE
C        DIVIDE ROW BY PIVOT
      KJ=K-N
      DO 75 J=1,N
      KJ=KJ+N
      IF(J-K) 70,75,70
   70 A(KJ)=A(KJ)/BIGA
   75 CONTINUE
C        PRODUCT OF PIVOTS
      D=D*BIGA
C        REPLACE PIVOT BY RECIPROCAL
      A(KK)=1.0/BIGA
   80 CONTINUE
C        FINAL ROW AND COLUMN INTERCHANGE
      K=N
  100 K=(K-1)
      IF(K) 150,150,105
  105 I=L(K)
      IF(I-K) 120,120,108
  108 JQ=N*(K-1)
      JR=N*(I-1)
      DO 110 J=1,N
      JK=JQ+J
      HOLD=A(JK)
      JI=JR+J
      A(JK)=-A(JI)
  110 A(JI) =HOLD
  120 J=M(K)
      IF(J-K) 100,100,125
  125 KI=K-N
      DO 130 I=1,N
      KI=KI+N
      HOLD=A(KI)
      JI=KI-K+J
      A(KI)=-A(JI)
```

```
  130 A(JI) =HOLD
      GO TO 100
  150 RETURN
      END

            FORTRAN DIAGNOSTIC RESULTS  FOR    MINV


      SUBROUTINE PRIM (AA,MA,NA)
      DIMENSION AA(6,6)
CPRINT-PRINT MATRIX-LIMIT ON PAGE≠,20,K)
      DIMENSION JP(5)
      LM=20
      LN=5
      J1=1
      N=NA
C NEW VIERITCAL BLOCK
   11 I1=1
      M=MA
C SET N LIMITS
      IF (N-LN) 13,13,14
   13 NP=N
CSET M LIMITS
      GO TO 15
   14 NP=LN
   15 IF (M-LM) 16,16,17
   16 MP=M
      GO TO 18
   17 MP=LM
C SET HEADING
   18 DO 19 L=1,NP
   19 JP(L)=J1+L-1
      PRINT 20,(JP(L),L=1,NP)
   20 FORMAT (///4H    ,5I20/)
CPRINT BLOCK
      I2=I1+MP-1
```

```
      J2=J1+NP-1                                                  HL 01102
      DO 23 L=I1,I2                                               HL 01103
      PRINT 22,L,(AA(L,KJ),  KJ=J1,J2)                            HL 01104
   23 CONTINUE                                                    HL 01105
   22 FORMAT (I10,5E20.7)                                         HL 01106
      M=M-MP                                                      HL 01107
      IF (M) 24,24,15                                             HL 01108
   24 N=N-NP                                                      HL 01109
      IF (N) 25,25,26                                             HL 01110
   26 J1=J1+NP                                                    HL 01111
      GO TO 11                                                    HL 01112
   25 PRINT 60                                                    HL 01113
   60 FORMAT (/////4H    )                                        HL 01114
      RETURN                                                      HL 01115
      END                                                         HL 01116

             FORTRAN DIAGNOSTIC RESULTS  FOR    PRIM


      SUBROUTINE GMPRINT6 (AA,MA,NA)                              HL 01118
      DIMENSION AA(6,6)                                           HL 01119
      DIMENSION JP(5)                                             HL 01120
      LM=20                                                       HL 01121
      LN=5                                                        HL 01122
      J1=1                                                        HL 01123
      N=NA                                                        HL 01124
C NEW VIERITCAL BLOCK                                              HL 01125
   11 I1=1                                                        HL 01126
      M=MA                                                        HL 01127
C SET N LIMITS                                                    HL 01128
      IF (N-LN) 13,13,14                                          HL 01129
   13 NP=N                                                        HL 01130
CSET M LIMITS                                                     HL 01131
      GO TO 15                                                    HL 01132
   14 NP=LN                                                       HL 01133
   15 IF (M-LM) 16,16,17                                          HL 01134
```

```
   16 MP=M                                                              HL 01135
      GO TO 18                                                          HL 01136
   17 MP=LM                                                             HL 01137
C  SET HEADING                                                          HL 01138
   18 DO 19 L=1,NP                                                      HL 01139
   19 JP(L)=J1+L-1                                                      HL 01140
      PRINT 20,(JP(L),L=1,NP)                                           HL 01141
   20 FORMAT (///4H    ,5I20/)                                          HL 01142
CPRINT BLOCK                                                            HL 01143
      I2=I1+MP-1                                                        HL 01144
      J2=J1+NP-1                                                        HL 01145
      DO 23 L=I1,I2                                                     HL 01146
   21 PRINT 22,L,(AA(L,KJ), KJ=J1,J2)                                   HL 01147
   23 CONTINUE                                                          HL 01148
   22 FORMAT (I10,5E20.7)                                               HL 01149
      M=M-MP                                                            HL 01150
      IF (M) 24,24,15                                                   HL 01151
   24 N=N-NP                                                            HL 01152
      IF (N) 25,25,26                                                   HL 01153
   26 J1=J1+NP                                                          HL 01154
      GO TO 11                                                          HL 01155
   25 PRINT 60                                                          HL 01156
   60 FORMAT (/////4H    )                                              HL 01157
      RETURN                                                            HL 01158
      END                                                               HL 01159

          FORTRAN DIAGNOSTIC RESULTS  FOR    GMPRINT6
```

```
      SUBROUTINE IMAGES(FRED)
      COMMON TEE(6,6),ESM(6,6),TEETRAN(6,6),PRODUCT(6,6),
     *BIGK(48,48),MM(48),DEF(48),R(48),M(48),PROP(48,10),LL(48)
     *,ZEE(48)
      DIMENSION IMAGE(48,48),HEAD(48)
      INTEGER FRED,HEAD
      PRINT 3
    3 FORMAT (10X,≠THE IMAGE OF THE MASTER STIFFNESS MATRIX IS≠,////)
      DO 4 I=1,FRED,5
    4 HEAD(I)=I
      PRINT 5,(HEAD(I),I=1,FRED,5)
    5 FORMAT (5X,12(I2,8X),/)
      STAR=2H *
      ONE=2H 1
      DO 1 I=1,FRED
      DO 1 J=1,FRED
      IMAGE(I,J)=STAR
      IF (BIGK(I,J) .NE. 0.) IMAGE(I,J)=ONE
    1 CONTINUE
      DO 6 J=1,FRED
      PRINT 7,(IMAGE(J,K),K=1,FRED)
    7 FORMAT (5X,60A2)
      DO 8 L=1,FRED,5
      IF (J .EQ. L) PRINT 10,L
    8 CONTINUE
   10 FORMAT (2H+ ,I2)
    6 CONTINUE
      PRINT 11
   11 FORMAT (5X,////)
      RETURN
      END
```

FORTRAN DIAGNOSTIC RESULTS FOR IMAGES

Problems

Problem 1. A uniaxial rod (bar) element is used in a 3-d structure. It is connected to nodes 16 and 21. Obtain the following:

a. The element stiffness matrix in element coordinates.
b. The element [T] matrix.
c. The expression for (do not multiply out) the element stiffness matrix in master coordinates.

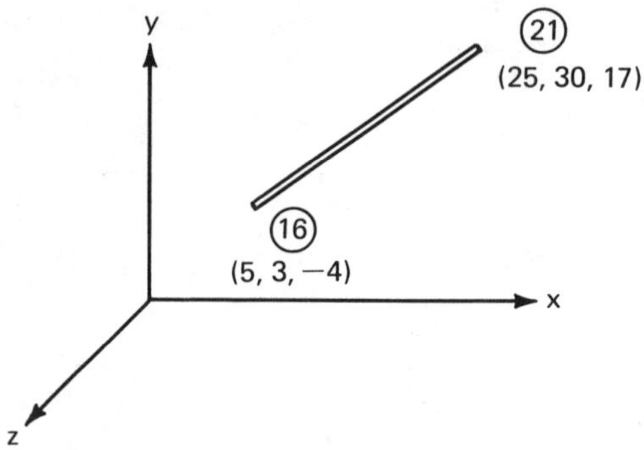

Problem 2. In problem 1, assume you had done part c. The form of $\bar{K}$ would be

$$\bar{K} = \begin{bmatrix} \bar{K}_{11} & \bar{K}_{12} & \cdots & \bar{K}_{16} \\ \vdots & & & \\ \bar{K}_{61} & \cdots & \cdots & \bar{K}_{66} \end{bmatrix}_{6 \times 6}$$

Show which of the elements of the master stiffness matrix are to be incremented by each of the terms of $\bar{K}$; i.e. $\bar{K}_{11}$ increments $K_{master??}$, for all 36 elements.

Problem 3. A 2-d frame is built up of beam elements. The nodes are numbered as shown. Use the notation and sign convention of this chapter and those shown in the figure:

a. Discuss which elements of the assembled master stiffness matrix are affected by the supporting of nodes 1 & 2. Describe how you would alter the master stiffness matrix.

b. Show the complete (24 x 1) load vector for this case. P = 10,000 lb; M = 10,000 lb-in.

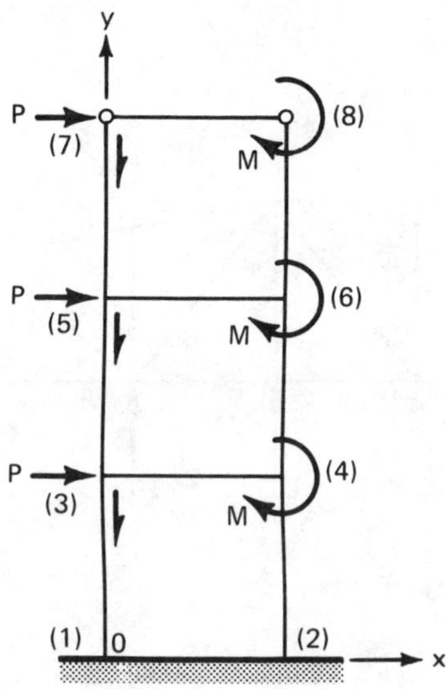

Problem 4. In a Fortran program, we are doing matrix operations. A skeleton outline is:

```
PROGRAM 408
DIMENSION -------, A(4,4),B(4,4),C(4,4)
  __________
  __________
  __________
CALL MTXM (-----------)
  __________
  __________
```

A SUBROUTINE MTXM is needed which will take [A] and [B] and their size (4 x 4) from the main program, perform the calculation $[A]^T \times [B]$ and return it to the main program in [C].

Do the following:

a. Write the <u>complete</u> subroutine.

b. Complete the CALL statement in accordance with your subroutine.

Problem 5. A wing rib is approximated by the following truss: (NTS)

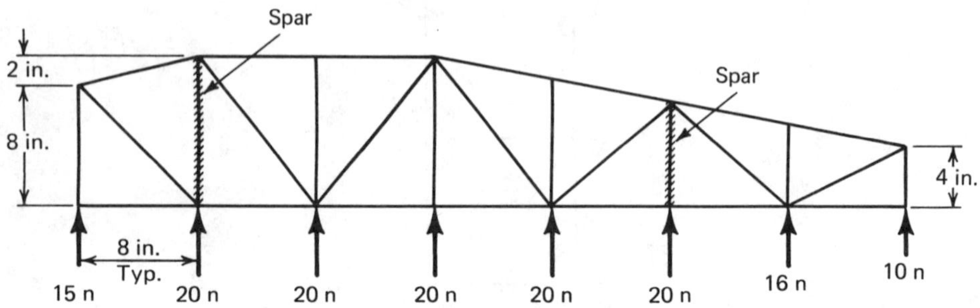

Material - 1/4" x 1/4" Douglas Fir; $E = 2.0 \times 10^6$ psi; spars = 1/4 x 3/4 fir; n = load factor. Loadings are given in pounds times load factor.

a. Determine most highly stressed elements for n = 1, and rearrange truss to decrease the stress in these members. (Keep truss internally statically determinate.)

b. Calculate maximum load factor allowable for the truss in the given configuration if $\sigma_{yield} = 5{,}000$ psi.

c. Repeat (a) for loads applied downward on top of elements (n = -1).

d. Repeat (b) for loads applied downward on top of rib elements.

Problem 6. A 2-d triangular element with in-plane forces is used in a 3-d structure as shown. Calculate its T matrix.

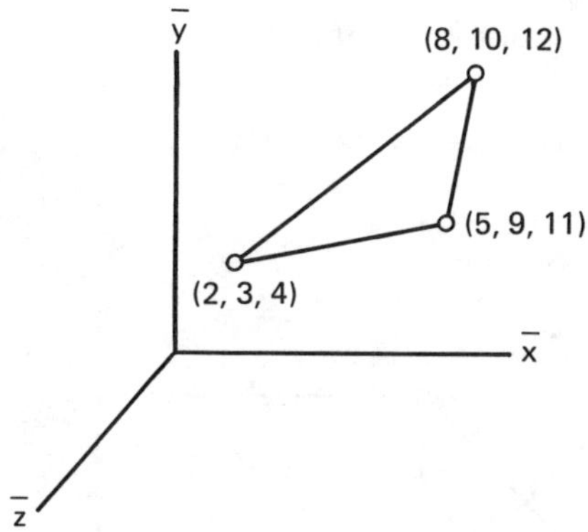

Problem 7. A pin-jointed bar element used in a 3-d structure has end coordinates (2, 4, 6) and (8, 10, 12). It has known values of E and A. Set up its $\left[K_{M_e}\right]$ matrix in structure coordinates. Have only numbers and E and A in your answer. Do not multiply any matrices.

Problem 8.

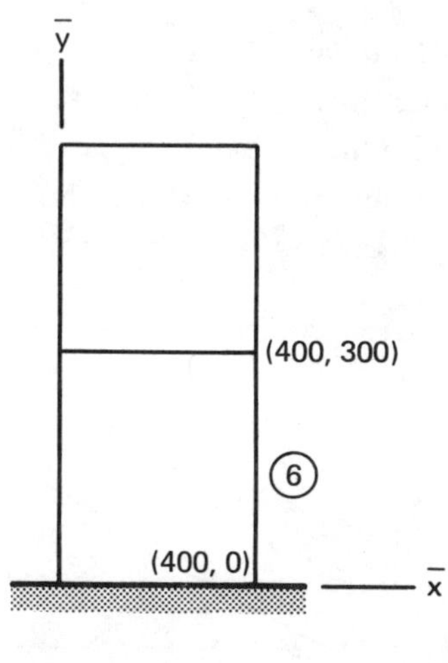

The beam element number 6 is used in a 2-d frame.

a. Calculate its T matrix.

b. Calculate its $\left[K_e\right]$ matrix if $I_Z = 15$; $A = 2$; $E = 30 \times 10^6$.

All units consistent. Neglect shear.

Problem 9. Element number 8 has area of 0.5 in^2 and modulus of elasticity, E, 10.5 x 10^6 psi. Write its stiffness matrix $[K_e]$ in local (element) coordinates.

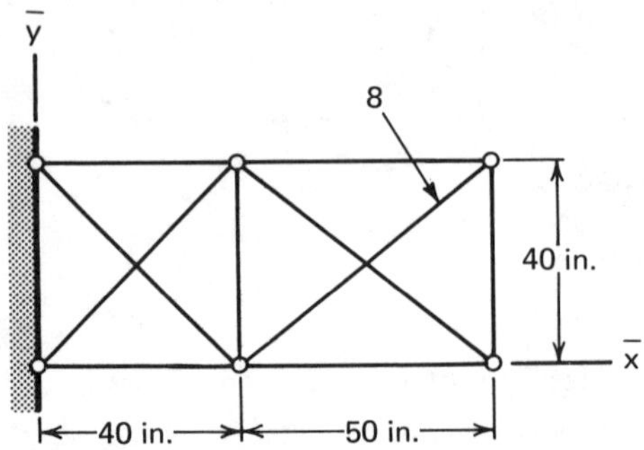

Problem 10. Write the transformation matrix, [T], for element number 8.

Problem 11. Show how, in complete numerical set up, you would get $[K_{M_e}]$, the element stiffness matrix in master (structure) coordinates. Have matrices only in numbers but do not multiply out.

Problem 12. For the element shown, if A = 0.25 in^2 and E = 30 x 10^6 psi then find the element $[K_e]$. Label element coordinates. Coordinates are in inches.

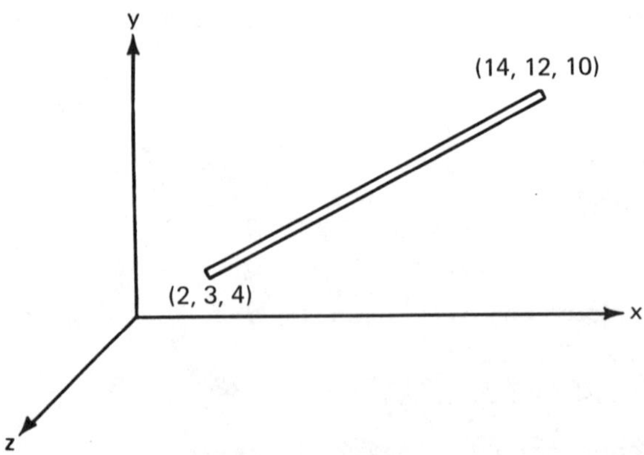

Problem 13. For the element shown, if t = 0.10 in and E = 10 x 10^6 psi, label the element axes and find the element $\left[K_e\right]$.

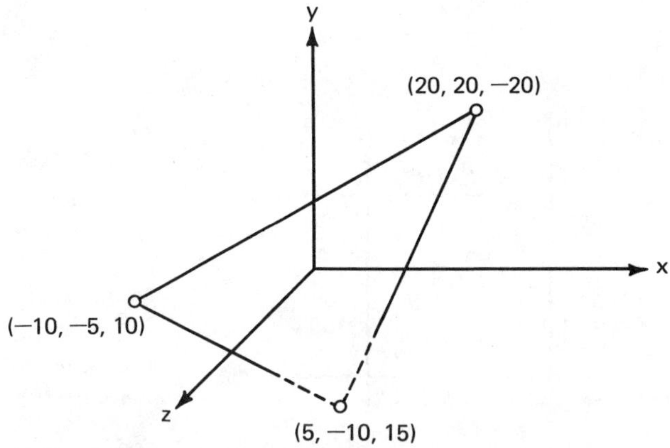

Problem 14. For the truss shown:

a. Solve for forces in this truss by statics.

b. Obtain a numerical, step-by-step solution for the displacements of the loaded node, use the stiffness method and explain each step. (One or two sentences.)

c. Run this problem on Hilsen (or equivalent) and compare with 1 and 2.

A = 0.1 in^2
E = 30 x 10^6 psi
L = 24 in
P = 400 lb

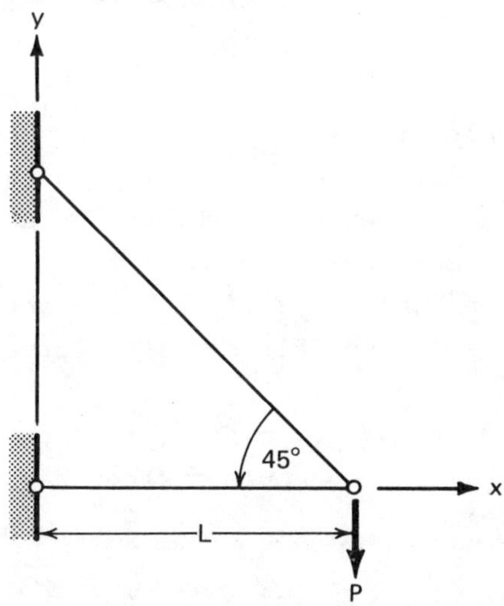

Problem 15. Use Hilsen to solve for structural displacements and elements of truss shown:

a. Loads are in pounds

b. Dimensions are in inches

c. $AE = 10^7 \frac{lb}{in^2} \cdot in^2$

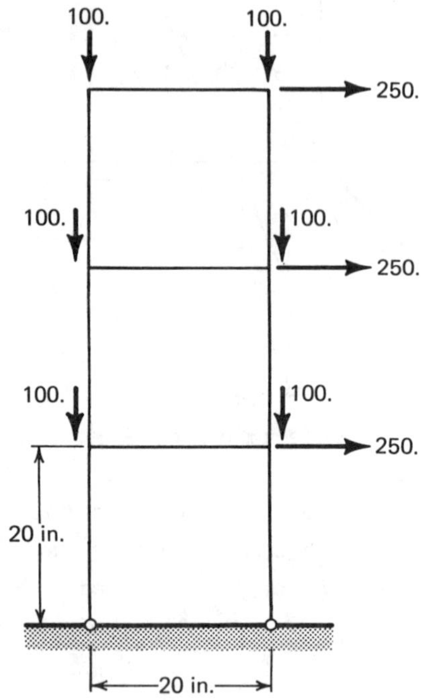

Problem 16. The guyed tower shown is a typical structure. The lateral loads are carried as a redundant truss, with additional supports provided by the guy wires. This is a two-dimensional view and since the wires are only good in tension, only half of them are shown. Select areas for members and guy wires, using actual tables available in civil or mechanical handbooks. Study the effects of varying the number of guy wires and/or their connection points. Discuss how you would further consider or improve the problem.

$A = 1\ in^2 \qquad E = 30 \times 10^6$ psi

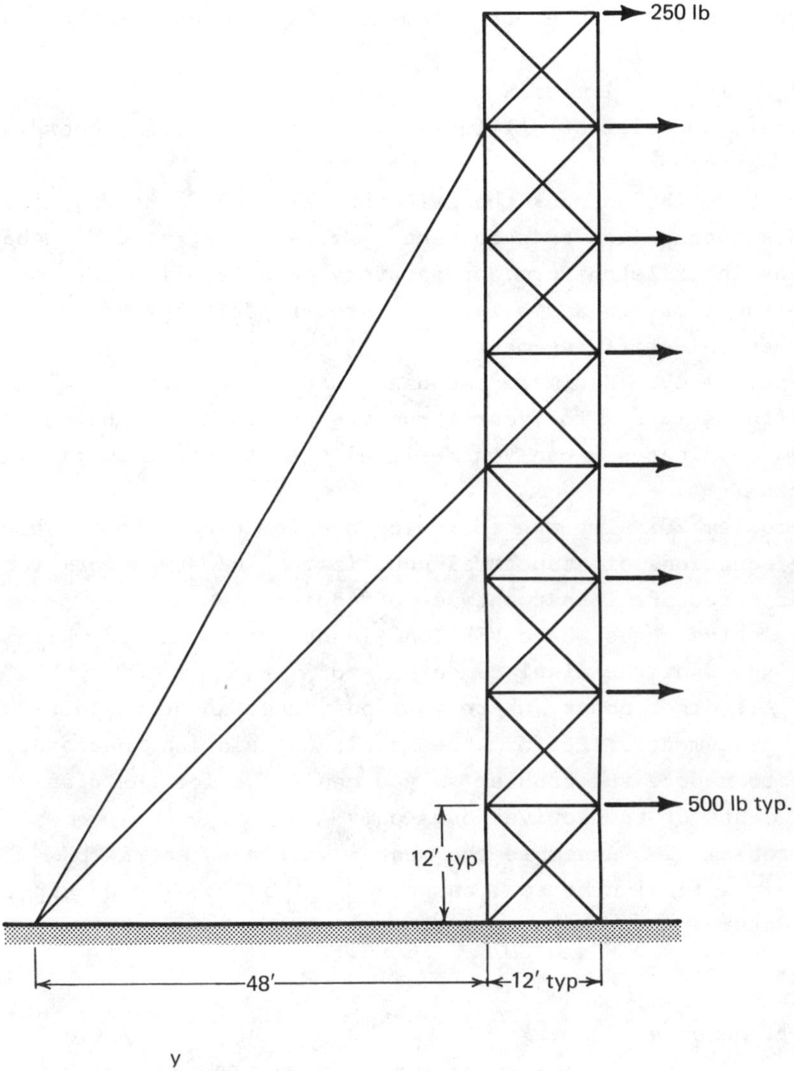
250 lb
500 lb typ.
12′ typ
48′
12′ typ

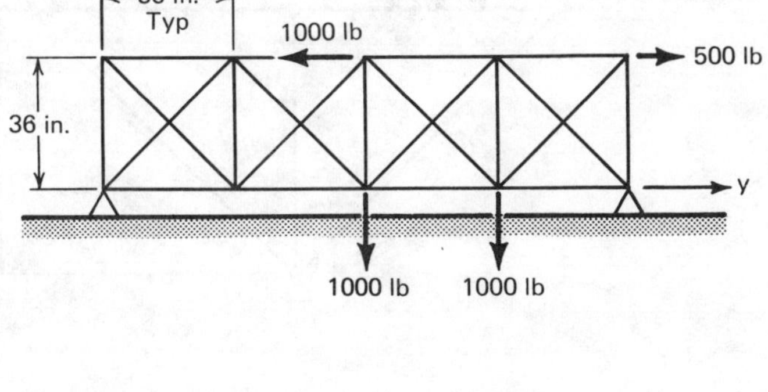
y
36 in.
Typ
1000 lb
500 lb
36 in.
y
1000 lb
1000 lb

Problem 17. Number the elements. Select any element and do the following:

a. derive the T matrix

b. get the element stiffness matrix in structural coordinates.

Problem 18. Number the grid points. Label the displacements appropriate to each node. Write the structural equations in skeleton form, preparatory to assembling the stiffness matrix terms into the proper positions of the structural stiffness matrix.

Problem 19. Estimate the bandwidth of the structural stiffness matrix for your structure as you have labeled the coordinates. Can you re-label them to get a smaller bandwidth?

Problem 20. Try the following problem closed book. Divide the equations of structural equilibrium in matrix form for your structure into two types of nodes:

a. Active nodes where VERTICAL loads occur, and on which you want the displacements, and

b. All other nodes, where you don't care about the displacements. Perform the matrix calculations necessary to reduce the problem so you can solve for the displacements of the active nodes only.

Problem 21. Assemble the master stiffness matrix $\left[K_M\right]$. Obtain a final numerical answer. $\left[K_M\right]$ is for the unsupported structure.

AE = 10^5 lb

All Members

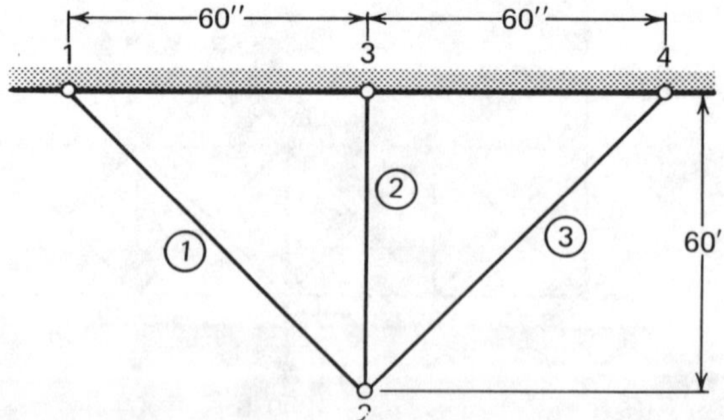

References

1. The Finite Element Method in Engineering and Science, O. C. Zienkiewicz, Mc-Graw Hill, London, 1971.
2. Introduction to the Finite Element Method, C. S. Desai and J. F. Abel, Van Nostrand Reinhold, New York, 1972.
3. NASTRAN Theoretical Manual, Richard H. MacNeal, Ed., NASA SP-221, Washington, D. C., 1970.
4. Survey of Advanced Structural Design Techniques, Z. Ziedans, ASME Paper 69-DE-13, New York, 1969.
5. Theory of Matrix Structural Analysis, J. S. Pryemueniecki, McGraw-Hill, New York, 1968.
6. Formulas for Stress and Strain, R. J. Roark, McGraw-Hill, New York, Sixth Edition, 1968.
7. Mark's Mechanical Handbook, T. Baumeister, Ed., McGraw-Hill, New York, Sixth Edition, 1968.
8. Matrix and Digital Computer Methods in Structural Analysis, Int. J. Num. Meth. Engr., Irons, B. M., Vol. 2, No. 1, Jan-Mar, 1970.

Chapter 7

THE METHOD OF VIRTUAL WORK AND STATICALLY INDETERMINATE STRUCTURES

7.1 VIRTUAL WORK AND THE DEFLECTION OF TRUSSES

The method of virtual work has wide application in all areas of structural analysis. Virtual work is one of a number of energy methods which forms an intriguing approach to static, dynamic and stability analysis of structures. References (1) and (2) provide an interesting summary of this topic.

The analytical power of the virtual work method is fascinating to observe. To do this, look first at Figure 7.1.1. A truss, 1-2-3, with members 1-3 and 2-3, is loaded by load P at node 3. The question arises as to what is the structural deflection at node 3. That is, how does the structure, the assembly of elements, displace?

It is very important to distinguish between structural displacements and element deformations. The structural displacements in this case is the movement of 3 to 3' during the loading. The element deformations are the respective lengthening or shortening of the two elements due to their own internal loads which result from the application of load P.

It is also very important to distinguish between external, or structural, loads and internal member forces. In Figure 7.1.1, P is an external load. The two members develop internal forces which are due to P. The two internal member forces may be found by considering the static equilibrium of node 3. For members one and two, we will denote the internal forces, due to P, as f_1 and f_2.

The question of interest to the stress analyst is: where does node 3 end up when the structure is loaded? (Where is point 3'?) This question can be answered without using virtual work--by using elementary strength of materials--but this is the hard way. To do the problem the hard way, we would calculate:

$$\delta_1 = \frac{f_1 L_1}{A_1 E_1} \quad \text{(extension)} \qquad (7.1.1a)$$

and

$$\delta_2 = \frac{f_2 L_2}{A_2 E_2} \quad \text{(compression)} \qquad (7.1.1b)$$

Then point 3' could be located by swinging arc C-D with center at node 2 and arc B-D with center at node 1. Simple logic tells us, that if the structure is going to remain joined at node 3, then point 3' must be at the intersection of these two arcs.

From this simple example we can see some very important things. One is that it is simple to calculate the element forces and deformations. The second is that it is harder to calculate the structural deflection. A third observation is very important. This is the fact that the structure must still "fit together" after deformation. This is called structural compatability.

The beauty of the method of virtual work is that it permits the calculation of the position of point 3' and automatically ensures structural compatability. In a simple, clean manner, without any additional fooling around with geometry or swinging of arcs, or nonsense like that. Virtual work is the easy way. We shall now show how it's done.

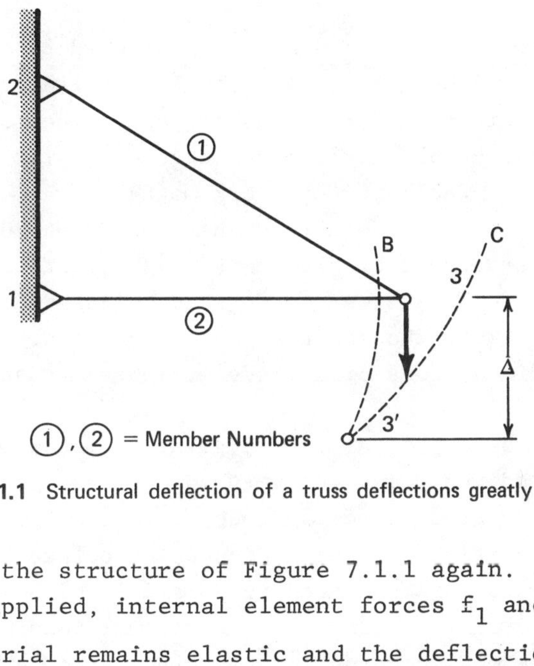

Figure 7.1.1 Structural deflection of a truss deflections greatly exaggerated

Consider the structure of Figure 7.1.1 again. When the load P is applied, internal element forces f_1 and f_2 result. If the material remains elastic and the deflections we discuss are "small," then Equations (7.1.1a) and (7.1.1b) will give the element deformations. The element force-deformation relationship is linear. If the structure is composed of linear elements, it is true (but is not proved here) that the structural load-deflection relationship will also be linear.

Look now at Figure 7.1.2. The element force-deformation relationship is shown graphically. The element stress-

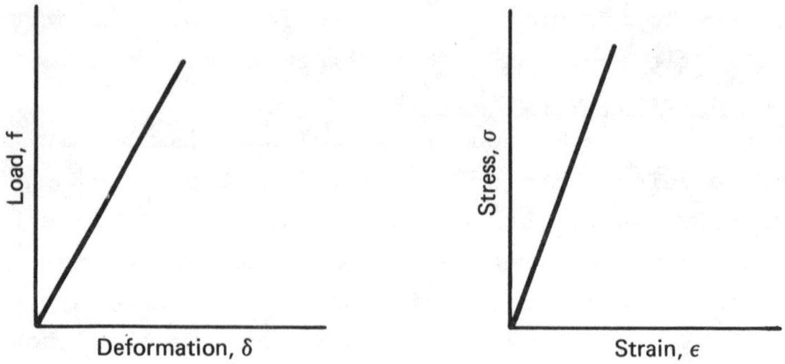

Figure 7.1.2 Element force-deflection and stress-strain relations

strain relationship is also shown. The load-deflection (P - Δ) curve for the structure will also be linear.

Now consider Figure 7.1.1 again and follow the logic in the next few paragraphs carefully, because it will provide an insight to the whole virtual work method.

We are going to apply load P to the structure and we wish to calculate structural displacement Δ--the vertical displacement of node 3. If we can do this calculation, we can generalize the method so that we can calculate the displacement of any node of any (determinate) truss in any direction.

To make this calculation, we are going to use what is generally called the dummy load method. We are going to put on a unit, virtual load, at node 3, vertically downward, and use the resulting element forces to calculate the displacement. To do this, we need to consider the work done, both during the real loading and the virtual loading.

First look at what happens both internally and externally when the real load P is applied.

The external work done by a slowly applied load is:

$$W_e = \frac{P\Delta}{2} \tag{7.1.2}$$

The internal work done (strain energy stored in members)

$$u_i = \sum_{i=1}^{2} \frac{f_i \delta_i}{2} \tag{7.1.3}$$

These two must be equal (external work done = internal strain energy stored).

$$\frac{P\Delta}{2} = \sum_{i=1}^{2} \frac{f_i \delta_i}{2} \tag{7.1.4}$$

Here's the tricky part. Suppose we had put on a unit load at node 3, vertically downward, before we put on load P. When we put on the unit load, we will develop internal member forces u_1 and u_2 which can be found from static equilibrium. At this point, the structure is sitting there with the unit load at node 3 and internal member forces u_1 and u_2 in the two members.

Now we apply load P and equate the external work done to the internal strain energy stored.

$$\underbrace{\underbrace{\frac{P\Delta}{2}}_{\text{Real Load}} + \underbrace{1 \cdot \Delta}_{\text{Virtual Load}}}_{\text{External}} = \underbrace{\underbrace{\sum_{i=1}^{2} \frac{f_i \delta_i}{2}}_{\text{Real Load}} + \underbrace{\sum_{i=1}^{2} u_i \delta_i}_{\text{Virtual Load}}}_{\text{Internal}} \tag{7.1.5}$$

Two of the terms in this equation

$$\left(1 \cdot \Delta \text{ and } \sum_{i=1}^{2} u_i \delta_i\right)$$

do not have a 1/2 in front. This is because they result from the unit load which was put on before load P was applied and, therefore, they were constant during the work process.

Now, Equations (7.1.4) and (7.1.5) tell us

$$1 \cdot \Delta = \sum_{i=1}^{2} u_i \delta_i \tag{7.1.6}$$

But

$$\delta_i = \frac{f_i L_i}{A_i E_i}$$

So that

$$\boxed{\Delta = \sum_{i=1}^{2} \frac{u_i f_i L_i}{A_i E_i}} \qquad (7.1.7)$$

It is essential to understand the meaning of each term in Equation (7.1.7) precisely and clearly. These should be memorized.

SYMBOL	MEANING
Δ	The structural deflection at the desired node and in the desired direction. In this case, the vertical deflection of node 3.
u_i	The force in the ith member resulting from the application of a unit load at the desired node and in the direction of the desired deflection.
f_i	The force in the ith member due to the applied external load. In this case, P.
L_i, A_i, E_i	The ith member properties.

Equation (7.1.7) can be generalized to calculate the deflection of any truss. The unit load is always put on at the node where the deflection is wanted and in the direction of the desired deflection. In Figure 7.1.3, for example, if we want the vertical deflection of node 3, we put on a unit load in the vertical direction at node 3 and calculate the resulting u_i.

Generally, then, the equation for calculating the deflection of any determinate truss is

$$\Delta = \sum_{i=1}^{n} \frac{u_i f_i L_i}{A_i E_i} \qquad (7.1.8)$$

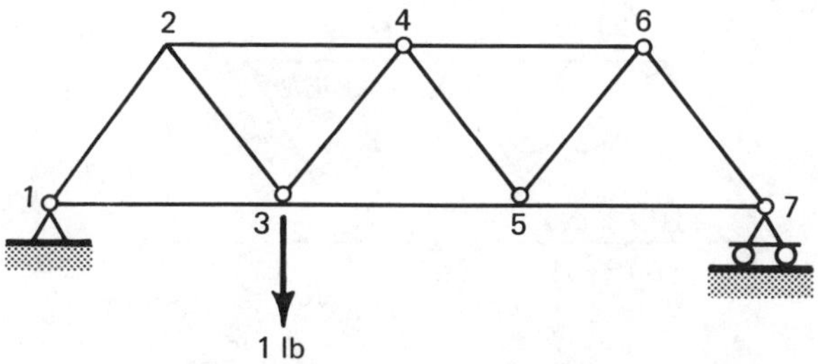

Figure 7.1.3

where the symbols have the same meaning as before, and n is the number of members.

It is now interesting to reflect on why the method of virtual work ensures compatability without explicitly requiring the consideration of members shortening or extension, or overtly requiring any geometry or trigonometry. The reason is that the geometry of the structure is reflected in the calculation of the f_i and the u_i. But it is much easier to calculate internal member forces than it is to work out the geometry of deflections. Therefore, the method of virtual work is very useful.

Example Problem:

Calculate the displacement of point C in a direction 45° with the horizontal as shown below

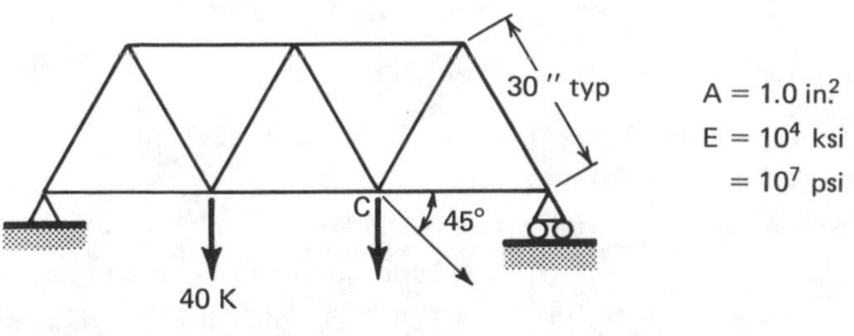

Solution:

The member forces S resulting from the applied loads are calculated from equilibrium at each joint, values shown are in kips (1,000 lb.).

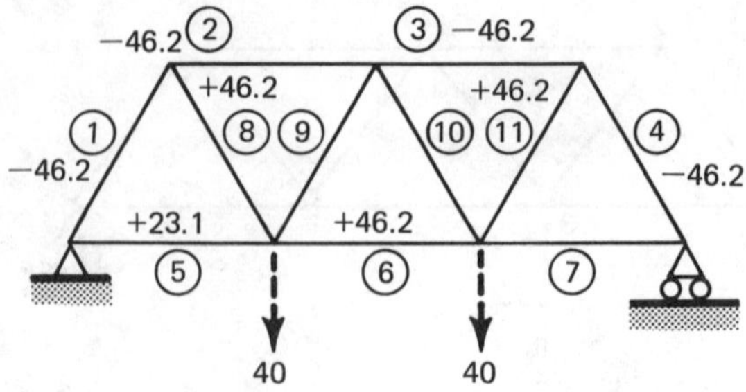

The forces, u, (in the members) resulting from a unit virtual load at point C in the direction of the desired deflection are given below.

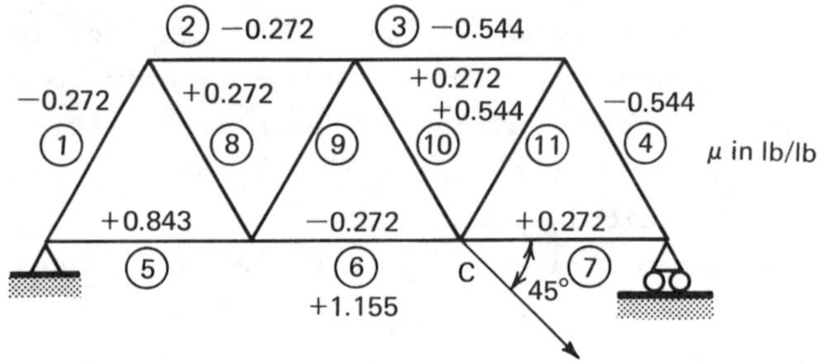

These results are summarized and final calculations are made in the table on the next page.

Objectives

(OB = Open Book; CB = Closed Book)

1. Be able to state Equation (7.1.8) and define all the terms. (CB)
2. Be able to use the method of virtual work to find any specified deflection of a given determinate truss. (OB)
3. Be able to state Equation (7.2.13) and define all the terms. (CB)
4. Be able to use the method of virtual work to find any specified deflection of a given determinate beam, arch or frame. (OB)
5. Be able to state Equation (7.3.1) and define all the terms. (CB)

Member No.	S [kip]	$\frac{L}{AE}$ [in/kip]	$\Delta = \frac{SL}{AE}$ [in.]	μ [lb/lb]	$\mu \cdot \Delta$ [in.]
1	−46.2	.003	−0.1386	−0.272	+0.0377
2	−46.2	↓	−0.1386	−0.272	+0.0377
3	−46.2	↓	−0.1386	−0.544	+0.0754
4	−46.2	↓	−0.1386	−0.544	+0.0754
5	+23.1	↓	+0.0693	+0.843	+0.0584
6	+46.2	↓	+0.1386	+1.155	+0.1600
7	+23.1	↓	+0.0693	+0.272	+0.0188
8	+46.2	↓	+0.1386	+0.272	+0.0377
9	0	↓	0	−0.272	0
10	0	↓	0	+0.272	0
11	+46.2		+0.1386	+0.544	+0.0754
$\delta = \Sigma \frac{S\mu L}{AE} = \Sigma\mu \cdot \Delta = +0.5765$					

6. Be able to use the method of virtual work to solve for any specified deflection of a given member in torsion. (OB)
7. Given any structure, the student should be able to determine the degree of indeterminancy (the number of redundants) by inspection, closed book. The distinction between external and internal redundants should be realized. (This is a review topic from the prerequisite subject of statics.) (CB)
8. Given an indeterminant structure, the student should be able to explain the procedure of enforcing compatibility by releasing a redundant (or redundants) and calculating a deflection (or deflections) and then developing enough redundant(s) to close the gap. When presented with a one-degree indeterminant structure, the student should be able to produce a written explanation of this procedure, with appropriate equations, illustrations and definitions. (CB)
9. Given a statically indeterminant truss, the student should be able to release the redundants and calculate

the f_i forces. He should be able to apply the appropriate unit load and calculate the u forces. He should be able to calculate δ_{aa} and δ_{ao}, realizing that the calculation for δ_{aa} may involve a redundant member. He should be able to perform the calculation: (OB)

$$X_a \delta_{aa} + \delta_{ao} = 0$$

10. Be able to solve for internal forces in a determinate truss.
11. Be able to calculate the deflection of an axially loaded member.
12. Be able to apply the dummy load and calculate the member force.
13. Be able to derive Equation (7.1.7).
14. Be able to solve for the deflection in any direction of any joint of any determinant truss.
15. Be able to write out the procedure for number 13 above.

7.2 BEAM DEFLECTIONS BY THE METHOD OF VIRTUAL WORK

Before going further and treating beam, frame and torsional deflections by the method of virtual work, it may be helpful to look at a simple illustration of the relation between virtual work and Castigliano's theorem.

Recall that the calculation for the structural deflections of a truss (which is composed of axial elements) involves two separate calculations. One for the real external loads and one for the appropriate unit load. Let us show the link between this operation and the use of Castigliano's theorem.

Castigliano's theorem says that the deflection of a structure at, and in the sense of, a load P is equal to the first partial derivative of the strain energy with respect to the load.

$$\Delta_p = \frac{\partial U}{\partial P} \tag{7.2.1}$$

A simple example of this is the deflection of an axial force element as shown in Figure 7.2.1.

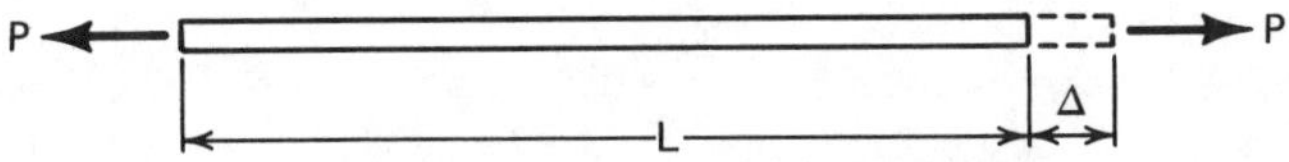

Figure 7.2.1 Deflection of axial force element

For this type of structural element, the strain energy is

$$U = \frac{P^2L}{2AE} \tag{7.2.2}$$

and

$$\Delta = \frac{\partial U}{\partial P} = \frac{PL}{AE} \tag{7.2.3}$$

a result which we already know. Looking at this result a little more closely, it is instructive to look at the first partial of Equation (7.2.2) in the following form

$$\frac{\partial U}{\partial P} = \frac{2P\left(\frac{\partial P}{\partial P}\right)L}{2AE} \tag{7.2.4}$$

The term $\frac{\partial P}{\partial P} = 1$ is the response of this simple structure to a unit load at the proper node, and in the sense of the desired deflection. Thus

$$\Delta = \frac{PL}{AE} \tag{7.2.5}$$

In general, when a structure such as a truss is composed of many elements, the strain energy is the sum over all the elements.

$$U = \sum_{i=1}^{n} \frac{f_i^{\,2}L_i}{2A_iE_i} \tag{7.2.6}$$

and Castigliano's theorem says

$$\Delta = \frac{\partial U}{\partial P} = \sum_{i=1}^{n} \frac{f_i \left(\frac{\partial f_i}{\partial P}\right) L_i}{A_i E_i} = \sum_{i=1}^{n} \frac{f_i u_i L_i}{A_i E_i} \qquad (7.2.7)$$

The term $\frac{\partial f_i}{\partial P}$ represents the u_i for each member. Thus we see the relation between Castigliano's theorem and virtual work.

Strain Energy of a Beam in Bending

A beam in bending stores strain energy. The energy relationship can be derived with the aid of Figure 7.2.2.

We know from previous study that

$$y'' = \frac{d\theta}{dx} = \frac{M}{EI}$$

$$d\theta = \frac{M}{EI}\,dx \qquad (7.2.8)$$

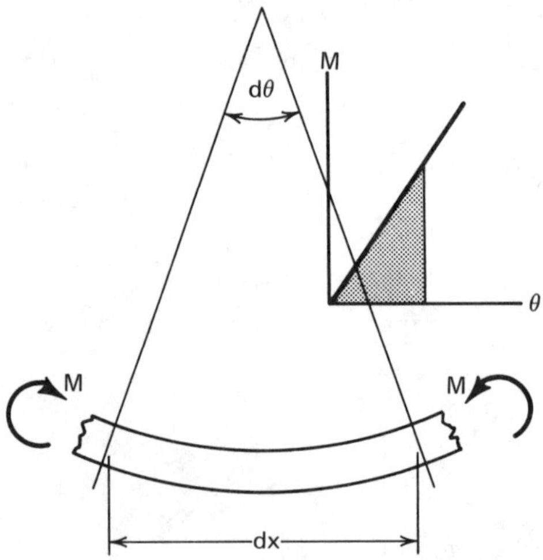

Figure 7.2.2 Beam deformation geometry second moment-slope relationship

We know that the moment-slope relation for a beam in pure bending is

$$\theta = \frac{Mx}{EI} \qquad (7.2.9)$$

and that for the linear relation shown in Figure 7.2.2 the work done (strain energy stored) in loading from zero to M is

$$dU = \frac{Md\theta}{2} \tag{7.2.10}$$

Integrating over the length, the energy stored in the beam is

$$U = \int_o^L \frac{M}{2}\, d\theta = \int_o^L \frac{M^2 dx}{2EI} \tag{7.2.11}$$

Castigliano's theorem then says that the deflection is the first partial derivative of the strain-energy with respect to the loading.

$$\Delta = \frac{\partial U}{\partial P} = \int_o^L \frac{2M\frac{\partial M}{\partial P}\, dx}{2EI} = \int_o^L \frac{M\frac{\partial M}{\partial P}}{EI}\, dx \tag{7.2.12}$$

Where P is the load which causes the moment.

If we associate the term $\frac{\partial M}{\partial P}$ with the response of a structure due to a unit load, we can write

$$\Delta = \int_o^L \frac{Mmdx}{EI} \tag{7.2.13}$$

To generalize this expression and relate it to the method of virtual work, consider a beam loaded by an external load P as shown in Figure 7.2.3.

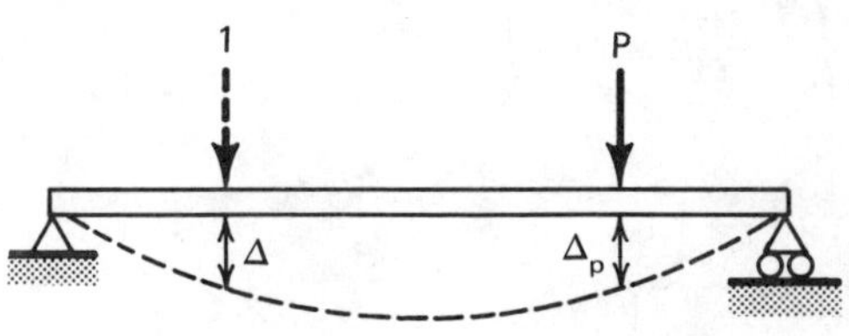

Figure 7.2.3 Deflection of a beam

The external work done is

$$\frac{P\Delta_P}{2} \tag{7.2.14}$$

The internal strain energy stored is

$$U = \int_o^L \frac{M^2}{2EI}\,dx \tag{7.2.15}$$

Then

$$\frac{P\Delta_p}{2} = \int_o^L \frac{M^2}{2EI}\,dx \tag{7.2.16}$$

Now, like the case of the truss deflection, consider that the unit load is applied to the beam at the point of desired deflection and in the direction of desired deflection before the load P is applied. At this point the unit load will be on and the beam will have a moment, m, due to the unit load. See the unit load and the Δ in Figure 7.2.3.

As the unit load is applied, the external work done is

$$\frac{1}{2}\,P\Delta_p + 1 \cdot \Delta \tag{7.2.17}$$

One term due to P and one due to the unit load. And the strain energy stored is

$$\int_o^L \frac{M^2}{2EI}\,dx + \int_o^L \frac{Mmdx}{EI} \tag{7.2.18}$$

Then

$$\frac{P\Delta_p}{2} + 1 \cdot \Delta = \int_o^L \frac{M^2}{2EI}\,dx + \int_o^L \frac{Mmdx}{EI}$$

And, from Equation (7.2.16)

$$\boxed{\Delta = \int_o^L \frac{Mmdx}{EI}} \tag{7.2.19}$$

One term is the strain energy due to the external load and one is due to the product of the constant internal moment, m, due to the internal load, riding through the deflections due to P.

It is important to know precisely the meaning of each term of the equation. It is also instructive to see the relation between each of the terms.

$$\Delta = \int_0^L \frac{Mmdx}{EI}$$

Symbol	Meaning
Δ	Deflection of a beam or frame.
M	Bending moment in beam due to applied external loads.
m	Bending moment in beam due to unit load at, and in the sense of, the desired deflection.
EI	Section properties of beam.

Note that the terms in the equation for beam deflection have the same meaning, in the most general sense, as the terms for truss deflection. M represents the response of the structure to external loading, and corresponds to the f_i of the truss. The response of the structure to the unit load is given by m, which corresponds to the u_i of the truss. A pattern is emerging whereby we begin to see that the response of the structure to the "real" external loading and also to the appropriate unit load are both important.

Example:

For the beam shown, find the horizontal displacement of point B.

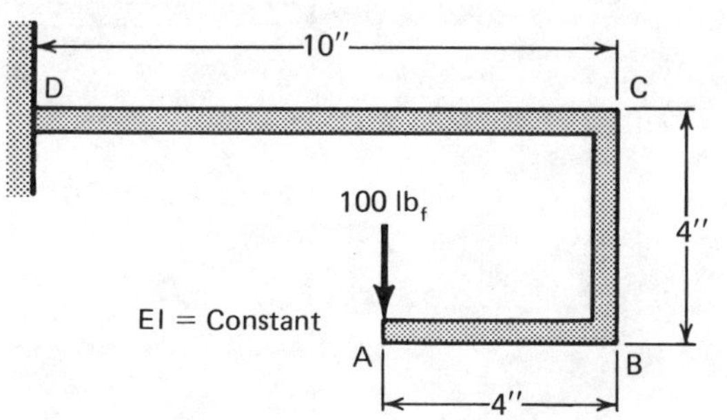

Solution:

The moment distribution on the frame due to the 100 lb_f load at point A is given below:

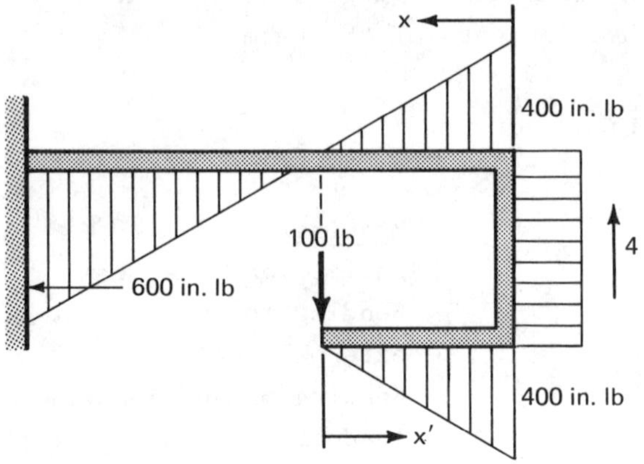

or

M = 100x'	400	400-100x
A to B	B to C	C to D

The moment distribution due to the horizontal virtual load f_e is given by:

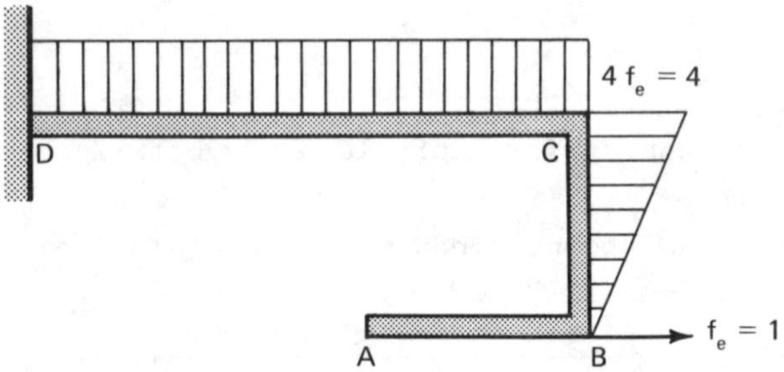

or

m = 1y	m = 4	m = 0
B to C	C to D	A to B

Now

$$\delta = \int_A^B \frac{M_{A-B}\, m_{A-B}\, ds}{EI} + \int_B^C \frac{M_{B-C}\, m_{B-C}\, ds}{EI} + \int_C^D \frac{M_{C-D}\, m_{C-D}\, ds}{EI}$$

$$\delta = \int_0^4 \frac{(100x')}{EI}(0)\,dx' + \int_0^4 \frac{400(1y)}{EI}\,dy + \int_0^{10} \frac{(400-100x)}{EI}\,4dx$$

$$\delta = \frac{400}{EI}\frac{y^2}{2}\Bigg|_0^4 + \frac{1600x}{EI}\Bigg|_0^{10} - \frac{400}{EI}\frac{x^2}{2}\Bigg|_0^{10}$$

$$\boxed{\delta = -\frac{800}{EI}\text{ in.}}$$

Sign indicates deflection in opposite direction of applied unit load.

Example:

A circular "arch" is loaded horizontally as shown. Find the horizontal deflection, Δ_H, of the support.

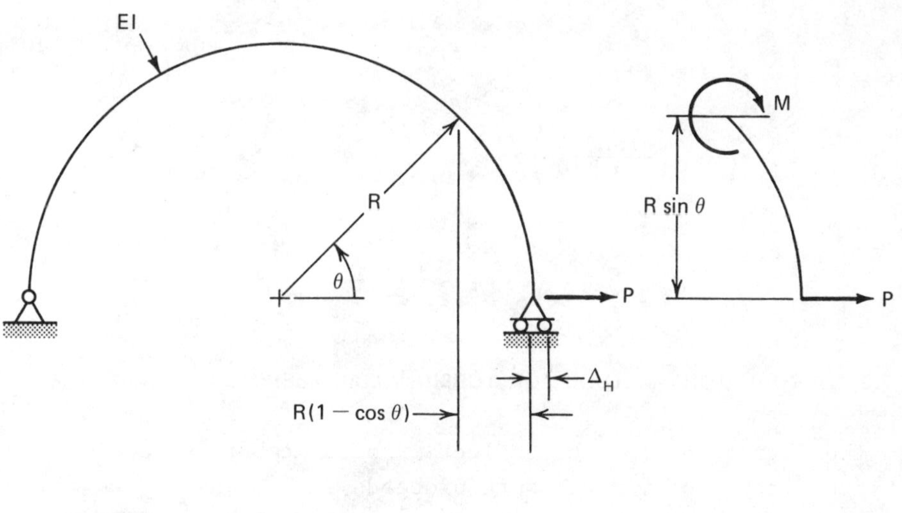

Solution:

The appropriate equation for this deflection is

$$\Delta_H = \int_L \frac{Mmds}{EI}$$

The bending moment M is the internal moment due to P. The free body diagram shown in the samll figure above shows

$$M = PR \sin \theta \quad \text{(compression outside = +)}$$

The bending moment m is the moment due to a unit load at, and in the sense of Δ_H. In this problem, this is the same as setting P = 1.

Then

$$m = R \sin \theta$$

In polar coordinates

$$ds = R\,d\theta$$

So that

$$\Delta_H = \int_0^{\pi} \frac{PR\sin\theta \; R\sin\theta \; Rd\theta}{EI}$$

$$= \frac{PR^3}{EI}\int_0^{\pi} \sin^2\theta\,d\theta$$

$$= \frac{PR^3}{EI}\left(\frac{\theta}{2} - \frac{\sin 2\theta}{4}\Big|_0^{\pi}\right)$$

$$\Delta_H = \frac{\pi PR^3}{2EI}$$

7.3 VIRTUAL WORK METHOD FOR TORSIONAL MEMBERS

The derivation for the deflection of a member in torsion by the method of virtual work proceeds in a manner which is analogous to that for beam bending. The derivation will not be given here. The result has the form:

$$\delta = \frac{TM_t}{GJ}\,ds \qquad (7.3.1)$$

where:

δ = the desired structural deflection

T = the internal torque distribution due to the external loading

M_t = the internal torque distribution due to a unit load applied at, and in the sense of, the desired deflection.

GJ = the member properties (need not be constant)

The similarity between these terms and those for truss and beam deflections should be noted. In particular note the following:

a. There is always a term which represents the internal response of the structure to external loading. For a truss: f_i. For a beam: M. For torsion: T.

b. There is always a term which represents the internal response of the structure to the appropriate unit load. For a truss: u_i. For a beam: m. For torsion: M_t.

c. A unit load is always put on the structure at, and in the sense of, the desired deflection. Remember that "load" is used in the general sense and may mean a force, moment, or torque.

d. Terms which give the member properties are always included.

For a truss: AE. For a beam: EI. For torsion: GJ.

Example:

Find the deflection of point A in the direction of the x-axis by the virtual work method. Set up integral only.

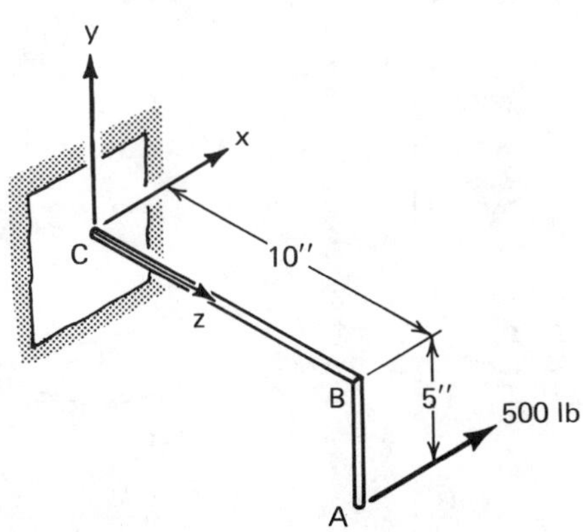

Solution:

The bending moment, torsional, unit bending, and unit torsion diagrams are as shown below for x-axis deflection of point A

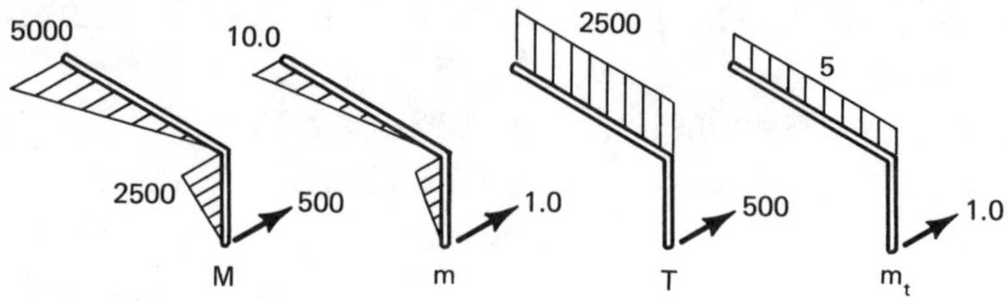

$$\delta_{Ax} = \int \frac{Mm}{EI} ds + \int \frac{TM_t}{JG} ds$$

$$= \int_0^5 \frac{\left(\frac{2500}{5}\right) y \ (y) dy}{EI} + \int_0^{10} \frac{\left(\frac{5000}{10}\right) z \ (z) dz}{EI} + \int_0^{10} \frac{2500(5) dz}{JG}$$

Example:

Find the deflection of point A in the direction of the y-axis by the method of virtual work. Set up the integrals and then integrate using the section properties given below.

Section Properties

$E = 29(10^6)$ psi
$G = 11(10^6)$ psi
$I = 1$ in.^{4}
$J = 2$ in.^{4}

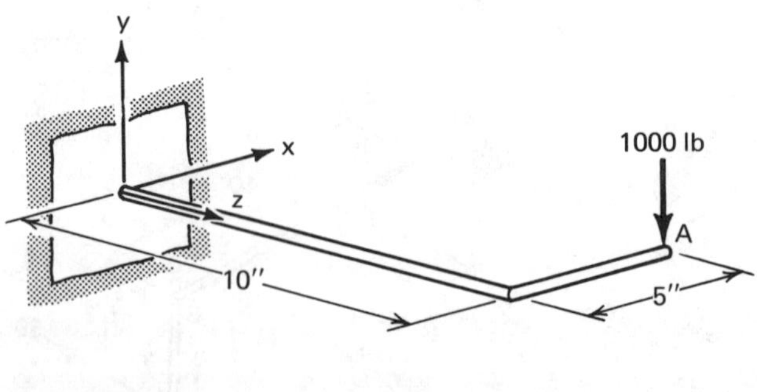

Solution:

$$\delta_{y_{TOTAL}} = \int \frac{Mmdx}{EI} + \int \frac{Mmdz}{EI} + \int \frac{Ttdz}{GJ}$$

$$= \int_0^5 \frac{1000x^2\,dx}{EI} + \int_0^{10} \frac{1000z^2\,dz}{EI} + \int_0^{10} \frac{5000(5)\,dz}{GJ}$$

$$= \frac{1000x^3}{3EI}\Bigg|_0^5 + \frac{1000z^3}{3EI}\Bigg|_0^{10} + \frac{25000z}{GJ}\Bigg|_0^{10}$$

Since $E = 29(10^6)$psi, $I = 1$ in^4, $G = 11(10^6)$psi, and $J = 2$ in^4, we have

$$\delta_{y_{TOTAL}} = \frac{1000(5)^3}{3(29)10^6}(1) + \frac{1000(10)^3}{3(29)10^6}(1) + \frac{25000(10)}{11(10^6)(2)}$$

$$= .001437 + .011494 + .011364$$

$$= .024295 \text{ inch}$$

$\therefore\ \delta_{y_{TOTAL}} \approx$ .0243 inch ↓ @ POINT A

7.4 STATICALLY INDETERMINATE STRUCTURES

In the previous section we saw how the method of virtual work is used to calculate the deflections of various types of structures. We shall now extend this skill to a systematic way of calculating the forces in statically indeterminate structures. Most "real" structures are statically indeterminate. Thus, we are learning an important skill.

A structure is statically indeterminate if the number of unknown forces (either internal member forces or reactions) exceeds the equations of static equilibrium which are available to calculate them. Thus the beam of Figure 7.4.1a is indeterminate because it has three redundants, R_1, R_2, and R_3 and only two static equations of equilibrium. (Sum of forces in the horizontal direction doesn't help in this case.) Therefore we have one excess force, or one redundant. The first step in understanding and solving statically indeterminate structures is to be able to determine the number of redundants. The number of redundants is the number of excess forces, internal or external, which the structures possess over those which can be solved for by using static equilibrium.

The basic plan for solving statically indeterminate structures can be illustrated with the beam of Figure 7.4.1. The procedure goes as follows:

1. Remove the redundant forces by physically freeing the structure such as shown in Figure 7.4.1(b). In this figure R_3 has been chosen as the redundant force and R_3 is removed from the structure. The structure will now be <u>statically determinate</u>. This statically determinate structure is called the <u>primary structure</u>.
2. Calculate the amount that the primary structure displaces in the sense of the (removed) redundant. This is illustrated as displacement Δ in Figure 7.4.1(b).
3. Calculate the amount of redundant force which is required to restore the structure to its original configuration. In the example of Figure 7.4.1, this is the magnitude of R_3 which is required to restore Δ to zero. Note that this calculation is also for the primary (determinate) structure. This step enforces structural compatibility.

General Comments

The above example has been used to illustrate a procedure which will be used for all statically indeterminate structures problems. First the redundants are removed, thus forming a primary structure. The deflections of the primary structure due to the external loading are calculated. Then the magnitude of redundant forces which are required to restore those displacements back to the original structure configuration are calculated. We shall see that the method of virtual work will be useful for these calculations.

If we have a structure which has n redundants (is n degrees statically indeterminate) then we will end up with a system of n linear algebraic equations in n unknowns, with the n redundant forces as the unknowns. Because it is the redundant forces which are the unknowns in this method of analyzing indeterminate structures , the method is called the force method. The force method is an ancestor of the matrix force method and the matrix force method was one of the first computer methods for stress analysis.

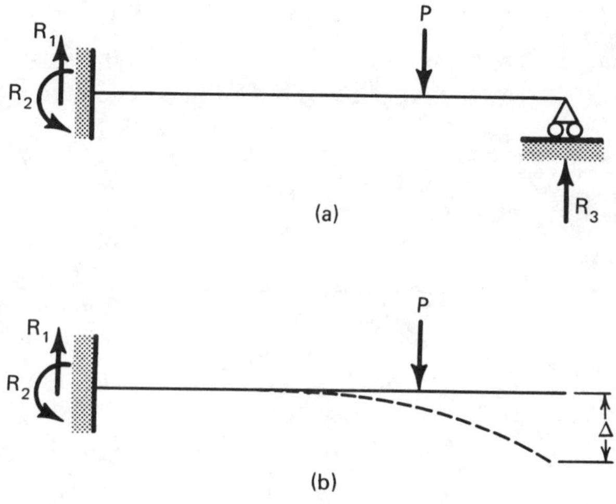

Figure 7.4.1

We can now generalize the notation for deflection.

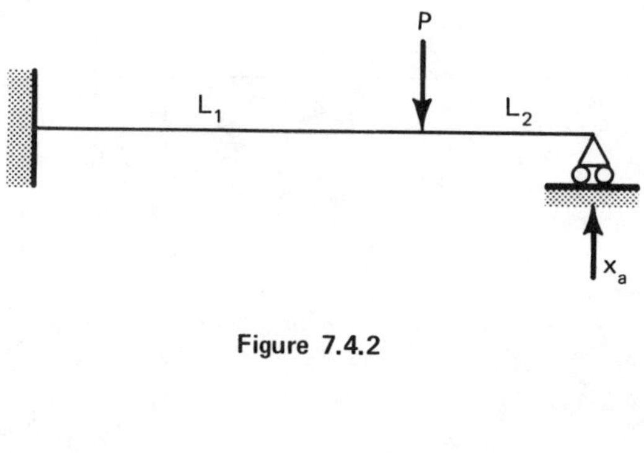

Figure 7.4.2

Recall the steps used to obtain reaction X_a

1. Cut the structure back to a determinant structure.

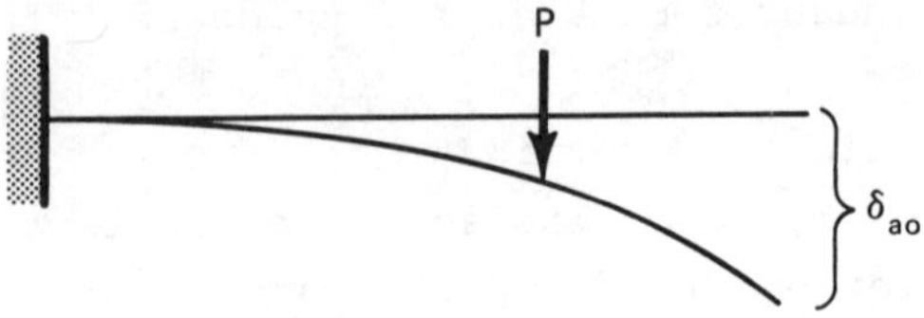

Figure 7.4.3

2. Calculate deflection, at point of redundant, in direction of redundant, for cut-back structure.

$$\delta_{ao} = \frac{PL_1{}^3}{3EI} + \frac{PL_1{}^2L_2}{2EI} \tag{7.4.1}$$

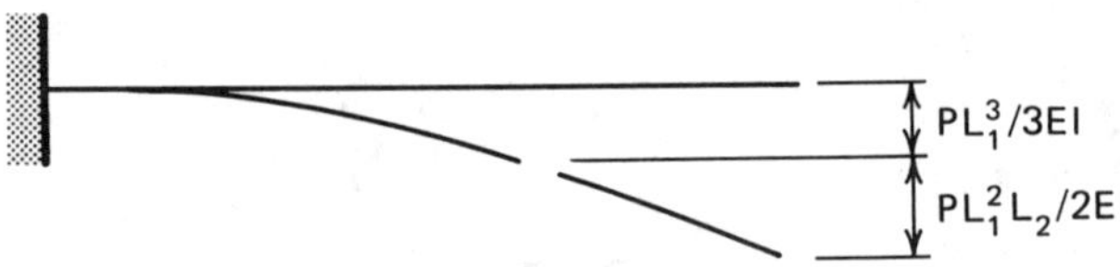

Figure 7.4.4

3. Apply unit load on cut-back structure, in direction of redundant, at redundant.

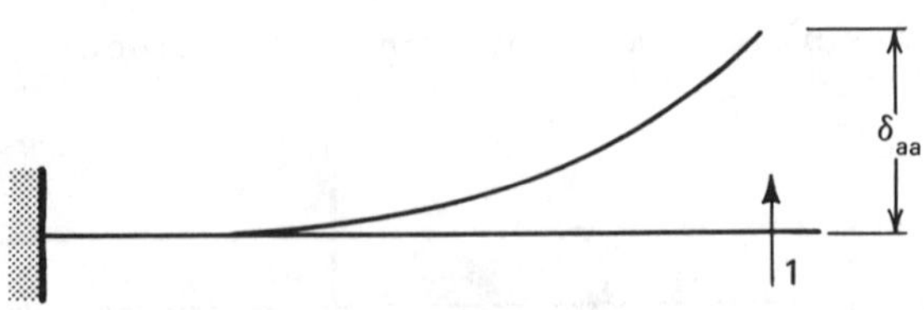

Figure 7.4.5

$$\delta_{aa} = \frac{1(L_1 + L_2)^3}{3EI} \tag{7.4.2}$$

4. X_a must be just enough to push δ_{ao} back to zero.

$$\delta_{ao} + X_a\,\delta_{aa} = 0$$

$$X_a = \frac{-\delta_{ao}}{\delta_{aa}} \tag{7.4.3}$$

Now in general, we denote

$\delta_{a\ o}$

- o — Due to external loads
- a — Displacement at the redundant X_a and in the direction of X_a

$\delta_{a\ b}$ $(\delta_{aa}, \delta_{ba}, \ldots \delta_{mn})$

- b — Deflection due to a unit load at redundant X_b and in the sense of X_b.
- a — Deflection is at the redundant X_a and in the direction of X_a.

One-degree Redundant Trusses

A truss with one internal redundant is shown in Figure 7.4.6. We wish to find the distribution of forces (the internal member forces) for this truss. It is important to realize that--in contrast to a determinate truss--the distribution of internal forces in the truss members is a function of the size and material properties of each truss member. That is: for a determinate truss, the distribution of internal member forces is not affected by member size, but only by static equilibrium. For an indeterminate truss, the distribution of internal forces is affected by member size. (Really, by member stiffness.) To emphasize this, the truss members have been given different properties, as shown in Figure 7.4.6.

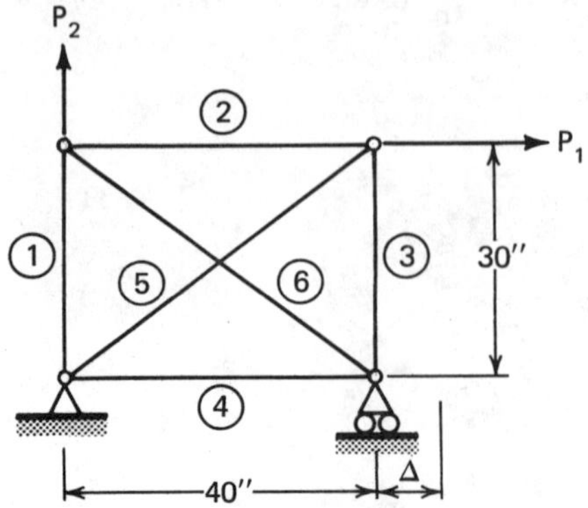

Member No.	Area in.²	E 10⁶ psi
1	1.0	29
2	0.9	30
3	0.8	10
4	0.7	10
5	0.6	29
6	0.5	30

P_1 = 707 lb
P_2 = 666 lb

Figure 7.4.6

The primary structure is shown in Figure 7.4.7. Member number 5 has been selected as the redundant and is shown as being cut, but still being present. We solve for the reactions and internal forces and apply the method of virtual work to calculating the "gap" developed in the primary structure; this is δ_{ao}, the deflection of the primary structure due to the external loading. The results are shown in Table 7.4.1.

The results of Table 7.4.1 show that the primary structure deflection, δ_{ao} to be -77.46×10^{-4} inches. This is the amount of gap, or structural incompatibility, which would develop if the determinant primary structure were loaded by the real external structural loads. Note that in using the method of virtual work, we have calculated the u_i as the set of internal forces in the primary structure resulting from the application of a unit, virtual load which is applied at, and in the sense of, the desired deflection. This amounts to the prescription of a unit force in member five. This is shown in Figure 7.4.7.

It is important to note that the appropriate unit load in this case is a value of unity in the truss member, and not a 1-lb force put on the nodes of the truss as external loads, as shown in Figure 7.4.8.

Note: Tension = (+), Compression = (-)

Member No.	f_i (lb)	u_i (lb)	$\frac{L_i}{A_i E_i}$ (10^{-6} in/lb)	$\frac{f_i U_i L_i}{A_i E_i}$ (10^{-4} in)
1	+1196	-0.6	1.03	-7.39
2	+707	-0.8	1.48	-8.37
3	0	-0.6	3.75	0
4	+707	-0.8	5.71	-32.30
5	0	+1.0	2.87	0
6	-883	+1.0	3.33	-29.40

$$\delta_{ao} = \sum \frac{f_i U_i L_i}{A_i E_i} = -77.46 \times 10^{-4} \text{ in}$$

Table 7.4.1

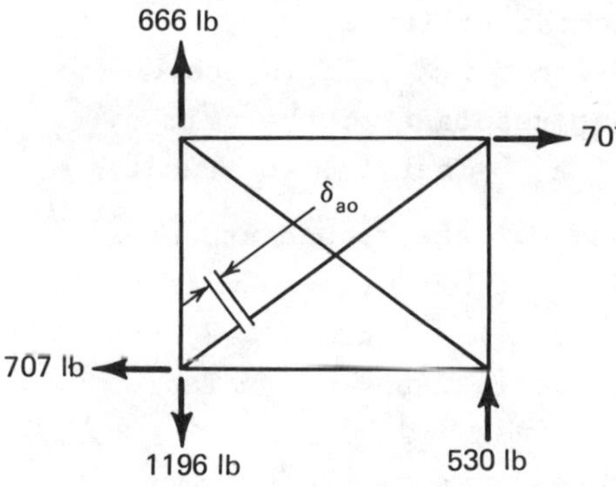

(a) Response of primary structure to real loads

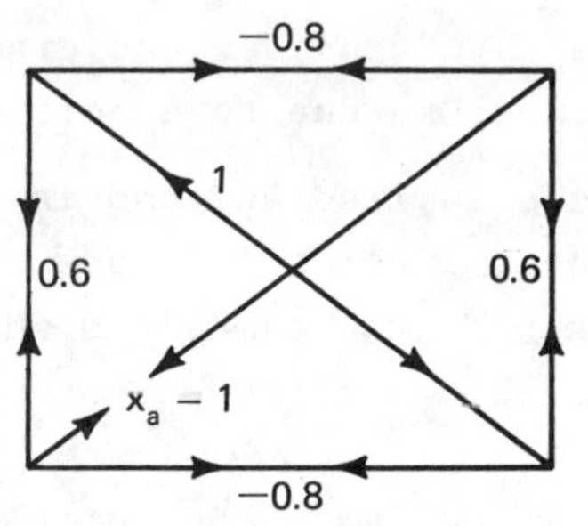

(b) Response of structure to unit x_a

Figure 7.4.7

Let us now return to our original objective, which was to calculate the value of the redundant force in member 5. Let us call this value of force X_a. The question is: How much is X_a? To answer this question, we must reflect on the meaning of structural compatibility. The primary structure has member 5 cut, and develops displacement δ_{ao}.

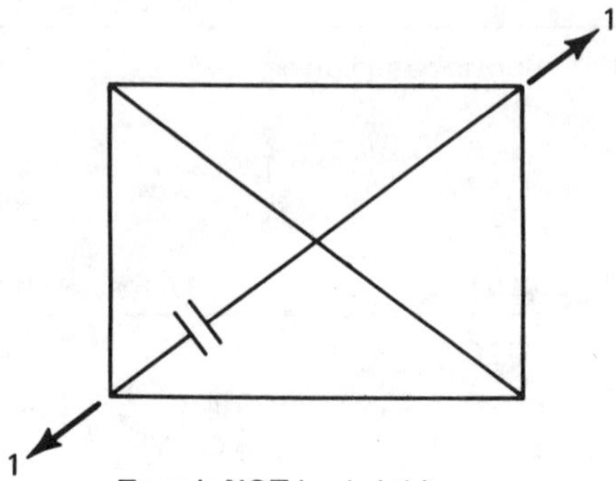

Truss is **NOT** loaded this way

Figure 7.4.8

The real structure does not have member 5 cut and develops internal force X_a in member 5 instead. The amount of X_a in the real structure is therefore, logically, just the amount of force required to pull δ_{ao} back to zero in the primary structure. (X_a is the force required to close the gap, or to restore structural compatibility.)

To perform the actual calculation for X_a, it is convenient to consider the displacement response of the primary structure due to a unit value of X_a. That is, to a unit load applied at, and in the sense of, the redundant force. It is interesting, and it saves a lot of work, to note that this is the same load which is shown in Figure 7.4.7, and which has already been used to calculate the u_i of Table 7.4.1.

If we denote the displacement response of the primary structure to this unit load as δ_{aa}, then the logic of structural compatibility may be expressed mathematically as the condition

$$X_a\delta_{aa} + \delta_{ao} = 0 \qquad (7.4.4)$$

That is to say that the redundant force X_a, times the displacement due to a unit redundant force, δ_{aa}, just cancels δ_{ao}, the "gap" which developed in the primary structure due to the external loads.

Now, let us think for a moment. We want to calculate the response of a determinate truss (the primary structure) due to a unit applied load. This, again, is depicted in Figure 7.4.7.

If we consider this as just another determinate truss calculation, and forget for a moment about indeterminate trusses and the rest of our worries, we will note that this is a problem where the real load and the virtual load are exactly the same, namely, a unit load as shown in Figure 7.4.7. Therefore, the deflection δ_{aa} is given by

$$\delta_{aa} = \sum_{i=1}^{6} \frac{u_i^2 L_i}{A_i E_i} \qquad (7.4.5)$$

where the u_i have already been calculated in Table 7.4.1. If we perform the sum indicated in Equation (7.4.5) we get

$$\delta_{aa} = 1.252 \times 10^{-5} \ \frac{\text{in}}{\text{lb}}$$

Note the units of in/lb. δ_{aa} is deflection per unit load. Now, using Equation (7.4.4) we get

$$X_a = \frac{-\delta_{ao}}{\delta_{aa}} = 618.6 \text{ lb}$$

Note that we have had to be careful about signs throughout the calculations, with tension being plus and compression being minus.

Once we know redundant force X_a, we can calculate the rest of the member forces from statics. But it is much easier to note that the force in each member is just

$$X_a u_i + f_i \qquad (7.4.6)$$

since the u_i are the response of the structure to a unit value of redundant. The final forces are shown in Table 7.4.2.

Member No.	Final Force, F_i (lb)
1	824.84
2	212.12
3	-371.16
4	212.12
5	618.6
6	-264.4

Table 7.4.2

This concludes the calculation for the member forces in the truss. All trusses of one-degree redundancy may be treated with the same general approach. Let us now turn our attention to a different, but related, problem: that of calculating structural deflection, Δ, of the real structure. An example is shown in Figure 7.4.9, where Δ is the horizontal structural displacement of the right support due to the external loads on the real structure. We note that δ_{ao} and δ_{aa} were not real structure displacements of the primary structure which were calculated for the purpose of determining X_a.

Considering now the structure of Figure 7.4.9 and the calculation of Δ, we may treat this as another problem of truss deflection by the method of virtual work with

$$\Delta = \sum_{i=1}^{n} \frac{f_i u_i L_i}{A_i E_i}$$

where:

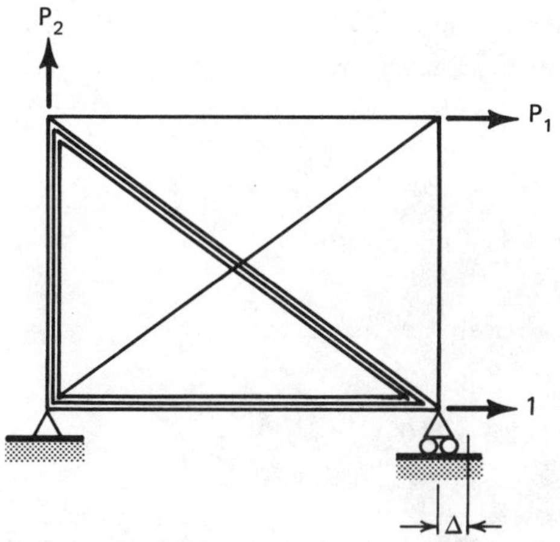

Figure 7.4.9

f_i are the internal member forces in the <u>real</u> structure due to the external load. (Just the solution to our original statically determinant problem.) Results are in Table 7.4.1.

u_i are the internal member forces in the <u>real</u> structure due to a unit load applied at the point of desired displacement and in the direction of desired displacement.

L_i, A_i, and E_i are member properties.

Looking at Figure 7.4.9, the situation at first looks grim. The unit load is shown in the figure and it appears that we will have to solve another statically indeterminate problem, since we must have the u_i for the real structure. However, one feature of the method of virtual work will save us much effort. It is presented here without proof:

For the u_i (the internal response to the external unit load) <u>ANY</u> set of internal forces which are geometrically compatible with the real structure and which are in equilibrium with the external unit load may be used.

This is extermely important because it permits us to select a geometrically compatible sub-truss which is statically determinate by selecting particular members from our real truss. An example is shown in Figure 7.4.9, where the sub-truss is indicated by heavy lines.

The u_i for this sub-truss is simply calculated as

$u_4 = 1$
All others = 0

So the displacement Δ is given by

$$\Delta = \frac{F_4 u_4 L_4}{A_4 E_4} = 1.22 \times 10^{-3} \text{ in}$$

It is necessary to elaborate slightly on what is meant by a geometrically compatible sub-truss. This means that a stable, determinant sub-truss must be formed from actual members of the real truss. We cannot introduce new members or create a set of internal equilibrating forces which is not consistent with the real truss.

Example:
Truss with single redundancy

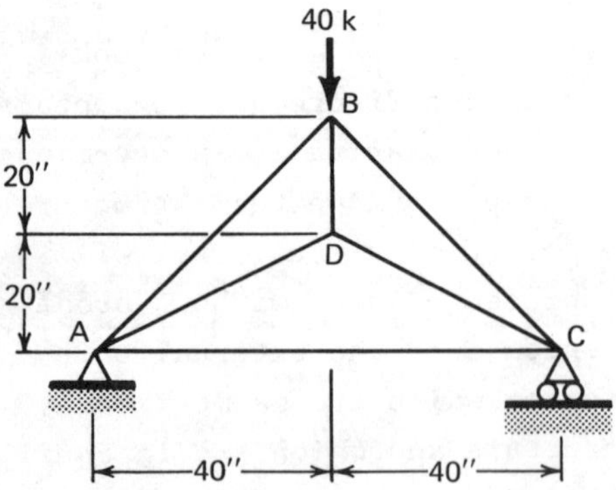

E = 10,000 psi

Member	Area, in.^{2}
AB	4
BC	4
BD	4
AD	3.6
DC	3.6
DC	3.6
AC	3.0

Select as a redundant the member AC. Cut this member.

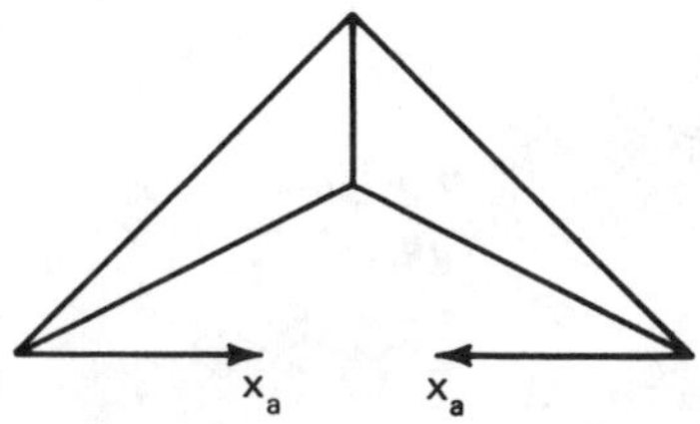

Real loads on cut back structure may be found from statics. Deflection δ_{ao} may be found from virtual work.

$$\delta_{ao} = \sum \frac{SuL}{AE}$$

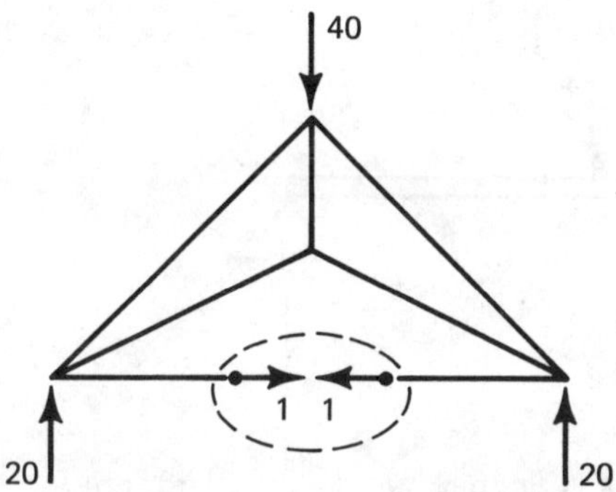

Member	S, kip	L/AE
AB	−56.6	0.001414
BC	−56.6	0.001414
AD	44.8	0.001245
DC	44.8	0.001245
BD	40.0	0.000500

Now the deflection δ_{aa} may be calculated, the appropriate unit force is the one already shown. But the deflection of member AC must be included

$$\delta_{aa} = \frac{u \cdot uL}{AE} = \frac{u^2L}{AE}$$

Member	u	$\frac{SuL}{AE}$	$\frac{u^2L}{AE}$
AB	1.414	−0.1130	0.00283
BC	1.414	−0.1130	0.00283
AD	−2.233	−0.1245	0.00621
DC	−2.233	−0.1245	0.00621
BD	−2.0	−0.0400	0.00200
		−0.5150	0.00267 AC !!!!!
		⇨	0.02275

Now

$$X_a = -\frac{\delta_{ao}}{\delta_{aa}} = \frac{0.5150}{0.02275} = 22.6 \text{ kips}$$

The total member forces are

$$S_T = S_o + X_a u$$

Problems

Problem 1. Use the method of virtual work to find the deflection of the center of the simply supported beam.

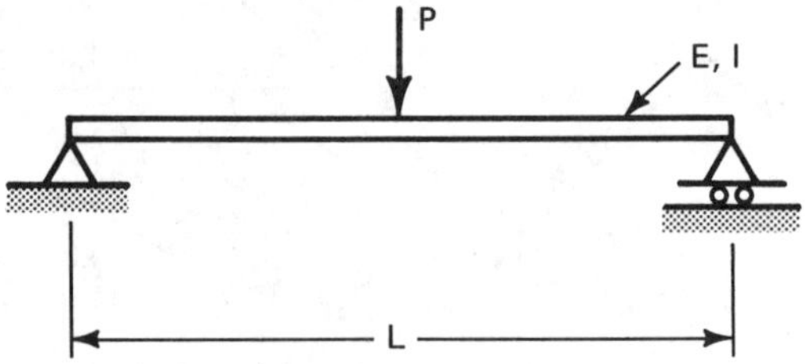

Problem 2. Use the method of virtual work to find the deflection of the center of the cantilevered beam.

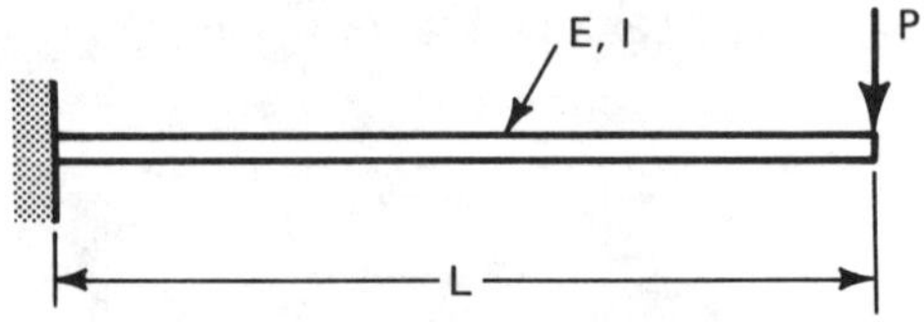

Problem 3. Use the method of virtual work to find the deflection of the center of the simply supported beam.

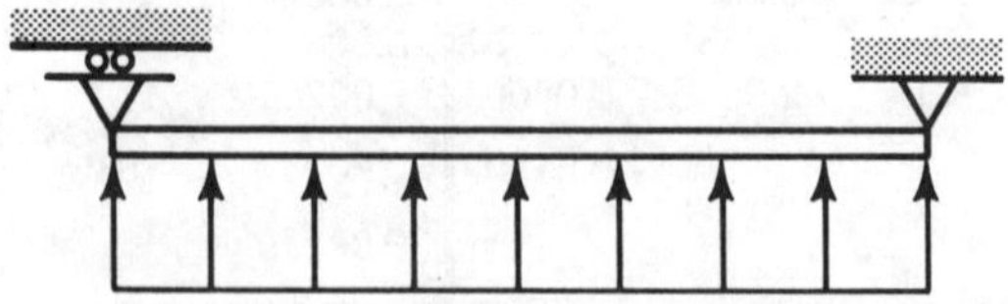

Problem 4. Find the vertical deflection at Point A for this truss

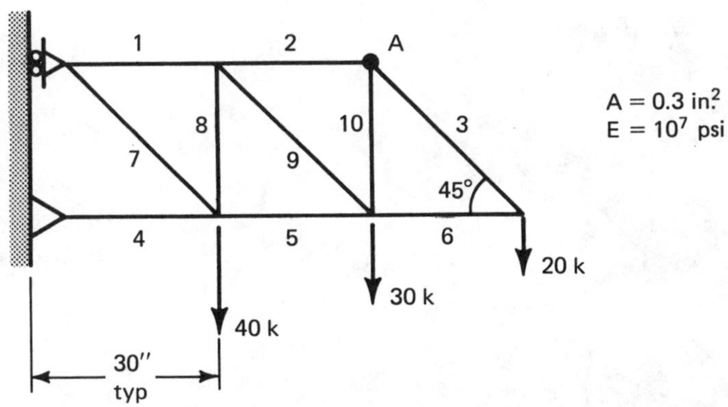

Problem 5. Find the horizontal deflection of point A on the truss

Member 1 Area = 0.5 in^2

Member 2 Area = 0.25 in^2

E = 10^7 psi

P = 1,000 lb.

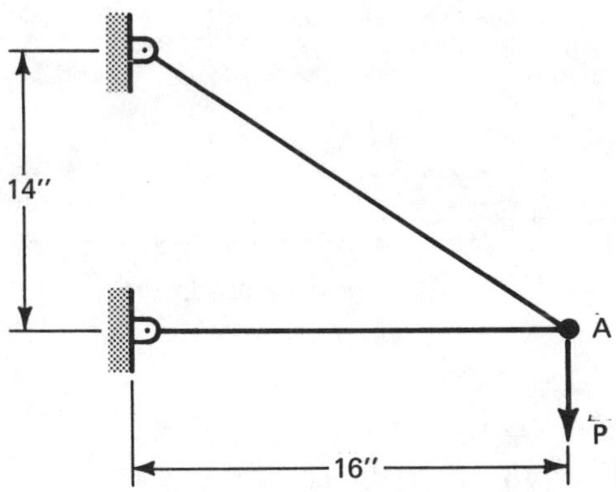

What will be the effect of the following on this deflection?

a. doubling P?

b. doubling the member areas?

c. doubling the size of the truss (member areas remain constant)?

Problem 6. Find the angle θ through the end of this structure will twist.

Combined Bending and Torsion

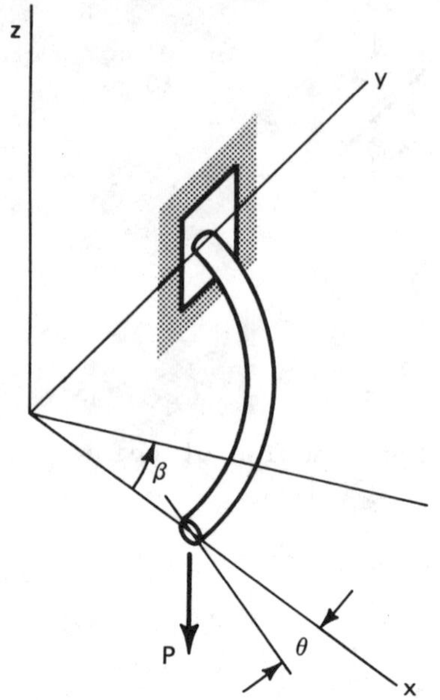

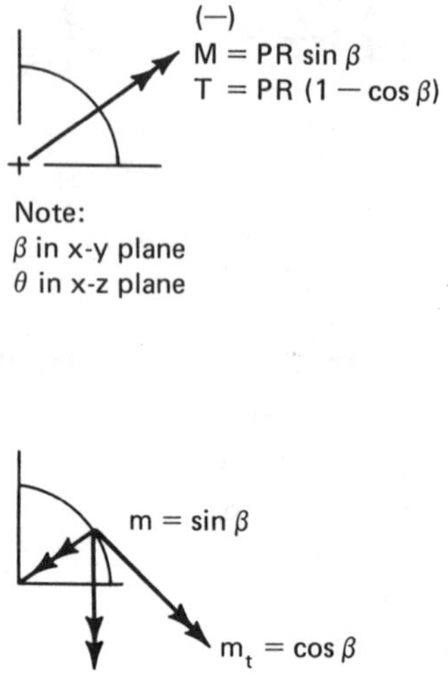

Problem 7. Give the degree of indeterminancy (redundancy) of the structures shown. Indicate whether these are external or internal redundants.

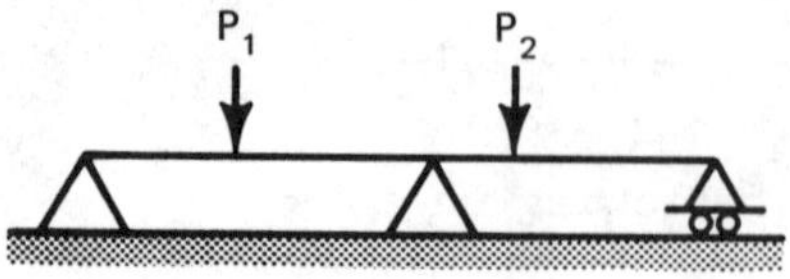

(a) ______

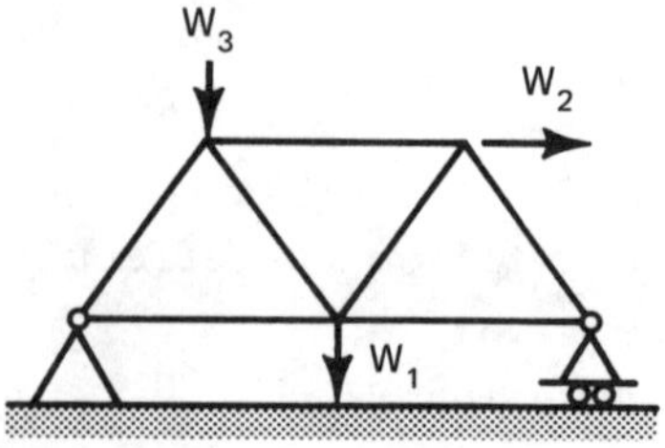

(b) ______

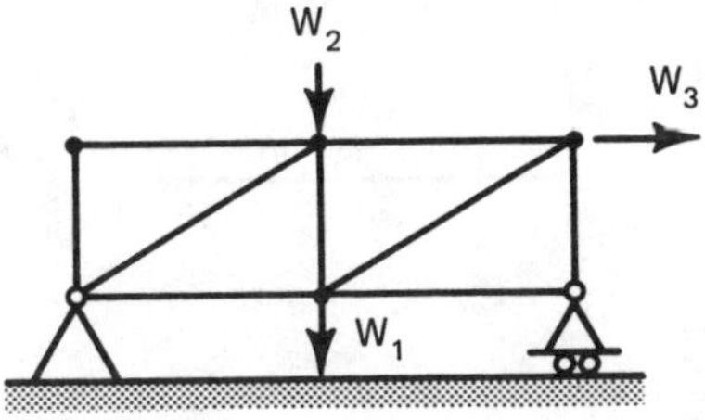

(c) ______

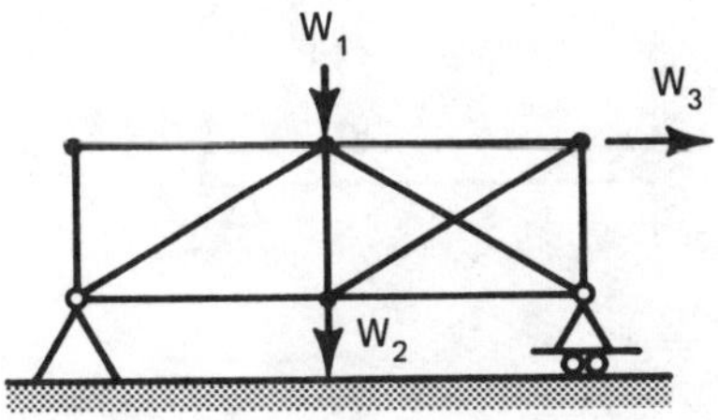

(d) ______

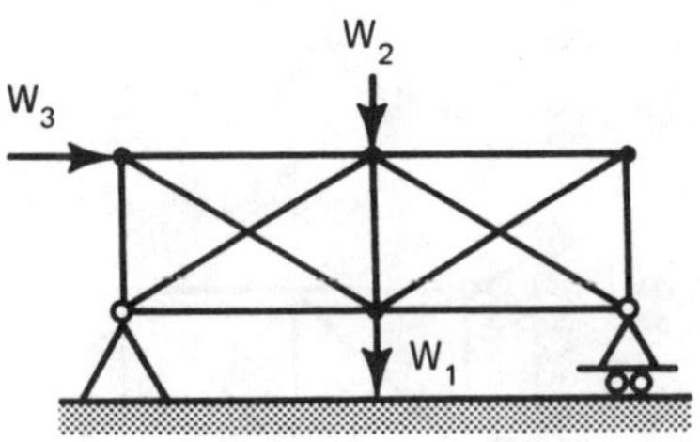

(e) ______

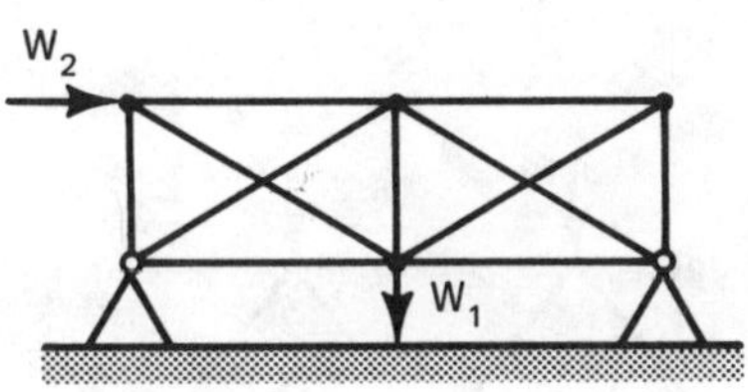

(f) ______

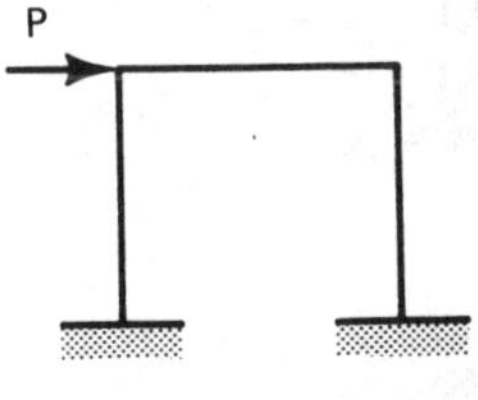

(g) ______

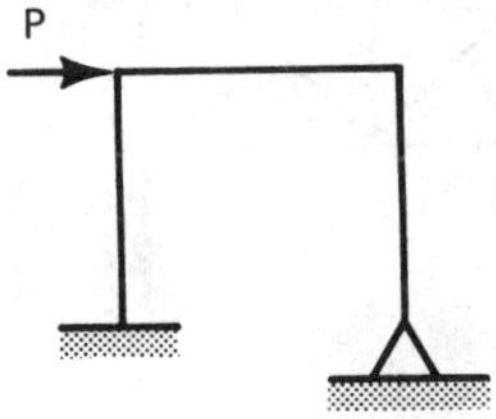

(h) ______

Problem 8. Give the degree of indeterminancy (redundancy) of the structures shown. Indicate whether these are external or internal redundants.

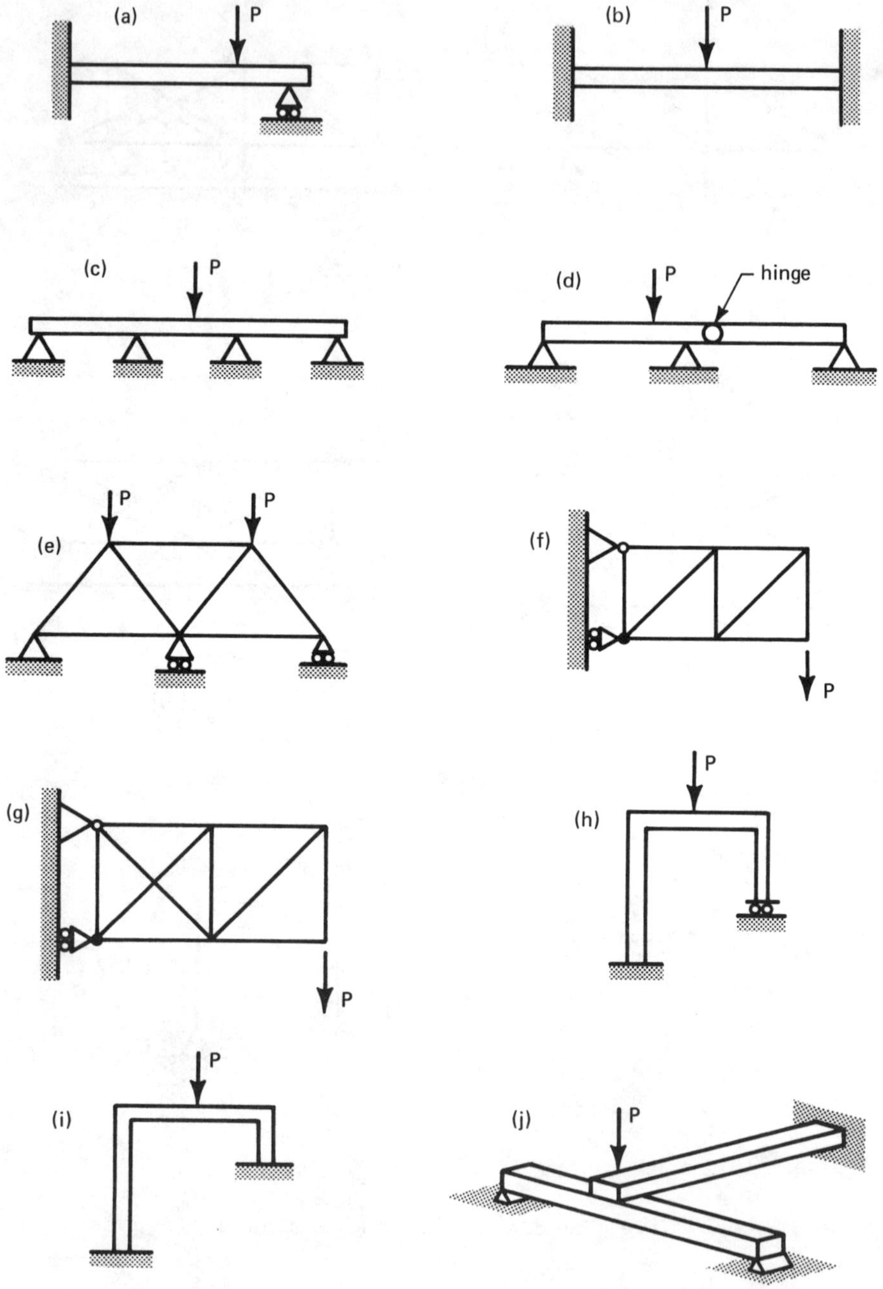
(a)
P
(b)
P
(c)
P
(d)
P
hinge
(e)
P
P
(f)
P
(g)
P
(h)
P
(i)
P
(j)
P

Problem 9. Give the degree of redundancy of the structures shown.

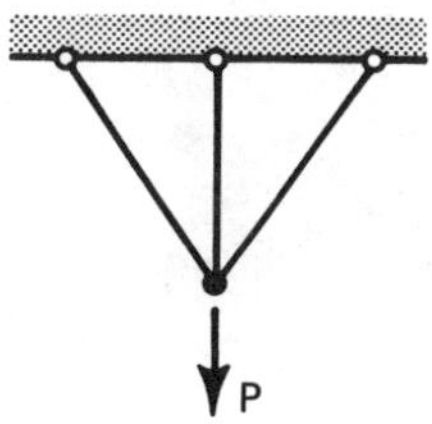

(a) ______

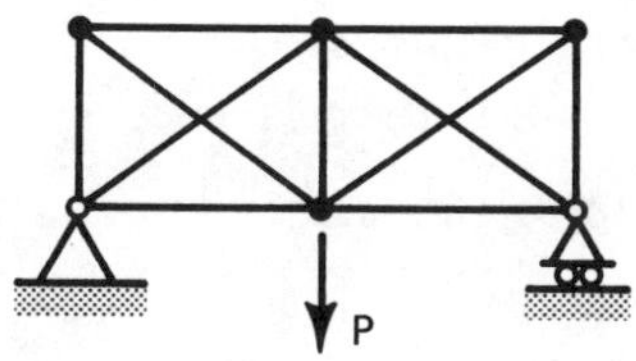

(b) ______

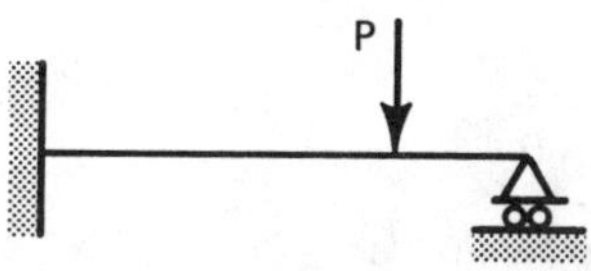

(c) ______

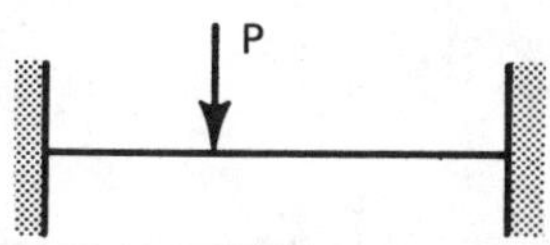

(d) ______

Problem 10. In the equation $\delta_{ao} + X_a \delta_{aa} = 0$, give the precise definition of each term: δ_{ao}; δ_{aa}; X_a.

Problem 11.

1. For the truss shown, calculate the internal forces, f_i, due to the applied load.
2. Show on the truss the unit load which you would apply to find the vertical deflection at point A.
3. Calculate the internal forces, u, resulting from this unit load.

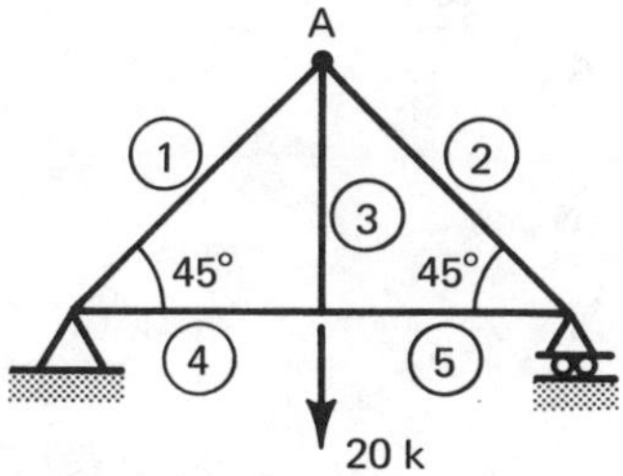

ANSWERS

Member	f_i, kip	u
1		
2		
3		
4		
5		

Problem 12. Solve for the forces in all members. Get a numerical result.

For all members

Area = 0.1 in^2

E = 30 x 10^6 psi

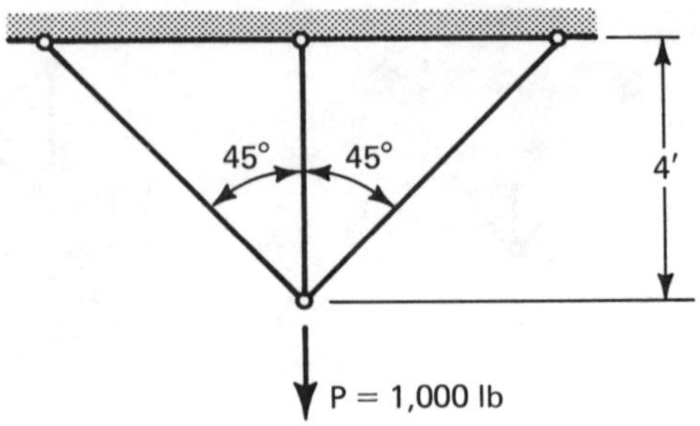

Problem 13. Find the forces in each member of the trusses shown.

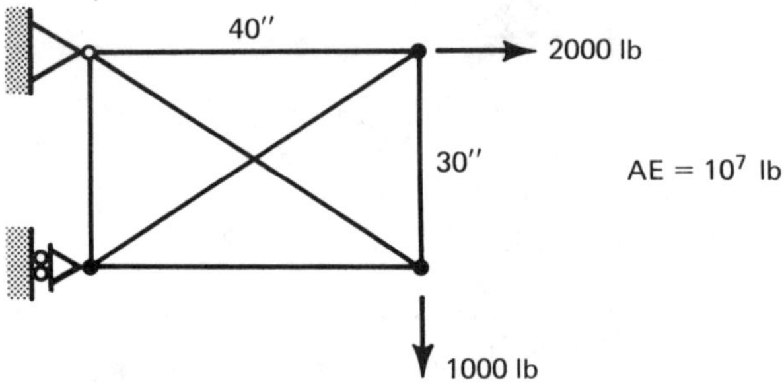

Problem 14. Find the loads in each member.

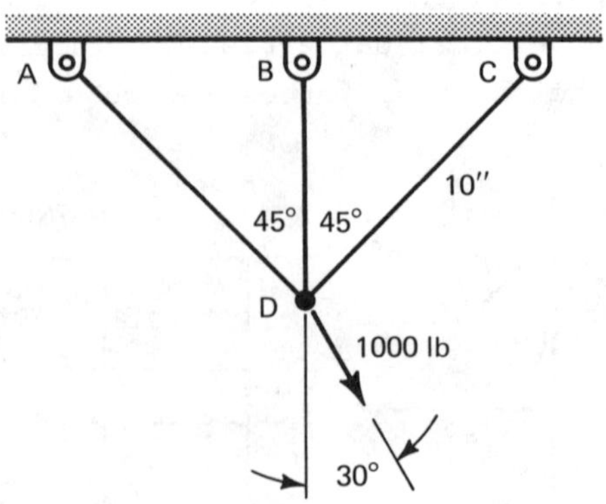

Problem 15. Give a written description of the process of geometric compatibility for this singly redundant beam.

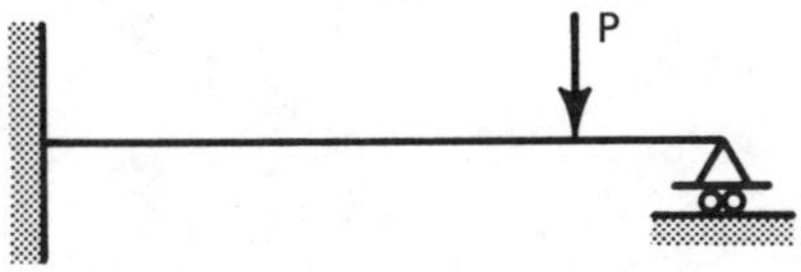

Problem 16. Derive the equations for the solution of the singly redundant structure shown.

Work out a numerical solution.

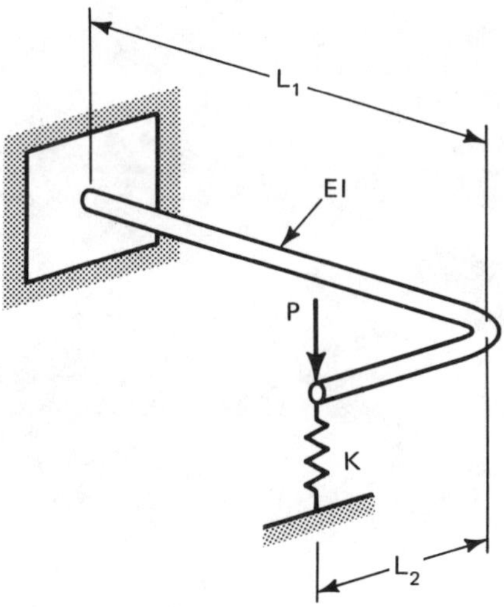

References

1. Aircraft Structures, David J. Perry, McGraw-Hill, New York, 1950, Chapter 16.
2. Energy Methods in Applied Mechanics, H. L. Langhaar, John Wiley and Sons, New York, 1962.

Chapter 8

FLANGE-WEB STRUCTURES (SEMI-MONOQUE STRUCTURES)

8.1 INTRODUCTION AND BASIC EQUATIONS

Whenever aerodynamic shape and forces are important, and, at the same time, appreciable volumes must be enclosed, we find that the flange-web structure becomes efficient. Examples of this application are: aircraft, missiles, boosters, urban transit vehicles, racing cars, etc.

The flange-web structure is characterized by thin, flat panels (webs) which carry shear, and long, thin stringers (flanges) which carry only tension or compression. An example of this is depicted in Figure 8.1.1. The idealized structure is also shown in Figure 8.1.2. The significant variables which appear in the idealization are the flange areas and the web thickness.

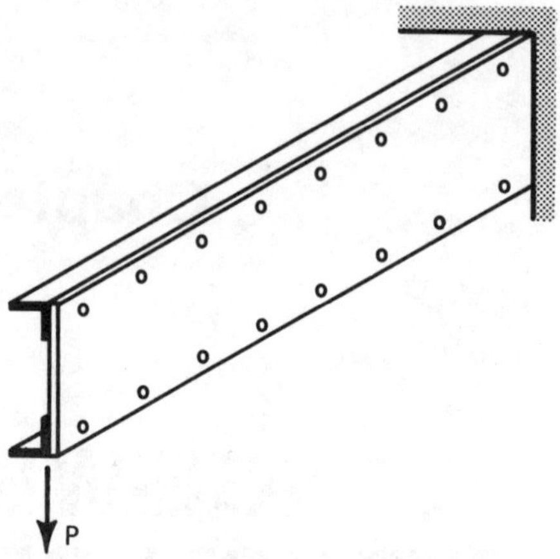

Figure 8.1.1 Actual flange-web structure riveted up from flat sheet and angle

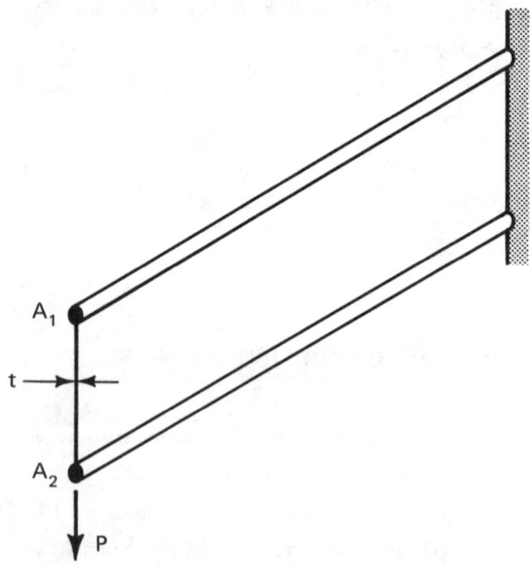

Figure 8.1.2 Structural idealization of Figure 8.1.1.
Only flange areas and web thickness are significant

The structural idealization made for the analysis of flange-web type structures appears in the following statements:

A. The webs carry all the shear and only shear.

B. The flanges carry all the bending (by going in tension and compression) and only bending.

These two assumptions simplify the analysis of this kind of structure.

The deformation and bending behavior of the structure of Figure 8.1.2 is depicted in Figure 8.1.3.

If we ignore the wall effects, we observe the following in Figure 8.1.3:

1. The top flange extends in tension.
2. The bottom flange contracts in compression.
3. The panel deforms in pure shear.

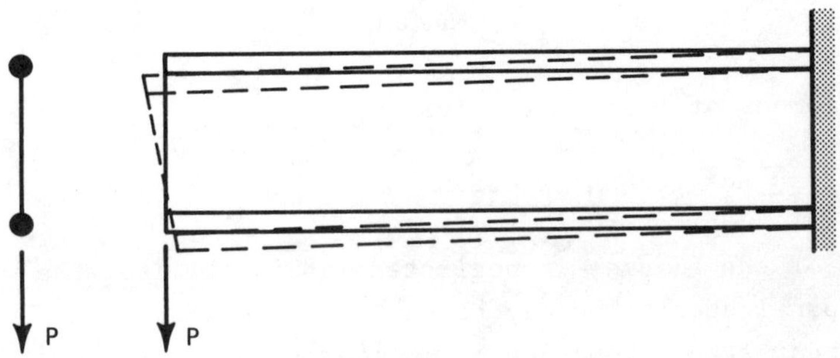

Figure 8.1.3 Deformation pattern (exaggerated) for flange-web structure

We may use the beam bending stress and beam shear stress equations to calculate the flange forces and stresses and the panel shear flow as illustrated in the following example. (See Figure 8.1.4.) Let us take

$A_1 = A_2 = 1.0 \text{ in}^2$

$L = 48''$

$h = 12''$

$t = 0.10$ in

$P = 1{,}000$ lb

Our first observation is that all the significant area (for purposes of bending stress) has been lumped in the flanges. (Web area is negligible.)

Therefore the neutral axis (centroid) is at the center of the cross section. Six inches up from the bottom.

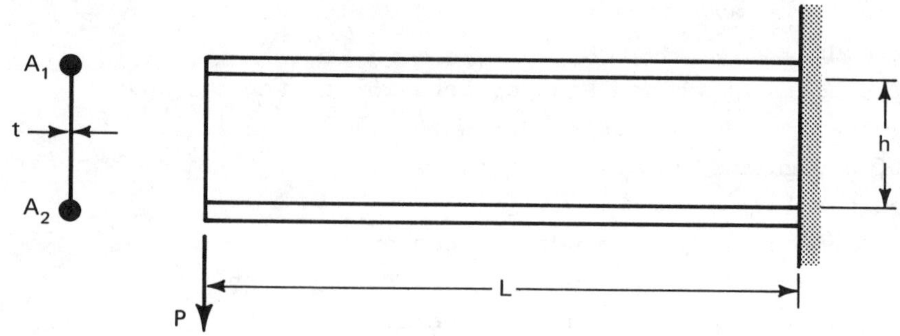

Figure 8.1.4

The moment of inertia is given by

$$I_{N.A.} = \sum Ad^2 \text{ for the flanges} \tag{8.1.1}$$

The I_o of the flanges is neglected, as is the I of the web.

The panel shear flow may be easily calculated from the beam shear stress equations. We know that

$$\tau = \frac{VQ}{It} \tag{8.1.2}$$

$$q = \tau \cdot t = \frac{VQ}{I} \tag{8.1.3}$$

We observe that, from the bottom of the panel to the top, we do not cross any (significant) area. All the $Q = A\bar{y}$ is in the flanges. The panel shear flow is everywhere constant, both depthwise and spanwise.

With respect to the neutral axis,

$$Q = A\bar{y} = 1(6) = 6 \text{ in}^3$$

$$q = \frac{VQ}{I} = \frac{1{,}000(6)}{72} = \frac{1{,}000}{12} \frac{\text{lb}}{\text{in}}$$

The shear flow in the panel is shown in Figure 8.1.5. The shear flow in the panel can also be calculated by considering the depthwise equilibrium of the front face.

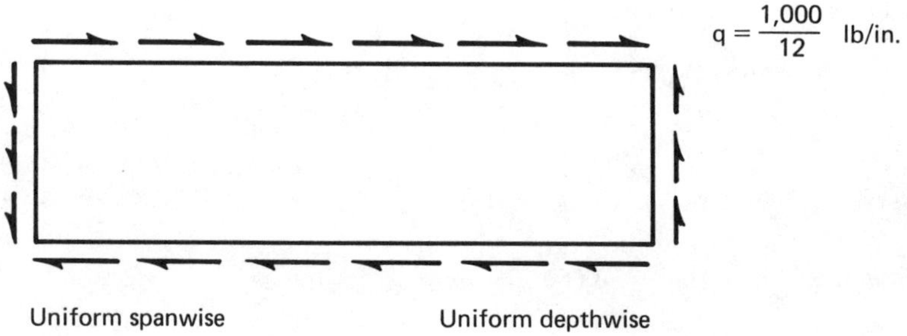

Figure 8.1.5 Shear flow in panel

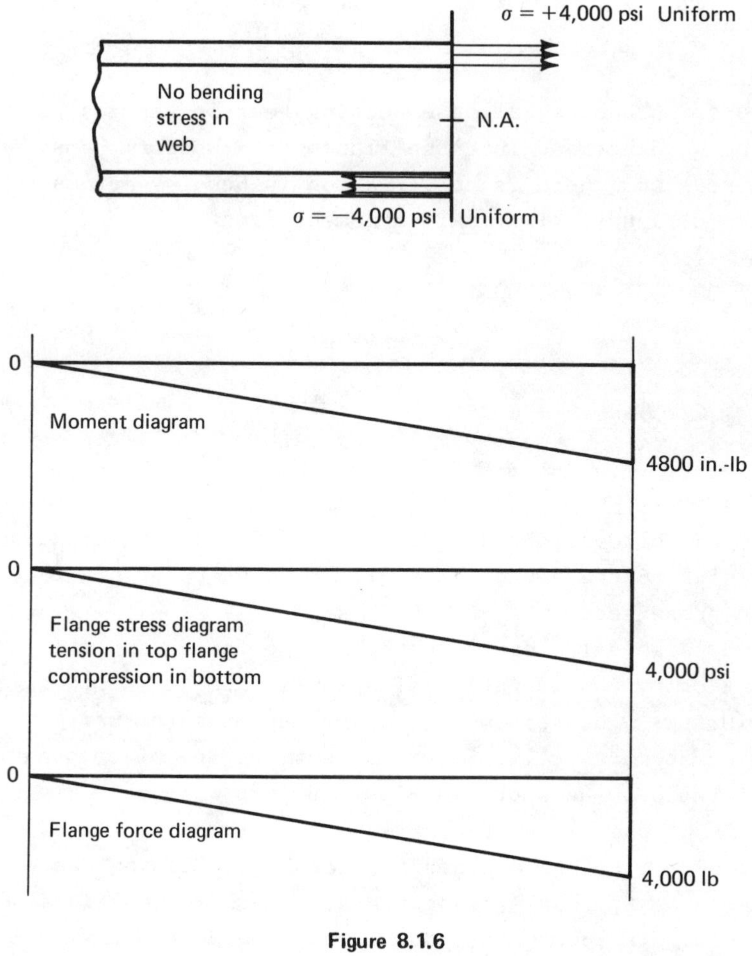

Figure 8.1.6

From this we can see that

$$q = \frac{V}{h} = \frac{1,000}{12} \frac{lb}{in}$$

We have

$$I = A_1 d_1^{\ 2} + A_2 d_2^{\ 2} = 1(6)^2 + 1(6)^2$$

$$I = 72 \text{ in}^4$$

Now in

$$\sigma = \frac{Mc}{I} \qquad (8.1.4)$$

we have c = <u>6</u> because all the bending goes in the flanges. Also the bending goes into the flange as a uniform tension or compression, there is no variation in bending across the lumped flange. The <u>maximum</u> moment is at the wall

$$M = 1,000(48) = 48,000 \text{ in lb}$$

and

$$\sigma = \pm \frac{48,000(6)}{72} = \pm 4,000 \text{ lb}$$

(at the wall).

This can be considered as an application of the equation of statics to the depthwise equilibrium of the front face of the cross section.

We may also calculate both the flange forces and the panel shear flow by consideration of spanwise moment and force equilibrium. For this purpose, any one inch (or unit) spanwise segment will do; let us consider the piece near the end of the beam as shown in Figure 8.1.7.

We first observe the following:

1. According to the engineering theory of bending, deformations, Δ, of the flanges are linearly proportional to the distance up (or down) from the neutral axis. See Figure 8.1.7.

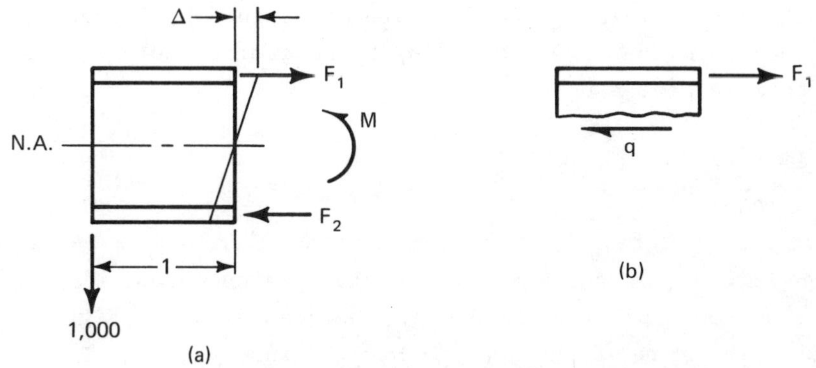

Figure 8.1.7 Spanwise moment and force equilibrium

2. Therefore, in a unit span, flange strains are linearly proportional to the distance from the neutral axis.
3. Since stress is proportional to strain, we can say that stresses are linearly proportional to the distance from the neutral axis.
4. Since $F = A\sigma$, the flange forces are proportional to the flange areas.

Let us now rework our example problem by reference to statics alone. Look at Figure 8.1.7. The equilibrium of spanwise moments says that

$$P(1) = M, \text{ or } M = 1{,}000 \text{ in-lb per in. of span}$$

Now, since the flanges are equidistant from the neutral axis and have equal areas, the flange forces are equal,

$$F_1 = F_2$$

and, from equilibrium of internal forces

$$F_1 d_1 + F_2 d_2 = M \qquad (8.1.5)$$

$$F_1(6) + F_2(6) = 1{,}000$$

$$F_1 = \frac{1{,}000}{12} \text{ lb/in of span}$$

This equals 4,000 lb at the wall for a check.

The signs of forces F_1 and F_2 are determined by inspection.

We may then use Figure 8.1.7(b) to determine the value of q, since we have a unit span,

$$q(1) = F_1$$

$$q = 1000/12 \text{ lb/in}$$

which is in agreement with our previous calculation.

Let us now apply this method to the beam shown in Figure 8.1.8. The span of the beam is L = 100 in.

We first calculate the centroid. Using the bottom flange as a reference, the centroid, $\bar{y}$, is

$$\bar{y} = \frac{\sum A\bar{y}}{\sum A} = \frac{1(12)}{3} = 4'' \text{ up from bottom} \qquad (8.1.6)$$

Since the flange forces are proportional to their areas and their respective distance from the neutral axis, we can write

$$F_1 = \left(\frac{d_1}{d_2}\right)\left(\frac{A_1}{A_2}\right)F_2 \qquad (8.1.7)$$

or,

$$F_1 = F_2$$

as before.

Note that this is necessary for spanwise equilibrium. We have

$$P(1) = M$$

$$M = 1{,}000 \text{ in-lb per in. of span}$$

Then the in-plane equilibrium will give

$$F_1 d_1 + F_2 d_2 = M$$

$$F_1(8) + F_2(4) = M$$

but

$$F_1 = F_2$$

thus

$$12F_1 = M$$

$$F_1 = \frac{1,000}{12} \text{ lb per in. of span}$$

Using now either the upper or lower portion of Figure 8.1.9, spanwise equilibrium is given by

$$q = 1000/12 \text{ lb/in}$$

We can observe some general things regarding this two-flange, cantilevered beam with a concentrated load at the tip:

1. Flange forces vary linearly from zero at tip to max at wall. (± sign determined by inspection.)
2. Flange stresses are flange forces divided by flange areas and vary as in number 1.
3. Shear flow in the web is uniform throughout.

<u>Example Problem</u>: Three flanges.

The beam of Figure 8.1.10 is loaded as shown. Let us find the flange forces per unit span, F_i, and the shear flows q. The method of statics will be used:

The centroid is found to be at

$$\bar{y} = \frac{\sum A\bar{y}}{\sum A} = \frac{1(12) + 2(6)}{6} = 4''$$

up from the bottom flange.

Considering the equilibrium of a unit span about point A, the neutral axis, we have

$$F_1(8) + F_2(2) + F_3(4) = 1{,}000\ (1) \tag{8.1.8}$$

From Equation 8.1.7, we have

$$F_1 = \left(\frac{8}{2}\right)\left(\frac{1}{2}\right) F_2 = 2F_2$$

and

$$F_2 = \left(\frac{2}{4}\right)\left(\frac{2}{3}\right) F_3 = \frac{1}{3} F_3$$

Substituting into Equation (8.1.8) gives

$$(2/3)\ F_3\ (8) + 1/3\ F_3\ (2) + F_3\ (4) = 1{,}000$$

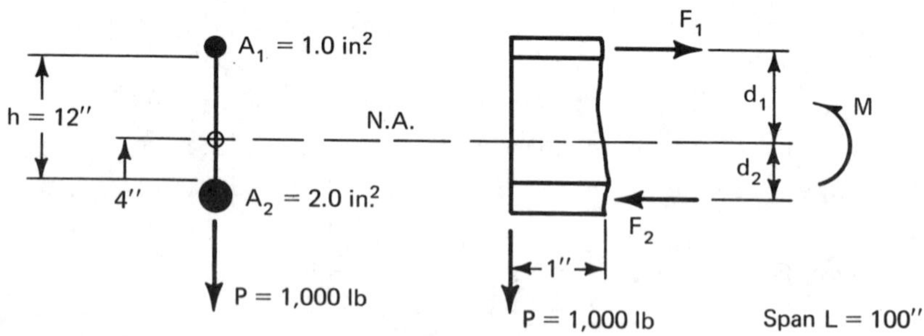

Figure 8.1.8

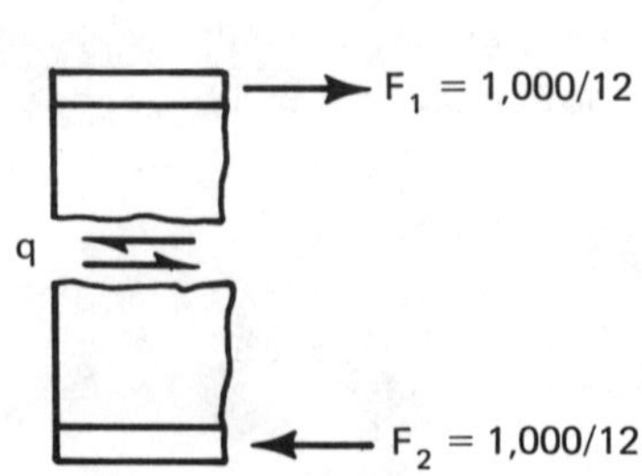

Figure 8.1.9

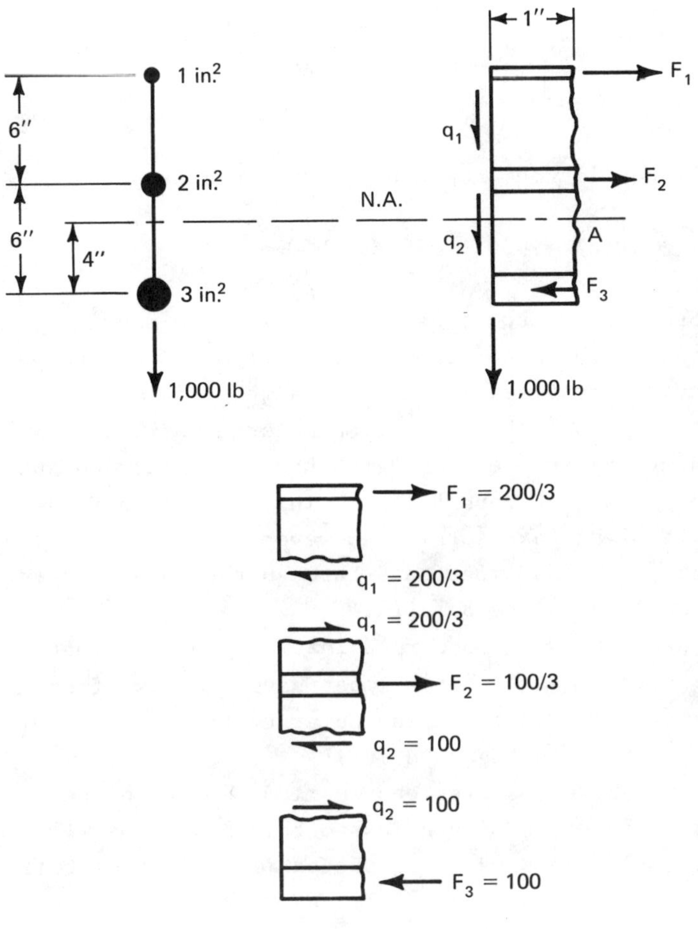

Figure 8.1.10

$$F_3 = 100 \ \frac{\text{lb}}{\text{in}} \qquad \text{(Compr.)}$$

$$F_2 = \frac{100}{3} \ \frac{\text{lb}}{\text{in}} \qquad \text{(tens.)}$$

$$F_1 = \frac{200}{3} \ \frac{\text{lb}}{\text{in}} \qquad \text{(tens.)}$$

Considering the span-wise free body diagrams of Figure 8.1.10 we see that

$$q_1 = \frac{200}{3} \text{ lb/in}$$

$$q_2 = 100 \text{ lb/in}$$

Checking Equilibrium down the front face, we see

$$q_1(6) + q_2(6) = 400 + 600 = 1{,}000$$

An observation from these results will be useful to us in the future. The shear flows, q, undergo a jump change in magnitude every time we cross a flange. The magnitude of the jump is equal to F, the flange force per unit span.

The next case to be considered has cross section such as the one shown in Figure 8.1.11. This beam appears as though it might take torsion; however, in theory it cannot. Therefore, the load must pass through the shear center, c, as depicted in Figure 8.1.11.

We are going to be concerned initially with symmetric bending. This means that we must have an axis of symmetry (of the flange areas) perpendicular to the bending moment vector as shown in Figure 8.1.11.

With the above observations in mind, we may proceed to apply the equations of statics to this beam. We will follow the same principles used for the earlier beams of this chapter.

The centroid is located:

$$\bar{y} = \frac{\sum A\bar{y}}{\sum A} = \frac{(1)(6) + (1)(6)}{(6)} = 2 \text{ in.}$$

up from the bottom flanges.

The moment of a spanwise unit section about point A is

$$F_1(4) + F_2(4) + F_3(2) + F_4(2) = 1{,}000$$

By use of Equation (8.1.7), we can relate the four forces. This results in

$$F_1 = F_2 = F_3 = F_4$$

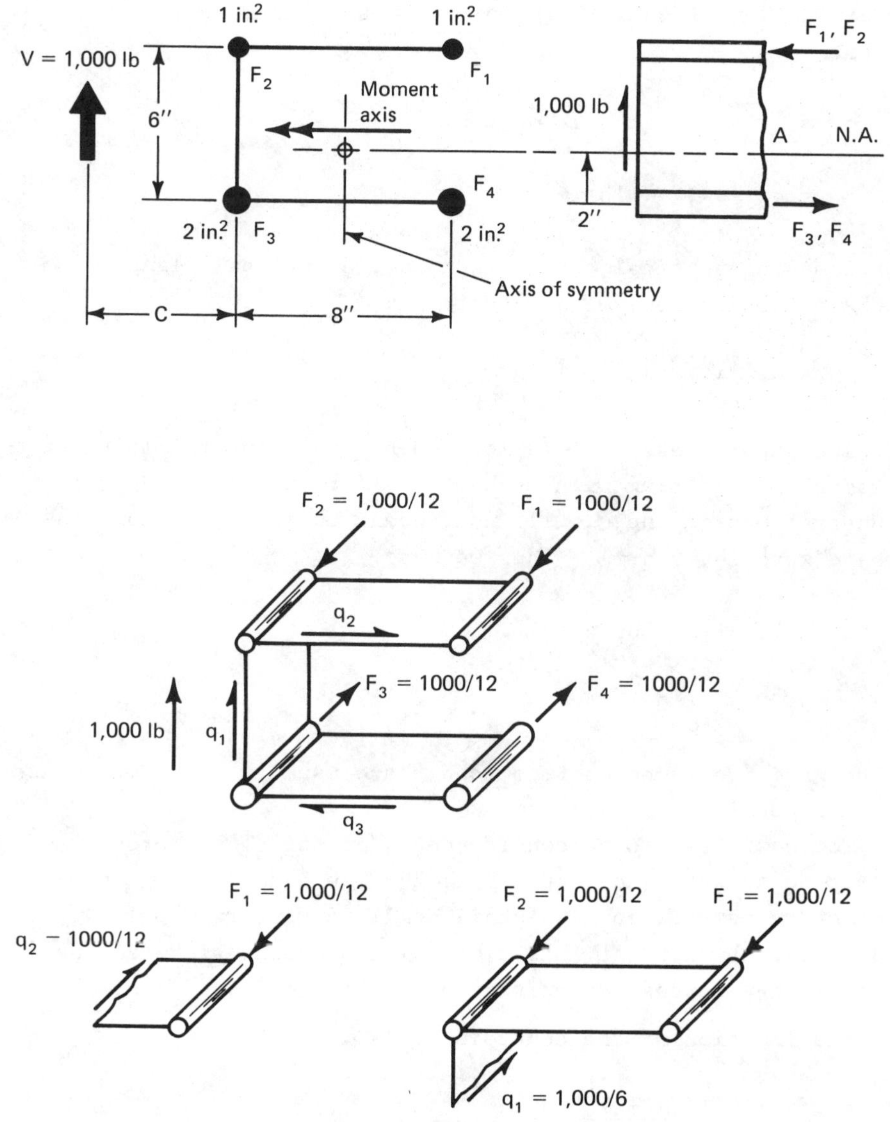

Figure 8.1.11

and thus

$$12F_1 = 1{,}000$$

$$F_1 = \frac{1000}{12} \text{ lb/in of span}$$

Examination of equilibrium of spanwise slices will show that

$$q_1 = \frac{1,000}{6} \text{ lb/in}$$

$$q_2 = q_3 = \frac{1,000}{12} \text{ lb/in}$$

Checking vertical equilibrium in the vertical flange of the cross section, we see that

$$q_1(6) = \frac{1,000}{6}(6) = 1,000$$

The shear center, c, is located by examining the in-plane moment equilibrium with respect to (for example) flange 3. Shear flows q_1 and q_3 will not appear in this equation (they have zero lever arm). Thus we have

$$q_2\,(8)(6) = 1,000\ c$$

$$c = 4''$$

The positive signs of F, q, and c are usually just determined by inspection.

The next beam to be considered is one of closed cross section as shown in Figure 8.1.12. The beam is now capable of carrying torsion. We shall still be concerned only with symmetric bending at this time. The procedure for finding the flange forces per unit span is the same as before.

The location of the centroid is

$$\bar{y} = \frac{\sum A\bar{y}}{\sum A} = \frac{1(5) + 1(5)}{5} = 2 \text{ in.}$$

The sum of moments on a unit span about point A gives

$$F_1(3) + F_2(3) + F_3(2)\ F_4(2) = 1,000\ (1)$$

We note that

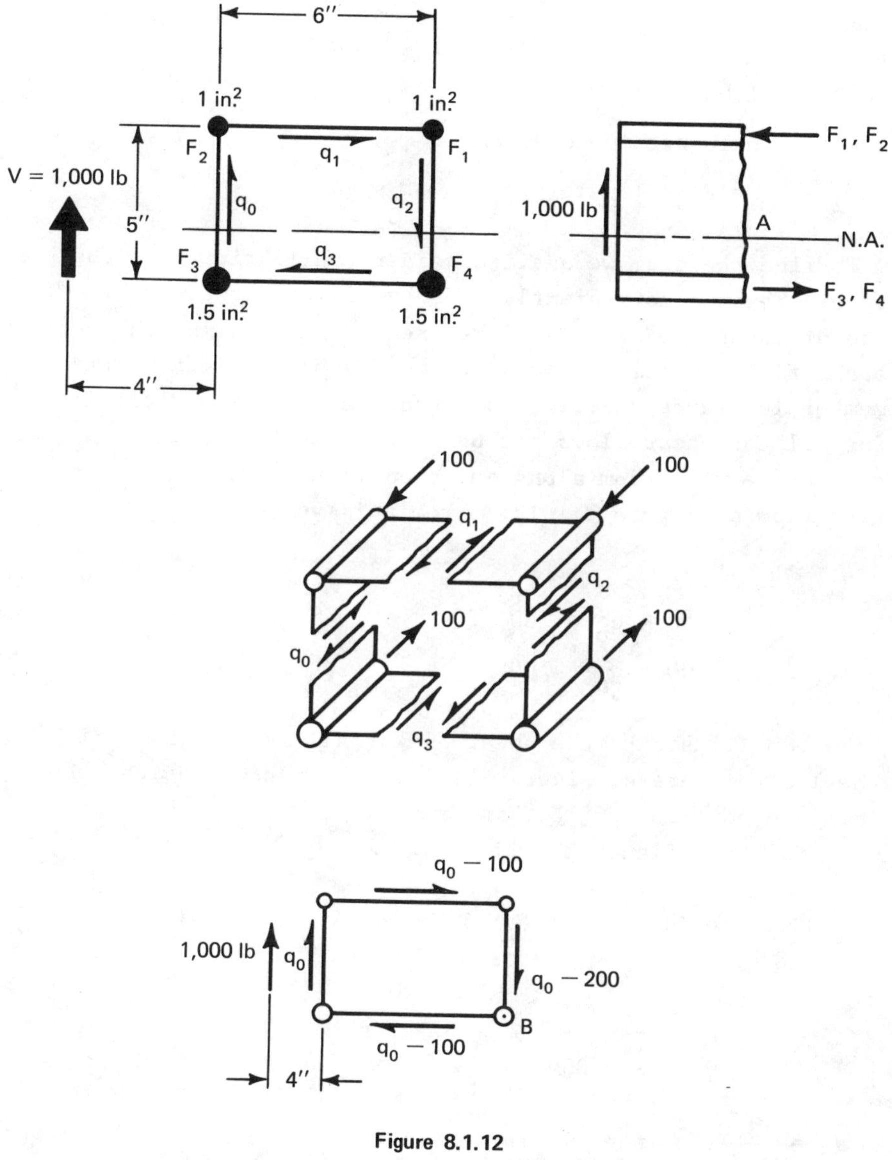

Figure 8.1.12

$$F_1 = (3.2)(1/1.5)F_2 = F_2$$

and similarly

$$F_1 = F_2 = F_3 = F_4$$

Thus

$$10F_1 = 1{,}000$$

$$F_1 = 100 \text{ lb per in. of span}$$

$$F_2 = F_3 = F_4 = 100 \text{ lb per in. of span}$$

To find the q's, we again consider equilibrium down the span. This is conveniently done by taking (arbitrarily) one of the shear flows as a reference and relating the other shear flows to it. q_o is shown acting on the left vertical member in Figure 8.1.12. Note that the positive direction for all the shear flows has been arbitrarily assumed. We consider equilibrium along the span in the exploded perspective view of Figure 8.1.12. From this we see:

$$q_1 = q_o - 100$$

$$q_2 = q_1 - 100 = q_o - 200$$

$$q_3 = q_1 + 100 = q_o - 100$$

We can illustrate the shear flows, all related to q_o in the last picture of Figure 8.1.12. If we now consider the in-plane equilibrium of this front face, we can, by taking moments about flange B, write

$$1{,}000(4) = (q_o - 200)\ (5)(6) + (q_o - 100)(6)(5)$$

or

$$4{,}000 = 60\ q_o - 9{,}000$$

$$q_o = \frac{1300}{6} \text{ lb/in} = 216 \text{ lb/in}$$

Let us check the sum of the vertical forces on the front face; we have:

$$1{,}000 = q_o(5) - (q_o - 200)(5)$$

$$1{,}000 = 1{,}000$$

The final shear flows are

$$q_o = 216 \text{ lb/in}$$
$$q_1 = 116 \text{ lb/in}$$
$$q_3 = 16 \text{ lb/in}$$
$$q_4 = 116 \text{ lb/in}$$

All as shown in Figure 8.1.12.

Let us now consider another way of solving this problem. Examination of Figure 8.1.12 shows that q_o is flowing around the front face and appears in every panel. The final resultant shear flow can be considered as the superposition of the shear flow situation for which $q_o = 0$ plus one in which the webs contain only $q_o = \frac{M}{2A}$. The value of M is equal to the moment of the external force about a flange such as number 3 minus the moment of the q's in Figure 8.1.13(a). This is written as M = T - Mo.

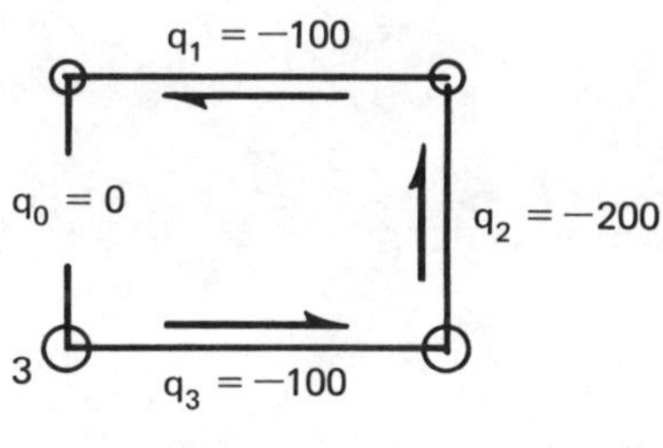

$$M_o = -300(6)(5) = -9{,}000 \text{ in-lb}$$
$$T = 4{,}000 \text{ in-lb}$$

(a)

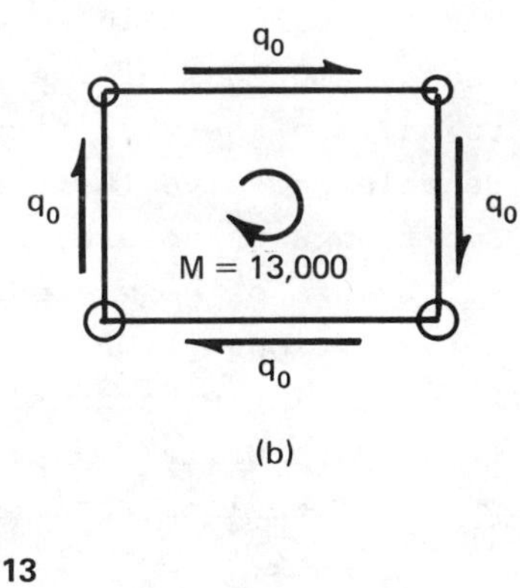

(b)

Figure 8.1.13

An example will be worked out to clarify this discussion. We will use the same problem as before. Refer to Figure 8.1.12 for flange forces and sign conventions. It is very important to keep the signs straight.

From Figure 8.1.13

$$Mo = 9000 \text{ in-lb (CW+)}$$

From Figure 8.1.12

$$T = +4{,}000 \text{ in-lb}$$
$$M = 4{,}000 - (-9{,}000) = 13{,}000 \text{ in-lb}$$
$$2A = 60 \text{ in}^2$$

Then we have, Figure 8.1.13(b)

$$q_o = \frac{13{,}000}{60} = 216 \text{ lb/in}$$
$$q_1 = 116 \text{ lb/in}$$
$$q_2 = 16 \text{ lb/in}$$
$$q_3 = 116 \text{ lb/in}$$

as before. The similarities between the two methods should be studied.

Objectives

(OB = Open Book; CB = Closed Book)

1. Be able to state in words and illustrate with sketches the assumptions of the flange-web structure. Be able to explain the resulting structural behavior. (CB)
2. Be able to solve for the flange forces and shear flow in two-flange structures.
   A. By use of beam bending stress and beam shear stress equations. (OB)
   B. By using equations of statics. (OB)
3. Be able to write down strain energy for a shear panel of given dimensions, and, for the same panel, be able to

apply the method of virtual work to the calculation of the deformation due to shear. (OB)

4. Be able to calculate flange forces, panel shear flow and the shear center for open sections of flange-web type by application of the equations of statics. (OB)
5. Be able to calculate beam deflections and the angle of twist in torsion for the structure of objective number 4 by application of the method of virtual work. (OB)
6. Be able to calculate the flange forces and panel shear flows for closed sections of four flanges or more. (OB)
7. Be able to calculate the bending deflection and the angle of twist in torsion for structures of the type given in objective 6 by application of the method of virtual work. (OB)
8. Be able to solve for the flange force and panel shear flow in one-and two-redundant multicell flange-web structures by application of the method of virtual work. Be able to discuss the extension of the method to more redundants. (OB)

8.2 TORSION OF FLANGE-WEB STRUCTURES—ANGLES OF TWIST

We can easily apply the method of virtual work to the calculation of the angle of twist due to the torsion of flange-web structures. Let us consider the problem of Figures 8.1.12 and 8.1.13 of the previous section. We consider only the torsion, since the torsion and bending effects can be separated.

The q resulting from the application of the load to the structure of Figure 8.1.12 may be considered to be the response of the structure to the applied external load. This is shown in Figure 8.2.1.

The equation for the change of angle of a shear panel with shear flow q and shear flow q_u, due to unit load is, following the previous equations for unit-load deflection:

$$\Delta = \frac{qq_u ab}{Gt} \qquad (8.2.1)$$

where:

q = shear flow due to real loads
q_u = shear flow due to a unit load at, and in the sense of, the desired deflection
a = the panel length
b = the panel width
G = the shear modulus of elasticity
t = the thickness of the panel

The unit shear flows, q_u, are easily calculated for the problem of angle of twist. Since the appropriate unit load is a unit moment as shown in Figure 8.2.1, the q_u are simply

$$q_u = \frac{1}{2A}$$

where A is the area enclosed by the webs. Note that the flange forces per unit span, F, are not affected by a pure torque.

Let us now work a numerical example for this problem by taking:

a = 100 in
G = 12×10^6 psi
t = 0.025 in. for all four panels

To get the angle of twist, θ, we simply apply the unit torque and sum over all the panels

$$\theta = \sum_{i=1}^{4} \frac{a_i q_{ui} a_i b_i}{G t_i} \tag{8.2.2}$$

The q_{ui} are all equal to $\frac{1}{2A}$, so we may write

$$\theta = \frac{1}{2AG} \sum_{i=1}^{4} \frac{q_i a_i b_i}{t_i} \tag{8.2.3}$$

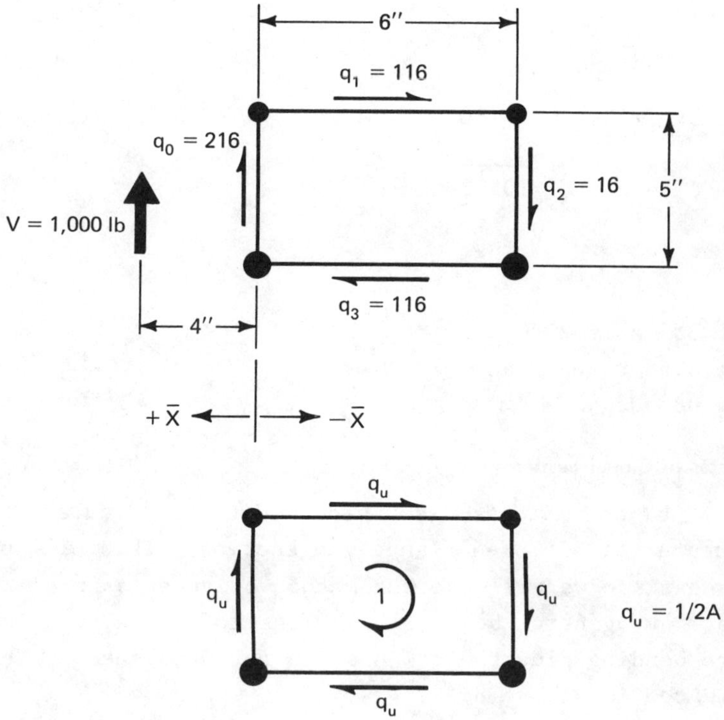

Figure 8.2.1

Since the panel lengths, a_i, are all equal, we may write

$$\theta = \frac{a}{2AG} \sum_{i=1}^{4} \frac{q_i b_i}{t_i} \tag{8.2.4}$$

Let us consider the term in front of the sum

$$\frac{a}{2AG} = \frac{100}{2(5)(6) \times 12 \times 10^6} = 1.388 \times 10^{-8}$$

Then, summing the remaining terms over the four panels:

panel 0: $\frac{216(5)}{0.025} = 43200$

panel 1: $\frac{116(6)}{0.025} = 27840$

$$\text{panel 2:} \quad \frac{16(5)}{0.025} = 3200$$

$$\text{panel 3:} \quad \frac{116(6)}{0.025} = 27840$$

$$\text{sum} = 102080$$

and

$$\theta = 1.388 \times 1.02080 \times 10^{-3}$$
$$\theta = 1.4168 \times 10^{-3} \text{ rad}$$
$$\theta = 0.083 \text{ deg}$$

Shear Center of Closed Section

We may use the methods discussed to date to calculate the shear center of a closed flange-web section. This will be important since we will use the method of superposition to transform our general beam calculations to the calculation for pure bending plus the calculation for pure shear. This is indicated in Figure 8.2.2.

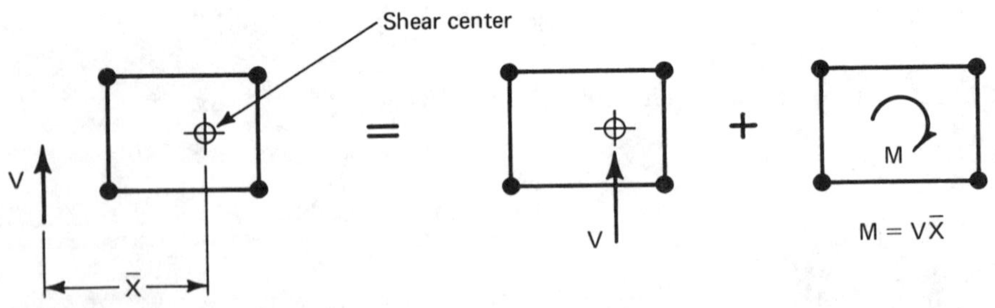

Figure 8.2.2

Let us return to the problem of Figure 8.2.1. Now, instead of the load V acting 4" to the left, let us call this distance the variable distance $\bar{X}$. We will calculate the position of $\bar{X}$ which will give a zero angle to twist to the structure. This will be the shear center.

Naturally, the q_o, q_1, q_2, q_3 of the structure will change. Figure 8.2.3 shows the application of the method of slicing

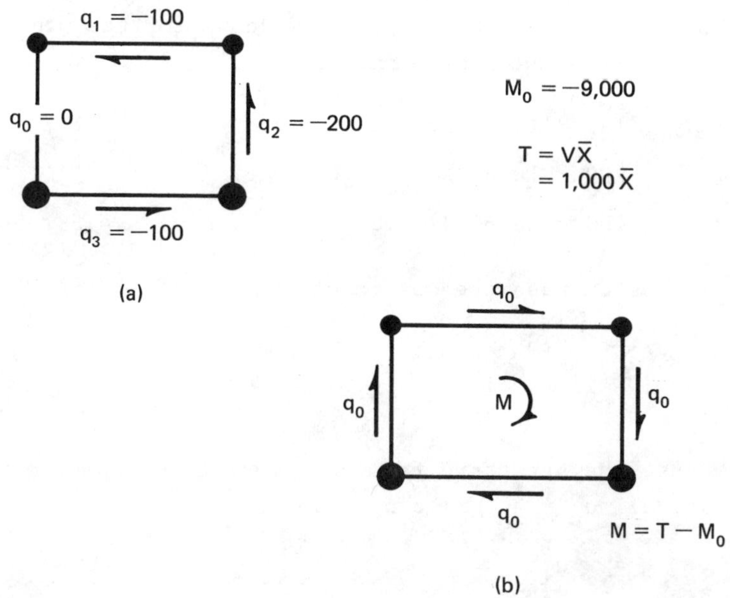

Figure 8.2.3

a web to this problem. The problem is the same as before except that the torque T is now equal to $V\overline{X}$. The location of the shear center is independent of the load, so we may take V = 1,000 lb as before.

The results of Figure 8.2.3a will be the same. The moment M, of Figure 8.2.3b will be the same in principle, but now $T = V\overline{X} = 1{,}000\overline{X}$. We now have

$$M = T - M_o$$
$$= 1{,}000\overline{X} - (-9{,}000)$$
$$M = 1{,}000\overline{X} + 9{,}000$$

and

$$q_o = \frac{M}{2A}$$
$$q_o = \frac{1{,}000\overline{X} + 9{,}000}{60}$$
$$q_o = 16.667\overline{X} + 150 \qquad (8.2.5a)$$

where $\overline{X}$ is still a variable. Now, we have for the other (final) flange forces, in terms of $\overline{X}$,

$$q_1 = q_o - 100 = 16.667\overline{X} + 50 \tag{8.2.5b}$$

$$q_2 = q_o - 200 = 16.667\overline{X} - 50 \tag{8.2.5c}$$

$$q_3 = q_o - 100 = 16.667\overline{X} + 50 \tag{8.2.5d}$$

Now let us express the condition that $\theta = 0$.

$$\theta = 0 = \sum_{i=1}^{4} \frac{q_i q_{ui} a_i b_i}{G t_i}$$

If we use the structure from our previous example, and recall that all $q_u = \frac{1}{2A}$, we can again write

$$\theta = \frac{a}{2AG} \sum_{i=1}^{4} \frac{q_i b_i}{t_i}$$

where the q_i are given by the results shown in Equations (8.2.5).

Substituting numbers from previous calculations, we have

$$\theta = 0 = 1.338 \times 10^{-8} \sum_{i=1}^{4} \frac{q_i b_i}{t_i}$$

Actually, all $t_i = 0.025$, so we can factor t out also, and

$$\theta = 0 = \frac{1.338 \times 10^{-8}}{t} \sum_{i=1}^{4} q_i b_i = 0$$

$$= 5.352 \times 10^{-7} \sum_{i=1}^{4} q_i b_i = 0 \tag{8.2.6}$$

Writing out the terms in the sum, we have:

panel 0: $(16.667\overline{X} + 150)(5) = 38.335\overline{X} + 750$
panel 1: $(16.667\overline{X} + 50)(6) = 100.002\overline{X} + 300$

$$\text{panel 2:} \quad (16.667\overline{X} - 50)(5) \quad = 83.335\overline{X} - 250$$

$$\text{panel 3:} \quad (16.667\overline{X} + 50)(6) \quad = 100.002\overline{X} + 300$$

$$\text{sum} = \quad 366.674\overline{X} + 1100$$

Then $\theta = 5.352 \times 10^{-7}\ (336.674\overline{X} + 1100) = 0$

Notice that 5.352×10^{-7} cancels out, so all terms which will factor out of the summation do not need to be calculated. Thus

$$366.674\overline{X} + 1100 = 0$$

$$\overline{X} = \frac{-1100}{366.674} = -2.99 \text{ in.}$$

or, 2.999 in. to the right of web q_o.

8.3 STATICALLY INDETERMINATE (MULTI-CELL) FLANGE-WEB STRUCTURES

We now turn our attention to the solution of a structure such as shown in Figure 8.3.1. This structure is one-degree statically indeterminate due to the fact that there are more shear flows, q, than the equations of static equilibrium available for solution.

This problem is solved most conveniently by separating it into two separate problems as shown in Figure 8.3.1(b) and (c). The problem of "pure" bending and "pure" torsion are treated separately. We shall use the beam of Figure 8.3.1 as an example problem.

Bending at Shear Center

The solution for flange forces per unit of span, F_1 through F_6, is the same as before. The flange forces may be solved for even though the beam is statically indeterminate.

The centroid is located above the bottom flange line as follows:

$$y = \frac{\sum A\overline{y}}{\sum A} = \frac{(1)(6) + (1)(6) + (1)(6)}{9} = 2 \text{ in.}$$

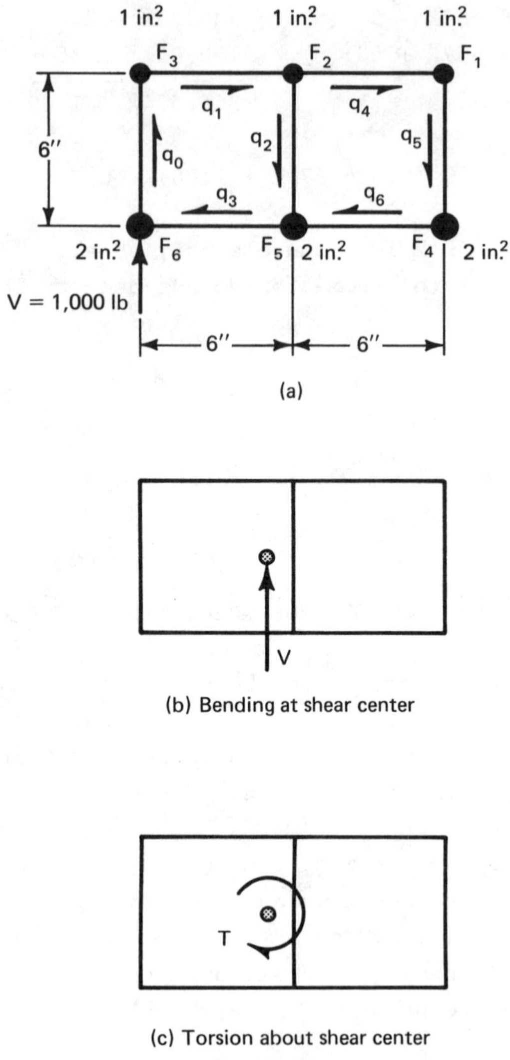

Figure 8.3.1

Using our previous procedure for relating the flange forces, Equation (8.1.7), we have

$$F_1 = F_2 = F_3 = F_4 = F_5 = F_6$$

Consideration of the moment equilibrium of a unit span gives the following equation

$$F_1(4) + F_2(4) + F_3(4) + F_4(2) + F_5(2) + F_6(2) = 1,000(1)$$

$$18\ F_1 = 1,000$$

$$F_1 = \frac{1,000}{18} = 55.55 \text{ lb/in}$$

The value of all flange forces per inch of span is 55.5 lb/in.

Calculation of the shear flows for the portion of the problem we are now dealing with (bending only) is now accomplished as before. A sufficient number of webs may be sliced to render the problem statically determinate. These will yield shear flows which we shall denote by $\bar{q}$. This is shown in Figure 8.3.2.

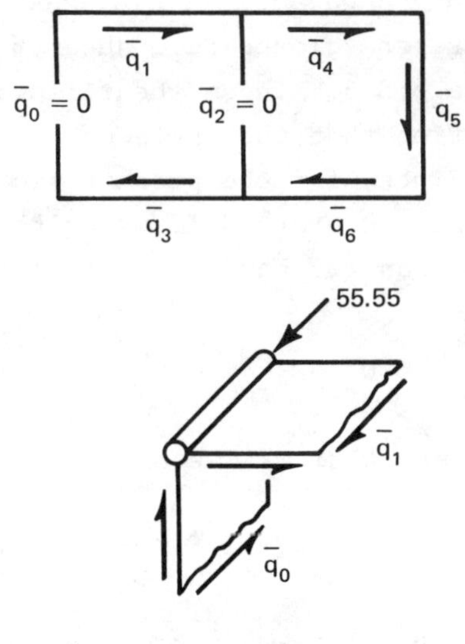

Figure 8.3.2

We see that

$$\bar{q}_o - 55.55 - \bar{q}_1 = 0$$

$$\bar{q}_1 = \bar{q}_o - 55.55 = 0 - 55.55 = -55.55 \text{ lb/in}$$

From this result, by inspection, we get

$q_4 = -111.1$ lb/in

$q_5 = -166.65$ lb/in

$q_6 = -111.1$ lb/in

$q_3 = -55.55$ lb/in

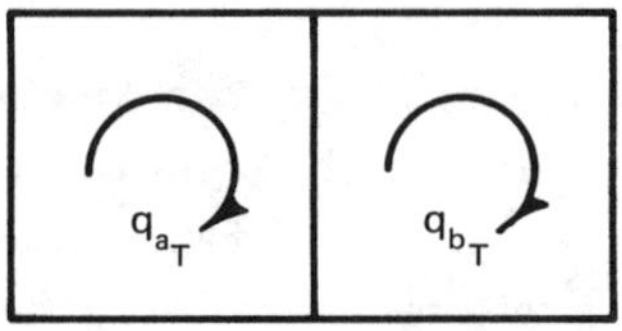

Figure 8.3.3

Superimposed on the $\bar{q}$ shear flow for the structure with the flanges cut, are the closed-cell shear flows q_{aT} and q_{bT} shown in Figure 8.3.3. These shear flows are calculated from the condition that the angle of twist of the two cells are zero. (Since, for the present problem, the beam is loaded at the shear center.) If we write this condition in terms of a unit span, we have

$$\sum \frac{q\ q_u\ ab}{Gt} = \sum \frac{q\ b}{2AGt} = 0 \qquad (8.3.1)$$

Canceling the constant terms, we have

$$\sum \frac{qb}{t} = 0$$

Where the q are the final shear flows for the bending portion of the problem, that is

$$q = \bar{q} + q_T$$

Writing the condition of Equation (8.3.1) for cell a, we have

$$\sum_a \frac{\bar{q}b}{t} + q_{a_T} \sum_a \frac{b}{t} - q_{b_T} \left(\frac{b}{t}\right)_{a-b} = 0 \qquad (8.3.2a)$$

and, doing this again for cell b, we have

$$\sum_b \frac{\bar{q}b}{t} + q_{b_T} \sum_b \frac{b}{t} - q_{a_T} \left(\frac{b}{t}\right)_{b\ -\ a} = 0 \qquad (8.3.2b)$$

If t is constant for each section, we will have for Equations (8.3.2)

$$\bar{q}_1\ (6) + \bar{q}_3\ (6) + q_{a_T}\ (6 + 6 + 6 + 6) - q_{b_T}\ (6) = 0$$

$$\bar{q}_4\ (6) + \bar{q}_5\ (6) + \bar{q}_6\ (6) + q_{b_T}\ (6 + 6 + 6 + 6) - q_{a_T}\ (6) = 0$$

or

$$(-55.55)(6) - (55.55)(6) + q_{a_T}\ (6 + 6 + 6 + 6) - q_{b_T}\ (6) = 0$$

$$(-111.1)(6) - (166.65)(6) - (111.1)(6) + q_{b_T}\ (24) - q_{a_T}(6) = 0$$

or

$$24\ q_{a_T} - 6\ q_{b_T} - 666.6 = 0$$

$$-6\ q_{a_T} + 24\ q_{b_T} - 2333.1 = 0$$

or

$$q_{a_T} = 55.55 \text{ lb/in}$$

and

$$q_{b_t} = 111.1 \text{ lb/in}$$

Superimposing our results to date, and recalling that we are still dealing with the bending portion of the beam problem only, we have

Cell a:

$$\bar{q}_o + q_{a_T} = 0 + 55.55 = 55.55 \text{ lb/in}$$

$$\bar{q}_1 + q_{a_T} = -55.55 + 55.55 = 0.0 \text{ lb/in}$$

$$\bar{q}_2 + q_{a_T} = 0 + 55.55 = 55.55 \text{ lb/in}$$

$$\bar{q}_3 + q_{a_T} = -55.55 + 55.55 = 0.0 \text{ lb/in}$$

Cell b:

$$\bar{q}_2 + q_{b_T} = 0 + 111.1 = 111.1 \text{ lb/in}$$

$$\bar{q}_4 + q_{b_T} = -111.1 + 111.1 = 0.0 \text{ lb/in}$$

$$\bar{q}_5 + q_{b_T} = -166.65 + 111.1 = -55.55 \text{ lb/in}$$

$$\bar{q}_6 + q_{b_T} = -111.1 + 111.1 = 0.0 \text{ lb/in}$$

These results are shown in Figure 8.3.4. Note that $q_2 = q_{b_T} - q_{a_T} = 55.55$ lb/in, up the web. Knowing the final shear flows for the bending portion of the problem, we can find the shear center by taking in-plane moments in the front face. From Figure 8.3.4 we have: $(\sum M_A)$

$$1{,}000\ \bar{x} = 55.55(6)(6) - 55.55(6)(6)$$

$$\bar{x} = 0$$

This says that the shear center is coincident with the center web. We could have determined this by inspection for this beam.

Torsion about Shear Center

With the shear center location known, we can calculate the response of the closed-cell structure to torsion about the shear center. This is shown in Figure 8.3.5.

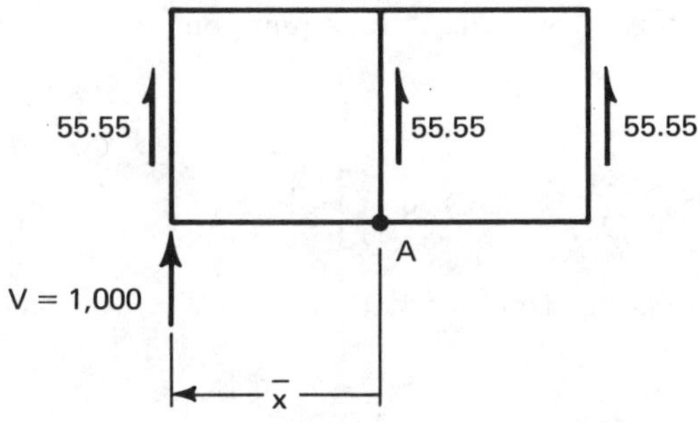

Figure 8.3.4

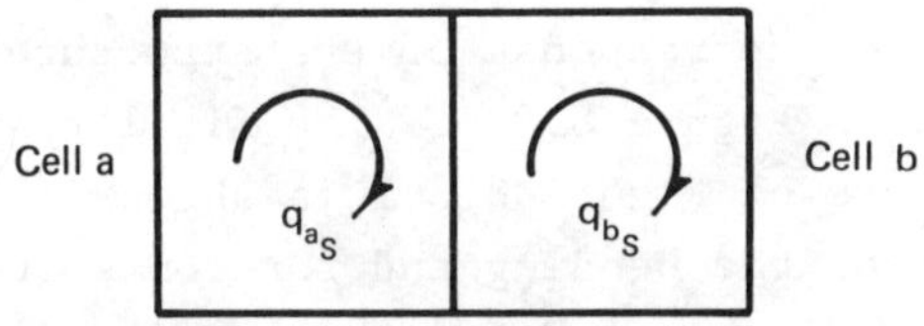

Applied torque $T = V\bar{X} = 1{,}000\ (6) = 6{,}000$ in.-lb

Figure 8.3.5

The conditions used to solve this portion of the problem are as follows:

1. The angle of twist of the two cells is equal.
2. The external torque, T, equals $\sum 2Aq$.

Writing condition 1 above, we have (for a unit span)

$$\sum_a \frac{q_{as} b}{2AGt} = \sum_b \frac{q_{bs} b}{2AGt} \tag{8.3.3}$$

Terms which are the same in each summation will cancel. Since we have taken t = constant for this problem, we will cancel all terms in the summation and have:

$$q_{as} \sum_a b = q_{bs} \sum_b b$$

Or, since each cell (in this problem) is a 6 in. square,

$$q_{as} = q_{bs} \tag{8.3.4}$$

Condition 2, above, yields the equation

$$T = \sum 2Aq$$

$$6{,}000 = 2\left[(36)\ q_{as} + (36)\ q_{bs}\right] \qquad (8.3.5)$$

Combining with Equation (8.3.4) we have

$$6{,}000 = 144\ q_{as}$$

$$q_{as} = q_{bs} = 41.66 \text{ lb/in}$$

The results for the response of the structure to pure shear are shown in Figure 8.3.6. The final result for the original structure of Figure 8.3.1 is shown in Figure 8.3.7. This is the sum of the bending and torsion solutions. It is also a superposition of Figures 8.3.4 and 8.3.6.

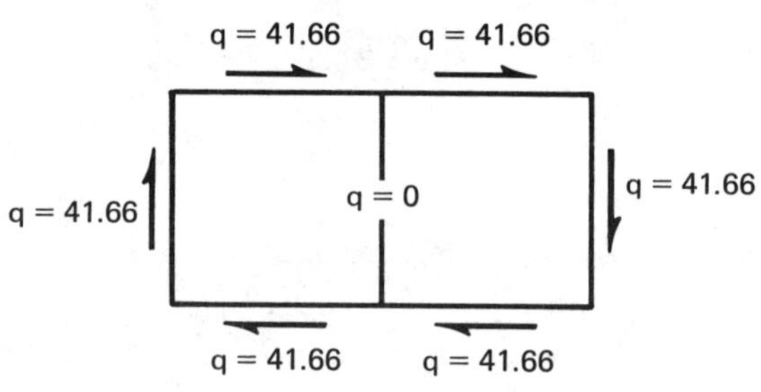

All units are lb/in.

Figure 8.3.6

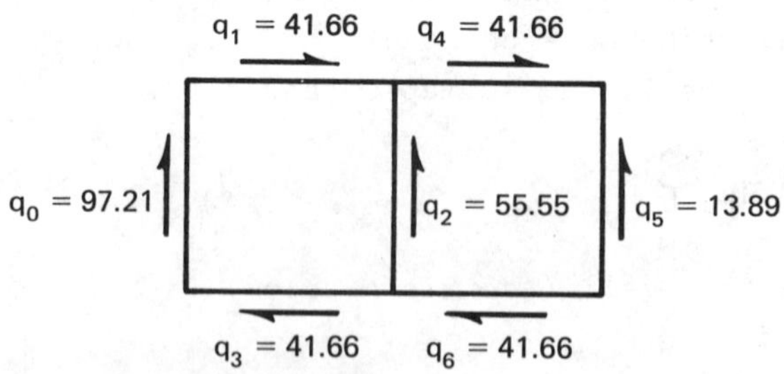

All units are lb/in.

Figure 8.3.7

Problems

Problem 1.

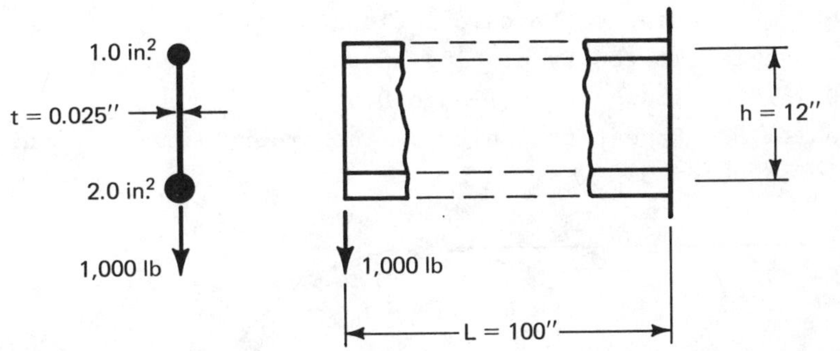

For the flange-web structure shown, find the following values by use of the beam bending and shear stress equations and by statics:

a. The flange stresses and forces at the wall.
b. The flange forces per unit of span.
c. The panel shear flow.

Problem 2. For the beam of problem 1, find the vertical downward deflection under the load.

Problem 3. The cross section of a flange-web beam is shown. The beam is cantilevered and has length L = 100 in. as shown in problem 1. What are the flange forces per unit of span and panel shear flows for this structure? What is the maximum flange stress. Where is the shear center?

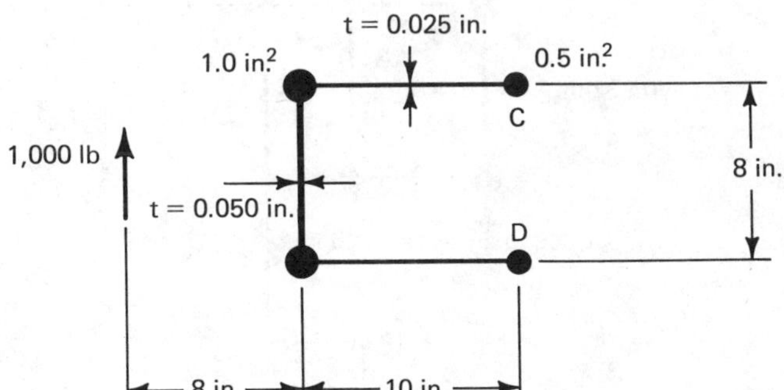

Problem 4. For the structure of problem 3, find the vertical deflection of the lower left flange. Include the effects of bending and torsion.

Problem 5. Repeat the calculations of problem 3 if the structure is altered by placing an additional web of t = 0.050 in from flange C to flange D.

Problem 6. Repeat the calculations of problem 4 for the beam of problem 5.

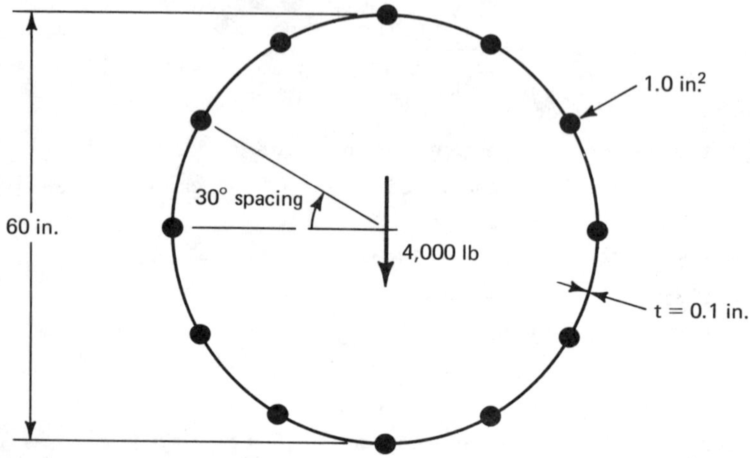

Problem 7. The cylinder shown has a length of 12 ft. and is loaded as shown. Find the flange forces in lb/unit span and find the panel shear flows.

Problem 8.

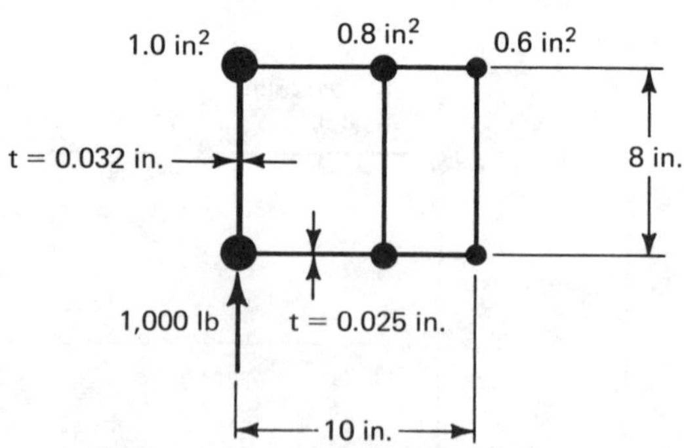

Find the flange forces per unit span and the panel shear flows for the structure shown.

Chapter 9

STRESS CONCENTRATIONS AND METAL FATIGUE

9.1 INTRODUCTION, STRESS CONCENTRATION

The problem of failures in structures which are subjected to repeated, cyclic or time-varying loads defines the important, but frequently neglected problem of metal fatigue. Metal fatigue is a cause of many structural failures in aerospace, civil and mechanical structures. Consider for example the findings of Holshouser and Mayner of the National Transportation Safety Board: (Reference 1).

"A review of records maintained by the National Transportation Safety Board showed that 16,054 civil aviation accidents occurred in the United States during the 3-year period ending December 31, 1969. Material failure was an important factor in the cause of 942 of these accidents. Fatigue was identified as the mode of the material failures associated with the cause of 155 accidents and in many other accidents the records indicated that fatigue failures might have been involved. There were 27 fatal accidents and 157 fatalities in accidents in which fatigue failures of metal components were definitely identified."

Very closely allied to fatigue is the subject of stress concentrations, which are the local peak stresses caused by any geometric discontinuity or irregularity in the structure. Stress concentrations are virtually inevitable in "real" structure due to grooves, fillets, oil holes, set screw holes, machining, and just about anything else. The worst stress concentrations are frequently those due to unforseen situations such as machinery errors, gravel nicks in propellors, microvoids in microstructure, non-metallic inclusions, poor workmanship, inadequate detection of damage, etc.

Objectives

1. Be able to define and illustrate stress concentration factors. (CB)
2. Be able to state Equation (9.1.3). (CB)
3. Be able to state the definition of, and discuss the meaning of "basic" fatigue properties. (CB)
4. Be able to use stress concentration charts to find K_t. (OB)
5. Be able to draw an estimated basic S-N diagram given only the F_{tu} of the material. (Restricted to steels.) (OB)
6. Be able to use the modifying factors to calculate the endurance limit S_{es} and draw an estimated service S-N diagram. (OB)
7. Be able to draw Goodman diagrams for desired life cycles. (OB)
8. Be able to use the Goodman diagram to determine if a part subjected to mean and alternating stresses will survive a given number of cycles. (OB)
9. Be able to use Miner's rule to determine whether or not a part will survive a given loading spectrum.

10. Be able to use charts such as Figures 9.6.1 and 9.6.2 to determine fatigue properties. (OB)

The combination of fatigue loading and a stress concentration of any kind is a potentially hazardous situation.

Figure 9.1.1 shows the distribution of stress around a circular hole in a flat panel. Let us consider carefully what is happening to the stresses in the vicinity of the hole, for the situation is worse than it seems at first.

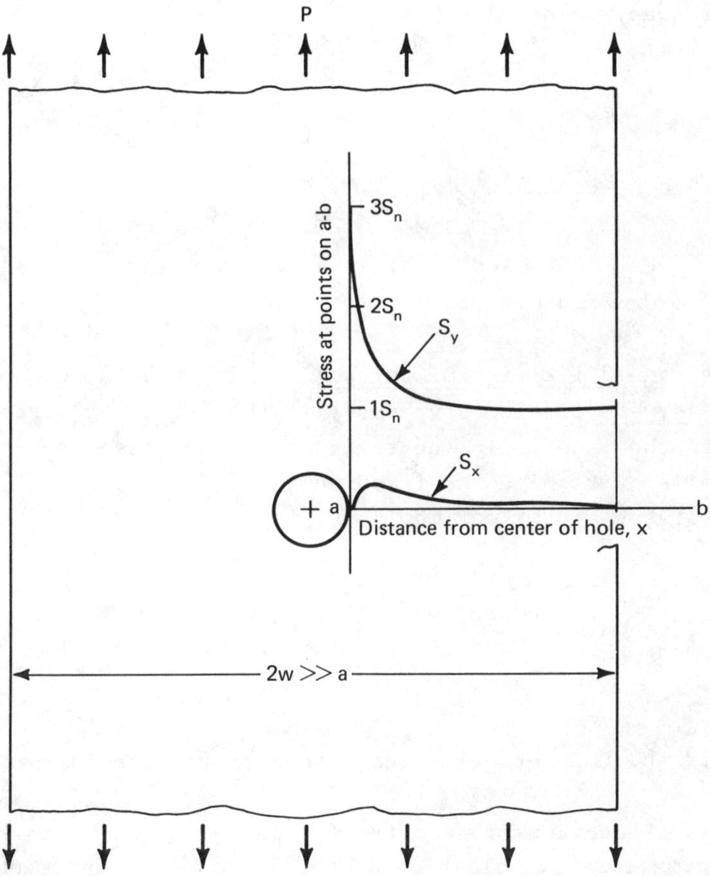

Figure 9.1.1 Stress distribution around a hole in a thin sheet under tension

First, consider a flat panel with no hole at all. The stress in the panel is

$$\sigma = \frac{P}{A} = \frac{P}{t x W} \qquad (9.1.1)$$

When a hole is drilled in the panel, it is natural to assume that the panel has a higher stress due to the fact that there is now less cross sectional area to carry the load.

Thus, one would think that

$$\sigma_{hole} = \frac{P}{A_{net}} \qquad (9.1.2)$$

where:

A_{net} = Net cross sectional area left when hole is drilled out.

Equation (9.1.2) is NOT correct. This is because the hole causes severe local disturbances in the "flow" of stress through the panel. In the region near the hole, the peak stress is quite high, as is shown in Figure 9.1.1.

The stress at the maximum point is given by

$$\sigma_p = \sigma_{peak} = K_t \frac{P}{A_{net}} \qquad (9.1.3)$$

where:

K_t is the theoretical stress concentration factor.

K_t is a function of geometry (shape, size) only, so long as the material is elastic. Thus, K_t is not a function of load or materials; it is the same for all magnitudes of the same loading state and for all different (elastic) materials.

The usual way of presenting stress concentrations is shown in Figures 9.1.2 - 5. Many strength of materials books, civil and mechanical handbooks, company stress manuals, etc., present such charts in great volume. Reference (2) is an entire book devoted to the subject of stress concentration. References (3 - 5) are also recommended.

The stress concentration factor K_t is usually found by electrical strain gages or photoelasticity. (See Chapters 11 and 12 of this text.) When K_t is known the peak stress in a given situation can be found from

$$\sigma_{peak} = K_t \sigma_{nom} \tag{9.1.4}$$

where:

$\sigma_{nominal}$ is the stress calculated from the usual theory ($\sigma = P/A$, etc.) for the case in question.

EXAMPLE:

The flat coupon shown is loaded in tension. What is the peak stress at the root of the fillet? D = 3/4 in; d = 5/8 in; ρ = 1/8 in; t = 0.10.

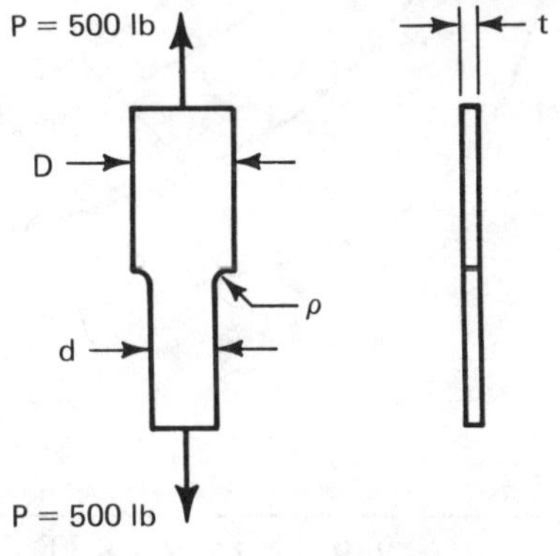

SOLUTION:

Refer to Figure 9.1.2.

$\rho/d = 0.2$

$D/d = 1.2$

from the figure:

$$K_t = 1.7$$

$$\sigma_{nom} = \frac{P}{td} = \frac{500}{(0.10)(.625)} = 8{,}000 \text{ psi}$$

$$\sigma_{peak} = K_t \sigma_{nom} = (1.7)(8{,}000)\text{psi} = \underline{\underline{13{,}600 \text{ psi}}}$$

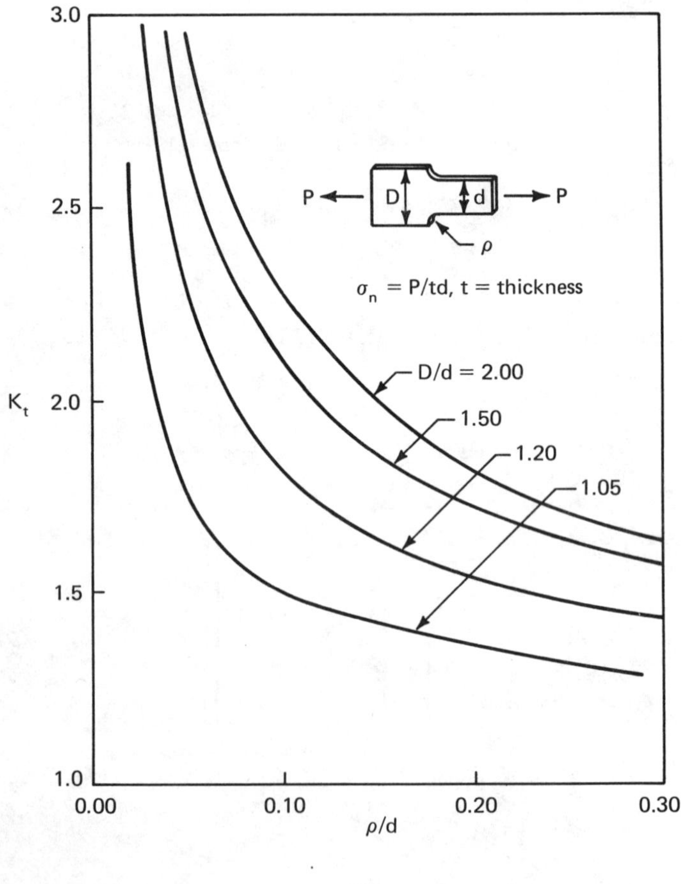

Figure 9.1.2 Stress-concentration factors for flat bar with fillet under tension (based on photoelastic data).

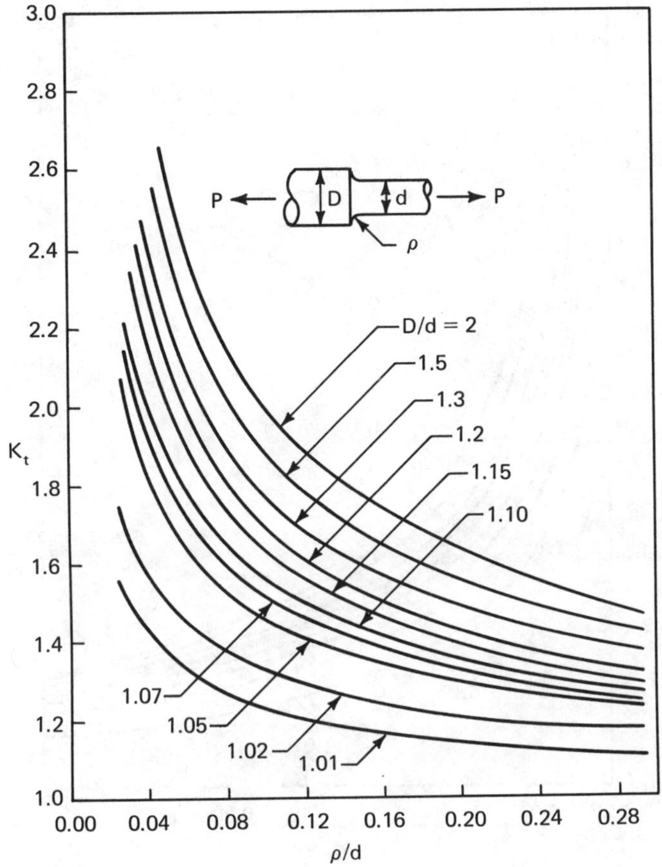

Figure 9.1.3 Stress-concentration factors for a round bar with fillet under tension

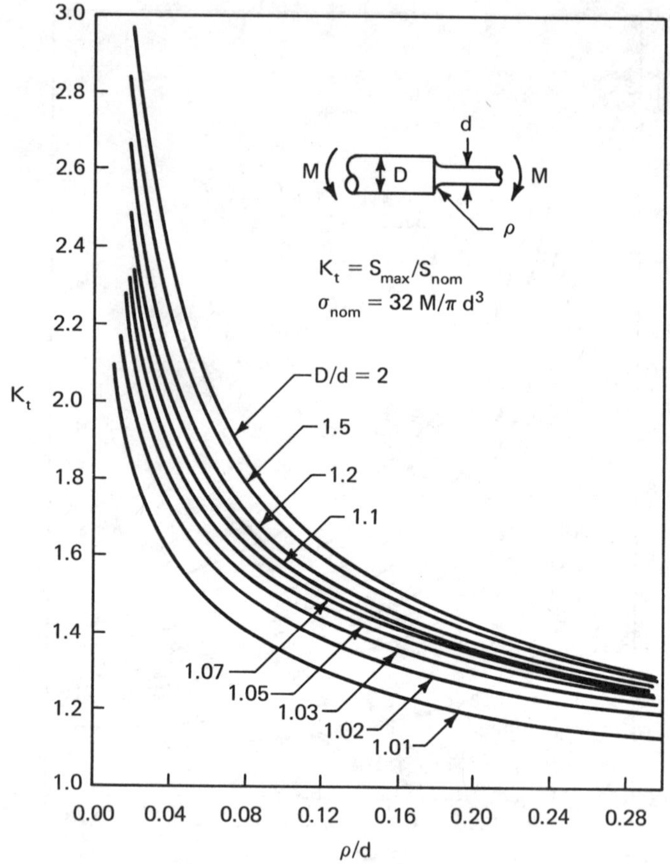

Figure 9.1.4 Stress-concentration factors for a round bar with fillet under bending

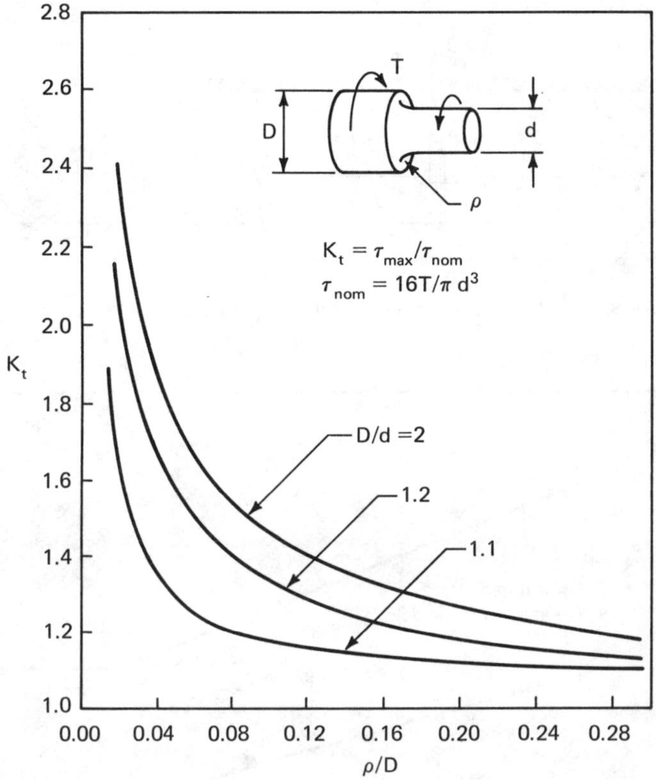

Figure 9.1.5 Stress-concentration factors for a round bar with fillet under torsion

9.2 METAL FATIGUE

The problem of metal fatigue has a long history. (References 6 - 8.) Wohler first studied the problem of fatigue in railroad axles in the 1860's. See Figures 9.2.1 and 9.2.2. Noteable fatigue failures, such as the Comet crashes, Reference (7), have been a prominent part of engineering history. Unfortunately, in spite of a huge volume of effort in this area, fatigue failures still occur too often and the fatigue problem cannot be considered to be "solved." Part of the problem lies in the very wide statistical spread in all of the variables which affect fatigue.

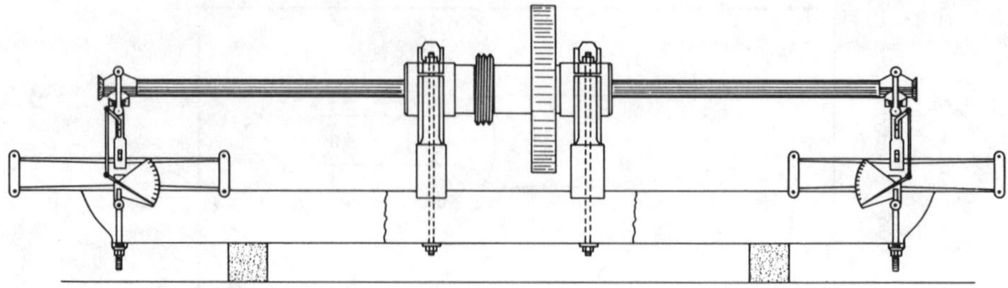

Figure 9.2.1 Wöhler's machine for fatigue testing of railway axles

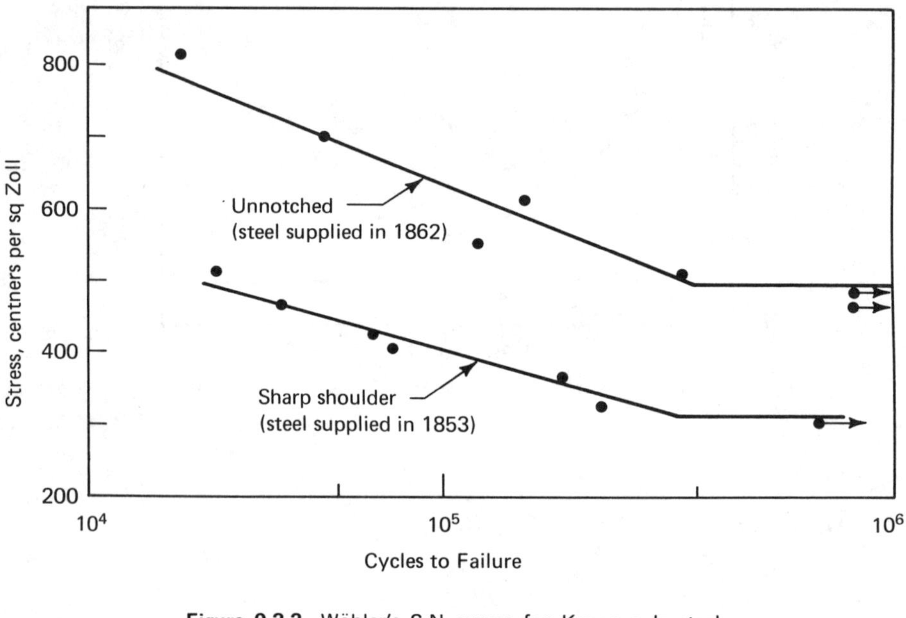

Figure 9.2.2 Wöhler's S-N curves for Krupp axle steel

The design methodology presented in this section can be used to design parts for fatigue. The results must be considered preliminary and the resulting part must be proved by testing.

We shall follow the procedure listed below as we develop our approach to fatigue. It is well to keep this sequence in mind as we go through this section and also when we study outside literature.

1. Load terminology.
2. The definition of the basic fatigue properties for standard specimens.
3. The effects of stress concentration, size, and other factors on the basic properties.
4. The effects of mean stress.
5. The effects of spectral loading.

Load terminology. The behavior of structures in fatigue is usually based on the assumption that the load is sinusoidal. The terminology for load and stress is shown in Figure 9.2.3.

The "basic" fatigue properties. The term "basic" with respect to fatigue has a very precise meaning in this chapter. The basic fatigue properties are given for:

- A standard specimen machined to exact specifications and surface finish. (Figure 9.2.4.)
- Zero mean stress
- Reversed bending

When the standard specimen is subjected to reversed bending, the stress amplitude σ_a is plotted vs. the number of cycles to failure, N and the results appear on plots such as Figure 9.2.5. (log-log plot.) This basic S-N diagram is sometimes available in the literature, although generally speaking fatigue data is relatively scarce.

Figure 9.2.6 shows that for steels the basic S-N diagram may be estimated by drawing a straight line from $0.9F_{tu}$ at 10^3 cycles to $0.5F_{tu}$ at 10^6 cycles, and horizontally thereafter. The estimates for steels for reversed axial loading and reversed torsion are shown in Figure 9.2.7.

The horizontal portion of the curves for steel define the endurance limit. Theoretically, there is a reversed stress amplitude, S_e, for which life is infinite for steels.

Note that we have already lost 50% of the ultimate tensile strength in long-life fatigue applications without considering any further reductions of fatigue strength due to other factors.

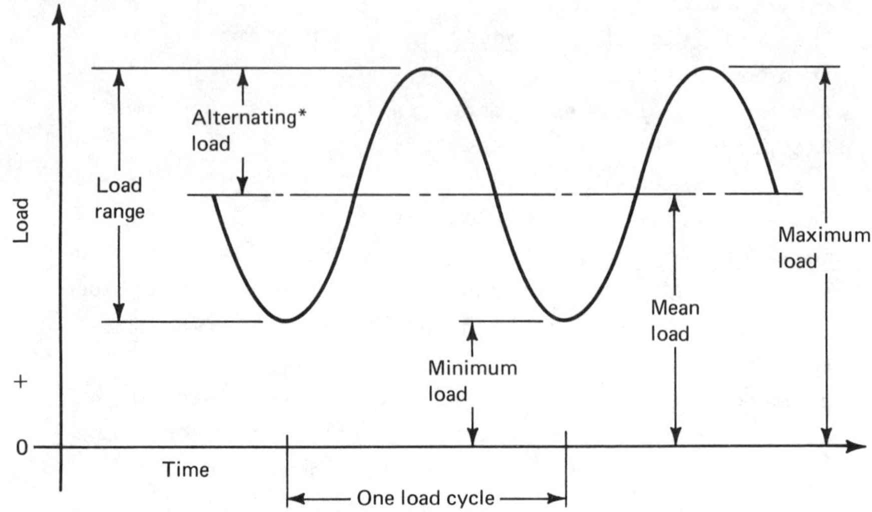

(a)

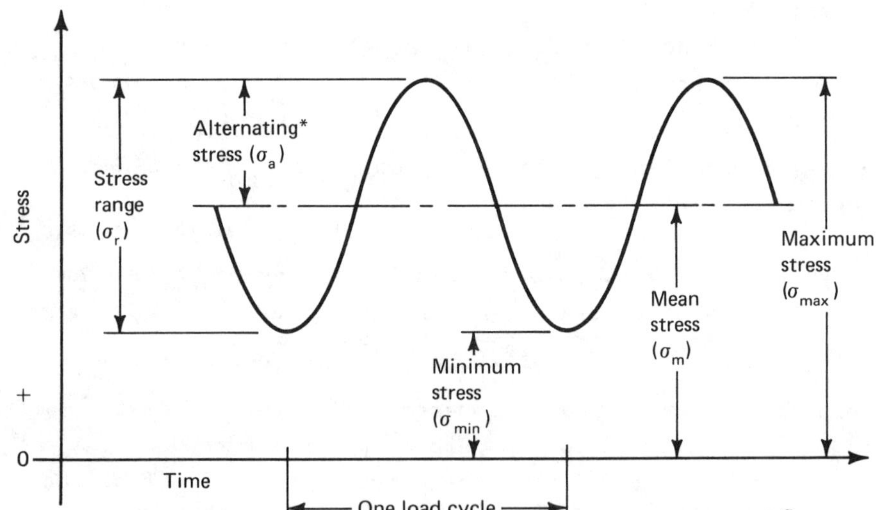

(b)

$$R = \frac{\sigma_{min}}{\sigma_{max}}$$

$$A = \frac{\sigma_a}{\sigma_m}$$

*Also frequently referred to as "variable" load or stress.

Figure 9.2.3 Fatigue load and stress nomenclature

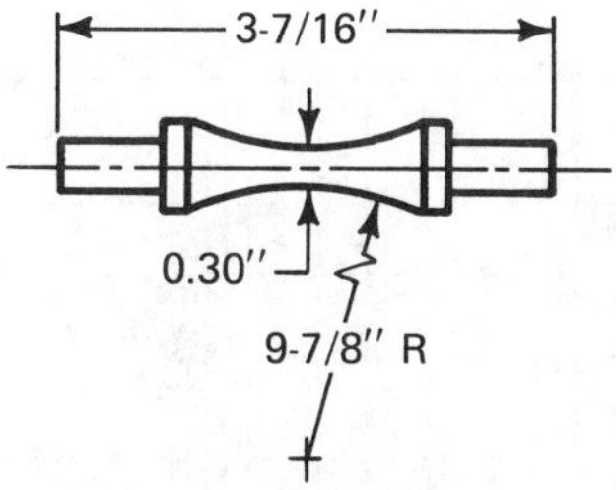

Figure 9.2.4 Standard specimen

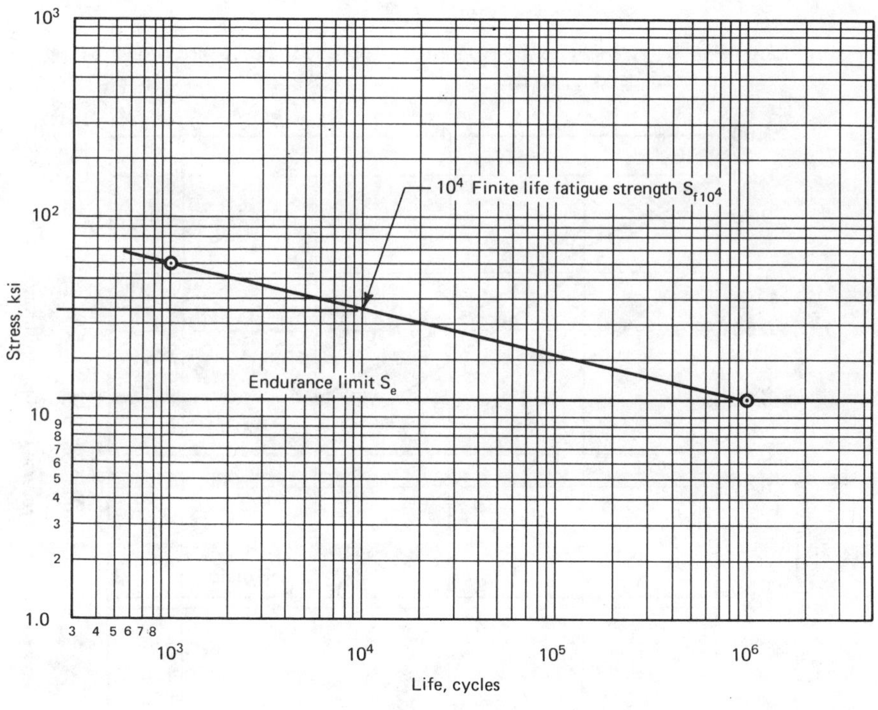

Figure 9.2.5

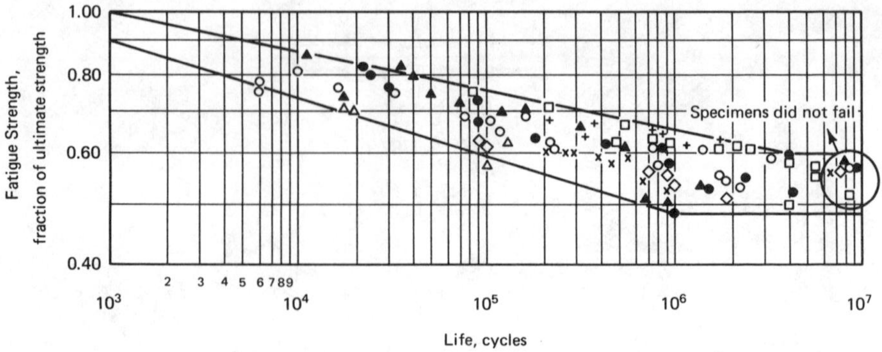

Figure 9.2.6 Fatigue strength vs. life, unnotched polished specimens, wrought steel, reversed bending loads

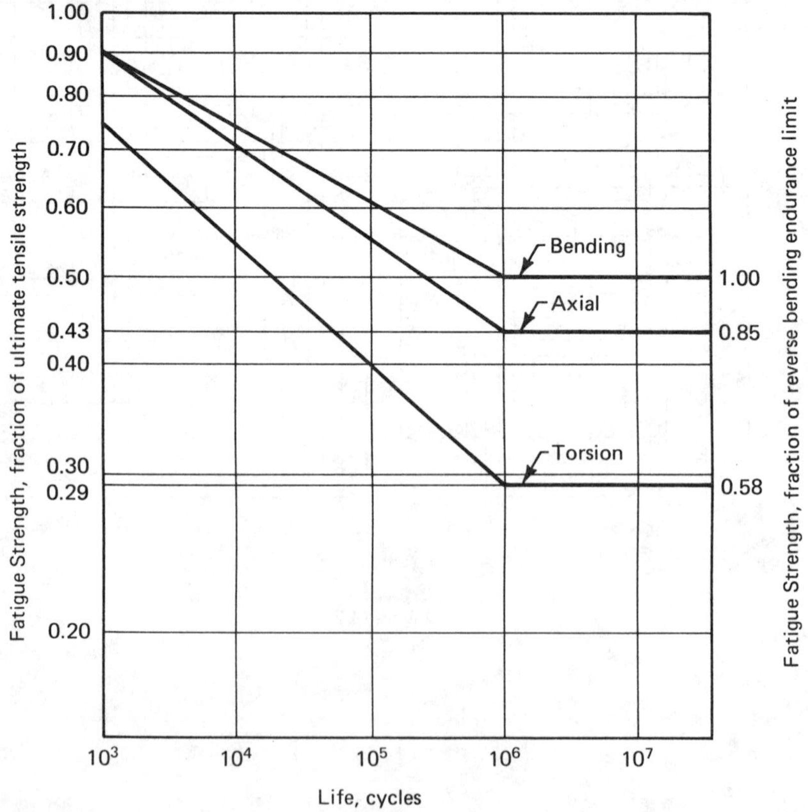

Figure 9.2.7 Generalized steel S-N curves reversed axial, bending, and torsional loading

This presents another dangerous aspect of the fatigue situation. Very high tensile strength steels are excellent in static applications but may be relatively poor in fatigue.

The S-N diagrams for an aluminum material are shown in Figure 9.2.8.

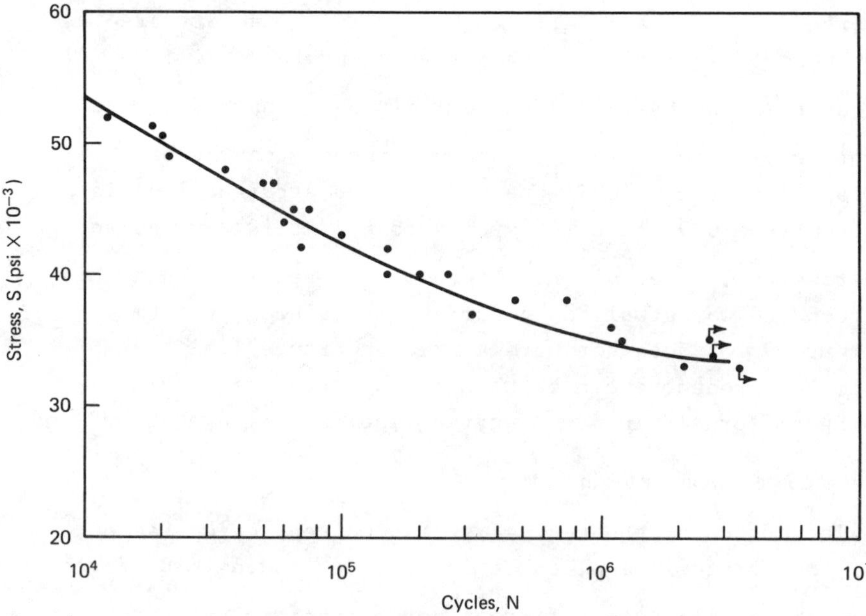

Figure 9.2.8

9.3 THE EFFECTS OF STRESS CONCENTRATION, SIZE AND OTHER FACTORS ON THE BASIC PROPERTIES

The effects of various factors in fatigue are conveniently handled by reducing the basic endurance limit, S_e, in accordance with

$$S_{es} = (K_a \times K_b \times K_c \times . . .)\, S_e \qquad (9.3.1)$$

where:

S_e = the <u>basic</u> endurance limit

S_{es} = the service endurance limit, modified for all the factors which affect the fatigue life

K_a, K_b, K_c. . . = modifying factors for service conditions

In reality, almost every factor in fatigue which is different from the basic test conditions makes the situation worse. Thus, nearly all of the modifying factors are less than one. As a result, the service endurance limit S_{es} is less than the basic endurance limit S_e. The S-N diagram for service conditions is drawn by modifying the basic S-N diagram as shown in Figure 9.3.1. The straight line is now drawn from $0.9F_{tu}$ at 10^3 cycles to S_{es} at 10^6 cycles and horizontally thereafter. It should be recalled that this procedure works well only for steels, since Aluminum, Figure 9.2.8 does not have a true endurance limit. A similar procedure can be used to estimate the service S-N diagram for Aluminum. A curved line can be "eyeballed" in.

Stress Concentration Factor in Fatigue

The effect of theoretical stress concentration factor, K_t, is not fully applicable in fatigue loading. Different materials will have different <u>notch-sensitivity factors</u>, q. Figure 9.3.2 (see, for example, Reference 9) shows typical plots of notch-sensitivity factor q. The <u>fatigue stress concentration factor</u>, K_f, can be found when K_t and q are known from

$$K_f = 1 + (K_t - 1)q \tag{9.3.2}$$

The stress concentration modifying factor, K_a, is then given by

$$K_a = \frac{1}{K_f} \tag{9.3.3}$$

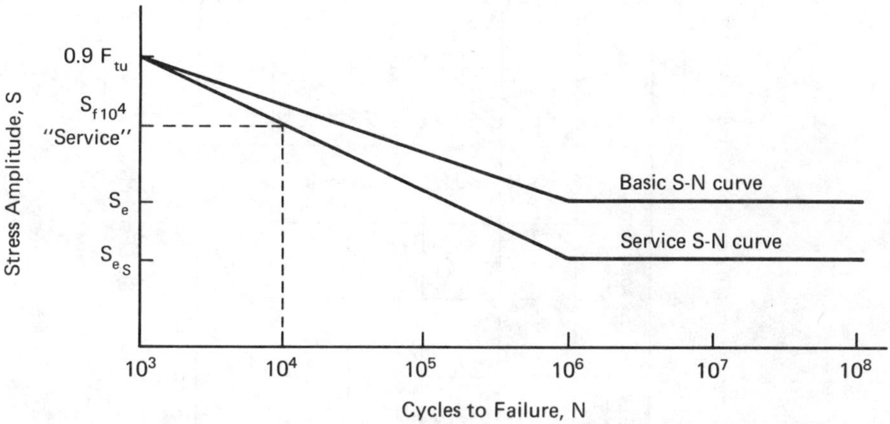

Figure 9.3.1

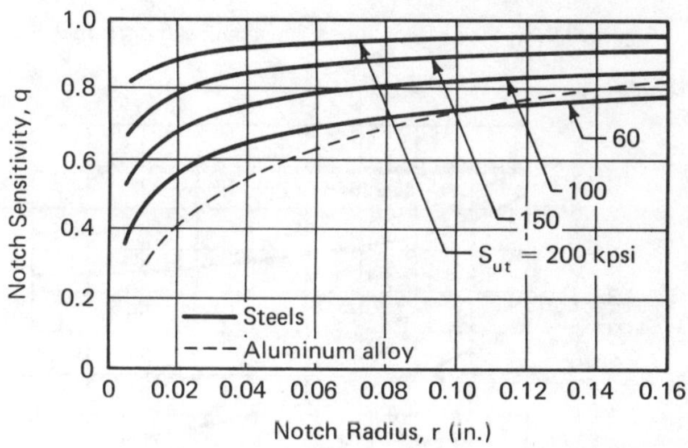

Figure 9.3.2

<u>Size Effect</u>. The size of a part affects its fatigue life. Up to a certain size, at least, a large part is more suceptible to crack initiation than a small one. Figure 9.3.3 (see, for example, Reference 10) shows the size effect, K_b, for bending, axial and torsional loads.

<u>Surface Finish</u>. The surface finish of a part has a great effect on its fatigue life. This is shown in Figure 9.3.4.

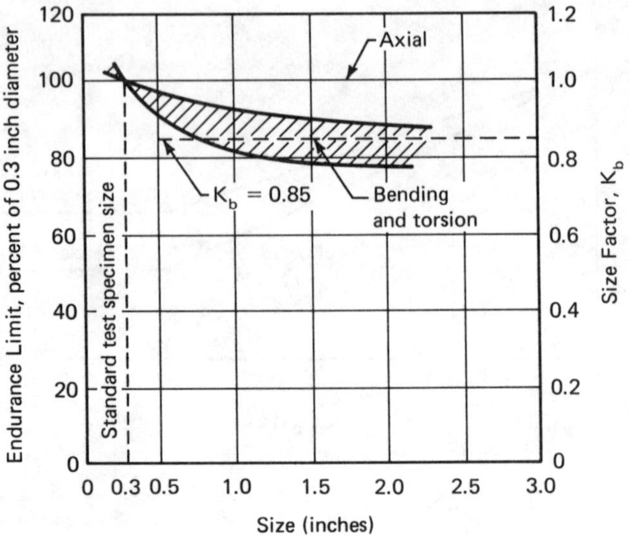

Figure 9.3.3 Size effect—bending, axial, and torsional loads

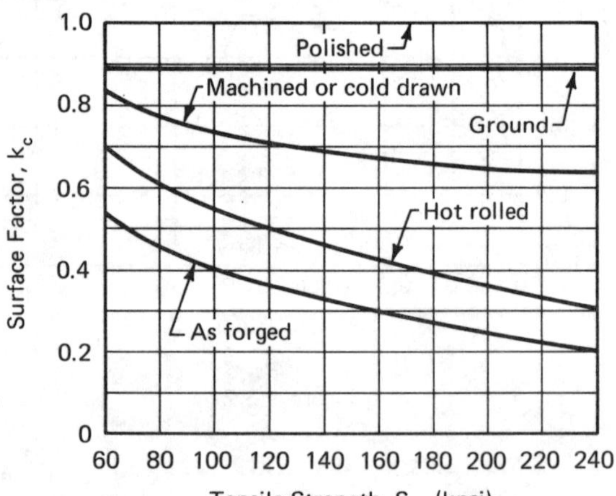

Figure 9.3.4 Modifying factors for surface finish

The surface finish modifying factor, K_c, is given in charts similar to this one.

Other Modifying Factors. Almost everything which happens to a part which makes it different than a standard test specimen will affect its fatigue life. The literature on fatigue, and the effects of these factors is voluminous and widespread. The reference at the end of this chapter will provide a starting point.

One of the few processes which improve fatigue life is nitriding. Another is shot peening. The effects of shot peening are shown in Figure 9.3.5. (After Reference 5.)

The procedure of this section provides a way of estimating the service S-N diagram for purposes of design. When the time is available (fatigue testing takes relatively huge amounts of time) and when the volume of production and/or money available is sufficient, the fatigue properties of specific parts made of specific materials for specific application can be determined by direct testing. Figure 9.3.6 is an example of this. This procedure provides more precise information but it is not always justifiable.

9.4 THE EFFECT OF MEAN STRESS

The mean stress on the part is a major variable in fatigue. This is especially true of tensile pre-loading. Fatigue is a brittle fracture phenomena, even in ductile materials and as such is most sensitive to tensile loading. Figure 9.2.3 shows the definition of stress ratios R and A. Information about the effects of mean stress on fatigue can be represented on S-N diagrams with the stress ratios R and A as parameters. It is usually more useful to plot a Goodman diagram, Figure 9.4.1. This diagram is a plot of pure alternating failure on the vertical axis and pure static failure on the horizontal axis. The endurance limit for the part in service forms the intercept on the vertical axis for the 10^6 life cycle line. The finite life fatigue strength, see Figure 9.3.1, forms the basis for other lives. A straight line of constant $A = \sigma_a/\sigma_m$ defines the stress ratio for a given situation. If the stress state for a given part falls outside a desired life cycle line, we predict a failure; otherwise a success. The mean and alternating failure points are connected by straight lines. A straight line is considered conservative for most cases.

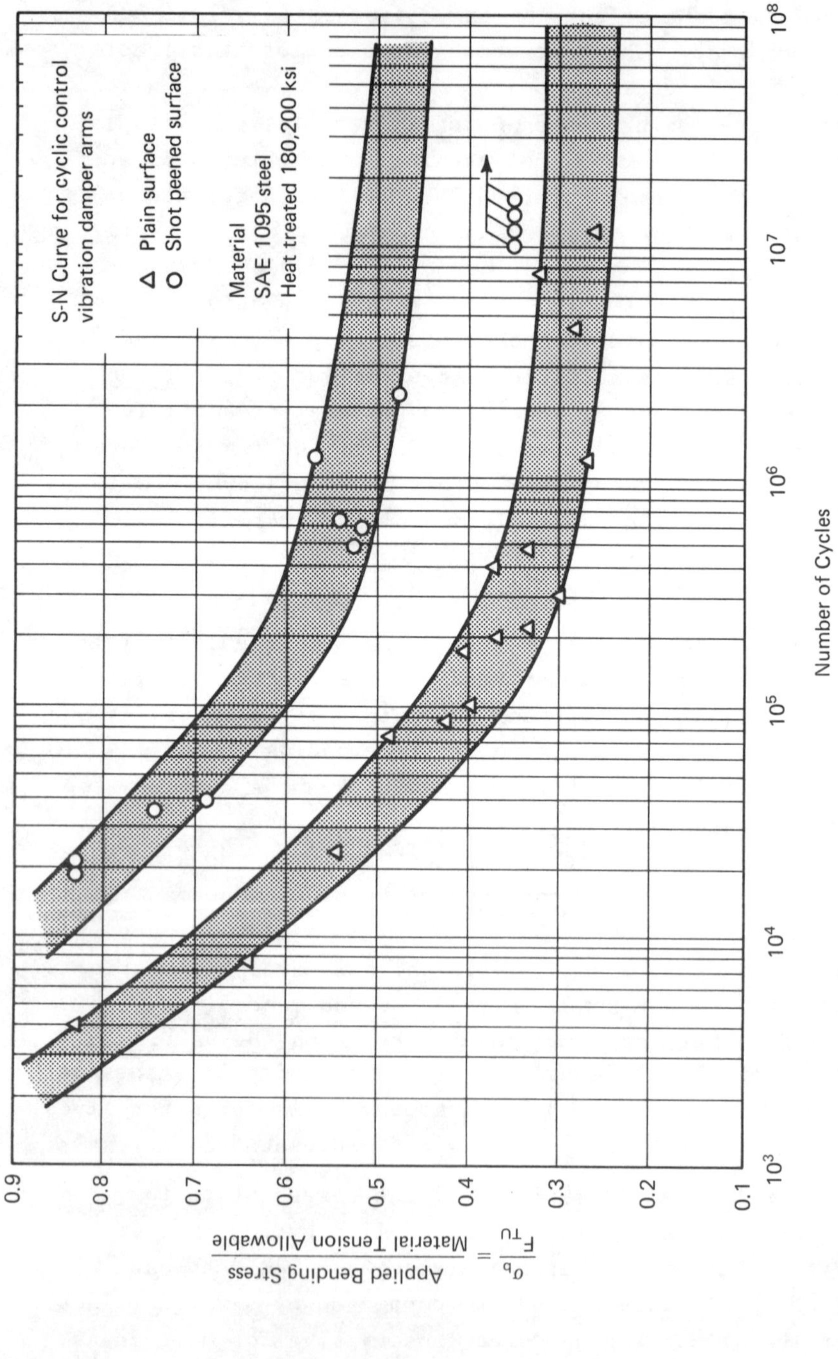

Figure 9.3.5 Improvement in fatigue life due to shot peening

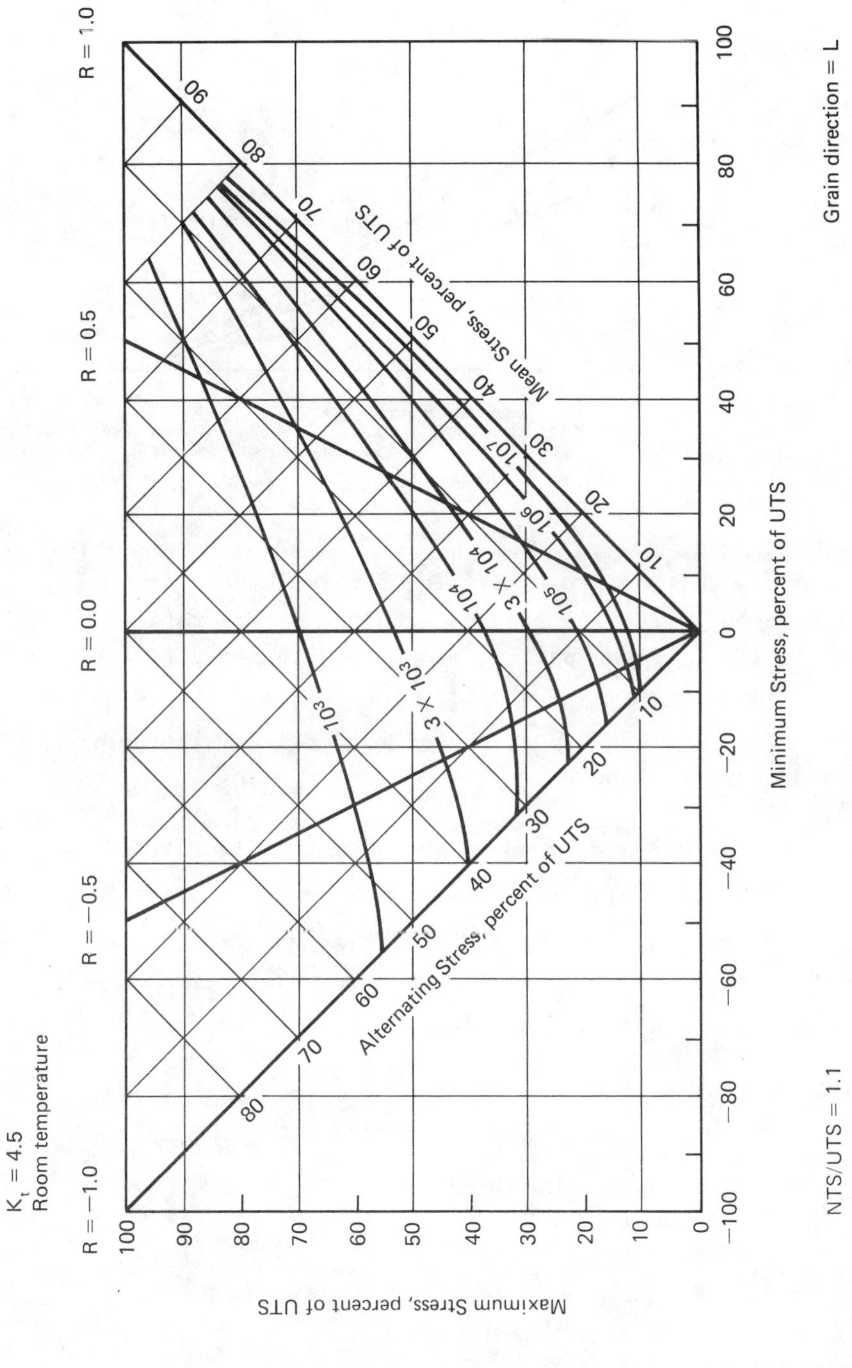

Figure 9.3.6 Modified Goodman diagram for Ti-6AL-4V plate diffusion-bond cycle

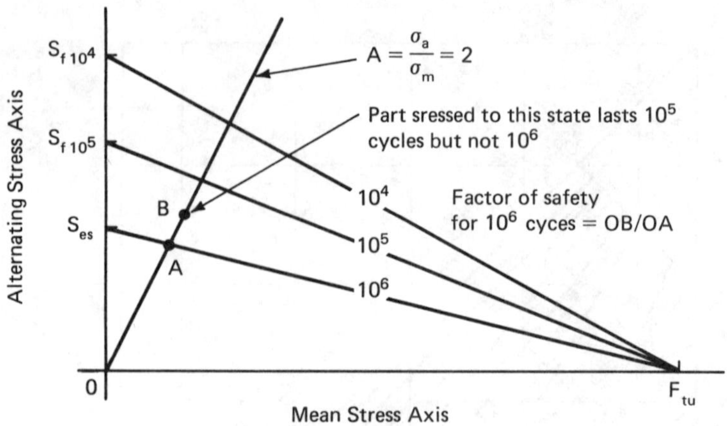

Figure 9.4.1

EXAMPLE:

The coupon of the first example of this chapter is to be subjected to fatigue loading with the stress ratio A equal to 1.0. If the part is made from 116 ksi ultimate steel, find the stress state (σ_a, σ_m) to which it can be subjected for a life expectancy of 10^4 cycles. The part is cold drawn finish.

SOLUTION:

1. The "basic" S-N diagram is shown in the figure. Lines are drawn from $0.9F_{tu} = 0.9\ (116{,}000)$ at 10^3 cycles to $.43F_{tu}$ at 10^6 cycles; horizontal thereafter.
2. The endurance limit is corrected to service by

$$S_{es} = K_a x K_b x K_c Se$$

Stress concentration: From the previous example, $K_t = 1.7$.

From Figure 9.3.2, $q = 0.82$.

From Equation (9.3.2)

$$K_f = 1 + (K_t - 1)\ q = 1 + (1.7 - 1)\ 0.82$$

$$K_f = 1.4$$

$$K_a = 1/1.4 = 0.72$$

Size Factor: From Figure 9.3.3, K_b = 1.0 for axial loading.
Surface Finish: From Figure 9.3.4, K_c = 0.73.

$$S_{es} = (0.72)(1.0)(0.73)50{,}000 = 26{,}300 \text{ psi}$$

The service S-N diagram is shown in the figure for this example.

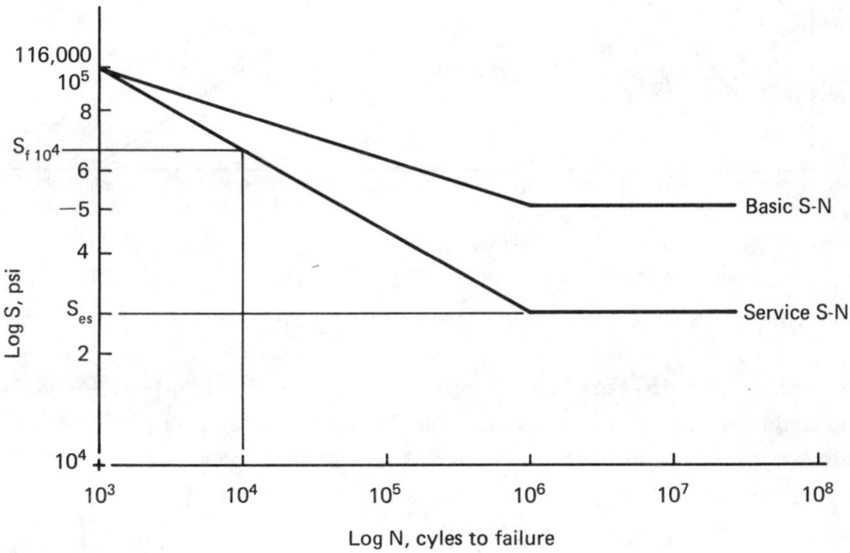

3. The 10^4 life cycle fatigue strength S_{f10^4} is taken from the service S-N diagram.

$$S_{f10^4} = 65 \text{ ksi}$$

The Goodman Diagram for this part is shown on the next page.

4. From the Goodman diagram, $\sigma_a = \sigma_m$ = 40 ksi is the permissible stress for 10^4 life cycles.

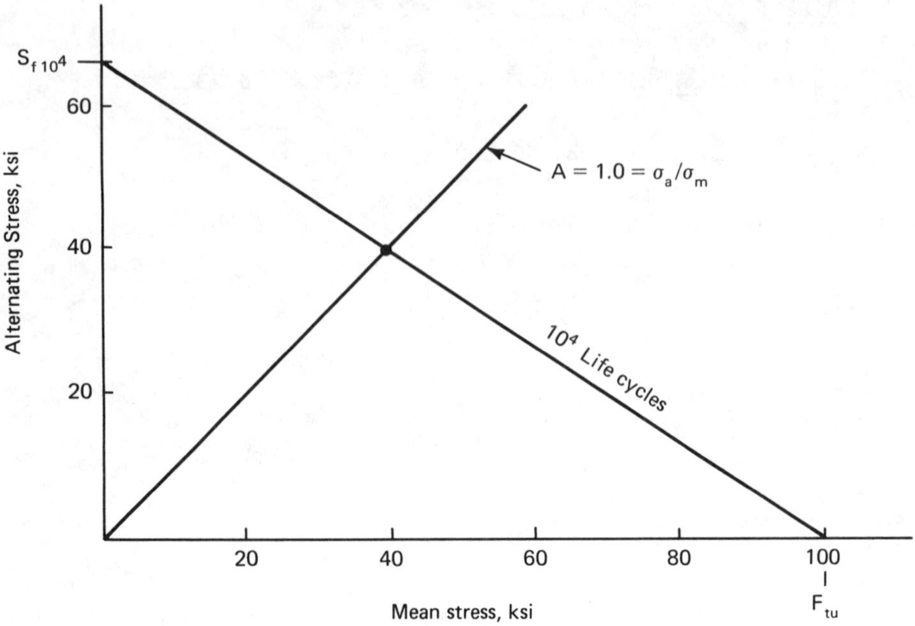

9.5 THE EFFECTS OF SPECTRAL LOADING (CUMULATIVE DAMAGE)

Actual parts in service are not usually subjected to sinusoidal loading of constant amplitude. There is, rather, a period of high amplitude, then low, then higher, etc. The situation is depicted in Figure 9.5.1. The n_i represent the number of applied cycles of the corresponding stress amplitude σ_i. From the S-N diagram for the part, the N_i correspond to the maximum number of cycles the part would take at any given σ_i. The part, according to a <u>linear</u> cumulative damage theory, is going to fail when

$$\sum_{i=1}^{\text{total}} \frac{n_i}{N_i} = 1 \qquad (9.5.1)$$

Equation (9.5.1) is called Miner's rule, or the Palmgren-Miner rule. It is not infallible, but it provides a reasonable estimate. There is, for example, some dependency on the order of stressing (high-low vs. low-high) on life, etc. The right hand side of Equation (9.5.1) is not always unity. (Reference 12.)

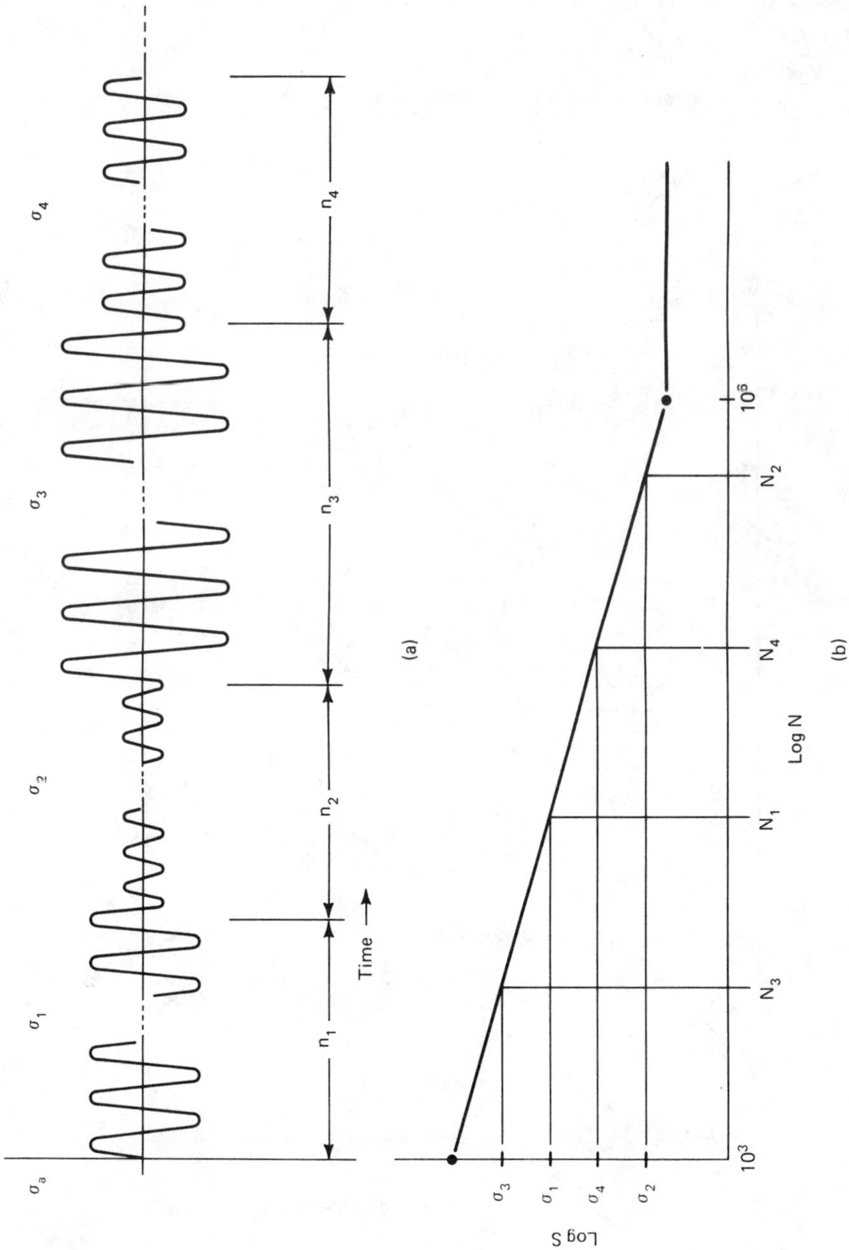

Figure 9.5.1

9.6 FATIGUE DATA FOR SOME IMPORTANT ALLOYS

Figures 9.6.1 and 9.6.2 present some fatigue data for steel and aluminum alloy. The Goodman diagrams are at an angle and stress ratios R and A are both plotted on the charts.

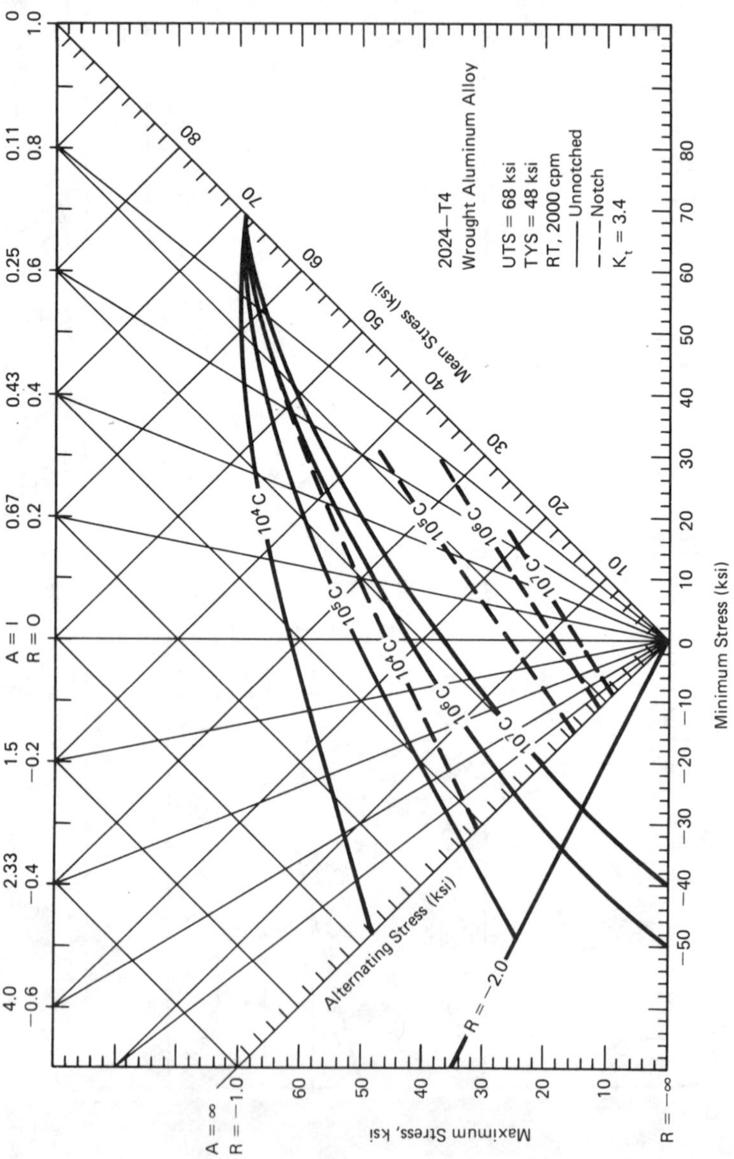

Figure 9.6.1 Typical fatigue properties, wrought 2024-T4 aluminum alloy at room temperature

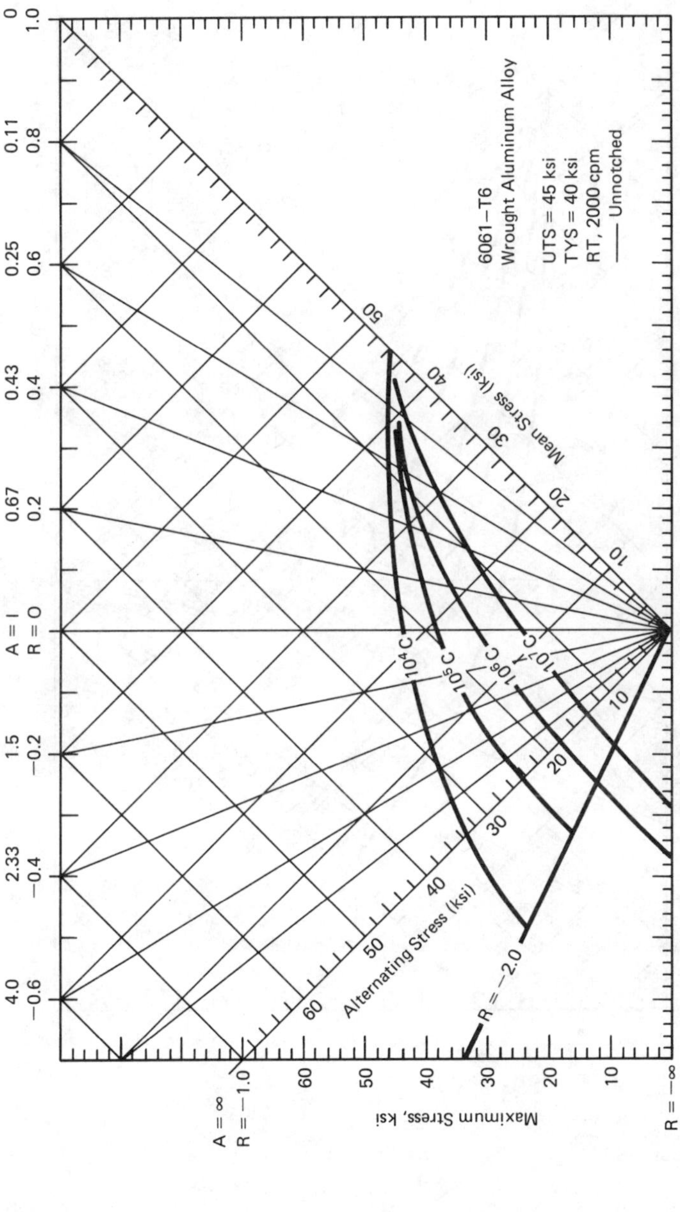

Typical fatigue properties, wrought 6061-T6 aluminum alloy at room temperature

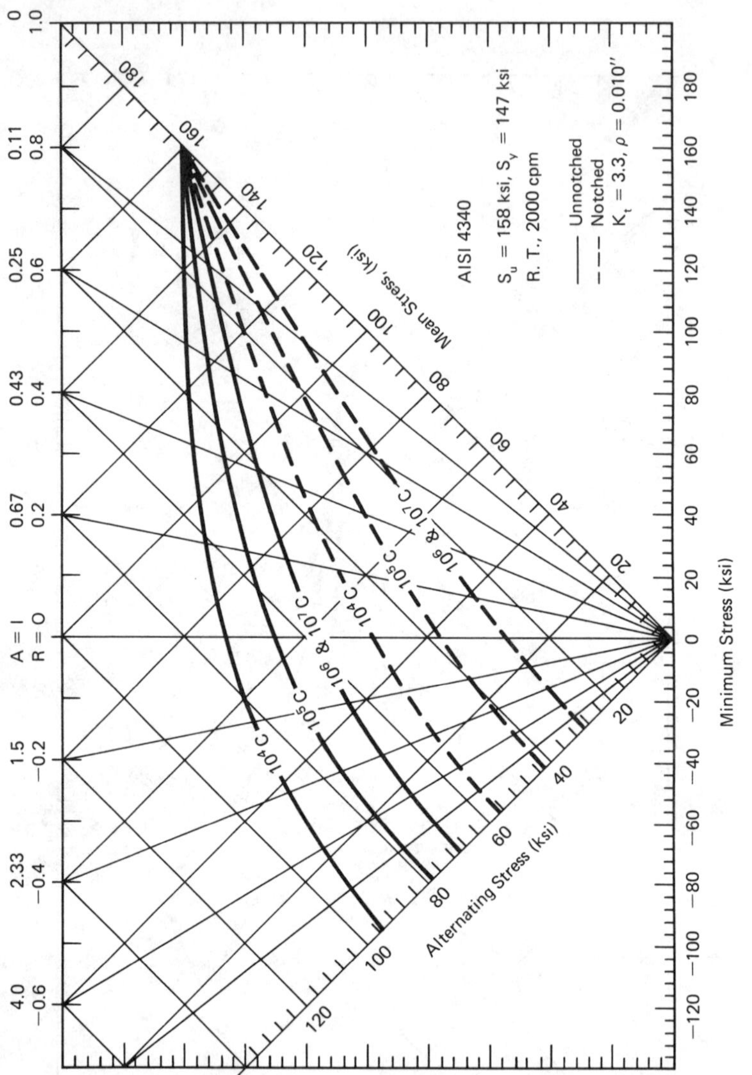

Figure 9.6.2 Typical fatigue properties, AISI-4340 steel at 158 ksi tensile ultimate

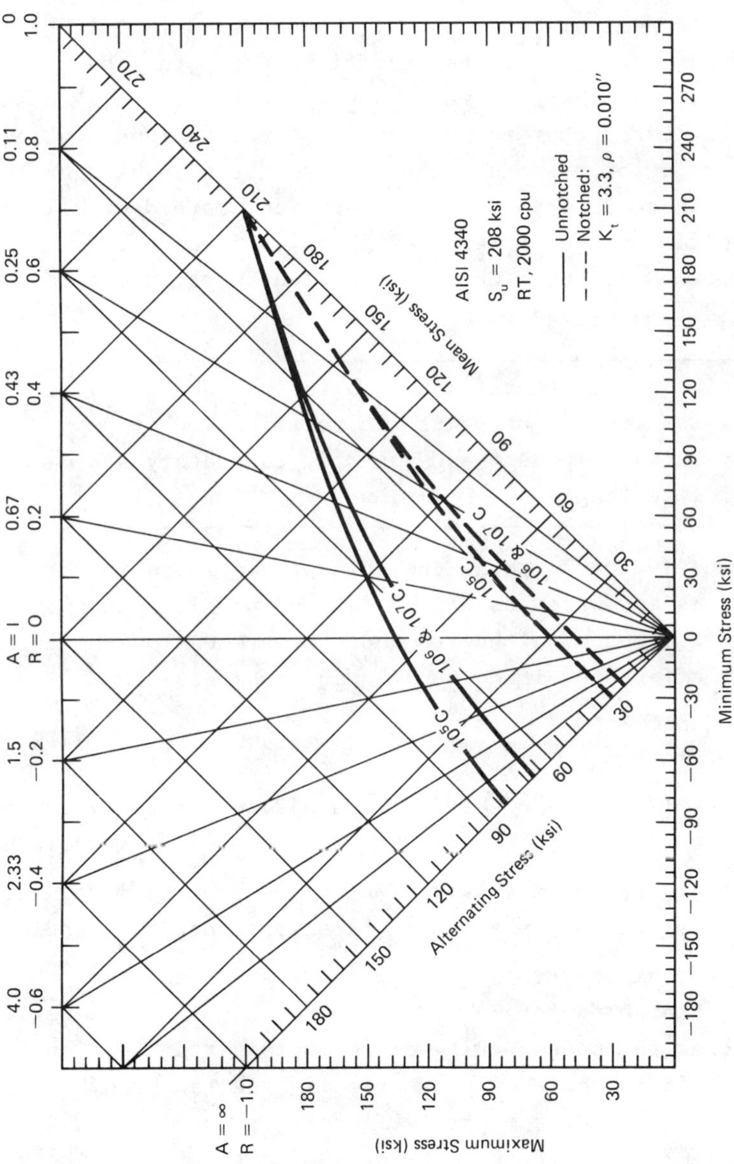

Typical fatigue properties, AISI-4340 steel at 208 ksi tensile ultimate

9.7 SUMMARY AND WARNINGS

The material in this Chapter presents some fundamentals of fatigue analysis and design. Students will still have very much left to learn about fatigue after they have completed this Chapter. Everyone who works in fatigue has a lot to learn, that is the nature of the fatigue problem. There are a lot of things about metal fatigue which are not known. There is also a wide statistical spread of all the data in all of the fatigue variables (Reference 13).

The methods of this Chapter represent ways to obtain design estimates, and:

There is no substantiation of a fatigue design by analysis alone. There must be verification of analysis by component and assembly laboratory and service testing.

The most valuable thing the student can learn in this chapter is an awareness of the fatigue problem, an awareness of its potential hazard and an awareness of the magnitude of effort required to safeguard against it.

Problems

Problem 1. Part in service is subjected to loading which gives rise to σ_a = 25 ksi and σ_m = 25 ksi. The material properties for the metal from which the part is made are: S_e = 45 ksi and S_{tu} = 90 ksi. Experience and data indicate that the fatigue stress conc. factor is 0.9 and that the surface finish factor is 0.8. Other factors may be neglected. Draw the Goodman Diagram for this part (based on infinite life) and show geometrically the factor of safety. Also state whether the part will fail or not.

Problem 2. A part is to be subjected to axial, fluctuating stresses, the stresses having the mean and alternating components shown below. The material from which the part is made has an ultimate tensile strength F_{tu} of 50,000 psi.

The theoretical stress concentration factor, K_t, is 1.75. The notch-sensitivity factor, q, is 0.7. The effects of all other factors on the fatigue strength is taken to be a factor of 0.7. Determine whether or not this part will survive 10^5 cycles.

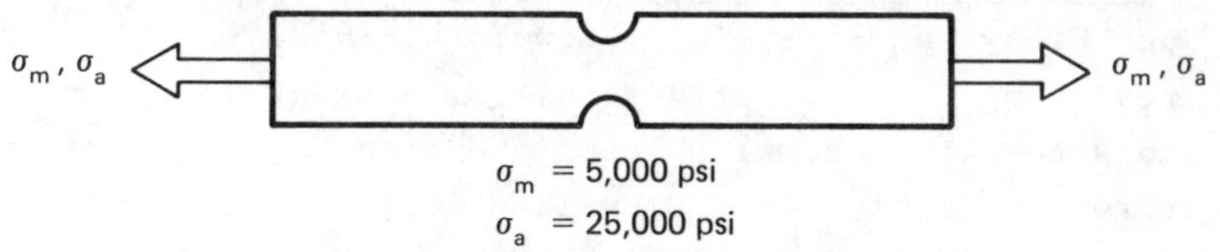

Problem 3. The shaft shown is subjected to mean and alternating bending such that $\sigma_a/\sigma_m = 0.75$. Find the mean and alternating stresses for failure in:

a. AlSl 4340 at 158 ksi tensile ultimate and 10^5 cycles;
b. 2024-T4 Aluminum alloy at 10^7 cycles. Use Figures 9.6.1, 9.6.2, and 9.1.4.

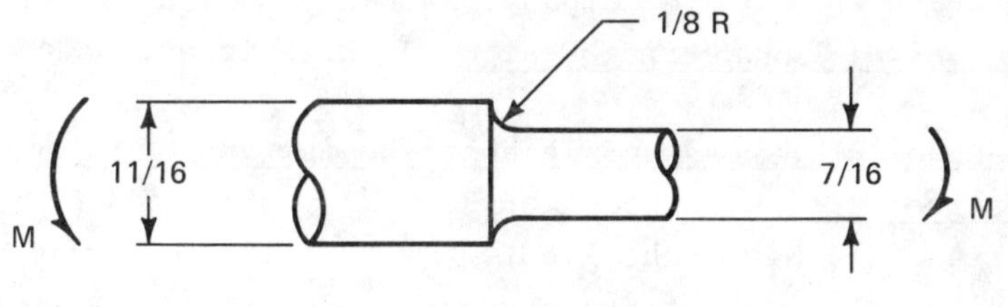

Problem 4. The part with S-N diagram of Figure 9.2.5 is subjected to

10^3 cycles at 50,000 psi
10^3 cycles at 30,000 psi
10^4 cycles at 20,000 psi

Will the part fail? If not, how many more cycles at 30,000 psi will cause it to fail?

References

1. Fatigue Failure of Metal Components as a Factor in Civil Aircraft Accidents, William L. Holshouser and Ruth D. Mayner, in Advanced Approaches to Fatigue Evaluation, NASA-SP-309, Washington, D.C., 1972.
2. Stress Concentration Design Factors, R. E. Peterson, John Wiley and Sons, Inc., New York, 1953.
3. Preventing Fatigue Failures, T. B. Stulen, H. N. Cummings and W. C. Schulte, reprinted from Machine Design, Penton Publishing Co., Cleveland, Ohio, 1961.
4. Analysis and Design of Flight Vehicle Structures, E. F. Bruhn, Tri-State Offset, 1973.
5. Metal Fatigue, G. Sines and J. L. Waisman, Eds., McGraw-Hill, New York, 1959.
6. Metal Fatigue, Frank R. Stone, Civil Aeronautics Board, Engineering Division Bulletin No. 63-1, U. S. Government Printing Office, 1963.
7. Fatigue of Metals, G. R. Gohn, H. F. Hardrath, R. E. Peterson, Materials Research and Standards, Vol. 3, No. 2, February, 1963.
8. Fatigue and Stress Corrosion Manual for Designers, Lockheed California Company, Burbank, California, 1968.
9. Mechanical Engineering Design, J. E. Shigley, McGraw-Hill, New York, 1963.
10. Handbook of Stress and Strength Design and Material Applications, C. Lipson and R. C. Juvinall, The Macmillan Co., New York, 1963.
11. Fatigue of Aircraft Structures, H. J. Grover, Department of the Navy, NAVAIR 01-1A-13, U. S. Government Printing Office, Washington, D. C., 1966.
12. Effects of Environment and Complex Load History on Fatigue Life, ASTM STP 462, ASTM, Philadelphia, Pennsylvania, 1970.
13. A Probalistic Method of Designing Specified Reliabilities into Mechanical Components with Time Dependent Stress and Strength Distributions, D. B. Kececioglu, J. W. McKinley and M. J. Saroni, NASA NGR 03-002-044, The University of Arizona, 1967.

Chapter 10

ELASTIC STABILITY

10.1 INTRODUCTION

The consideration of elastic stability, or buckling arises frequently in the stress analysis of structures in all fields of engineering. This topic is often the governing criterion in aircraft and missiles, building frames, ships, oil refinery storage tanks, and a large assortment of civil and mechanical structures.

Elastic buckling is a phenomenon which defines a failure mode in which the structure "collapses" without <u>necessarily</u> having any portion of the structure exceeding the yield strength of the material from which the structure is made. A familiar example is the "spring out" of a long slender object such as a fishing rod or wooden yardstick when a force is applied on the end. When the load is released, the structure returns to its unstressed condition and is undamaged. Nevertheless it has failed as a structural element when it buckles. In this chapter we will consider the buckling of columns in some detail. Then we will give a brief introduction to the instability of plates, beams and shells.

Objectives

1. Be able to state the definition of an Euler column. (CB)
2. Be able to state Equation (10.2.1) and derive Euler's equation. (CB)
3. Be able to state and use Euler's equation. (CB)
4. Be able to state, recognize and calculate a cross section radius of gyration. (CB)
5. Be able to plot and utilize plots such as Figures 10.2.2, 10.3.2, 10.5.2, 10.7.2, 10.10.2, etc. (OB)
6. Be able to look up the effective length of a column for end conditions other than pinned-pinned. (OB)
7. Be able to state and use Equation (10.3.1). (CB)
8. Be able to state Equation (10.3.3) and recognize it as the proper way to treat inelastic buckling. (CB). Be able to use this equation and the material stress-strain curves to plot σ_{cr} vs. L_e/p. (OB)
9. Be able to calculate core radius r_c and eccentricity ratio e/r_c. Be able to use Figure 10.5.2 for buckling of eccentric columns. (OB)
10. Recognize that some tubular columns must be checked for local crushing as well as column instability. Be able to use figures such as Figure 10.3.3 to perform such checks. (OB)
11. Be able to recognize and/or define and sketch the edge and end conditions for a flat plate, curved plate or cylinder. (CB)
12. Be able to calculate the critical torque for a given cylinder in torsion. (OB)

13. Be able to calculate the compressive or shear buckling stress for a given flat plate or curved panel. (OB)
14. Be able to describe and sketch the buckling behavior of a flat plate. (CB)
15. Be able to describe and sketch the behavior of a beam experiencing lateral buckling. (CB)
16. Be able to calculate the buckling load of a given beam. (OB)
17. Be able to calculate the buckling load or stress for a given circular cylinder in compression, bending or torsion. (OB)

10.2 ELASTIC STABILITY OF COLUMNS

A column is a member which carries compression along its longitudinal axis. The elastic stability of columns was first studied by Euler in approximately 1757, (Reference 10). The basic equation for column buckling is still called Euler's equation and the basic buckling load is often referred to as the Euler load for the column. We shall derive Euler's equation for the pinned, uniform, elastic, initially straight column. The column is shown in Figure 10.2.1. It is laid on its side, but that doesn't really matter. It's all in how you look at it. The weight of the column is neglected. The product of modulus of elasticity E and minimum moment of inertia I is bending rigidity EI, which is constant with length.

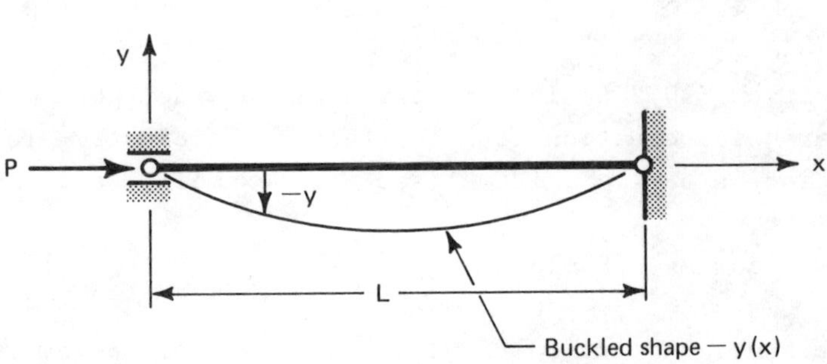

Figure 10.2.1

We assume that the buckled shape of the column, y(x), is shown by the curved portion of Figure 10.2.1. From Chapter 4, the differential equation governing such an elastic curve is

$$\frac{d^2y}{dx^2} = \frac{M}{EI} \qquad (10.2.1)$$

The bending moment for the column is given by

$$M = P(-y) \qquad (10.2.2)$$

So that

$$\frac{d^2y}{dx^2} = \frac{-Py}{EI} \qquad (10.2.3)$$

or

$$y'' + \frac{P}{EI}\, y = 0 \qquad (10.2.4)$$

The term P/EI is defined to be a new constant

$$k^2 = \frac{P}{EI} \qquad (10.2.5)$$

Then

$$y'' + k^2 y = 0 \qquad (10.2.6)$$

Equation (10.2.6) is a linear, second order, ordinary differential equation with constant coefficients. The solution is easily written (see, for example, Reference 7).

$$y = A \cos kx + B \sin kx \qquad (10.2.7)$$

The general solution, Equation (10.2.7) can be evaluated by using the boundary conditions for the column of Figure 10.2.1.

At the left end we have

$$y(x)\Big|_{x=0} = 0$$

$$A \cos 0 + B \overset{0}{\cancel{\sin}}\ 0 = 0$$

$$A = 0 \tag{10.2.8}$$

So that

$$y(x) = B \sin kx \tag{10.2.9}$$

At the right end

$$y(x)\Big|_{x=L} = 0$$

$$B \sin kL = 0 \tag{10.2.10}$$

B cannot be zero, if it were Equation (10.2.9) would be identically zero. Therefore we must have

$$\sin kL = 0 \tag{10.2.11}$$

The roots of Equation (10.2.11) are

$$kL = n\pi;\ n = 0, \pm 1, \pm 2, \pm \ldots \tag{10.2.12}$$

For the more technically oriented reader, the $n\pi$ are the eigenvalues of Equation (10.2.10).

The root $n = 0$ is trivial, the first buckling load is given by $n = 1$, for which

$$kL = \pi \tag{10.2.13}$$

Recalling the definition of k

$$\sqrt{\frac{P}{EI}}\,L = \pi$$

or

$$\boxed{P = P_E = \frac{\pi^2 EI}{L^2}} \qquad (10.2.14)$$

The buckled shape is

$$B\sin\sqrt{\frac{P}{EI}}\,x \qquad (10.2.15)$$

Equation (10.2.14) gives the lowest, or fundamental buckling load. Theoretically the column will buckle at the loads corresponding to $n = \pm 2, \ldots$ but in actuality will buckle at the load P_E unless artificially forced to a higher load.

We recall that the Euler load, P_E is for a column which is:

a) pinned
b) elastic
c) straight
d) uniform

In subsequent sections we shall consider what happens when the column no longer meets the above conditions.

The equation (10.2.14) can be rearranged to give the "buckling stress." We divide by A and use area property called radius of gyration, P, where

$$I = \rho^2 A \qquad (10.2.16)$$

Then

$$\sigma_E = \frac{\pi^2 E}{(L/\rho)^2} \qquad (10.2.17)$$

Equation (10.2.17) plots as a parabola with σ_E vs. L/ρ as shown in Figure 10.2.2.

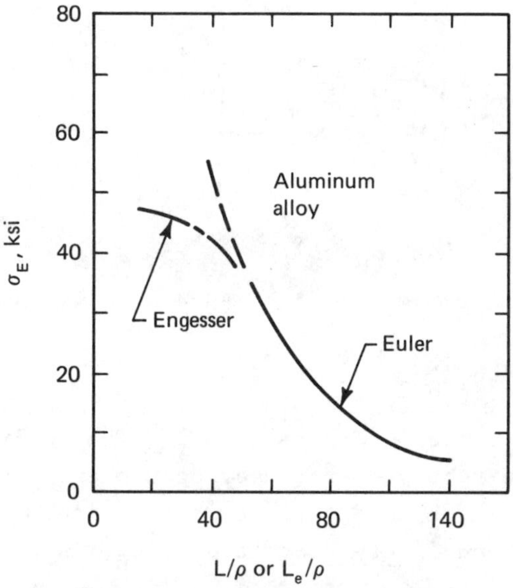

Figure 10.2.2

10.3 END CONDITIONS OTHER THAN PINNED

When the end conditions are different than the basic Euler column (pinned), the boundary conditions of the solution for Equation (10.2.7) are different. This problem has been treated extensively in the literature; for example, see References (2), (3), (7) and (8). The customary way to present the results is to define an equivalent length, L_e, for columns with other end conditions with the Euler column as the reference. This is illustrated in Figure 10.3.1.

Equation (10.2.17) may then be written

$$\sigma_{cr} = \frac{\pi^2 E}{(L_e/\rho)^2} \tag{10.3.1}$$

And Figure 10.2.2 will plot the same, but the abscissa is (L_e/ρ). Thus the concept of equivalent length takes care of different end conditions. All aluminum columns

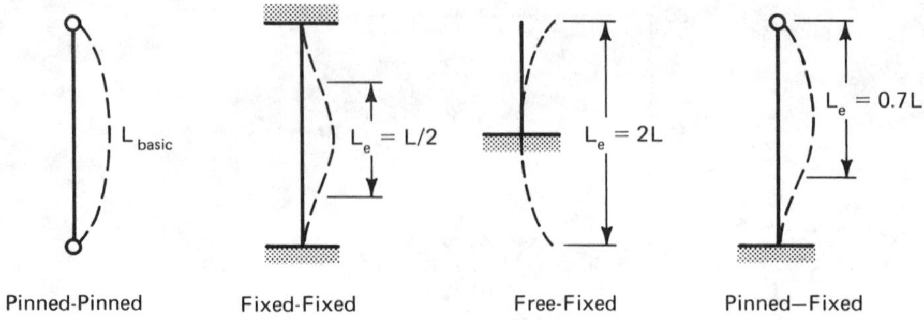

Figure 10.3.1

now fall on the curve of Figure 10.2.2. The equivalent lengths, L_e, are tabulated for many column end conditions. See Reference (4) for an excellent compilation.

Sometimes the end fixity coefficient, C_E, is tabulated where

$$C_E \equiv \frac{L_e}{L} \tag{10.3.2}$$

Samples of these curves for Steel, Aluminum and Magnesium are shown in Figures 10.3.2, 3 and 4. (Reference 9.)

10.4 THE EFFECTS OF INELASTIC BUCKLING

Many columns are designed to yield and buckle at about the same load. This can be shown to give an efficient column. (Reference 1.) There is some degree of uncertainty regarding inelastic column behavior. The best way to handle this problem is to use the tangent modulus, E_t, (Chapter 2, our text) in Equation (10.3.1) to get

$$\sigma_{cr} = \frac{\pi^2 E_t}{(L_e/\rho)^2} \tag{10.3.3}$$

Equation (10.3.3) is Engesser's equation, developed circa. 1898, (Reference 10). The effect of inelastic considerations on the Aluminum Alloy of Figure 10.2.2 is shown by the flattening of the curve at the higher stresses. Examination of Figures 10.3.2, 3 and 4 reveal that the curves are sometimes "chopped" or drawn horizontally at some upper limit stress.

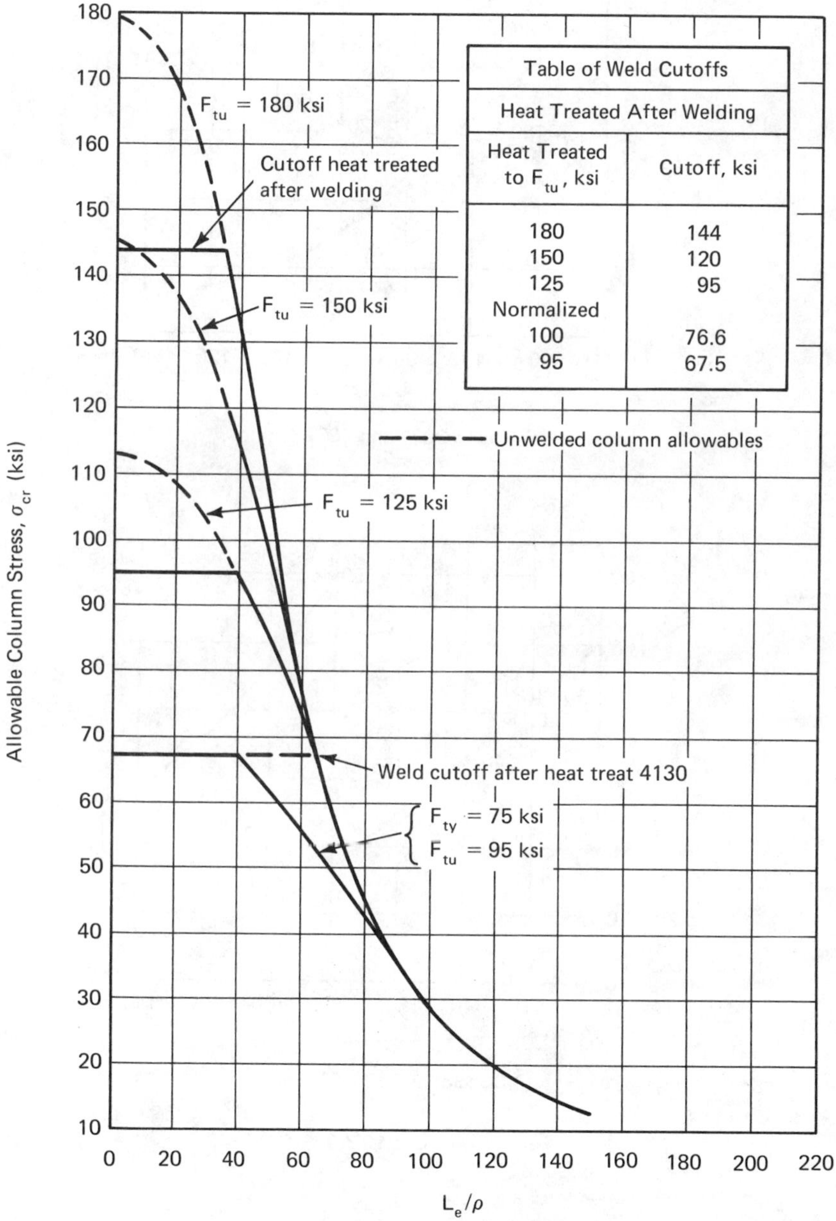

Figure 10.3.2

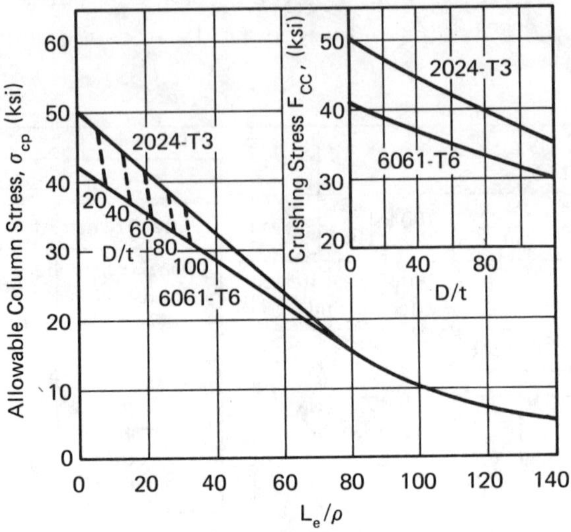

(a) Round 2024 and 6061 tubing

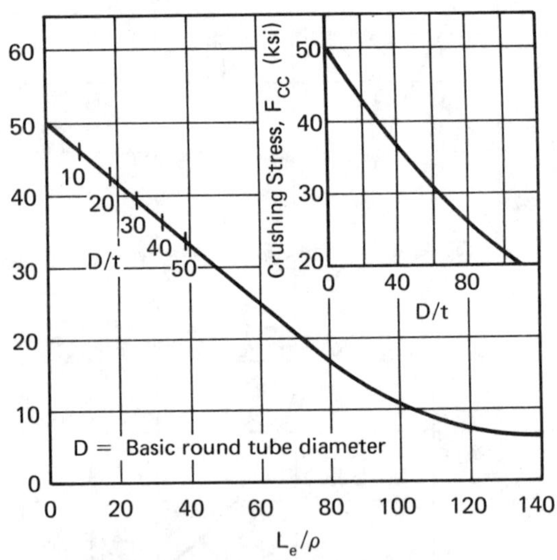

(b) Streamline 2024-T3 tubing

Figure 10.3.3

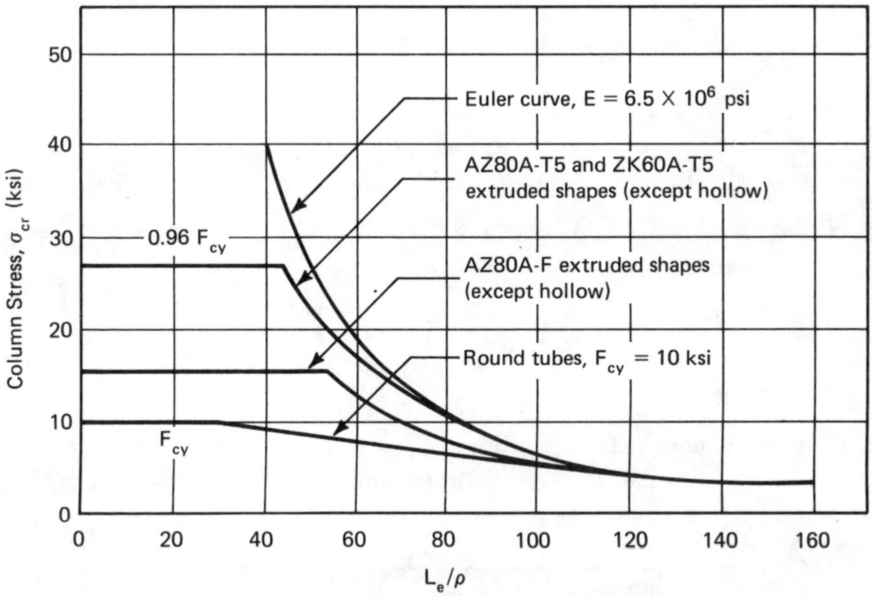

Figure 10.3.4

10.5 THE EFFECTS OF ECCENTRICITY

In actuality, columns are not initially straight. References (1), (3) and (7) develop the mathematics of the column with initial imperfection; the column which is not initially straight; and the column subjected simultaneously to lateral loading (the beam-column). We shall not treat the beam-column. The column with initial lack of straightness--intentional or accidental--is shown in Figure 10.5.1.

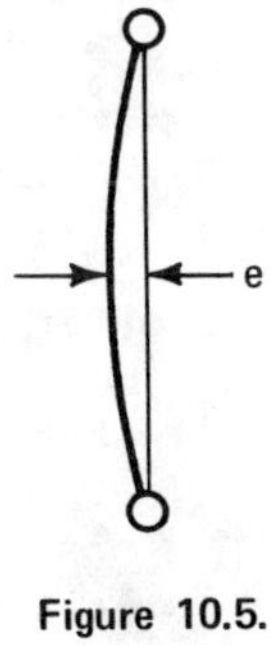

Figure 10.5.1

The inital maximum imperfection is denoted by e. We define the column <u>eccentricity ratio</u> as

$$e/r_c \tag{10.5.1}$$

where r_c is the core radius = ρ^2/c or Z/A; another cross section property. The critical stress for such a column is given implicitly by

$$\sigma_y = \sigma_{cr}\left(1 + \frac{ec}{\rho^2} \sec \frac{L}{2\rho}\sqrt{\frac{\sigma cr}{E}}\right) \tag{10.5.2}$$

A plot of the solution to this equation for a steel alloy is given in Figure 10.5.2. (Reference 1, 3, 7.)

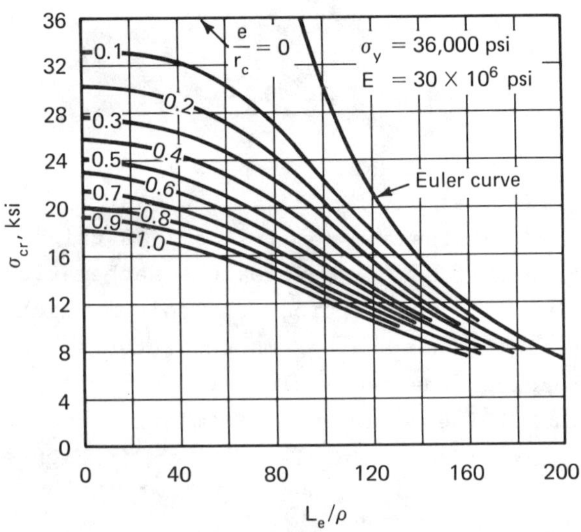

Figure 10.5.2

Note that the concept of plotting σ_{cr} vs. L_e/ρ allows all of the column conditions: end fixity, inelasticity, imperfections to be plotted on the same chart.

10.6 LOCAL CRUSHING OF COLUMNS

As hollow circular columns become thinner (the D/t ratio becomes higher) they may be subjected to local crushing or crippling. This is illustrated in Figure 10.6.1.

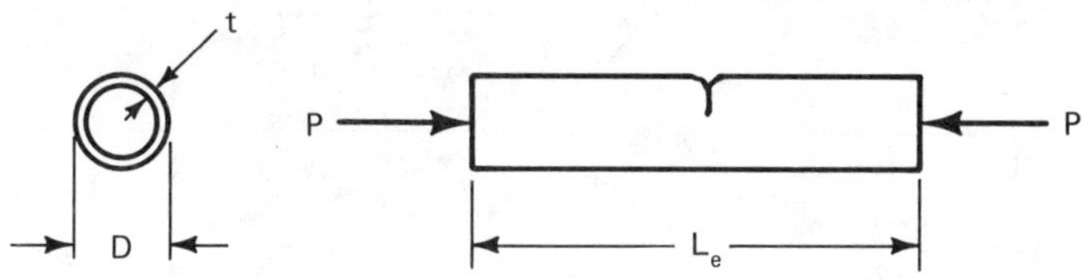

Figure 10.6.1

Figure 10.3.3 shows the curves for Aluminum Alloy local crushing stress vs. D/t ratio. These columns must be checked for Euler buckling and local crushing. The cut-off values of D/t are shown for these tubes on the σ_{cr} vs. L_e/ρ curves.

Figure 10.3.3 can also be used to determine local crushing for circular tubes in bending since the compressive bending stress has essentially the same effect as the direct compressive stress. Figure 10.6.2 gives the local buckling stresses for round tubes in torsion.

10.7 THE STABILITY OF FLAT PLATES

A thin, flat panel may be compressed along opposing edges while all four edges of the panel are supported in some fashion. (Figure 10.7.1.) The plate which is free to rotate along the edges but is restrained against z direction displacements is said to be <u>simply supported</u> or <u>clamped</u>. If the edges of the plate are built in the plate is <u>fixed</u>. Various combinations of these conditions are also possible. Flat plates buckle by "springing out" in the center, or in multiple sine-waves m as shown in Figure 10.7.1.

The equation for critical stress of such a panel is:

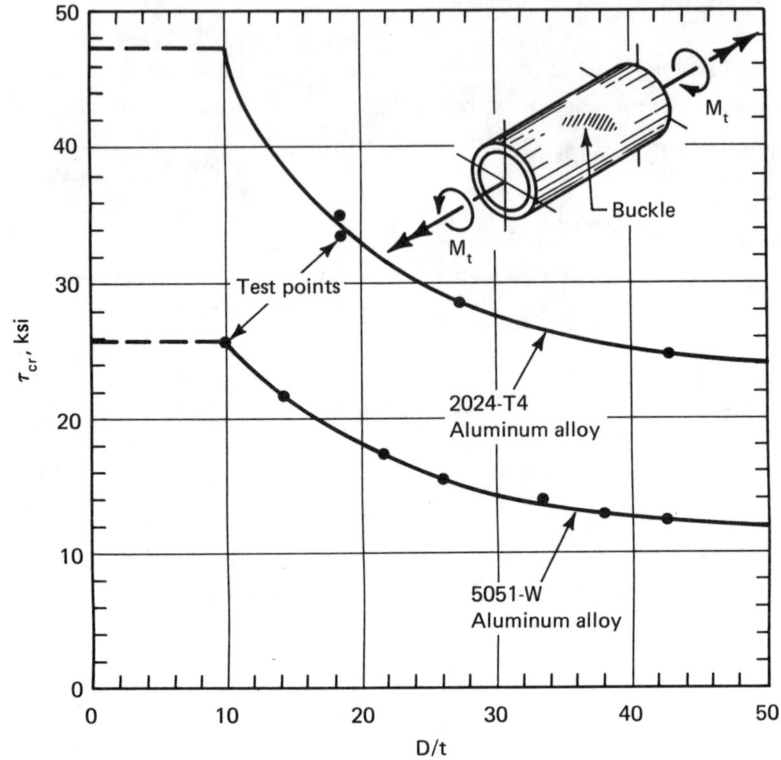

Figure 10.6.2

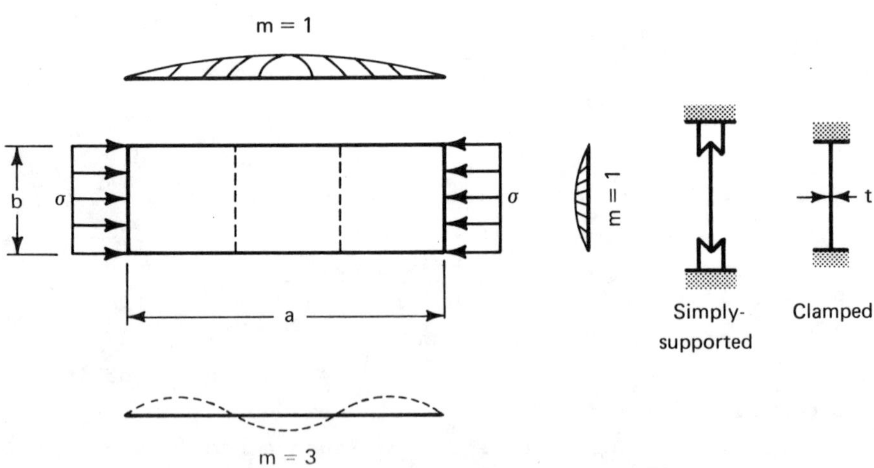

Figure 10.7.1

$$\sigma_{cr} = \frac{\pi^2 K_c E}{12(1 - \nu^2)} \left(\frac{t}{b}\right)^2 \qquad (10.7.1)$$

Figure 10.7.2 shows the buckling coefficient, K_c, for various edge conditions. (Reference 3, et al.) Many other kinds of loading and edge conditions may be considered. See especially Reference (4).

A similar equation governs the buckling of panels in shear (Reference 3, et al),

$$\tau_{cr} = \frac{\pi^2 K_s E}{12(1 - \nu^2)} \left(\frac{t}{b}\right)^2 \qquad (10.7.2)$$

Values of K_s are shown in Figure 10.7.3.

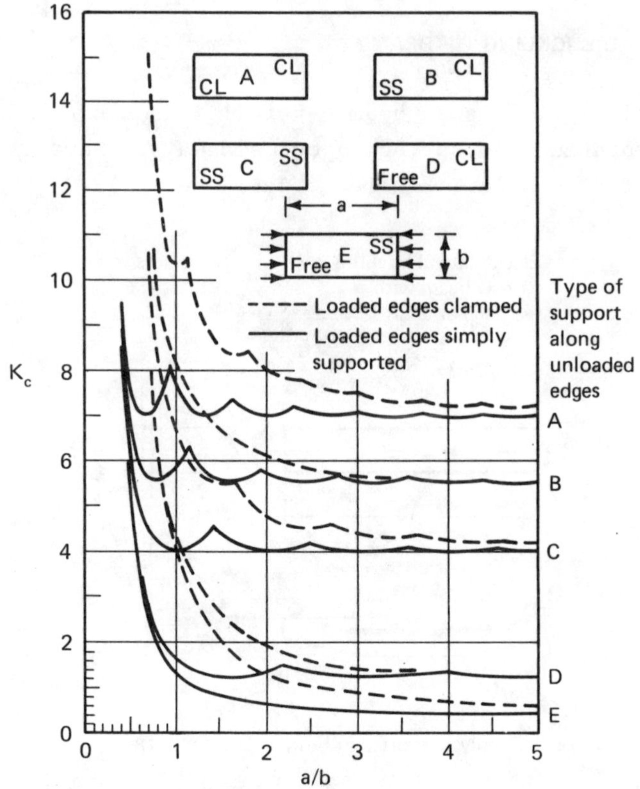

Figure 10.7.2

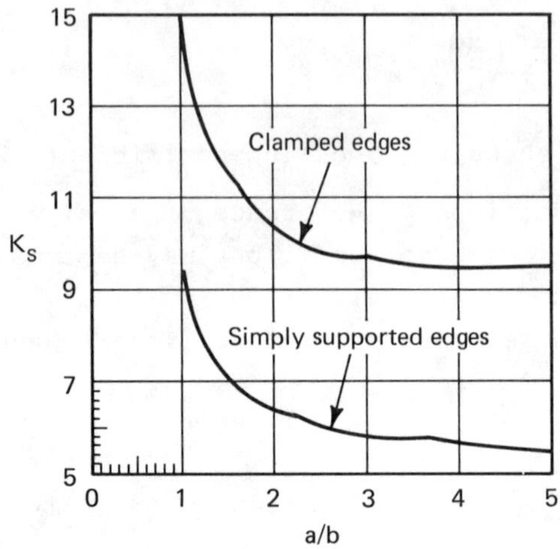

Figure 10.7.3

10.8 LATERAL BUCKLING OF BEAMS

A relatively long-span beam can fail in an instability mode by rotating and springing out sideways. Several cases are illustrated below (Reference 4, 10, et al).

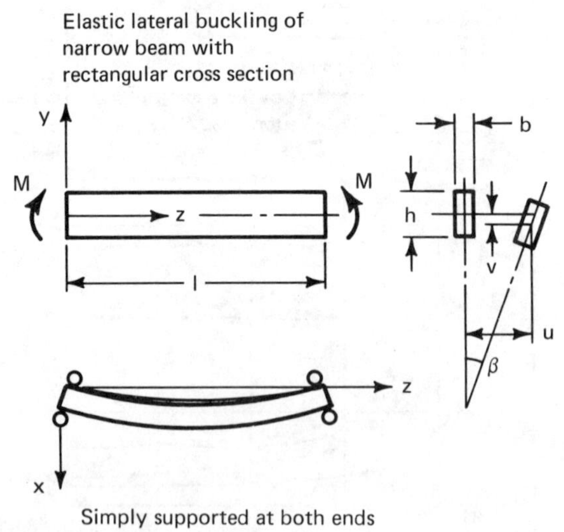

$$M_{cr} = \frac{\pi\sqrt{EI_y GJ}}{l}$$

where

$$Iy = \frac{b^3 h}{12}E, \quad J = \frac{b^3 h}{3}\left(1 - 0.630\ \frac{b}{h}\right)$$

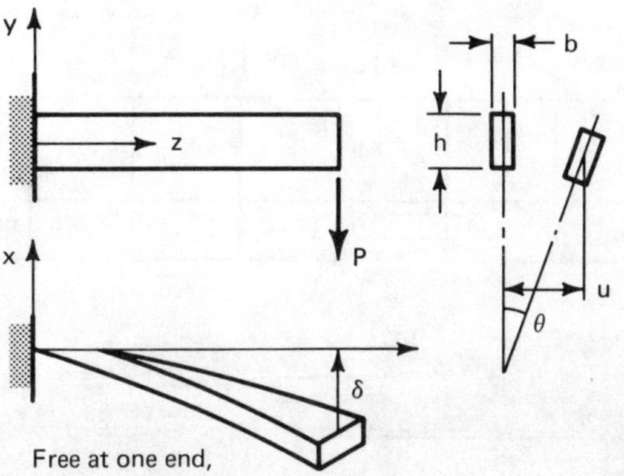

Free at one end,
built-in at another

$$P_{cr} = \frac{k\sqrt{EI_y GJ}}{l^2}$$

where

$$I_y = \frac{hb^3}{12}, \quad J = \frac{hb^3}{3}\left(1 - 0.630\frac{b}{h}\right).$$

b/h	$<\frac{1}{10}$	$\frac{1}{10}$	$\frac{1}{5}$	$\frac{1}{3}$
k	4.013	4.085	4.325	5.030

k is calculated by the following eqùation when the load P is applied at distance g above the centroid of cross section,

$$k = 4.013\left(1 - \frac{g}{l}\sqrt{\frac{EI_y}{GJ}}\right)$$

$$P_{cr} = \frac{k\sqrt{EI_y GJ}}{l^2}$$

Values of k

$\frac{l^2 GJ}{EC_w}$	0.1	1	2	3	4	6	8	10
k	44.3	15.7	12.2	10.7	9.76	8.69	8.03	7.58

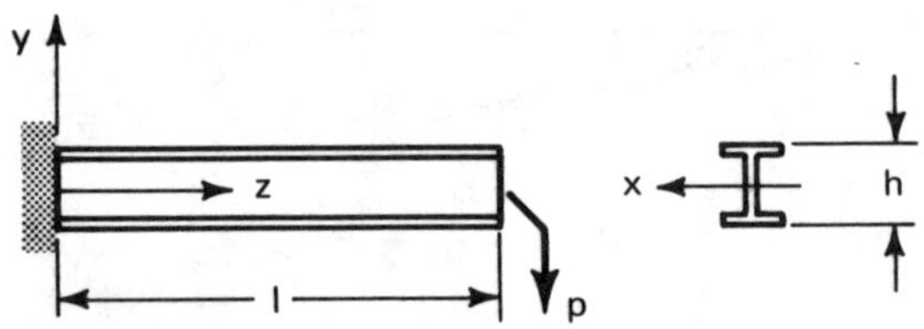

Free at one end,
built-in at another

$\frac{l^2 GJ}{EC_w}$	12	14	16	24	32	40	>40
k	7.20	6.96	6.73	6.19	5.87	5.64	$\frac{4.01}{(1-a/l)^2}$

10.9 THE BUCKLING OF CURVED PANELS

A sample of the critical shear stress relationships for curved panels is shown in Figures 10.9.1 and 2. (Reference 11.)

10.10 THE STABILITY OF CYLINDRICAL SHELLS

Cylindrical shells are closed, thin-walled structures which develop their strength from their shape. They are structurally similar to a beer can. Such structures, or

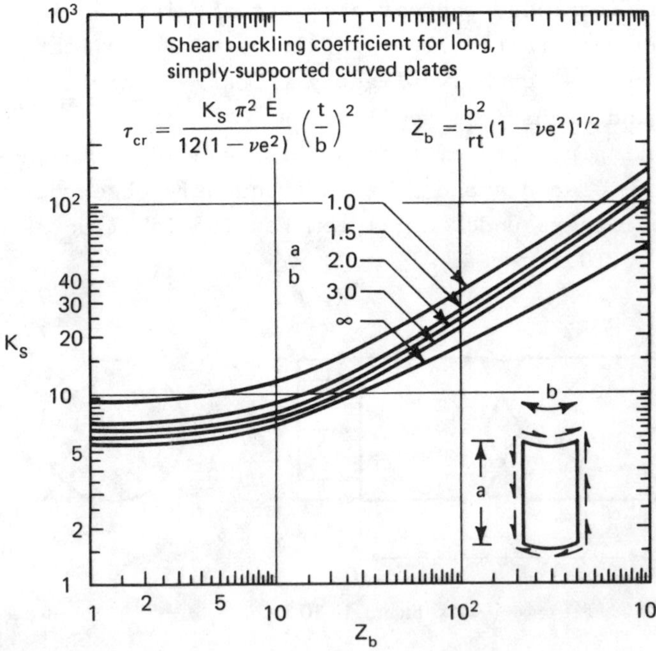

Figure 10.9.1

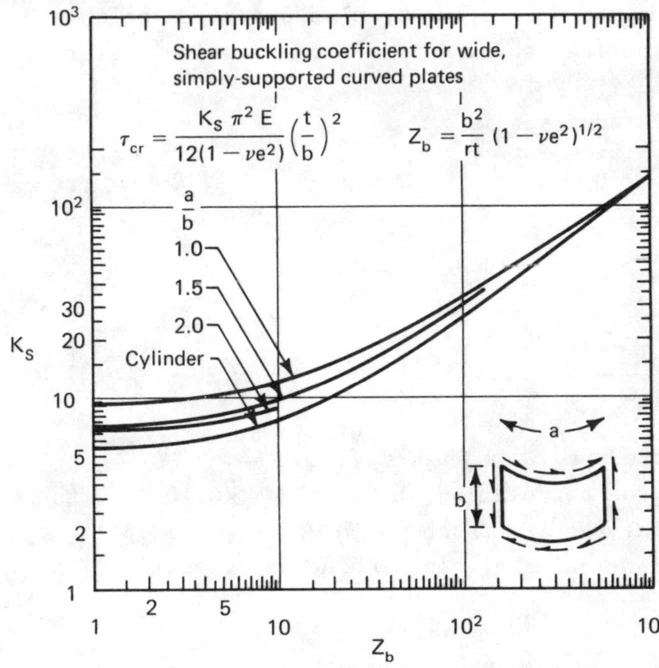

Figure 10.9.2

slight modifications, thereof are extensively used in missiles and aircraft. There is also a wide application of shell structures, usually slightly thicker in civil, architectural and mechanical applications.

The buckling pattern of thin shells is a local in-and-out pattern. The diamond pattern frequently observed in circular cylinders under axial compression is illustrated in Figure 10.10.1.

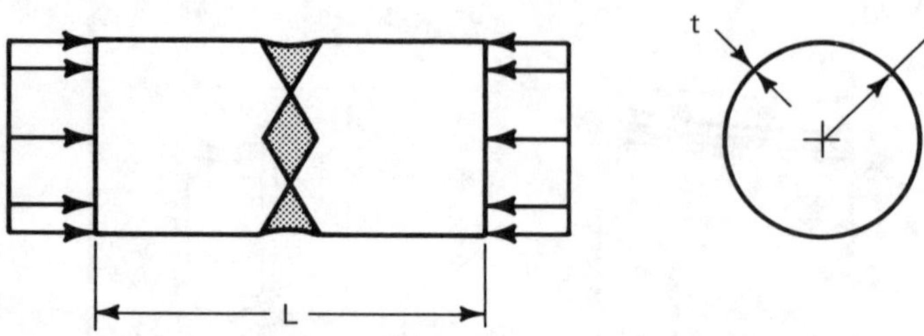

Figure 10.10.1

The theory of shells is far beyond the scope of this text. Some representative results are given below.

The cylindrical shell in axial compression, References (4) and (13):

The basic equation is:

$$\sigma_{cr} = K_c \frac{\pi^2 D}{L^2 t} \tag{10.10.1}$$

where:

$$D = \frac{Et^3}{12(1 - \nu^2)}$$

K_c is plotted vs. Z in Figure 10.10.2.

Some of the correlation of Equation (10.10.1) with experimental results is shown in Figure 10.10.3.

The cylindrical shell in bending is illustrated in Figure 10.10.4.

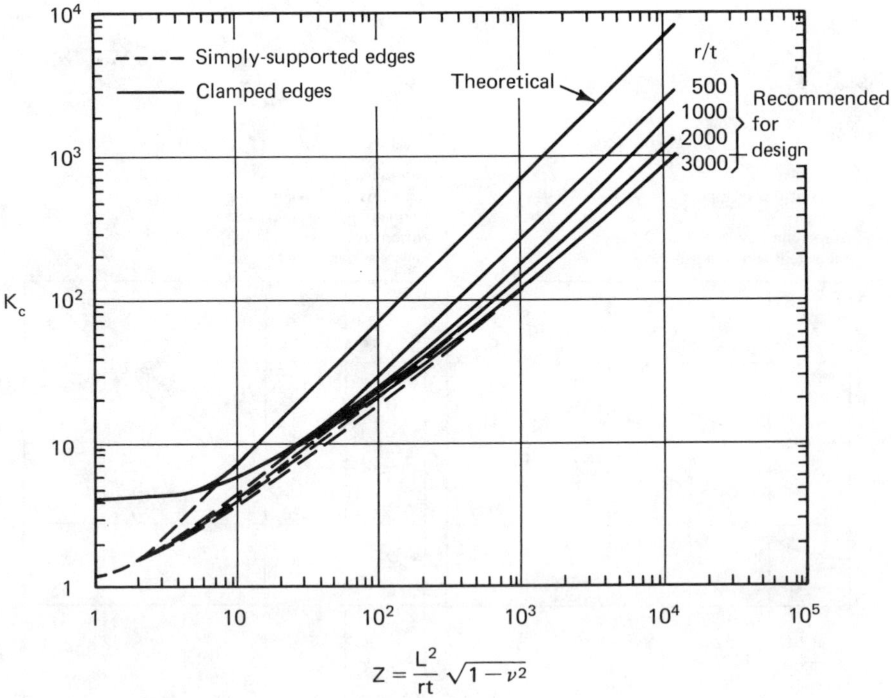

Figure 10.10 2

The critical bending moment, References (9), (14), and (15), is:

$$M_{cr} = \alpha \frac{E}{(1 - \nu^2)} Rt^2 \tag{10.10.2}$$

where.

For a long tube,

$$\alpha = \frac{2\sqrt{2}\pi}{9} = 0.99$$

The value of $\alpha = 0.99$ is theoretical. The average value of test results is reported as $\alpha = 1.14$, with a minimum (conservative) value being $\alpha = 0.72$.

The cylindrical shell in torsion, References (9) and (16) is depicted below in Figure 10.10.5.

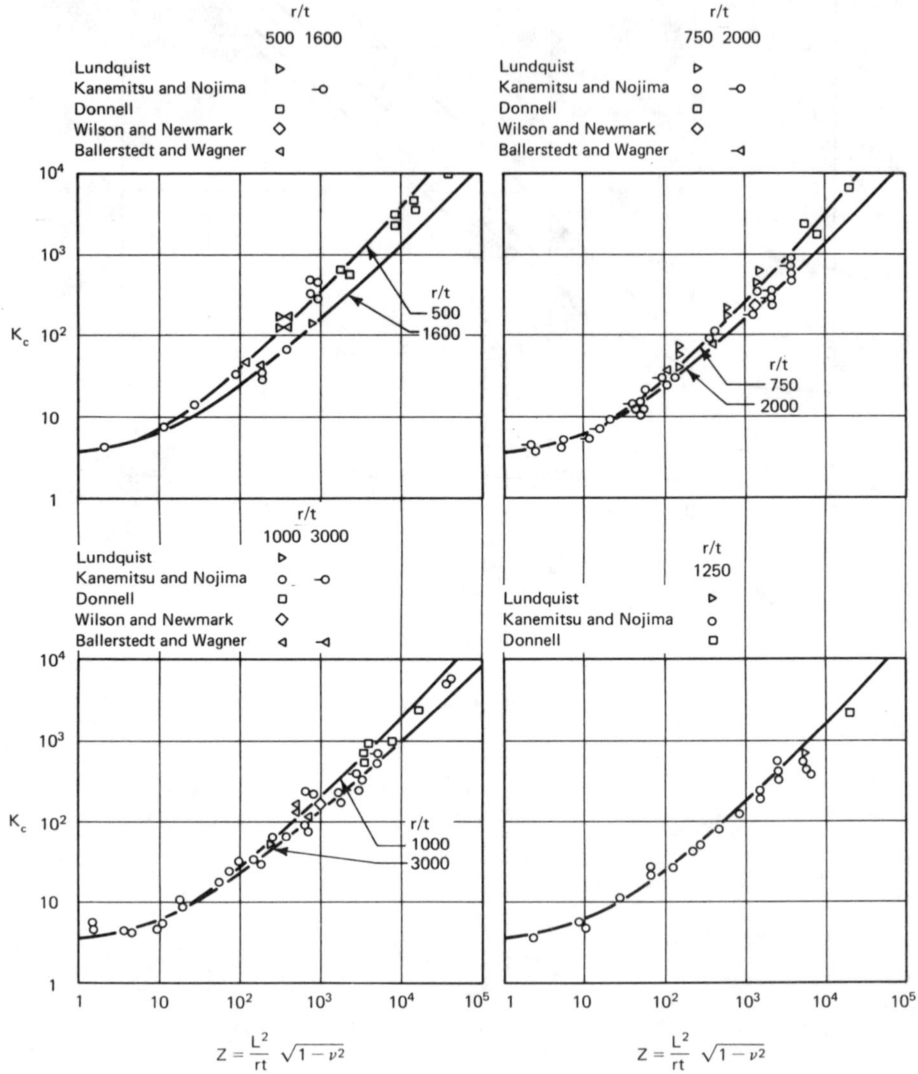

Figure 10.10.3

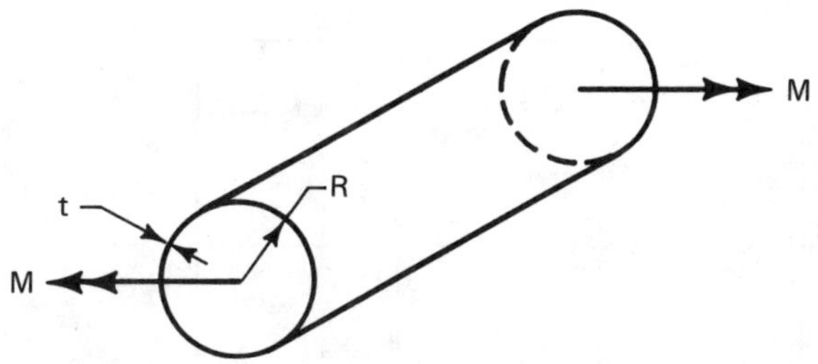

Figure 10.10.4

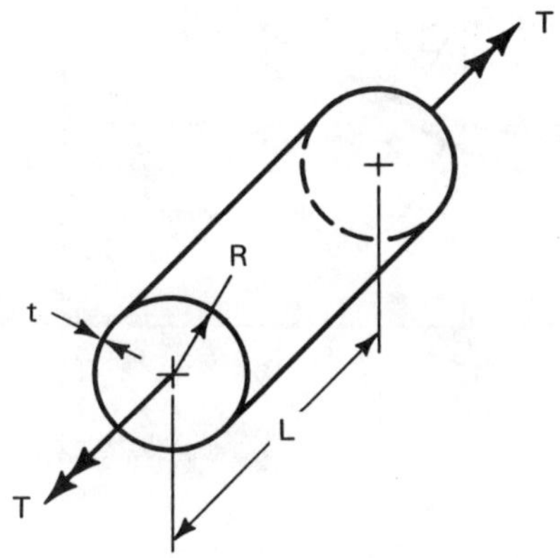

Figure 10.10.5

The critical shear stress is given by

$$\tau_{cr} = K_s \frac{\pi^2 D}{L^2 t} \tag{10.10.3}$$

where

$$D = \frac{Et^3}{12(1 - \nu^2)}$$

Some results for K_s vs. Z are shown in Figure 10.10.6. A comparison with experimental results is shown in Figure 10.10.7.

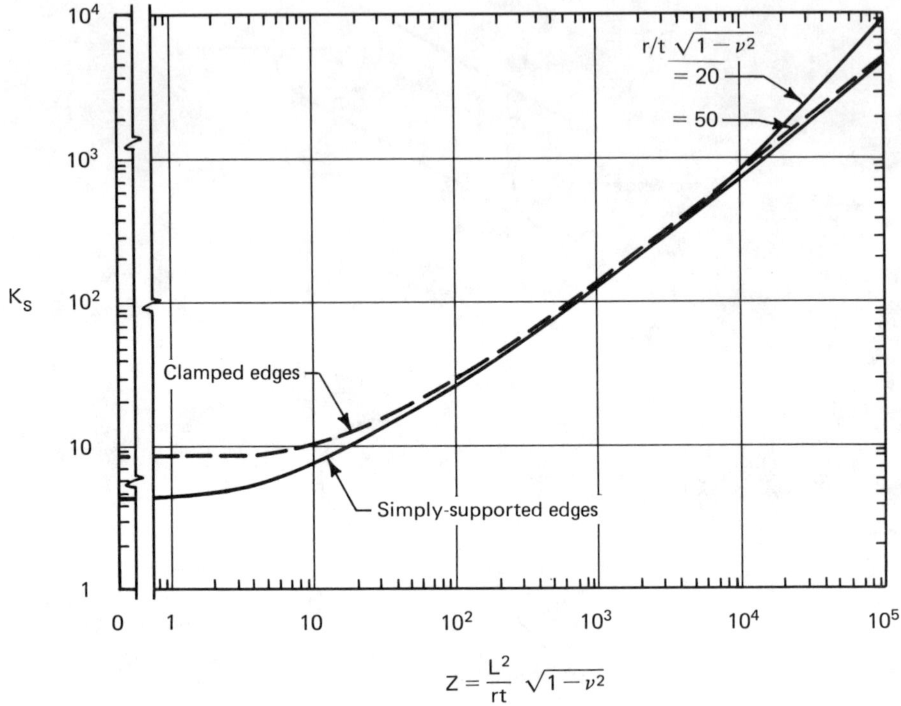

if $50 < Z < 10\left(\frac{r}{t}\right)^2 (1-\nu^2)$

$$K_s = 0.85\, Z^{3/4}$$

Figure 10.10.6

10.11 SUMMARY COMMENTS

The body of literature on the elastic stability of columns, tubes, beams, plates, stiffened plates, shells, stiffened shells, etc., etc., etc., is quite staggering and the presentation of this present chapter represent an extremely small sample of the available formulas. Similarly, the references at the end of this chapter represent only a beginning sample.

Reference (4) is to be especially recommended for a handbook of results. It is extremely thorough and comprehensive, imminently practical and it contains an unbelievably extensive bibliography. For those interested in a more

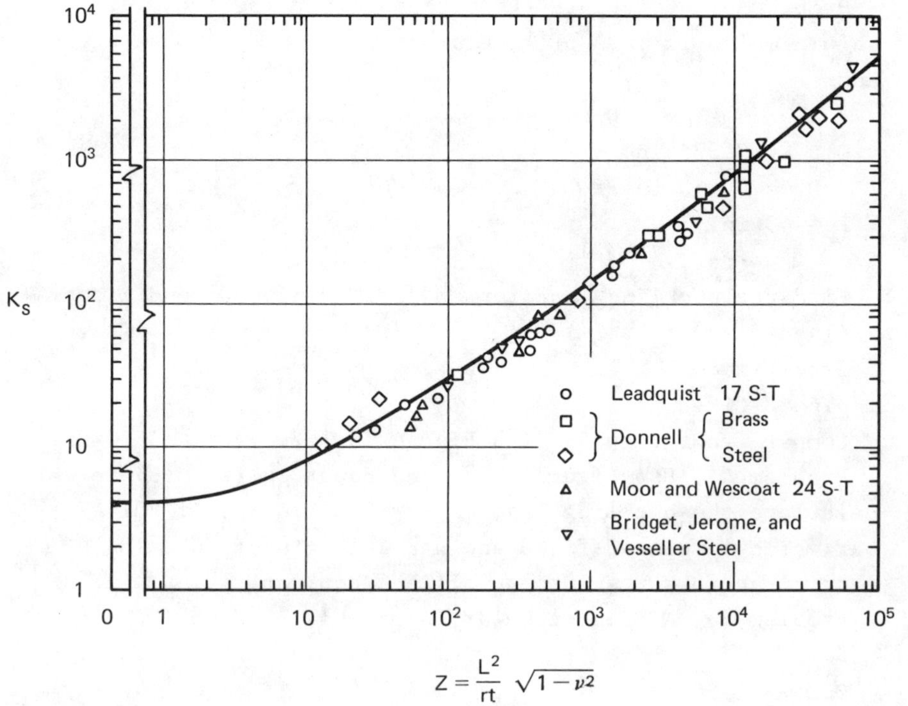

Figure 10.10.7

thorough study of derivations and theory, the classic, Reference (7) is still hard to beat.

Problems

Problem 1. Derive the equation for elastic buckling of the fixed-form column and show that

$$P_{cr} = \frac{4\pi^2 EI}{L^2}$$

and that $L_e = 2L$. Follow the method of this text, Equation (10.2).

Problem 2. Plot the Euler and Engesseur curves of Figure 10.2.2 for the magnesium of Figure 2.3.4; 75°F AZ63A-T6.

Problem 3. Calculate the buckling load for an Euler column which is made of tubular steel with $F_{ty} = 36$ ksi, and:

O.D. = 0.875 in.
I.D. = 0.635 in.
L = 48 in.

Find the buckling load for the following end conditions:
fixed-fixed
free-fixed
pinned-fixed
Compare your results with Figure 10.3.2.

Problem 4. Use Figure 2.3.3 and Equation (10.3.3) to calculate the buckling load of a 1.0 x 1.0 in. Al 2014-T6 bar which is pinned-fixed and has a length of 20.0 in. A trial-and-error solution will be required.

Problem 5. A steel tubular column has

O.D. = 1.50 in.
Wall thickness = 0.049 in.

If the column is loaded as shown, find σ_{cr}. Use Figure 10.5.2.

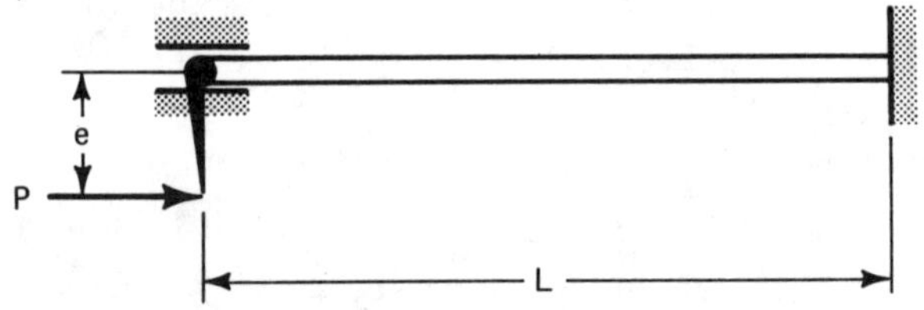

e = 0.176 in.
L = 30 in.

Problem 6. A 1/2 O.D. x 0.049 in. wall 2024-T3 Aluminum alloy tubing is used as an Euler column. If its length is 4 in., compare its column buckling and its local crushing stress. See Figure 10.3.3.

Problem 7. Find the torsional buckling stress for the tube in Problem 6.

Problem 8. A 7075-T6 Aluminum alloy plate has dimensions of 8 in. x 10 in. by 0.080 in. thickness. The narrower end is loaded in compression. Calculate the panel buckling stress for:

loaded edges clamped, and

(a) unloaded edges simply supported

(b) unloaded edges clamped

loaded edges simply supported and

(c) unloaded edges simply supported

(d) unloaded edges clamped.

Problem 9. A 12 x 10 x 58 WF I beam is loaded in cantilever fashion. Calculate the lateral buckling load, P_{cr}. Check for elastic buckling if the beam is AISC 36 ksi yield material.

Problem 10. An Aluminum cylinder has simply supported ends and is subjected to compression. Calculate the critical stress if:

$D = 100$ in.

$t = 0.125$ in.

$L = 96$ in.

Calculate the critical axial load also.

Problem 11. For the cylinder of Problem 10, calculate the critical bending moment.

Problem 12. For the cylinder of Problem 10, calculate the critical torque.

Problem 13. A thin-walled tube has a mean diameter of 4 inches, a wall thickness of .016 inches and a length of 100 inches. The tube is composed of 2014-T6 aluminum and is pinned at both ends. Determine whether column or local buckling will occur first.

(a) Local Buckling

(b) Column Buckling

(c) Must consider both (a) and (b) because they are within 100 psi of each other

(d) Lateral Buckling

(e) None of the above

Problem 14. An aluminum alloy bar with a square cross section (4.84" x 4.84") is to be used as an Euler column. Its length is 100 inches with fixed-pinned end constraints.

If Figure 10.3.3 describes the material, at what load will the column fail? (Pick answer ± 10%.)

(a) 212 kips
(b) 169 kips
(c) 35 kips
(d) 44 kips
(e) None of the above

Problem 15. Consider three separate square metal panels. All edges are simply supported.

Panel #1	Panel #2	Panel #3
6" x 6"	5" x 5"	4" x 4"
AISI 4130 Steel	2024-T3 Al	M1A Magnesium
t = .03"	t = .04"	t = .05"

Determine which panel will buckle first.

(a) Panel #1
(b) Panel #2
(c) Panel #3
(d) Both Panel #1 and Panel #2 will buckle first within ± 50 psi of each other
(e) None of the above

Problem 16. A solid bar is to be used as an Euler column. (Assume material is steel with pinned-pinned end constraints.) The maximum allowable stress is 40,000 psi.

If the bar is to be 100 inches long, pick the smallest diameter bar.

(a) OD = 4-1/2"
(b) OD = 3"
(c) OD = 2-1/4"
(d) OD = 1-3/4"
(e) None of the above

Problem 17. Determine σ_{max} and $\delta_{midspan}$ for the circular column.

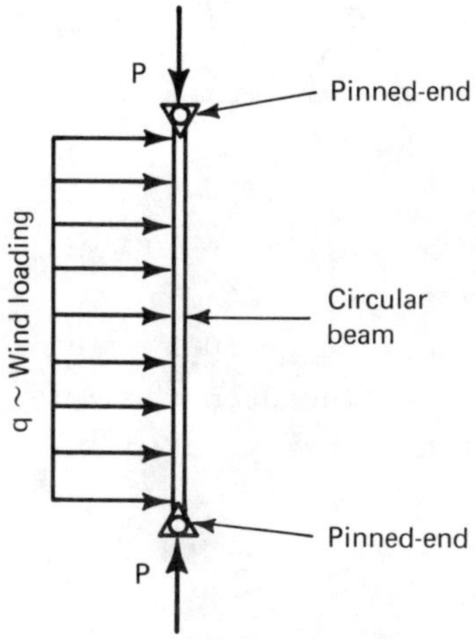

Column Specifications:

Cross section ~ 2 inch O.D. Tube

t = .065 inch

L = 72 inches

E = $10(10^6)$ psi

Cross Section Properties:

A = .3951 in^2 I = .1851 in^4

ρ = .6845 in Z = .1851 in^3

Loading Condtions:

q = 0.5 lb_f/in. P = 1750 lb_f

Problem 18. Two columns and their properties are shown below. Calculate the critical load and the critical stress for each case. Which column has the highest critical load? Which column has the highest critical stress?

Case A. Pinned-Pinned

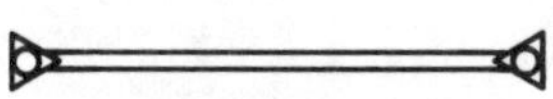

Magnesium Column

$$\left\{ \begin{array}{l} E = 6.5\,(10^6)\text{ psi} \\ \alpha = 1 \\ I = 30.68\text{ in.}^4 \\ A = 19.63\text{ in.}^2 \\ \rho = 1.25\text{ in.} \\ L = 120\text{ in.} \end{array} \right\}$$

Case B. Pinned-Fixed

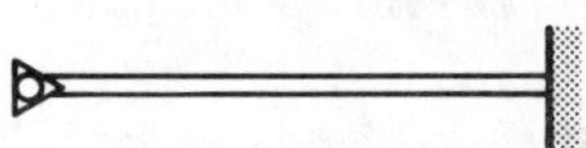

Aluminum Column

$$\left\{ \begin{array}{l} E = 10\,(10^6)\text{ psi} \\ \alpha = 1/2 \\ I = 3.976\text{ in.}^4 \\ A = 7.069\text{ in.}^2 \\ \rho = .75\text{ in.} \\ L = 120\text{ in.} \end{array} \right\}$$

Problem 19. Consider an aluminum compression member which is supported at both ends by pinned constraints. An engineer is faced with the task of selecting one of the three cross sections shown below. Choose the cross section that has the highest critical stress if buckling is assumed to occur about the x-axis. The column length is 120 inches.

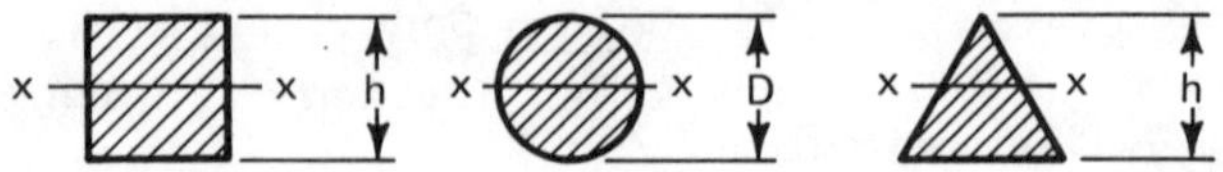

h = 3 in. D = 3 in. h = 3 in.

Problem 20. A circular cylinder has a radius r = 100 in., a length L = 200 in., and a wall thickness t = .05 in. The material is an aluminum alloy 7075-T6 whose $E_c = 10.5 \times 10^6$ psi. If an internal pressure of 5 psi is placed on the cylinder, what compressive axial stress can it carry based on 90% probability?

References

1. Mechanics of Materials, F. R. Shanley, McGraw-Hill, New York, 1967.
2. Analysis and Design of Flight Vehicle Structures, E. F. Bruhn, Tri-State Offset Co., 1973.
3. Introduction to Structural Stability Theory, George Gerard, McGraw-Hill, New York, 1962.
4. Handbook of Structural Stability, Column Research Committee of Japan, Ed., Corona Publishing Co. Ltd., Tokoyo, 1971.
5. The Stability of Flat Plates, P. S. Bulson, American Elsevier Publishing Co., New York, 1969.
6. Formulas for Stress and Strain, R. J. Roark, McGraw-Hill, New York, 1965. (A later edition has recently been printed.)
7. Theory of Elastic Stability, S. P. Timoshenko and J. M. Gere, McGraw-Hill, New York, 1961.
8. Buckling Strength of Metal Structures, F. Bleich, McGraw-Hill, New York, 1952.
9. Metallic Materials and Elements for Aerospace Vehicle Structures, Dept. of Defense, Washington, D.C., 1966.
10. History of the Strength of Materials, S. P. Timoshenko, McGraw-Hill, New York, 1953.
11. Handbook of Structural Stability. Part III. Buckling of Curved Plates and Shells, Gerard and Becker, NACA TN 3883.

12. Torsional Strength of Aluminum Alloy Round Tubing, R. L. Moore, NACA TN 879, 1943.

13. Critical Stress of Thin-Wall Cylinders in Axial Compression, S. B. Batdorf, M. Shildcrout and M. Stein, NACA Report 887, 1947.

14. On the Flexure of Thin Cylindrical Shells and Other Thin Sections, L. G. Brasier, Proc. Roy. Soc. London, Ser. A. Vol. 116, pp. 104-114, 1927.

15. The Bending Stability of Thin-Walled Unstiffened Circular Cylinders Including the Effects of Internal Pressure, H. S. Suer, L. A. Harris, W. T. Skene and R. J. Benjamin, J. Aero Sci., Vol. 25, No. 5, pp. 281-287, May 1958.

16. A Simplified Method of Elastic-Stability Analysis for Thin Cylindrical Shells, S. B. Batdorf, NACA Report 874, 1947.

Chapter 11

THE BONDED ELECTRICAL RESISTANCE STRAIN GAGE IN EXPERIMENTAL STRESS ANALYSIS

11.1 INTRODUCTION

The bonded electrical resistance strain gage works on the principal that a metal element, such as a wire, will change resistance when it is elongated. Lord Kelvin first discovered this effect in 1856. If a wire, insulated by a backing material, is cemented to the structure whose strain is desired, and the resistance change of the wire during loading is measured, this change in resistance can be converted to strain measurements.

A simple illustration demonstrates this effect.

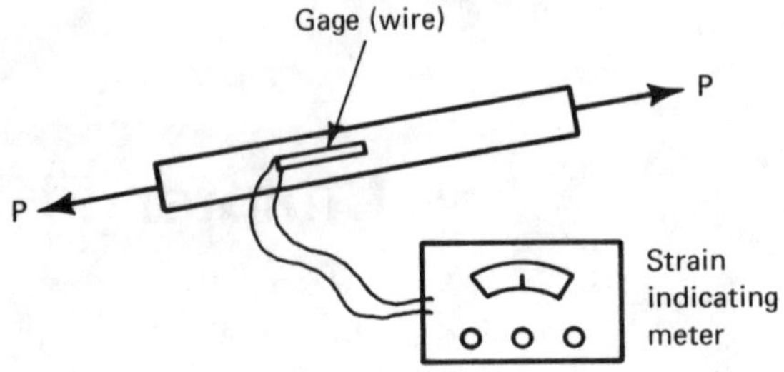

Figure 11.1.1 Illustration of a basic test

The three basic steps in using this gage for experimental stress analysis are:

1. Selection of gage and cementing in proper location.
2. Using instrumentation to convert the resistance change to a strain signal.
3. Translations of strains to stresses.

The types of gages available and some of their characteristics will be discussed next.

11.2 MANUFACTURE AND TYPE OF GAGES

The basic types of gages available are:

1. Wire
2. Foil
3. Semiconductor

The semiconductor gage will not be discussed at present.

The foil gage is currently very popular and is made by photographically etching a desired pattern on a thin metal foil. Some typical shapes which are available are illustrated below. Other shapes are shown in Figure 11.2.3.

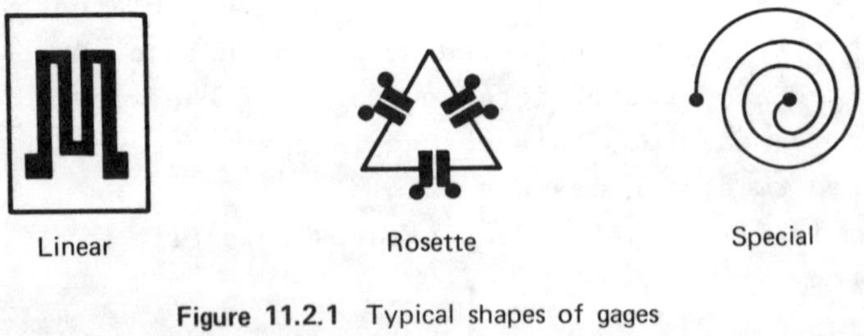

Figure 11.2.1 Typical shapes of gages

The wire wound gage is made by pulling the wire between hooks in a grid pattern. The wire grid is then mounted on paper and covered with felt; the felt covering protects the gage.

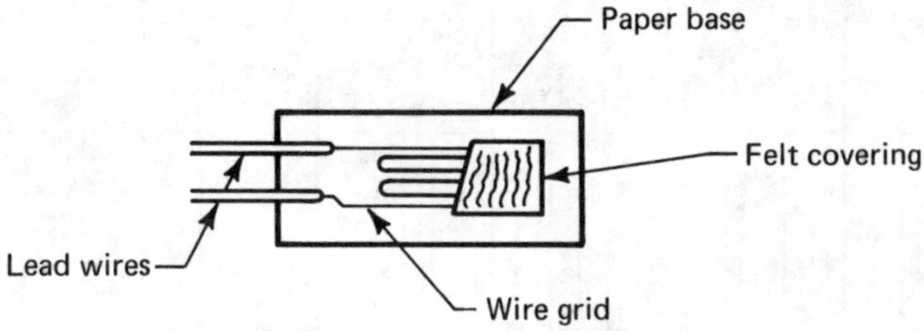

Figure 11.2.2 Typical wire wound gage

Gages of all types are available in lengths from 1/64 in. up. Special orders are available from all manufacturers. It must be remembered that these gages will measure the average strain over their gage length; therefore, if strain gradients are high at a particular location, the gage length must be small to obtain accurate readings.

Resistances of strain gages vary from 60Ω to 5,000Ω. A typical value is 120Ω.

Gages are available in many forms. Among these are the following:

1. Single element. For uniaxial tension.
2. Two-element rosette. For two-dimensional stresses when the principal stress directions are known.
3. Three-element rosettes. For general two-dimensional stress work.
4. Special Purpose. For a wide variety of special applications.

Gage backing materials are usually paper or epoxy. The paper gage is used for room temperature short time tests, the epoxy for high temperatures and high stability.

A number of adhesives are available for bonding gages to the test structure. Among these are:

1. Nitrocellulose cements (Airplane glue) - for room temperature short time applications.

GAGE PATTERN (Actual Size) Grids are Vertical Except as Noted	GAGE DESIGNATION Insert Desired S-T-C No. in Spaces Marked XX	STOCK STATUS	RES. IN OHMS	LIST PKG. PRICE (Five Gages)	OPTIONS AVAILABLE Add Indicated Price to Package Price. Note: Stock Status Symbols Do Not Apply to Gages with Optional Features: W	E	S	SE	L	LE	DIMENSIONS IN INCHES / *MILLIMETRES* (M = Matrix; S = Section (e.g., S1 = Sec. 1); ES = Each Section; CP = Comp. Patt.; GD = Grid Dia.; PD = Patt. Dia.; APD = Act. Patt. Dia.): GAGE LENGTH	OVERALL LENGTH	GRID WIDTH	OVERALL WIDTH	OTHER SERIES AVAILABLE: GAGE DESIGNATION	Stock Status	LIST PKG. PRICE
125UT 'C' FEATURE 	CEA-XX-125UT-120	A	120 ± 0.4%	50.00	*CEA Series Strain Gages feature large integral copper solder tabs and a completely encapsulated grid. Refer to the front of this Catalog for details.*						0.125 ES	0.325 CP	0.165 ES	0.356 CP			
	CEA-XX-125UT-350	A	350 ± 0.4%	60 00							*3.18 ES*	*8.26 CP*	*4.19 ES*	*9.04 CP*			
											Matrix Size (Pretrimmed) →		0.425L x 0.455W / *10.8L x 11.6W*				
					Pattern Description: 2-Element 90° 'tee' rosette for general-purpose use. Exposed solder tab area 0.10" x 0.07" *(2.5mm x 1.8mm).*												
125TD 	EA-XX-125TD-120	A	120 ± 0.2%	42.00	18.00	5.50	7.00	12.50	7.00	12.50	0.125 ES	0.328 CP	0.125 ES	0.392 CP	EP-08-125TD-120	C	52.00
	MA-XX-125TD-120	A	120 ± 0.2%	47.00			7.00				*3.18 ES*	*8.33 CP*	*3.18 ES*	*9.96 CP*	SA-XX-125TD-120	C	69.00
	ED-DY-125TD-350	C	350 ± 0.5%	53.00		5.50	7.00	12.50	7.00*	12.50*	*Matrix Size* →		0.43L x 0.53W / *10.9L x 13.5W*		SK-XX-125TD-350	C	75.00
	WA-XX-125TD-120	C	120 ± 0.4%	69.00	21.00*										SD-DY-125TD-350	C	73.00
	WK-XX-125TD-350	C	350 ± 0.4%	75.00	21.00*										WD-DY-125TD-350	C	73.00
					Pattern Description: 2-Element 90° rosette used for shear-strain measurement and in torque transducers.										TA-XX-125TD-120	C	53.00
															TK-XX-125TD-350	C	58.00
															TD-DY-125TD-350	C	57.00
125TR 	EA-XX-125TR-350	B	350 ± 0.2%	47.00	18.00	5.50	7.00	12.50	7.00	12.50	0.125 ES	0.273 CP	0.125 ES	0.392 CP	EP-08-125TR-350	C	58.00
	MA-XX-125TR-350	B	350 ± 0.2%	52.00			7.00				*3.18 ES*	*6.93 CP*	*3.18 ES*	*9.96 CP*	SA-XX-125TR-350	C	76.00
	ED-DY-125TR-10C	C	1000 ± 0.5%	60.00		5.50	7.00	12.50	7.00*	12.50*	*Matrix Size* →		0.39L x 0.52W / *9.9L x 13.2W*		SK-XX-125TR-10C	C	87.00
	WA-XX-125TR-350	C	350 ± 0.4%	76.00	21 00*										SD-DY-125TR-10C	C	85.00
	WK-XX-125TR-10C	C	1000 ± 0.4%	87.00	21.00*										WD-DY-125TR-10C	C	85.00
					Pattern Description: 2-Element 90° rosette. Higher resistance version of the 125TD pattern, with more compact geometry.										TA-XX-125TR-350	C	57.00
															TK-XX-125TR-10C	C	64.00
															TD-DY-125TR-10C	C	63.00
125TF 	EA-XX-125TF-120	A	120 ± 0.2%	38.00			5.25	10.25	6.00	13.00	0.125 ES	0.500 CP	0.150 ES	0.150 CP	EP-08-125TF-120	C	48.00
	MA-XX-125TF-120	B	120 ± 0.2%	45.00			5.25				*3.18 ES*	*12.70 CP*	*3.81 ES*	*3.81 CP*	SA-XX-125TF-120	B	65.00
	ED-DY-125TF-350	B	350 ± 0.4%	50.00			5.25	10 25			*Matrix Size* (Trim Marks) →		0.535L x 0.185W / *13.6L x 4.7W*		SK-XX-125TF-350	C	71.00
															SD-DY-125TF-350	C	70.00
					Pattern Description: General-purpose 2-element 90° 'tee' rosette with narrow pattern geometry. Sections have a common electrical connection.										TA-XX-125TF-120	C	49.00
															TK-XX-125TF-350	C	55.00
															TD-DY-125TF-350	C	55.00
125VB 	EA-XX-125VB-120	B	120 ± 0.2%	39.00			7.00	12.50			0.125 ES	0.500 CP	0.150 ES	0.150 CP	EP-08-125VB-120	C	49.00
	MA-XX-125VB-120	B	120 ± 0.2%	46.00			7.00				*3.18 ES*	*12.70 CP*	*3.81 ES*	*3.81 CP*	SA-XX-125VB-120	B	66.00
	ED-DY-125VB-350	C	350 ± 0.5%	51.00			7.00	12.50			*Matrix Size* (Trim Marks) →		0.570L x 0.220W / *14.5L x 5.6W*		SK-XX-125VB-350	C	72.00
															SD-DY-125VB-350	C	71.00
					Pattern Description: General-purpose 2-element 90° 'tee' rosette. Identical to 125TF preceding except sections are electrically independent.										TA-XX-125VB-120	C	50.00
															TK-XX-125VB-350	C	56.00
															TD-DY-125VB-350	C	56.00
125TG 	EA-XX-125TG-350	A	350 ± 0.2%	49.00			5.25	10.25	6.00	13.00	0.125 ES	0.500 CP	0.150 ES	0.150 CP	EP-08-125TG-350	C	58.00
	MA-XX-125TG-350	C	350 ± 0.2%	55.00			5.25				*3.18 ES*	*12.70 CP*	*3.81 ES*	*3.81 ES*	SA-XX-125TG-350	C	74.00
	ED-DY-125TG-10C	C	1000 ± 0.5%	58.00			5.25	10.25			*Matrix Size* (Trim Marks) →		0.535L x 0.185W / *13.6L x 4.7W*		SK-XX-125TG-10C	C	91.00
	WA-XX-125TG-350	C	350 ± 0.4%	74.00											SD-DY-125TG-10C	C	89.00
					Pattern Description: General-purpose 2-element 90° 'tee' rosette with high-resistance grid. Sections have a common electrical connection.										TA-XX-125TG-350	C	59.00
															TK-XX-125TG-10C	C	69.00
															TD-DY-125TG-10C	C	67.00

Figure 11.2.3 Typical foil gage configurations (Courtesy Miero Measurements, Inc.)

2. Epoxy cements - for high temperatures, good humidity resistance, and long-time stability.
3. Contact Cements - for rapid installation (Eastman 910).
4. Ceramic Cements - for very high temperatures.

A wide variety of gages and cements are available, and manufacturers catalogs should be consulted for current information. Cement and adhesive technology also changes rapidly. The recommendations of manufacturers should be followed.

Basic factors in gage selection include the following:

1. Selection of base (backing) material.
2. Selection of filament material and type.
3. Selection of grid and lead configuration.
4. Special applications.
5. Gage length.

The actual mechanics of gage application, cementing, and lead attachment will be discussed later. The measurement of strain will now be discussed.

11.3 MEASUREMENT OF STRAIN

Each gage is characterized by a gage factor, defined to be[1]

$$G.F. = \frac{\Delta R/R}{\Delta L/L}$$

or

$$G.F. = \frac{\Delta R/R}{\varepsilon} \tag{11.3.1}$$

where

G.F. = gage factor

R = gage resistance (unstrained)[2]

ΔR = change in gage resistance during elongation. This is the quantity actually measured.

ε = strain being measured.

---

1. This gage factor is supplied with each gage.
2. The gage resistance is also supplied by the manufacturer.

The gage factor permits the measured change in resistance to be converted to strain.

To measure the changes in resistance, which are very small, a Wheatstone bridge is used.

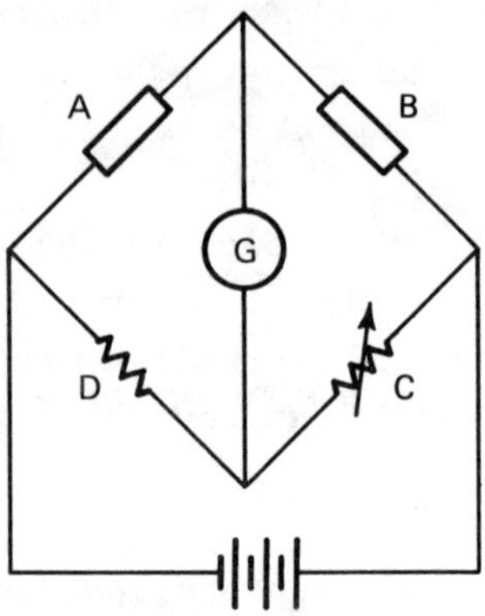

A = active, or strain-measuring gage.
B = the temperature compensating gages.
C = D = Internal resistances in the meter.
G = Galvanometer

Figure 11.3.1

It can be shown that for a balanced bridge,

$$\frac{R_A}{R_D} = \frac{R_B}{R_C} \tag{11.3.2}$$

It can also be shown that for maximum sensitivity,

$$R_A = R_B$$
$$R_C = R_D \tag{11.3.3}$$

In using the strain gage bridge, interesting adaptions of the basic setup can be made if two important rules are utilized.

1. Resistance changes in adjacent legs subtract algebraically.
2. Resistance changes in opposite legs add algebraically.

Some typical test setups are now discussed.

a. To double output in bending and eliminate temperature effects

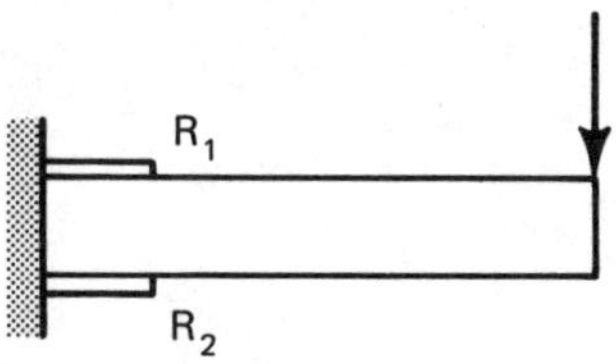

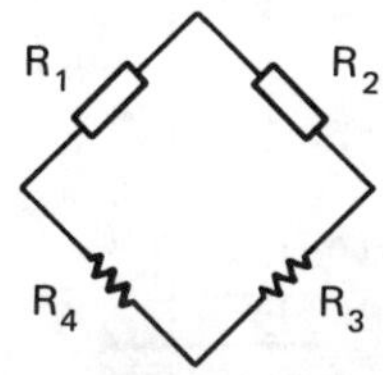

Bending strain: $R_1 \rightarrow +\varepsilon B$

$R_2 \rightarrow -\varepsilon B$

Output: $\varepsilon B - (-\varepsilon B) \rightarrow 2\varepsilon B$

Temperature: $R_1 \rightarrow +\varepsilon T$

$R_2 \rightarrow +\varepsilon T$

Output: $\varepsilon T - \varepsilon T \rightarrow 0$

b. To measure torque on a shaft

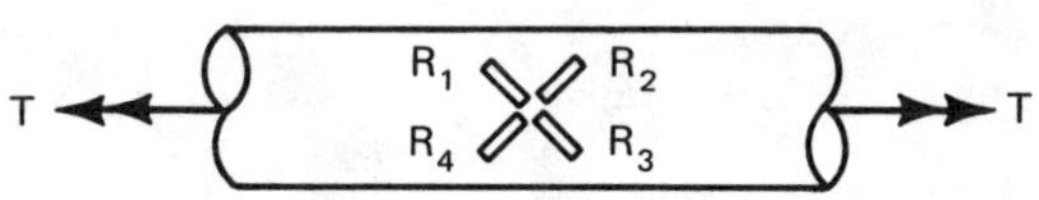

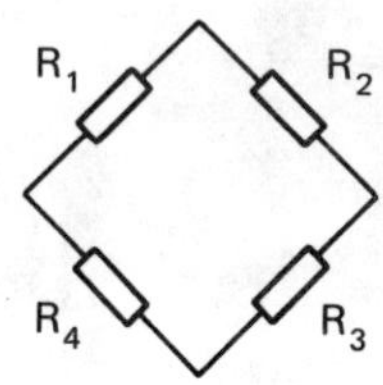

Due to torsion:

$R_1 = +\varepsilon_t$

$R_2 = -\varepsilon_t$

$R_3 = -\varepsilon_t$

$R_4 = +\varepsilon_t$

Output:

$\varepsilon_t - (-\varepsilon_t) + \varepsilon_t - (-\varepsilon_t) = 4\varepsilon_t$

Due to temperature:

$R_1 = R_2 = R_3 = R_4 = \varepsilon_T$

Output:

$\varepsilon_T - \varepsilon_T + \varepsilon_T - \varepsilon_T = 0$

c. To separate bending from axial strains. (Note loss of temperature compensation.)

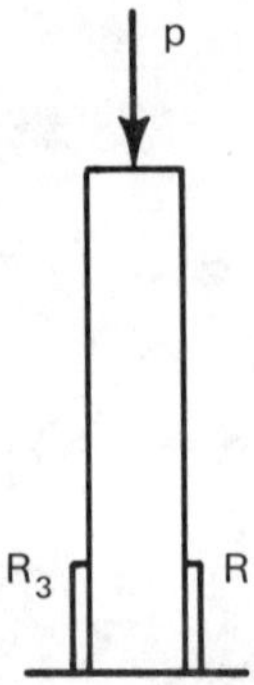

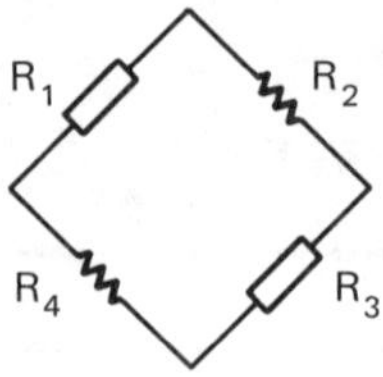

Due to axial strain:

$R_1 = \varepsilon_{1A}$

$R_3 = \varepsilon_{1A}$

Output:

$\varepsilon_{1A} + \varepsilon_{1A} = 2\varepsilon_{1A}$

Due to bending strain:

$R_1 = \varepsilon_{1B}$

$R_3 = -\varepsilon_{1B}$

Output:

$\varepsilon_{1B} + (-\varepsilon_{1B}) = 0$

Due to temperature:

$R_1 = \varepsilon_T$, $R_3 = \varepsilon_T$

Output $= \varepsilon_T$

A number of these variations can be used.

Bridges may be basically of two types

1. Balanced bridges (for static strain)

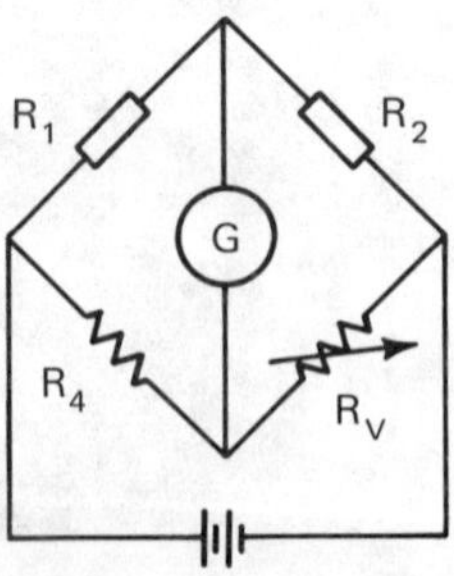

The bridge is re-nulled after each load increment.

The variable resistance R_V is calibrated to read strain directly in most commercially available meters. See Figures 11.3.2 and 3.

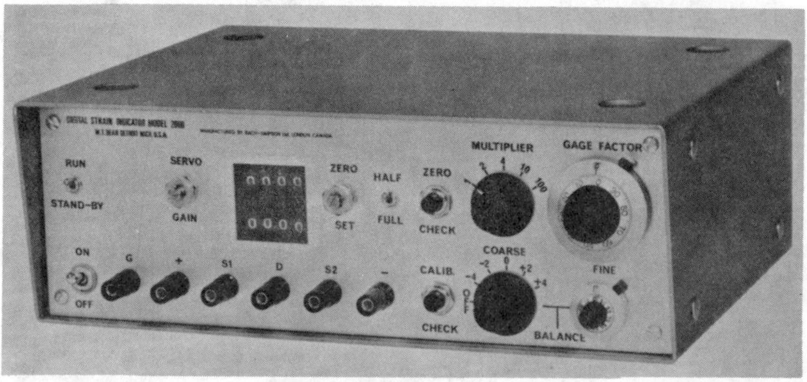

Figure 11.3.2 Typical commercial strain gage instrumentation

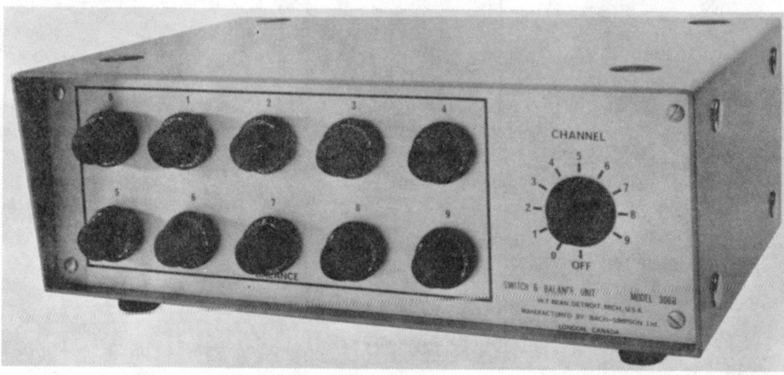

Figure 11.3.3 Typical commercial strain gage instrumentation

2. Unbalanced bridge (dynamic and stable strains)

A switch and balance unit may be used to provide multichannel capability.

It is also possible to measure transient strains and to telemeter strain signals by radio.

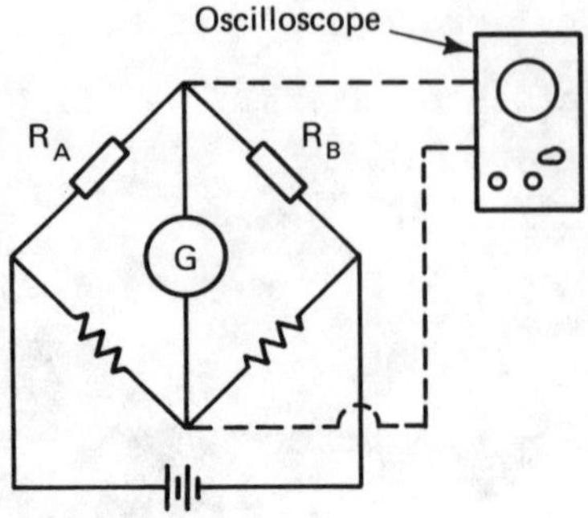

The dial G is calibrated to read in strain. An oscilloscope or other recorder may be inserted into circuit.

Calibration of the system may be built into the basic instrument or may be accomplished by inserting a calibration resistor R_p in parallel with the active gage.

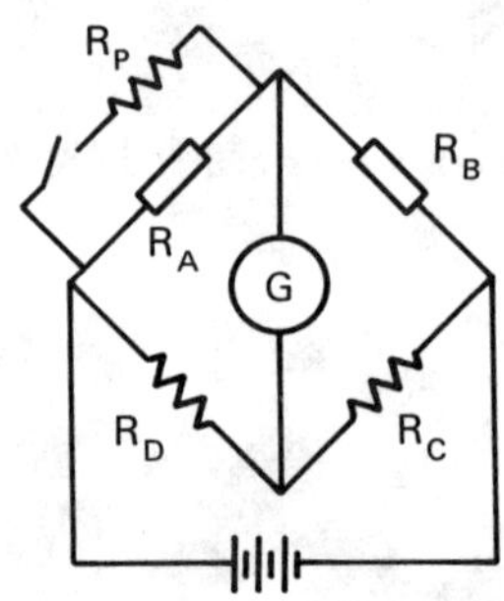

Figure 11.3.4 The basic Wheatstone Bridge with calibrating resistor R_p

We know

$$\frac{1}{R} = \frac{1}{R_A} + \frac{1}{R_p}$$

or

$$R = \frac{R_A R_p}{R_A} + R_p$$

Thus, during calibration

$$\Delta R_A = R_A - R = \frac{R_A{}^2}{R_A + R_p}$$

But also,

$$G.F. = \frac{\Delta R_A/R_A}{\varepsilon}$$

So that

$$\varepsilon = \frac{\frac{\Delta R_A}{R_A}}{G.F.} = \frac{R_A}{(R_A + R_p)\ G.F.}$$

Thus if the gage factor, gage resistance, and calibration resistance are known, the system can be given a known ε.

Once the strain measurements have been accomplished, they are converted to stresses on the test part. This conversion will be discussed in the next section.

11.4 CONVERSION OF STRAINS TO STRESSES

1. Uniaxial stress and strain. In some very simple cases, the experiment will be performed to determine uniaxial stress. An example of this is the simple tension test. If the test is being performed on a specimen which can be assumed to be in uniaxial tension, such as the one shown in Figure 11.1.1, then the uniaxial strain is converted to uniaxial stress by the simple equation

$$\sigma = E\varepsilon \tag{11.4.1}$$

where:

σ = the uniaxial stress, psi
E = Young's modulus, psi
ε = the measured uniaxial strain

2. Biaxial stress and strain with principal stress directions known. It is sometimes possible to run experiments on test structures when the directions, but not

the magnitudes, of the principal stresses are known in advance. For example, the principal stress directions may have been determined by performing an earlier test with brittle coating which will crack in a direction perpendicular to the maximum principal stress σ_1. If the principal stress directions are known in advance, as illustrated in Figure 11.4.1, a two-gage rosette may be used, and the following equations may be used to solve for σ_1 and σ_2.

$$\sigma_1 = \frac{E}{1 - \nu^2} (\varepsilon_1 + \nu\varepsilon_2)$$

$$\sigma_2 = \frac{E}{1 - \nu^2} (\varepsilon_2 + \nu\varepsilon_1)$$

$$\tau_{max} = \frac{E}{2(1 + \nu)} (\varepsilon_1 - \varepsilon_2) \qquad (11.4.2)$$

where:

ε_1 = max. strain along max. principal stress axis.

ε_2 = min. strain along min. principal stress axis.

ν = Poisson's ratio of test material.

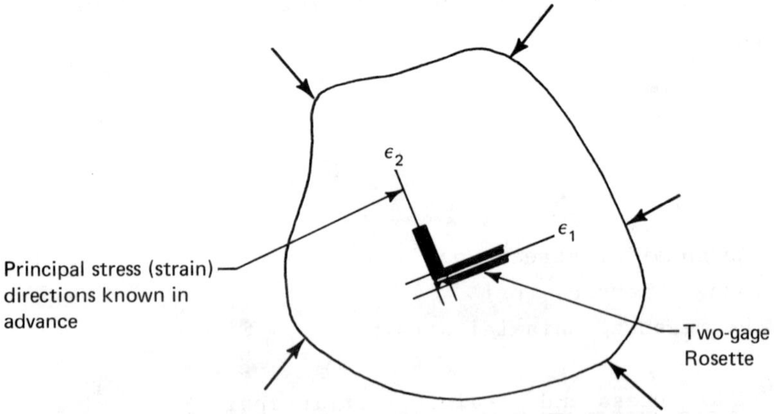

Figure 11.4.1 Measurement of strains for biaxial stress with principal stress directions known

3. Biaxial stresses with no information known in advance. Generally when biaxial stresses are measured in a two-dimensional test structure neither the directions or the magnitudes of the principal stresses are known in advance. In this case, measurement of strain in at least three different directions must be made with a three-element rosette. It is sometimes thought that meaningful information can be obtained by the use of a single gage or two gage rosette when dealing with a general case of biaxial stresses. This mistake must NEVER be made.
   The common forms of rosette are shown below:

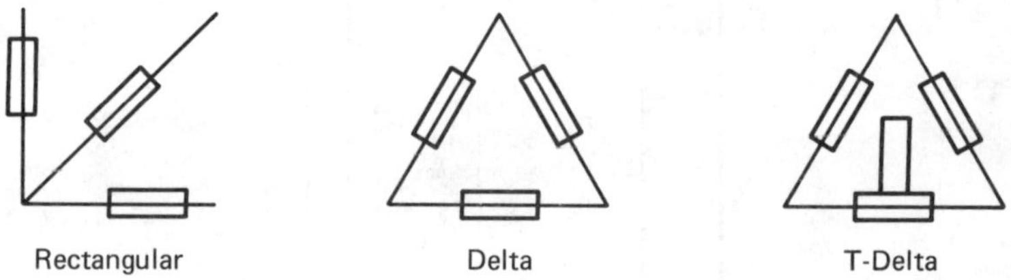

Figure 11.4.2 Commonly used strain gage rosettes

The T-Delta rosette supplies an excess of information. The extra data is used for checking results. The conversion of strain readings to stress readings for the common forms of rosettes is given in Table 11.4.1.

11.5 CEMENTING AND PROTECTION OF STRAIN GAGES. ATTACHMENT OF LEAD WIRES.

The bonded electrical resistance strain gage is an extremely versatile device whose applications in measuring strain are nearly unlimited. Strain gages can be mounted on virtually any surface where strain measurements are needed: underwater, in engine crankcases, attached to the legs of a Surveyor spacecraft landing on the moon, surgically implanted in the bones of living animals, high and low-temperature environments, etc.

Table 11.4.1

Gage arrangement 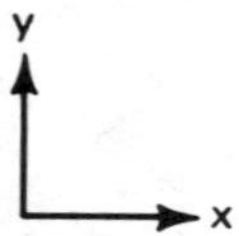 Type of rosette	Principal strain-principal stress	Principal angle	Identification $0 < \phi_1 < 90°$
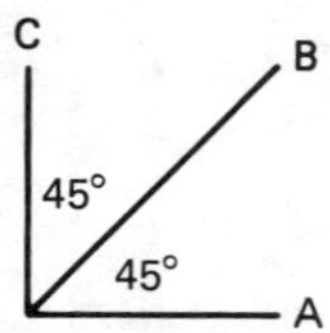 Rectangular, 3-element	$\varepsilon_{1,2} = \frac{\varepsilon_A + \varepsilon_C}{2} \pm \frac{1}{2}\sqrt{(\varepsilon_A - \varepsilon_C)^2 + (2\varepsilon_B - \varepsilon_A - \varepsilon_C)^2}$ $\sigma_{1,2} = \frac{E}{2}\left[\frac{\varepsilon_A + \varepsilon_C}{1 - \nu} \pm \frac{1}{1 + \nu}\sqrt{(\varepsilon_A - \varepsilon_C)^2 + (2\varepsilon_B - \varepsilon_A - \varepsilon_C)^2}\right]$	$\tan 2\phi_1 = \frac{2\varepsilon_B - \varepsilon_A - \varepsilon_C}{\varepsilon_A - \varepsilon_C}$	$\varepsilon_B > \frac{\varepsilon_A + \varepsilon_C}{2}$
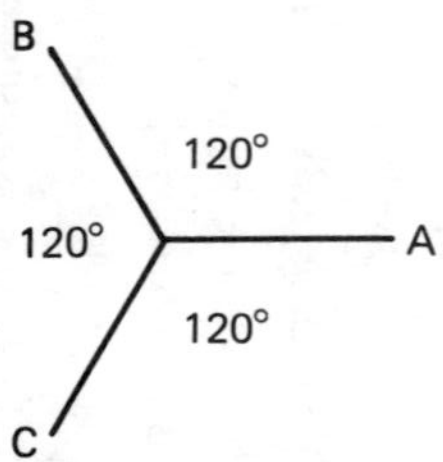 Delta, 3-element	$\varepsilon_{1,2} = \frac{\varepsilon_A + \varepsilon_B + \varepsilon_C}{3} \pm \sqrt{[\varepsilon_A - \frac{1}{3}(\varepsilon_A + \varepsilon_B + \varepsilon_C)]^2 + \frac{1}{3}(\varepsilon_C - \varepsilon_B)^2}$ $\sigma_{1,2} = E\left\{\frac{\varepsilon_A + \varepsilon_B + \varepsilon_C}{3(1 - \nu)} \pm \frac{1}{1 + \nu}\sqrt{[\varepsilon_A - \frac{1}{3}(\varepsilon_A + \varepsilon_B + \varepsilon_C)]^2 + \frac{1}{3}(\varepsilon_C - \varepsilon_B)^2}\right\}$	$\tan 2\phi_1 = \frac{\sqrt{3}(\varepsilon_C - \varepsilon_B)}{2\varepsilon_A - (\varepsilon_B + \varepsilon_C)}$	$\varepsilon_C > \varepsilon_B$

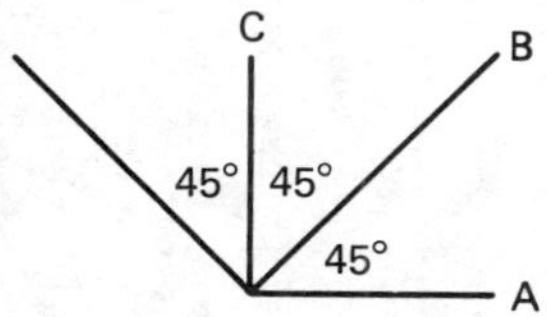Rectangular, 4-element	$\varepsilon_{1,2} = \frac{\varepsilon_A + \varepsilon_B + \varepsilon_C + \varepsilon_D}{4} \pm \frac{1}{2}\sqrt{(\varepsilon_A - \varepsilon_C)^2 + (\varepsilon_B - \varepsilon_D)^2}$ $\sigma_{1,2} = \frac{E}{2}\left[\frac{\varepsilon_A + \varepsilon_B + \varepsilon_C + \varepsilon_D}{2(1 - \nu)} \pm \frac{1}{1 + \nu}\sqrt{(\varepsilon_A - \varepsilon_C)^2 + (\varepsilon_B - \varepsilon_D)^2}\right]$	$\tan 2\phi 1 = \frac{\varepsilon_B - \varepsilon_D}{\varepsilon_A - \varepsilon_C}$	$\varepsilon_B > \varepsilon_D$
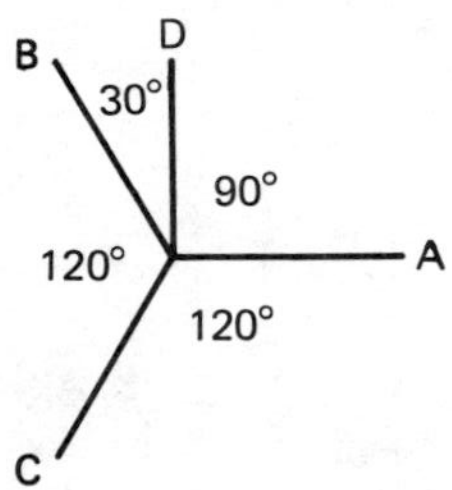Tee-delta, 4-element	$\varepsilon_{1,2} = \frac{\varepsilon_A + \varepsilon_D}{2} \pm \frac{1}{2} \quad (\varepsilon_A - \varepsilon_D)^2 + \frac{4}{3}(\varepsilon_C - \varepsilon_B)^2$ $\sigma_{1,2} = \frac{E}{2}\left[\frac{\varepsilon_A + \varepsilon_D}{1 - \nu} \pm \frac{1}{1 + \nu} \quad (\varepsilon_A - \varepsilon_D)^2 + \frac{4}{3}(\varepsilon_C - \varepsilon_B)^2\right]$	$\tan 2\phi 1 = \frac{2(\varepsilon_C - \varepsilon_B)}{\sqrt{3}(\varepsilon_A - \varepsilon_D)}$	$\varepsilon_C > \varepsilon_B$

In each of these cases, the gage installation can be no better than the cement attaching it to the specimen. It is for this reason that the selection of the proper cement, and the proper application of this cement, are critical steps in the procedure for using these gages.

When gages are used in a harsh environment, they must be protected. It is the usual practice to protect gages from moisture, temperature, pressure, wear, etc., by applying a protective cover of wax, shellac, or in some cases, metal cups.

The attachment and securing of lead wires is also an important step in strain gage practice. These steps will now be outlined. It must be emphasized that current practice is changing daily. The catalogs and bulletins of manufacturers of strain gages and cements should be consulted for current information.

Choice of Cement. The choice of cement depends principally on:

1. Materials used in the gage.
2. Environmental conditions of the test.
3. Time available for installing gage.

Gage backings fall in three major categories:

Backing	Appropriate Cement
1. Paper	Nitrocellulose (airplane glue)
2. Bakelite	Epoxy
3. Epoxy	Epoxy

Another cement, contact cement (Eastman 910) can be used for all gages. This is a cement with a drying time of about one minute, and is very useful if environmental conditions are not too severe.

Surface Preparation. Cleanliness is critical when a good cement bond is required. The surface must be not just clean, but super-clean.

Initially, all rust, scale, and gross contaminants must be removed. The surface is then polished with a fairly coarse emery cloth soaked in solvent. The goal here is to have a surface which is smooth in the overall sense, but not highly polished. Now the surface is painstakingly cleaned with solvent by wiping the surface alternately with fresh patches of first solvent-wetted tissue, then dry, then solvent-soaked, then dry. . ., ...

When the surface is clean, alignment marks may be scribed to aid in aligning the gage. If this step contaminates the area, it must be recleaned. The scribe lines should not pass under the gage, but merely up to the point where the edge of the gage will be placed.

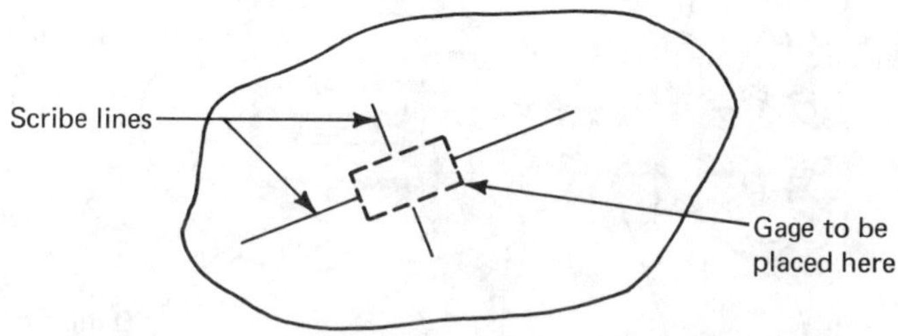

Figure 11.5.1

Cementing

Now the cementing is carried out. The actual steps involved depend on the cement to be used. Generally speaking, a minimum amount of cement to accomplish the job is spread on the cleaned surface. Then the cleaned gage is carefully aligned and pressed on the surface. A clamping pressure--from very light to very heavy depending on the cement used--is applied and the cement is cured.

Curing times vary from a few minutes to several hours. Heating may be employed to cure some of the epoxy cements.

Attachment of Lead Wires

The attachment of lead wires is another important step in strain gage usage. A clean solder joint of low resistance is essential. It is also important to provide some protective measures in case the wires are pulled. Before soldering, the wires should be firmly attached to the structure a short distance from the gage.

Terminal strips may be employed in the installation. Usually in this case a fine wire connects the gage tab to the terminal strip. Then the heavier lead wires (about 20-gage stranded wire) are run out from the terminal.

A typical installation is illustrated in Figure 11.5.2.

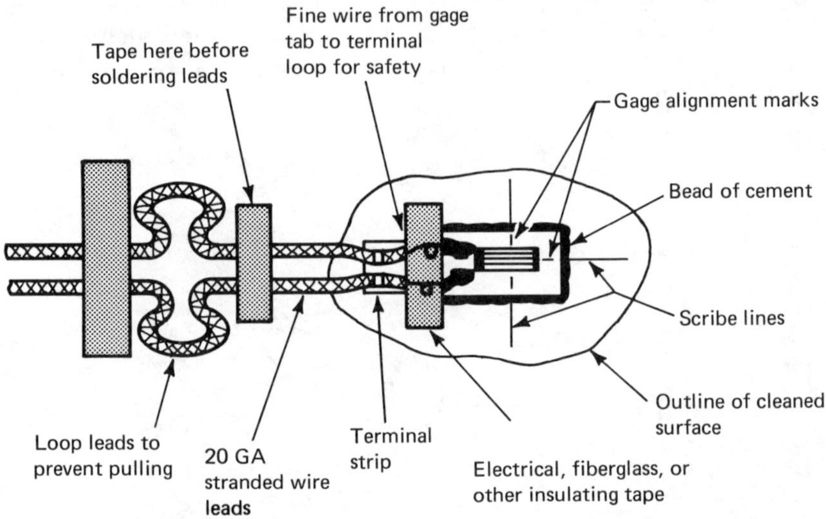

Figure 11.5.2 Typical gage and lead wire installation

Testing a Strain Gage Installation

The following checks can be performed and will usually indicate whether or not the gage installation is satisfactory.

1. Check the gage resistance with an ohmmeter. A change in resistance or an open circuit indicates a damaged gage.
2. Check the resistance from the gage to the structure. This should be very high. A minimum acceptable value is 50 megohms.
3. Connect the gage to the strain indicator and null the meter. Pressing lightly on the gage with the eraser end of a pencil should cause the meter to deflect slightly. Removing the pressure should cause the gage to return to zero. An unsatisfactory installation is indicated if the needle oscillates wildly when the pressure is applied.

Protecting Gages

A number of methods are employed for protecting gages. These vary from covering the gage with a few layers of electrical tape to elaborate arrangements with multiple layers of rubber sheet, metal caps, epoxy and tape.

One frequently required type of protection is moisture proofing. Gages must be protected from humidity, especially if they need to be stable over long periods of time. The gage can be covered with a variety of waxes, shellacs, or epoxy compounds. These are best applied in several thin layers rather than one thick one.

Protection from various atmospheric elements can be provided by the above, or by use of such items as Glyptal, liquid rubber, or ceramics. Protection from physical damage can be provided by using metal caps or other substantial protective devices.

For thorough discussions of protective coatings, the manufacturers bulletins and the references should be consulted. See the following table.

References

1. Perry, C. C., and H. R. Lessner, The Strain Cage Primer, (New York: McGraw-Hill Book Company, 1962), 2nd Ed., (A later edition is available).
2. Lee, G. H., Experimental Stress Analysis (New York: John Wiley and Sons, 1950).
3. Dove, R. C., and Paul H. Adams, Experimental Stress Analysis and Motion Measurement, (Columbus), Ohio: Charles E. Merrill, 1964).

11.6 SAMPLE EXPERIMENT

A sample laboratory report is included to further illustrate both stress analysis theory and experimental technique.

M-BOND STRAIN GAGE ADHESIVES

The success of a strain gage installation is measured by the validity and usefulness of the test results. Adhesive selection based on anticipated environmental test conditions and accuracy requirements contributes significantly to a successful installation.

To assist you in your selection, the chart below includes a number of typical testing requirements. If a test profile is not listed in the chart it is probably because performance limits are being approached or exceeded. The interaction of test characteristics is often too complex for simple chart presentation, in which case we suggest that you contact our Applications Engineering Department for recommendations.

On the following pages is a complete guide to M-Bond adhesives, detailing specific characteristics of each type and further application considerations.

TYPE OF TEST OR APPLICATION	OPERATING TEMPERATURE RANGE	TEST DURATION IN HOURS	ACCURACY REQUIRED[2]	CYCLIC ENDURANCE REQD		TYPICAL SELECTION	
				Maximum Strain, $\mu\epsilon$	Number of Cycles	Gage Series	M-Bond Adhesive
GENERAL STATIC OR STATIC-DYNAMIC STRESS ANALYSIS[1]	−50 to +150°F (−45 to +65°C)	$<10^4$	Moderate	± 1 500	$<10^6$	CEA, EA	200 or AE-10
		$>10^4$	Moderate	± 1 500	$<10^6$	CEA, EA	AE-10 or AE-15
		$>10^4$	Very High	± 1 500	$>10^6$	WA, SA	AE-15 or 610
		$>10^4$	High	± 2 000	$>10^6$	WK, SK	AE-15 or 610
	−50 to +400°F (−45 to +205°C)	$<10^3$	Moderate	± 1 500	$<10^6$	WA, SA	600 or 610
		$>10^3$	High	± 2 000	$<10^6$	WK, SK	600 or 610
	−452 to +450°F (−269 to +230°C)	$>10^3$	Moderate	± 2 000	$>10^6$	WK, SK	610
	<600°F (<315°C)	$<10^2$	Moderate	± 1 800	$<10^6$	WK	610
	<700°F (<370°C)	<10	Moderate	± 1 500	$<10^6$	WK	610
HIGH-ELONGATION (POST-YIELD)	−50 to +150°F (−45 to +65°C)	<10	Moderate	± 50 000	1	CEA, EA	200 or AE-10
		$>10^3$	Moderate	±100 000	1	EP	AE-15
		$>10^3$	Moderate	±200 000	1	EP	A-12
	0 to +500°F (−20 to +260°C)	$<10^2$	Moderate	± 15 000	1	SA, SK, WA, WK	610
	−452 to +500°F (−269 to +260°C)	$<10^3$	Moderate	± 10 000	1	SK, WK	600 or 610
DYNAMIC (CYCLIC) STRESS ANALYSIS	−100 to +150°F (−75 to +65°C)	$<10^4$	Moderate	± 2 200	10^7	ED	200 or AE-10
		$<10^4$	Moderate	± 2 500	10^7	WD	AE-10
	−320 to +500°F (−195 to +260°C)	$<10^4$	Moderate	± 2 500	10^7	WD	600 or 610
		$<10^4$	Moderate	± 2 000	$<10^5$	WD	600 or 610
TRANSDUCER GAGING	−50 to +150°F (−45 to +65°C)	$<10^4$	1 to 5%	± 1 500	$<10^6$	CEA, EA	200 or AE-10
		$<10^6$	1 to 5%	± 1 500	$<10^6$	EA	AE-15
	−50 to +200°F (−45 to +95°C)	$<10^4$	Better than 0.2%	± 1 500	10^6	MA	610 or 43-B
	−50 to +300°F (−45 to +150°C)	$<10^4$	0.2 to 0.5%	± 1 500	10^6	WA, SA, TA	610
	−320 to +350°F (−195 to +175°C)	$<10^4$	Better than 0.5%	± 1 800	10^6	WK, SK, TK	610

1. This category includes most testing situations where some degree of stability under static test conditions is required. For absolute stability with constantan gages over long periods of usage and temperatures above +150°F (+65°C), it may be necessary to employ half- or full-bridge configurations. Protective coatings may also influence stability in cases other than transducer applications where the element is hermetically sealed.

2. It is inappropriate to quantify "accuracy" as used in this table without consideration of various aspects of the actual test program and the instrumentation used. In general, "moderate" for stress analysis purposes is in the 2% to 5% range, "high" in the 1% to 3% range, and "very high" 1% or better.

References: Tech Note TN-132 *Strain Gage Selection — Criteria, Procedures, Recommendations.*
References to specific adhesive instruction are given with each adhesive description on the following pages.

Table 11.5.1 Sample table courtesy Micro Measurements, Vishay Intertechnology, Inc.

BIAXIAL STRESS CONCENTRATION EXPERIMENT

ARO 326
AEROSPACE STRUCTURES LABORATORY

by
Phillip McNamee
California State Polytechnic University, Pomona
November 21, 1977

1. SUMMARY

This experiment was performed to determine the magnitude and orientation of the principal stresses in the stress concentrated region near a circular hole in a biaxial test specimen undergoing uniaxial loading. These values were then compared to theoretically predicted values.

A modified delta strain gage rosette was placed approximately 0.1 inches from the edge of the hole oriented according to Figure 4.2 along the holes x-axis. The specimen was then loaded along the y-axis and appropriate values of strain recorded. These values of strain were used in conjunction with the equations included with Table 5.2 to determine the principal stresses and their orientation (Table 5.2). The values of the principal stresses were then compared to theoretical values calculated using equations also included with Table 5.2 (Table 5.3).

The values of the principal stresses were found to vary in a non-linear fashion with applied load. The maximum principal stress was found to be approximately 2.1 ± .25 times the applied normal stress (Table 5.4) at the experimental location (Figure 4.1). Experimental results correlated only generally with theoretical predictions differing by as much as 45%.

2. OBJECTIVES

The objectives of this experiment were to determine the magnitude and orientation of the principal stresses in the stress concentrated region near a circular hole in a biaxial test specimen undergoing uniaxial loading and to compare these values with theoretically determined values.

3. BACKGROUND

3.1 Biaxial Stress

The condition of biaxial stress is a highly complex one. A simple "common sense" approach to the analysis of a biaxial stress condition might be to orient a strain gage along the x or y axis of the sample. Then by using the stress-strain relationship 6 = $E\varepsilon$ one might determine the stress in the x or y direction by simply measuring the strain at a given point. From there the principal stresses and their orientation may be calculated. This method of analysis, however, may lead to serious errors.

The main flaw in the above approach is that this is a biaxial stress situation, and uniaxial methods of analysis are inapplicable. A more general method of analysis is required. Biaxial stress may be analyzed using a strain gage rosette. In a rosette a given number of strain gages (usually 3 or 4) are oriented in a given geometry. (More detailed information regarding strain gage rosettes may be obtained from Reference 8.3. A modified delta rosette configuration was used in this experiment.) By taking readings of the value of strain in the gages of the rosette and by the use of the formulas given in Reference 8.3 or Table 5.2, the principal stresses and their orientation may be determined.

3.2 Stress Concentrations

The presence of a hole or some other discontinuity in a stressed specimen will result in a concentration of stress in the neighborhood of the discontinuity. This experiment examines the specific case of the stress concentrations near a circular hole in a biaxial stress specimen undergoing axial loading.

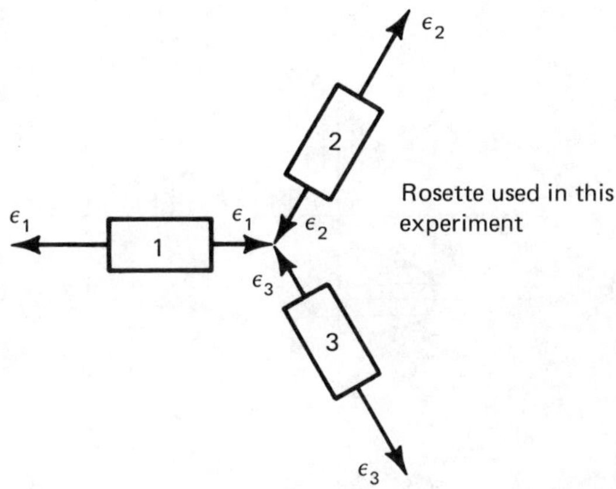

Figure 3.1

As a specimen is loaded the stress in the specimen will attempt to "flow" around any discontinuity. (Figure 3.2.) The resulting stress concentrations may be much greater than the actual applied normal stress, reaching a maximum theoretical value three times as great as the applied normal stress in the direction of the applied stress.

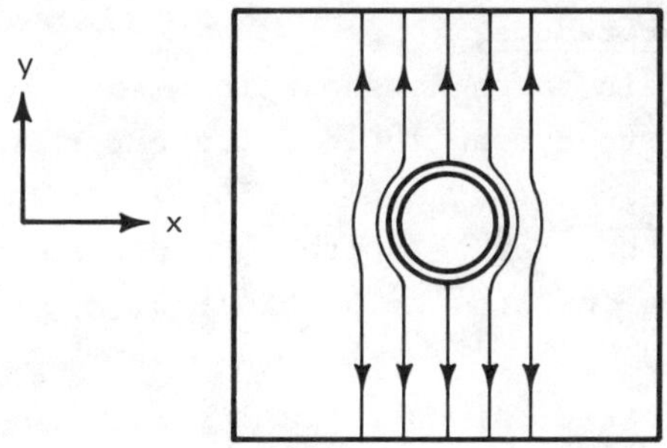

Figure 3.2 Diagram showing stress "flow" around a circular hole

4. EXPERIMENTAL PROCEDURE

4.1 <u>Required Apparatus and Equipment</u>

4.1.1 MTS testing system servo-hydraulic load frame.

4.1.2 MTS control console.

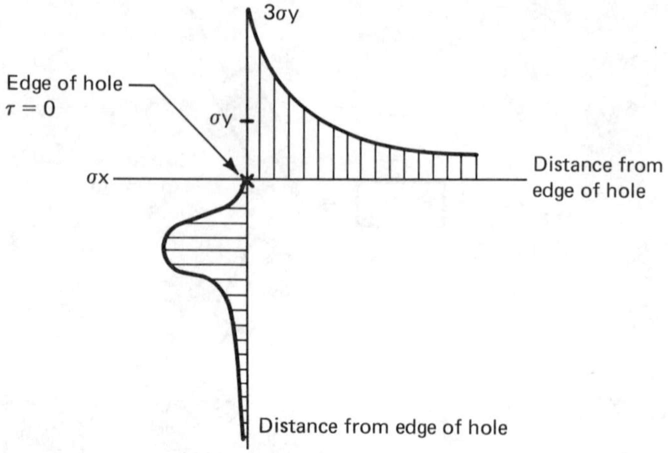

Figure 3.3 Stress distribution in the neighborhood of a circular hole

4.1.3 Baldwin Lima Hamilton (BLH) model 120 strain indicator.

4.1.4 BLH switching unit.

4.1.5 Biaxial stress test specimen (see Figure 4.1).

4.1.6 Three SR-4 120 ohm G.F. 2.05 strain gages.

4.2 Experimental procedure

4.2.1 Position strain gages as shown in Figure 4.2.

4.2.2 Calibrate servo-hydraulic load frame for 5000 lb_f load full scale. (Set point = 500.)

4.2.3 Connect strain gages to switching unit and connect the switching unit to the strain indicator as shown in Figure 4.3.

4.2.4 Record unstressed values for all strain gages.

4.2.5 Mount biaxial stress test specimen into servo-hydraulic load frame.

4.2.6 Load the test specimen. Starting at set point = 500 increase the set point value in increments of 50 up to a value of 750. Recording the strain of each gage at each increment.

4.2.7 Unload the test specimen and record the strain for each gage.

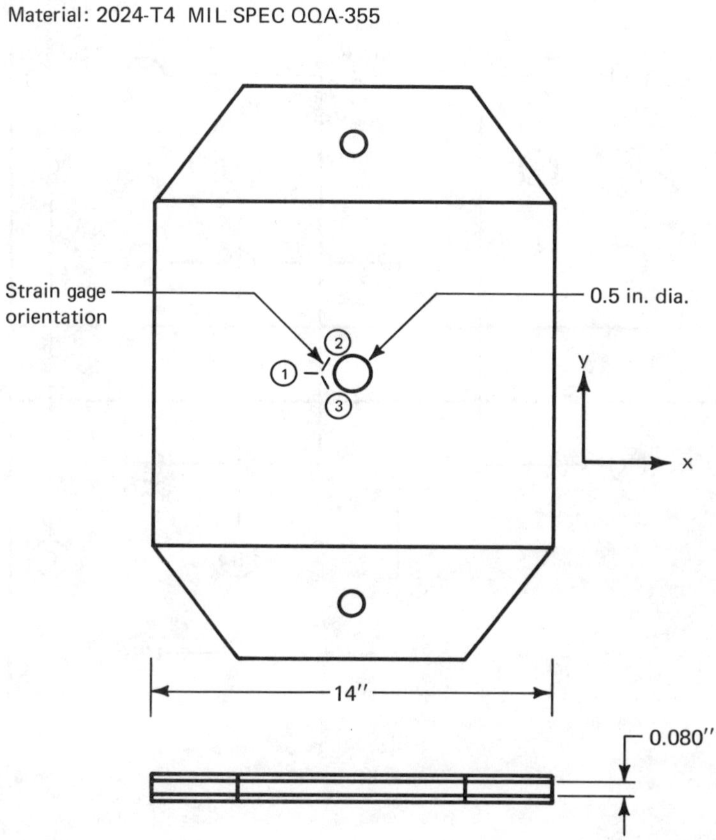

Figure 4.1 Biaxial stress test specimen

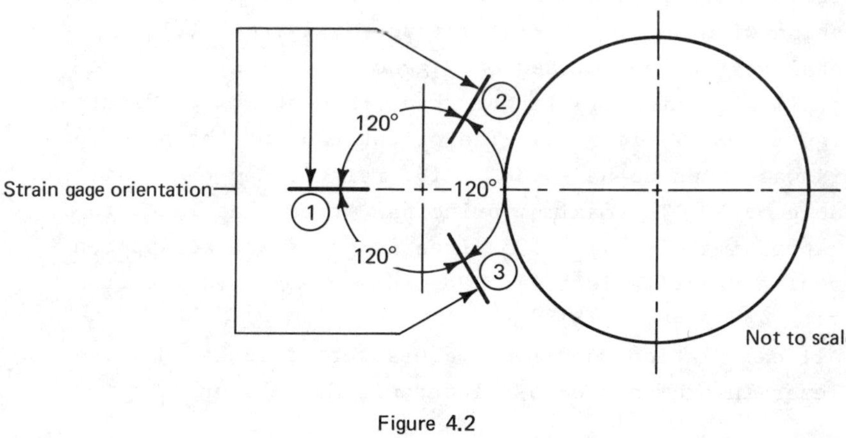

Figure 4.2

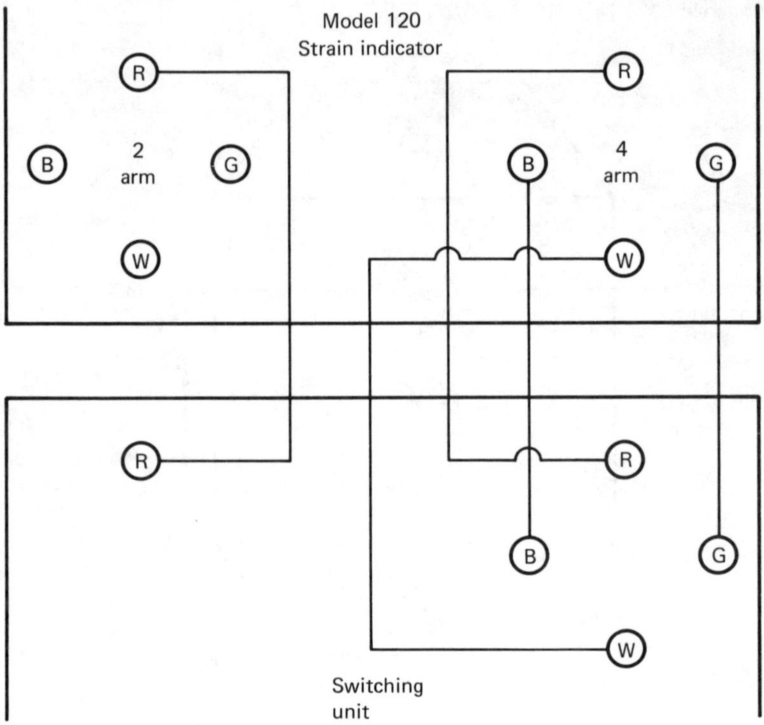

Figure 4.3

5. RESULTS

The principal stresses were found to vary in a non-linear fashion with applied normal stress. (Figure 5.1.) Experimental results correlated only generally with, and were overall significantly larger than, theoretical predictions. (Figure 5.2, Table 5.3.) The orientation of the principal axis was found to be $-3.38^{o} \pm 19^{o}$ relative to the x axis (Table 5.2). The maximum principal stress was found to be approximately $2.1 \pm .25$ times as great as the applied normal stress (Table 5.4). Percentage error varied from 2.7 to 45% (Table 5.3).

All calculations and mean values were calculated using a Texas Instruments SR-51 Electronic Calculator.

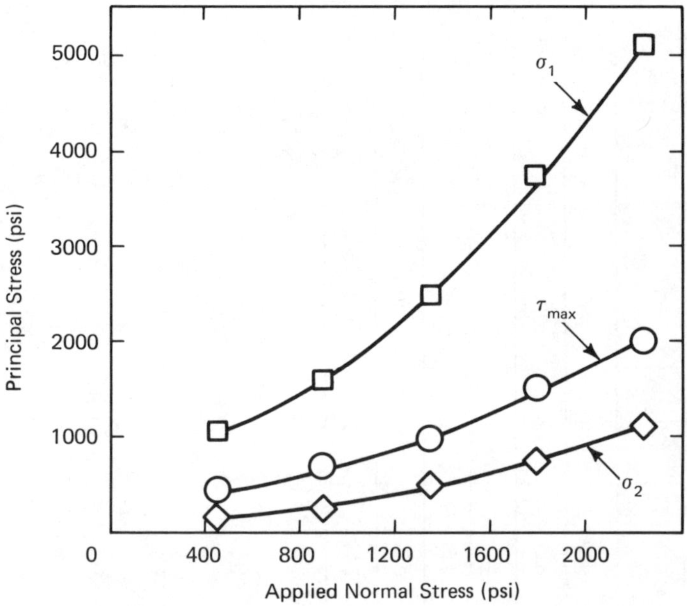

Figure 5.1 Graph showing applied normal stress σ versus principal stresses σ_1, σ_2, τ_{max}

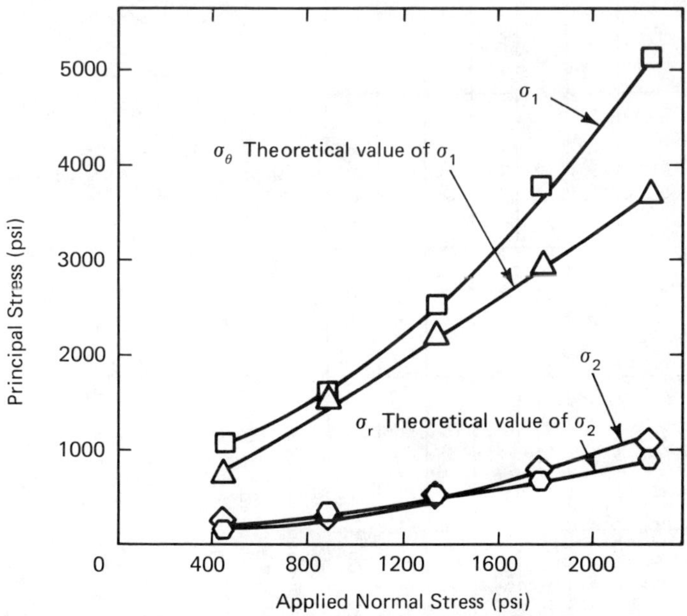

Figure 5.2 Graph showing applied normal stress versus stresses σ_1, σ_2 and theoretical values of principal stress σ_θ and σ_r

Table 5.1

Set Point	Strain Gage #1	Δ Strain #1	Strain Gage #2	Δ Strain #2	Strain Gage #3	Δ Strain #3	Applied Load in LBF	Applied Normal Stress in psi
(zero load) 500	42564	0	42858	0	41724	0	0	0
550	42550	-14	42936	78	41780	56	500	446.4
600	42538	-26	42970	112	41817	93	1000	892.9
650	42535	-29	43030	172	41870	146	1500	1339.3
700	42518	-46	43115	257	41945	221	2000	1785.7
750	42510	-54	43211	353	42024	300	2500	2232.1
(zero) 500	42546	-18	42848	-10	41708	-16	0	0

Cross sectional areas:
Plate area .080 in. x 14 in. = 1.12 in^2
Plate area net .080 in. x 13.5 in. = 1.08 in^2

Table 5.2

Applied Load in LBF	Principal Stress in psi			Principal Axis
500	1063.7	190.1	436.8	-3.31°
1000	1615.2	254.9	680.1	-3.63°
1500	2503.7	515.7	994.0	-3.41°
2000	3762.2	750.8	1505.7	-3.12°
2500	5140.8	1117.2	2011.8	-3.44°

FORMULAS USED IN DETERMINING THE MAGNITUDE AND ORIENTATION OF THE PRINCIPAL STRESSES

Experimental:

$$\delta_1 = E\left[\frac{\varepsilon_1 + \varepsilon_2 + \varepsilon_3}{3(1-\mu)} + \frac{1}{1+\mu}\sqrt{\left(\varepsilon_1 - \frac{\varepsilon_1 + \varepsilon_2 + \varepsilon_3}{3}\right)^2 + \left(\frac{\varepsilon_2 - \varepsilon_3}{\sqrt{3}}\right)^2}\right]$$

$$\delta_2 = E\left[\frac{\varepsilon_1 + \varepsilon_2 + \varepsilon_3}{3(1-\mu)} - \frac{1}{1+\mu}\sqrt{\left(\varepsilon_1 - \frac{\varepsilon_1 + \varepsilon_2 + \varepsilon_3}{3}\right)^2 + \left(\frac{\varepsilon_2 - \varepsilon_3}{\sqrt{3}}\right)^2}\right]$$

$$\tau_{max} = \frac{E}{1+\mu}\sqrt{\left(\varepsilon_1 - \frac{\varepsilon_1 + \varepsilon_2 + \varepsilon_3}{3}\right)^2 + \left(\frac{\varepsilon_2 - \varepsilon_3}{\sqrt{3}}\right)^2}$$

$$\theta = \frac{1}{2}\,\mathrm{TAN}^{-1}\left[\frac{\frac{1}{\sqrt{3}}\left(\varepsilon_2 - \varepsilon_1\right)}{\varepsilon_1 - \frac{\varepsilon_1 + \varepsilon_2 + \varepsilon_3}{3}}\right]$$

Theoretical:

$$\delta_r = \frac{S}{2}\left(1 - \frac{a^2}{r^2}\right) + \frac{S}{2}\left(1 + \frac{3a^4}{r^4} - 4\frac{a^2}{r^2}\right)\cos 2\theta$$

$$\delta_\theta = \frac{S}{2}\left(1 + \frac{a^2}{r^2}\right) - \frac{S}{2}\left(1 + \frac{3a^4}{r^4}\right)\cos 2\theta$$

E = modulus of elasticity = 10.5×10^6 psi
μ = Poisson's Ratio = .33
a = radius of hole = 0.25 in.
r = distance from center of hole to the center of the rosette = 0.35 in.
ε_1, ε_2, ε_3 = experimental strain values

Table 5.3

Applied Load in LBF	Measured Value σ_1	Theoretical Value σ_θ	% Difference Between σ_θ & σ_1	Measured Value σ_2	Theoretical Value σ_r	% Difference Between σ_r & σ_2
500	1063.7	734.2	45	190.1	167.3	13.6
1000	1615.2	1469.2	10	254.9	334.7	23.8
1500	2503.7	2203.9	13.6	515.7	502.1	2.7
2000	3762.2	2938.5	28	750.8	669.4	12.2
2500	5140.8	3673.	40	1117.2	836.7	33.5

Table 5.4

Applied Load in LBF	Maximum Principle Stress σ_1	Applied Normal Stress σ_0	σ_1/σ_0
500	1063.7	446.4	2.4
1000	1615.2	892.9	1.8
1500	2503.7	1339.3	1.9
2000	3762.2	1785.7	2.1
2500	5140.8	2232.1	2.3

Mean value of $\sigma_1/\sigma_0 = 2.1 \pm .25$

6. DISCUSSION OF RESULTS

The results of this experiment are generally acceptable. There are, however, several important problems. The presence of a non zero shear stress at the assumed edge of the hole indicates that the method used in this experiment is only approximate for this type of measurement. The presence of a non-zero principal angle indicates that the strain gage rosette is not properly oriented and may have contributed substantially to the large deviation between experimental and theoretical values.

7. CONCLUSIONS AND RECOMMENDATIONS

The value of the principal stresses was found to vary in a non-linear fashion with applied uniaxial loading (Figure 5.1). A high degree of stress concentration was present at the location measured. Principal stresses exceeded applied normal stress by as much as 2.4, times with an average concentration of 2.1 ± .25 times applied stress (Table 5.4).

8. REFERENCES

8.1 MILHbK5B, U. S. Government Printing Office.

8.2 Mechanics of Materials, F. R. Shanley, McGraw-Hill, 1967.

8.3 Bonded Electrical Resistance Strain Gage, J. W. McKinley, California State Polytechnic University, 1972.

Chapter 12

PHOTOELASTICITY IN EXPERIMENTAL STRESS ANALYSIS

12.1 INTRODUCTION

Photoelasticity is a widely used method for experimentally determining the stress magnitudes and directions in a model of a structural member. This model, which is made of a transparent material with certain stress-optic properties (to be discussed later) is put in a polariscope (also to be discussed in detail later). The polariscope transmits light through the model in a particular way, and the result is that the image of the model, when viewed on a screen, has alternating light and dark fringes. A particular example of this is shown in Figure 12.1.1. We see here a particular example of isochromatic fringes in a light field polariscope. These particular fringes reveal the principal stress difference according to the formula

$$\sigma_1 - \sigma_2 = \frac{C}{t} N \qquad (12.1.1)$$

where:

$\sigma_1 - \sigma_2$ is the principal stress difference
C is the stress-optic coefficient of the material (Photoelastic constant)
t is the model thickness
N is the fringe order

In Figure 12.1.1, the integral values of fringe order, N, are shown. Every white band is a fringe. Note the presence of fringes of order N = 0. These are areas of zero stress.

Photoelasticity is referred to as a <u>whole-field</u> method. By this we mean that the stress patterns are shown for the

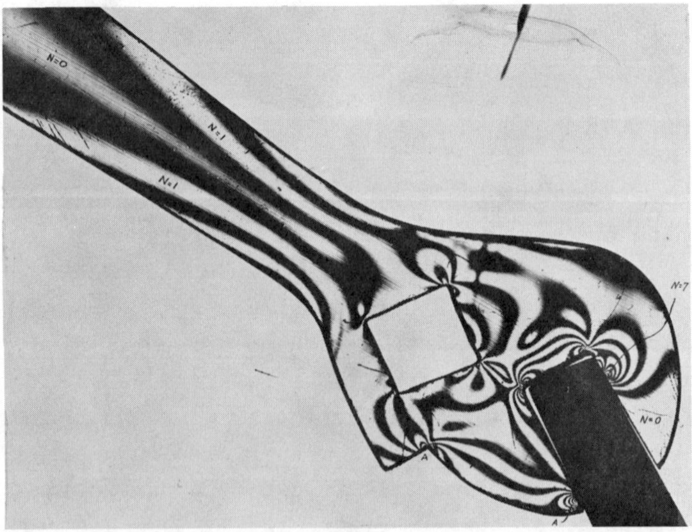

Figure 12.1.1 Isochromatics for a model of a crescent wrench. These have been done in a light field circular polariscope. The integer order fringes are white. Notice the presence of zero stress areas, N=0. Note also the stress concentrations. Material: CR-39.

whole model at one time. The primary information that is available from the photoelastic method is summarized below: (Not all of it is available from Figure 12.1.1.)

1. Principal stress difference, $\sigma_1 - \sigma_2$, for all points in the model.
2. Principal stresses at the model boundaries, since either σ_1 or σ_2 will be zero at a boundary.
3. Areas which are highly stressed; these are indicated by the presence of many closely-spaced fringes. See Areas A in Figure 12.1.1.
4. Areas of low stress. In such areas, material may be eliminated or rearranged to make a more efficient part.
5. The principal stress directions. These are given by a different pattern called the isoclinics.

Additional information can be obtained using the photoelastic technique, but it is harder to get compared to the five above. The advantages of the photoelastic method are that it gives a solution to elasticity problems which may be very difficult (or impossible) to solve theoretically. It is a whole field method. It is relatively easy to use. It gives valuable design information for real parts in service. Photoelasticity is a technique of popular current use, and it is utilized by many industries, research laboratories, and educational institutions. Current research is being carried out to extend this method to make it more useful (Reference 1).[1]

There are some disadvantages to the method. A model must be made and a polariscope must be available. If more information is needed than the five items mentioned above, the photoelastic method may become very tedious. Three-dimensional photoelasticity may be very complicated (but represents one of the few experimental approaches to 3-d work). Photoelasticity, like anything else, has its strengths and its weaknesses. Nevertheless it remains a powerful tool for the stress analyst.

---

1. References are listed at the end of the chapter.

We shall limit our present discussion to some elementary topics; included will be: some basic concepts of light, the elements and arrangements of polariscopes, calibration of materials, interpretation of results, model-to-prototype scaling, and materials selection. We will talk about two-dimensional work only. Not included in the present discussion will be some topics which are important to the serious photoelastician, but are beyond the scope of this effort; these include: three-dimensional photoelasticity and stress freezing, reflecting polariscopes, separation of principal stresses, etc. The references should be consulted for these topics.

12.2 PRINCIPLES OF PHOTOELASTICITY

Light may be treated with the wave theory for the purposes of photoelasticity. The particle theory of light will not be used. Light may be thought of as a time dependent vector which is uniformly oriented in all directions in space. The light vectors may be thought of as having amplitude a and frequency ω.

$$\underline{A} = a \sin \omega t \text{ (omnidirectional)} \qquad (12.2.1)$$

The frequency is $f = \omega/2\pi$. The group of vectors, all equal to $\underline{A}$, but uniformly distributed around a plane perpendicular to the line of transmission may be seen in Figure 12.2.1. The wavelength of the light, λ, is related to the frequency by the equation

$$f \lambda = c \qquad (12.2.2)$$

where c is the velocity of light in air. The color of light is determined by its wavelength. Ordinary white light is composed of a mixture of all visible wavelengths. Light of one color is called <u>monochromatic</u> light. In talking about photoelasticity, we will be concerned primarily with monochromatic light.

The <u>index of refraction</u> relates the velocity of light in any medium to the velocity of light in air. If the velocity of light in air is c, then the velocity in any medium, 1, is given by

$$c_1 = c/n_1 \qquad (12.2.3)$$

where n_1 is the index of refraction in medium 1.

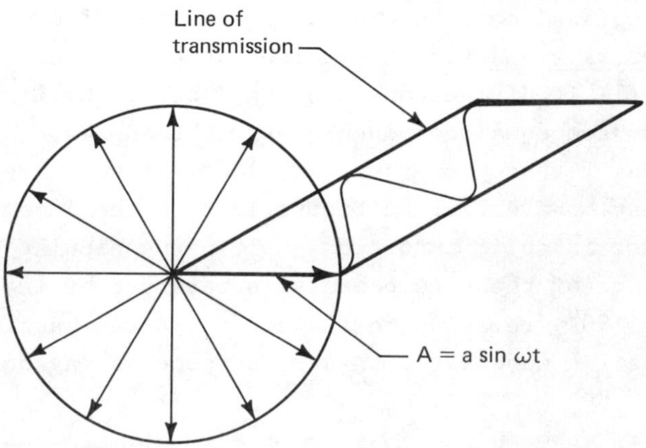

Figure 12.2.1 A light ray which is viewed perpendicular to the line of transmission. The vectors are uniformly distributed around the plane. The sinusoidal nature of one of the vectors is shown.

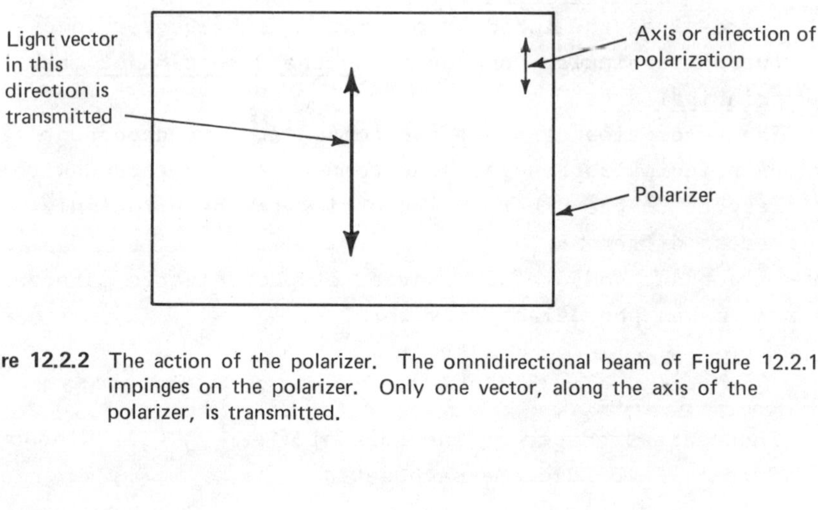

Figure 12.2.2 The action of the polarizer. The omnidirectional beam of Figure 12.2.1 impinges on the polarizer. Only one vector, along the axis of the polarizer, is transmitted.

Polarization of light is accomplished by passing the omnidirectional light vectors shown in Figure 12.2.1 through a polarizing filter.[2] Figure 12.2.2 depicts the result of this, only one of the vectors in the direction of the axis of the polarizer has been transmitted. We may represent this vector by

$$\underline{A} = a \sin \omega t \text{ (one direction)} \tag{12.2.4}$$

Another optical component essential to polariscopes is the quarter wave plate. The quarter wave plate has two effects: (1) It divides an incoming (polarized) light vector into two equal components at right angles, (2) It retards one of these components by 1/4 wavelength relative to the other. Referring to Figure 12.2.3, the incoming light vector is split into two equal, perpendicular components. One of these components is retarded by 1/4 wavelength (90°) relative to the other. After passing through the 1/4 waveplate, the two components may be represented by

$$\underline{A}_3 = 0.707 \, a \sin (\omega t + 90^o) \tag{12.2.5}$$

$$\underline{A}_4 = 0.707 \, a \sin \omega t$$

A photoelastic material behaves in two important ways:

1. It resolves the incident polarized light vector into two perpendicular components in the direction of the principal axes.
2. The velocities of these two components are dependent on principal stresses. One component is retarded with respect to the other in proportion to the principal stress difference.

This important and useful behavior of photoelastic materials is what makes photoelasticity work.

---

2. These are produced by The Polaroid Company, 119 Windsor Street, Cambridge, Massachusetts.

Let us now consider the plane polariscope, Figure 12.2.4. We will consider a single incoming light vector emerging from the polarizer and striking the model. At any one point on the model, the principal stress axes will be at angles θ and $(90^o - \theta)$ to the impinging light ray. We will consider only one point. Of course, the angle θ will vary from point to point in the model. The following equations are written for one point, but the same development applies to all points.

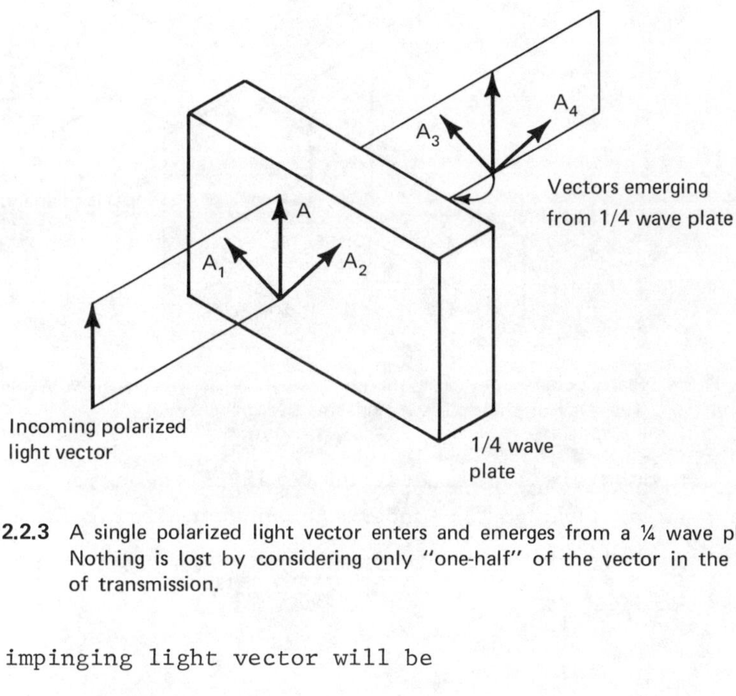

Figure 12.2.3 A single polarized light vector enters and emerges from a ¼ wave plate. Nothing is lost by considering only "one-half" of the vector in the plane of transmission.

The impinging light vector will be

$$\underline{A} = a \sin \omega t \qquad (12.2.7)$$

This vector will be resolved as follows (refer to Figure 12.2.4)

$$\underline{A}_1 = a \cos \theta \sin \omega t \qquad (12.2.8a)$$

$$\underline{A}_2 = A \sin \theta \sin \omega t \qquad (12.2.8b)$$

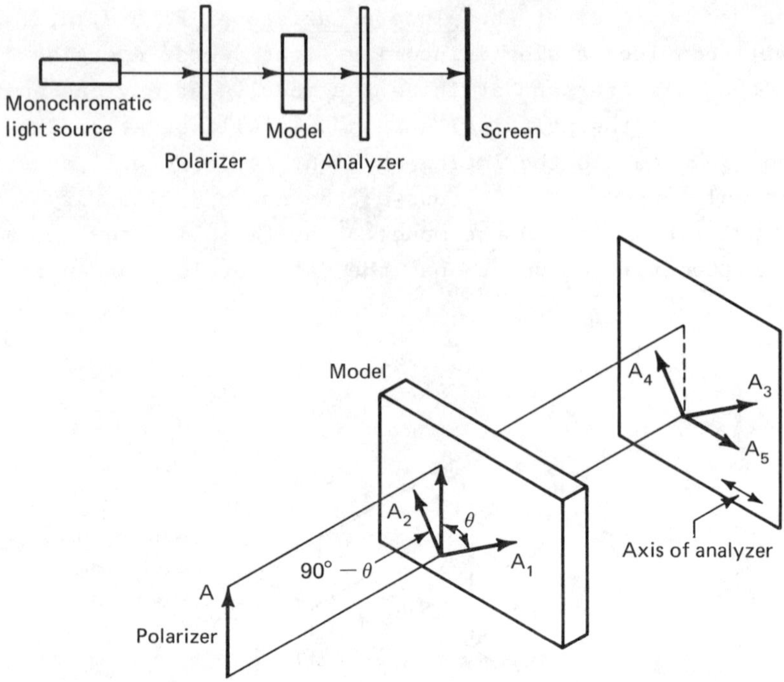

Figure 12.2.4 Plane polariscope. The analyzer is a polarizer with a different name. The axes of the polarizer and analyzer are crossed.

These components will be retarded relative to one another while passing through the model, and upon emerging from the model, will be

$$\underline{A}_3 = a \cos \theta \sin (\omega t + \omega \delta t) \tag{12.2.9a}$$

$$\underline{A}_4 = a \sin \theta \sin \omega t \tag{12.2.9b}$$

The time difference δt is a result of the difference in velocities of the two components in the model.

The analyzer is another polarizer with its axis of polarization perpendicular to the polarizer (crossed analyzer and polarizer). If the polarizer is vertical, the analyzer will be horizontal. The effect of the analyzer will be to add the horizontal components of the vectors $\underline{A}_3$ and $\underline{A}_4$. The result is

$$\underline{A}_5 = a \sin 2\theta \sin \frac{\omega\delta t}{2} \cos(\omega t + \omega\delta t) \tag{12.2.10}$$

This is the light vector which will be viewed on the screen.[3]

Let us now consider the time difference, δt. The photoelastic model causes the two components of light to travel through it with different velocity. This is because the indices of refraction along the two principal axes are different. The time of passage along each axis (1 and 2) will be

$$t_1 = \frac{n_1 t}{c}; \; t_2 = \frac{n_2 t}{c} \tag{12.2.11}$$

t is the thickness of material c is again, the velocity of light in air. Do not confuse time σt, t_1 and t_2 with thickness t. We have therefore

$$\delta t = \frac{t}{c}(n_1 - n_2) \tag{12.2.12}$$

Now let us note that

$$\omega\delta t = \frac{\omega t}{c}(n_1 - n_2) = \frac{2\pi t}{\lambda}(n_1 - n_2) \tag{12.2.13}$$

Since $f\lambda = c$ and $\omega = 2\pi f$. If N is the number of wavelengths of relative retardation, then (see Figure 12.2.5)

$$\omega\delta t = 2\pi N \tag{12.2.14}$$

From Equations 3.12 and 3.13

$$2\pi N = \frac{2\pi t}{\lambda}(n_1 - n_2)$$

or

$$N = \frac{t}{\lambda}(n_1 - n_2) \tag{12.2.15}$$

---

3. The intensity of the light is actually $\underline{A}_5^2$, but Equation (12.2.10) will serve our purpose.

It has been shown, primarily by experiment, that the difference in indices of refraction along the principal axes are proportional to the difference in principal stress, this allows us to write

$$N = \frac{t}{\lambda}(n_1 - n_2) = \frac{t}{c}(\sigma_1 - \sigma_2) \qquad (12.2.16)$$

where C is the stress-optic coefficient of the material, or the <u>photoelastic</u> <u>constant</u>. We will use photoelastic constant. This is a combined constant, and depends on the photoelastic material and the wavelength of light being used. For this reason, calibration is almost always necessary in photoelastic tests.

Rewriting Equation 12.2.10, we have

$$\underline{A}_5 = a \sin 2\theta \sin \frac{\pi t}{c}(\sigma_1 - \sigma_2) \cos(\omega t + \omega\delta t) \qquad (12.2.17)$$

When we view a model in a plane polariscope, we will see light and dark fringes similar to (but not exactly like) those in Figure 12.1.1. The dark bands we observe will occur when $\underline{A}_5$ in Equation 12.2.17 equals zero. This will occur under two conditions:

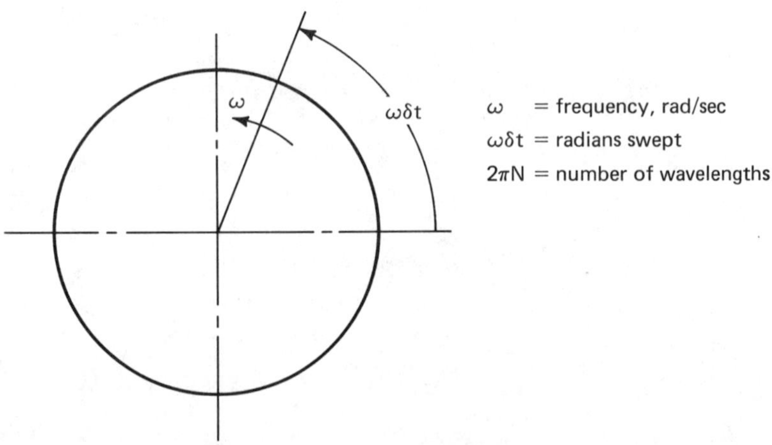

Figure 12.2.5 Illustrating that $\omega\delta t = 2\pi N$. If N=1, $\omega\delta t = 2\pi$, and one wavelength is swept out. (One cycle is completed.)

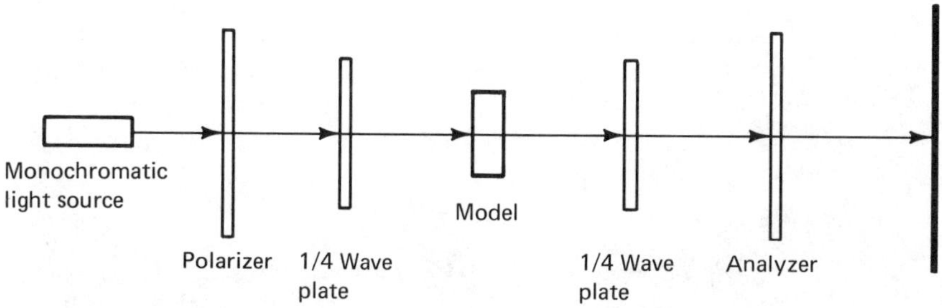

Figure 12.2.6 The elements of the circular polariscope

1. $\sin 2\theta = 0$

   $\theta = 0^o$ or 90°

   This is the case when the polarizer is aligned with one of the principal stress axes.

   These dark bands are called isoclinics. An isoclinic is a locus of all points where the polarizer is aligned with a principal stress axis. (The analyzer will be aligned with the other one.) The isoclinics do not show the principal stress directions directly. They show (at any particular orientation of polarizer) the points where the polarizer and a principal stress axis are in the same direction.

   Isoclinics for various angles, i.e., 0°, 10°, 20°, etc., can be generated by successively rotating the polarizer and analyzer through 10° increments. This operation will form the complete isoclinic pattern for the model. Isoclinics at 0° orientation (vertical polarizer) are shown in Figure 12.2.7.

2. $\sin \frac{\pi t}{C} (\sigma_1 - \sigma_2) = 0$

   $\frac{\pi t}{C} (\sigma_1 - \sigma_2) = N\pi$; N an integer, 0, 1, 2, ----

   These dark bands are the isochromatics. These are the fringes shown in Figure 12.1.1 as will be explained later. If the fringe order N is known, the difference in principal stresses can be determined from

$$\sigma_1 - \sigma_2 = \frac{NC}{t} \tag{12.1.1}$$

where:

N = 0 is a fringe of zero order

N = 1 is a fringe of order 1

N = 2 is a fringe of order 2

etc.

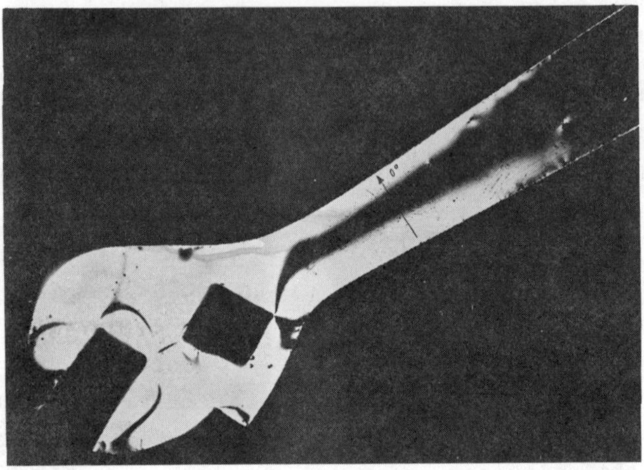

Figure 12.2.7 Zero degree isoclinics for the wrench of Figure 12.2.1. The dark band defines the areas where the principal axes are vertical and horizontal (0°).

This was given in Equation (12.1.1); we have now completed the derivation.

12.3 COMMENTS ON ISOCHROMATICS AND ISOCLINICS

Looking at Figure 12.1.1 again, we see that isochromatics look pretty much the same, regardless of their order, N = 0, 1, . . . integer values. In using the equation

$$\sigma_1 - \sigma_2 = \frac{NC}{t}$$

we must know the value of N in order to use this equation. There are several ways to evaluate N, among these are:

1. Increase the load slowly and count the fringes as they appear. The original unstressed model will have N = 0 throughout.
2. Locate a fringe of zero order, N = 0, and count up or down from it. Zero order fringes can be located in several ways.
   a. Points of known zero stress have N = 0.
   b. The zero order fringe does not move as the load is changed.
   c. <u>Isotropic points</u> are points where $\sigma_1 = \sigma_2$. It follows that N = 0 at these points. These points can be identified by examining the <u>isoclinic</u> patterns. All isoclinics meet at isotropic points.
   d. Free external corners cannot carry stress and will have N = 0.
   e. When white light is used for illumination, the zero order fringe will be black; other fringes will be colored.

It is ordinarily not too difficult to determine whether stresses are increasing and to decide whether to count up or down from the fringes of zero order.

The isoclinics are ordinarily viewed by using white light for illumination. They will then be black. An axis of symmetry of model load will be a zero degree isoclinic. All isoclinics will meet at an isotropic point. All isoclinics will go through a point of concentrated load application.

The main problem with the <u>plane</u> polariscope is that both isoclinics and isochromatics appear. This can lead to a rather confusing pattern. The isoclinics may be distinguished from the isochromatic by use of the following techniques:

1. A light load on the model will lead to few isochromatics and make the isoclinics stand out. However, a heavier load will make the isoclinics sharper.
2. The use of white light will make the black isoclinic stand out against the colored isochromatics (except

for the isochromatic of zero order).

3. The isoclinics will be unchanged and the isochromatics will be changed when the load is changed. When the polarizer and analyzer are rotated together, the isoclinics will move and the isochromatics will not.

Actually, this problem is avoided by using a circular polariscope. This is shown in Figure 12.2.5. The mathematical development of the circular polariscope will not be given. It can be shown that the circular polariscope will give only isochromatic fringes and no isoclinics. The elements of the circular polariscope can be arranged to give either a light field or dark field pattern.

In the light field polariscope, the integral order fringes will be light. Actually, they will be the color of the monochromatic light from the source, but the usual isochromatic photograph will show the light field fringes as white. Figure 12.1.1 shows a light field polariscope.

In the dark field polariscope, the integral order fringes will be black.

There is no simple way to make the isoclinics appear without the isochromatics. The isoclinics are found by using the plane polariscope as previously discussed. Figure 12.3.1, shows some typical commercial polariscopes.

12.4 CALIBRATION OF PHOTOELASTIC MATERIAL

There are several ways to calibrate photoelastic materials. All of these involve subjecting a specimen to known stress conditions in the polariscope, and using the light wavelength which is to be used in the test.

The simplest test which can be used for calibration is the tension test. The known stress in this case will be

$$\sigma_1 = \frac{P}{A}$$

$$\sigma_2 = 0$$

we therefore have

Figure 12.3.1

$$\sigma_1 - \sigma_2 = \sigma_1 = \frac{P}{A} = \frac{CN}{t}$$

The thickness of the model through which the light is transmitted is t. Let the cross sectional area of the model be t x W. Then

$$\frac{P}{A} = \frac{P}{t \times W} = \frac{CN}{T}$$

$$\frac{P}{W} = CN$$

or

$$C = \frac{(P/W)}{N} \qquad (12.4.1)$$

As the model is loaded, the integral order fringes will appear as the uniform central section turns dark (or light, depending on the polariscope) all at once. The values of P and N can be recorded, and if (P/W) vs. N is plotted, the slope of the line will give C. See Figure 12.4.1.

12.5 MODEL TO PROTOTYPE SCALING

The model and prototype must be geometrically similar in the 1-2 plane. The thickness, t, of model and prototype need not be the same. The loads must be applied in the same fashion, but the model load may be (and usually is) less than the prototype. Then the stresses in the prototype may be scaled from the stresses in the model according to the relation

$$\sigma = \frac{P_{L_m} t_m}{P_{m} L t} \sigma_m \qquad (12.5.1)$$

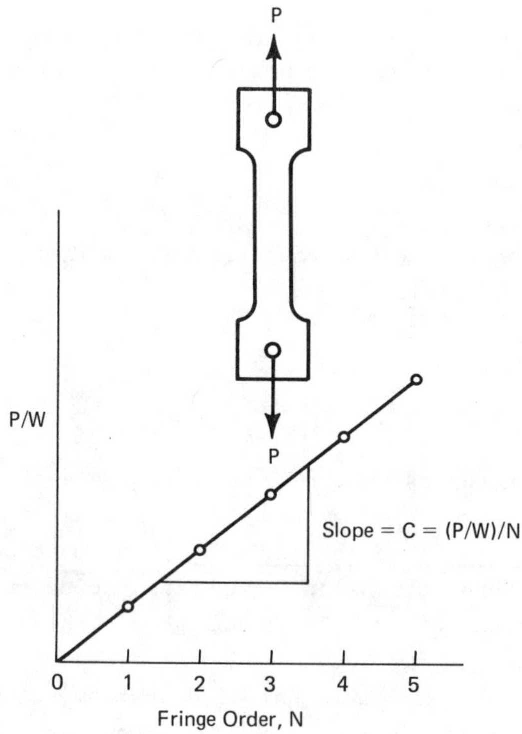

Figure 12.4.1 The simple tension test for calibration. If P/W is plotted vs. N, the slope will be C.

where:

m refers to the model
L is a characteristic length
t is the thickness
P is the load

12.6 PHOTOELASTIC MATERIALS

Material technology changes rapidly and material selection should always be guided by consulting current manufacturer's bulletins and current technical papers. See the following catalog excerpts.

Some important considerations in selecting model materials are:

1. Low value of C
2. Good machinability
3. Freedom from creep
4. Reasonable price
5. Freedom from time-edge effect (development of edge fringes with time)

A popular material is CR-39.[4]

12.7 ADVANCED TOPICS

Many advanced topics in the area of photoelasticity cannot be covered here. Among these are:

1. Three dimensional photoelasticity. The most popular method of three dimensional photoelasticity is the frozen stress method. In this method a three dimensional model is cast, and loaded at elevated temperature. When the model is unloaded at room temperature, the stresses are frozen in. Two dimensional slices are then removed from the model and examined by methods of this chapter.

PHOTOELASTIC MODEL MATERIALS

SHEET MATERIAL FOR TWO-DIMENSIONAL ANALYSIS OF MODELS

TYPE	THICKNESS						DESCRIPTION*
	⅛″	¼″	⅜″	½″	¾″	1″	
PSM-1	$17.00	$20.00	—	—	—	—	Polyester sheet—completely free of edge effect—highest sensitivity
PSM-5	14.00	20.00	—	—	—	—	Epoxy sheet—excellent machinability—high sensitivity
PSM-4	20.00	25.00	—	—	—	—	Polyurethane sheet—requires very low forces for straining
CR-39	—	10.00	$15.00	$20.00	$30.00	$40.00	Allyl polyester sheet—excellent transparency—medium sensitivity

4. Available from: The Homalite Corporation, 11-13 Brookside Drive, Wilmington, Deleware 19804.

2. Separation of principal stresses. At the interior points in the model, it is difficult to separate the principal stresses. There are a number of experimental and numerical techniques for doing this, all of them tedious.

For this and other advanced topics, the interested reader should consult the references.

References

1. Experimental Mechanics, Monthly Journal published by the Society for Experimental Stress Analysis, 21 Bridge Square, Westport, Conn.
2. Lee, G. H., An Introduction to Experimental Stress Analysis, John Wiley and Sons, Inc., New York, 1950.
3. Dally, J. W., and W. F. Riley, Experimental Stress Analysis, McGraw-Hill Book Co., New York, 1965.
4. Dove, R. C. and P. H. Adams, Experimental Stress Analysis and Motion Measurement, Charles E. Merrill, Inc., Columbus, Ohio, 1964.
5. SESA, Manual on Experimental Stress Analysis, W. H. Tupenny and A. S. Kobayashi, Eds., Society for Experimental Stress Analysis, Westport, Conn., 2nd Ed.

12.8 SAMPLE LAB REPORT

DETERMINATION OF STRESS CONCENTRATION FACTOR BY PHOTOELASTICITY

ETA 349

AEROSPACE STRUCTURES LABORATORY

Jerry Klaeser

California State Polytechnic University, Pomona

November 20, 1977

1. SUMMARY

The purpose of this experiment was to determine the stress concentration factor (K_t) of a photoelastic tensile specimen with a central hole. The ratio of the hole diameter over the width of the specimen as .44, see Figure 1. The method used to find K_t was to load the photoelastic specimen under the circular polariscope until the 4th fringe order (N = 4) appeared at the edge of the hole. K_t was then found by:

$$K_t = \frac{C\ N\ (W - d)}{P}$$

where C is the photoelastic constant, W is the specimen width, d is the diameter of the center hole, and P is the load.

The experimentally derived K_t was 1.88. According to reference 1, K_t in this type of situation is 2.2. The percentage difference between these two values is 8.8%.

2. OBJECTIVE

The objectives of this experiment were: (1) to determine the stress concentration factor (K_t) in a photoelastic tensile specimen with a central hole, and (2) compare the derived value of K_t with a published value.

3. BACKGROUND INFORMATION

A stress concentration is a condition in which high localized stresses occur at an abrupt change in the geometry of a structure. For example in Figure 1, the formula $\sigma = \frac{P}{A}$ is valid only for finding the average stress in the vicinity of the hole of the tensile specimen. In order to determine the maximum stress which occurs at the edge of the hole, the average stress must be multiplied by a stress concentration factor that applies to the specific geometry of the specimen:

$$\sigma_{max} = K_t \, \sigma_{average}$$

There are charts available in which graphs of empirically derived values of K_t are plotted against a geometry factor for different types of structural discontinuities. Figure 2 is a reproduction of a stress concentration chart from reference 1. In this situation, the geometry factor is equal to $\frac{d}{W}$, where d equals the hole diameter and W equals the width of the specimen.

The K_t can also be determined for a photoelastic specimen under the circular polariscope by the derivation

$$\sigma_1 - \sigma_2 = \frac{C}{t} N \quad \text{(from Reference 2)}$$

where:

C = the photoelastic constant of the material
t = the thickness of the specimen
N = the integer fringe order
σ_1 = the stress in the y direction
σ_2 = the stress in the x direction = 0

$$\sigma_{max} = \sigma_1 = \frac{C}{t} N = K_t \frac{P}{A_{net}}$$

$$\frac{C}{t} N = K_t \frac{P}{t(W - d)}$$

$$K_t = \frac{CN(W - d)}{P}$$

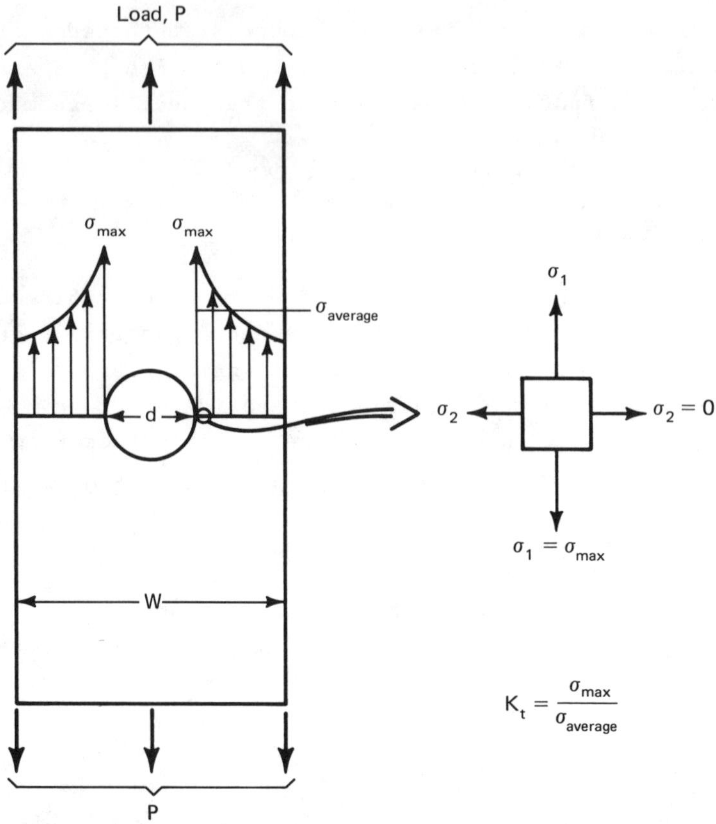

Figure 1. Stress concentrations in a tensile specimen with a center hole

4. EXPERIMENTAL PROCEDURE

4.1 Equipment List

1. A photoelastic tensile specimen (see Figure 3), made of PSM-1 material, C = 42.7 lb/in.
2. Circular polariscope with tensile tester and electrical load readout.

4.2 Procedure

1. Zero the electrical load indicator on the low scale setting (0-300 lbs).
2. Apply a tensile load on the specimen while observing the edge of the center hole through the telescope

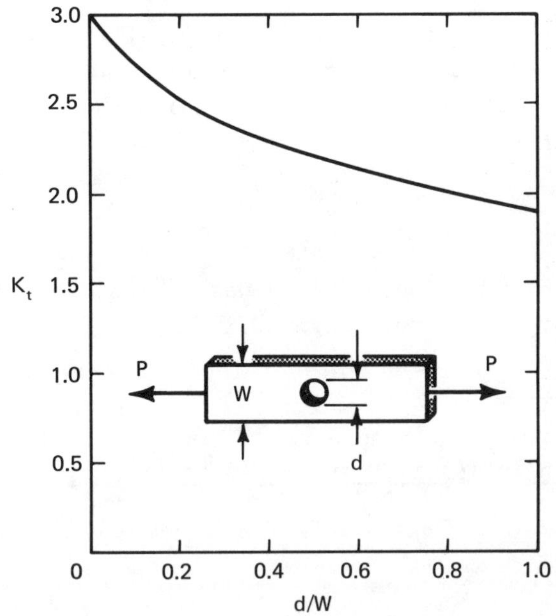

Figure 2. K_t versus d/W for a tensile specimen with a center hole

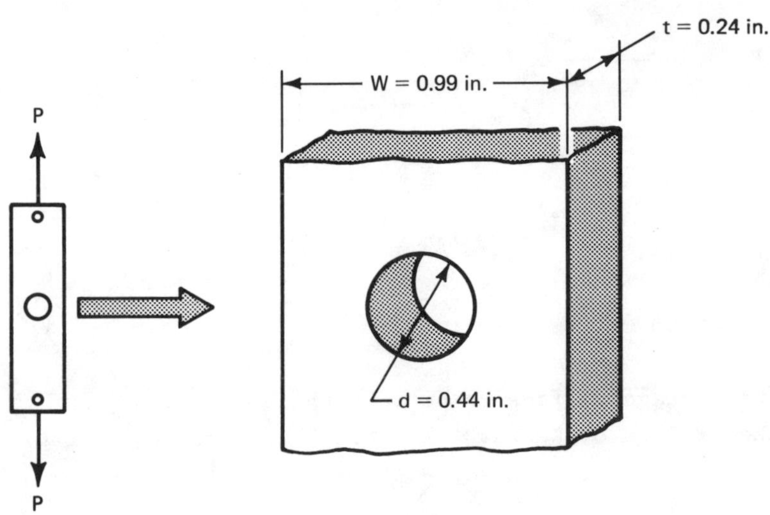

Figure 3. Tensile specimen

mounted in front of the polariscope. When the 4th fringe order appears, record the load indication.

3. Repeat step 2 several times to obtain an average value of load when N = 4.

5. RESULTS

The average load value obtained was 50 lbs. when N = 4. K_t is found by the formula derived in section 3:

$$K_t = \frac{C\ N\ (W - d)}{P}$$

$$K_t = \frac{42.7\ \text{lb/in}\ (4)\ (.99\ \text{in.} - .44\ \text{in.})}{50\ \text{lb.}}$$

$$K_t = 1.88$$

According to Figure 1, K_t is equal to 2.2 when $\frac{d}{W}$ is equal to .44 for the situation of the tensile specimen used in this experiment. The percentage difference between these two values of K_t is found by:

$$\%\ \Delta\ K_t = \frac{K_{t_1} - K_{t_2}}{K_{t_2}} \times 100\%$$

$$\%\ \Delta\ K_t = \frac{1.88 - 2.2}{2.2} \times 100\%$$

$$\%\ \Delta\ K_t = -14.5\%$$

6. DISCUSSION OF RESULTS

The percentage difference between the experimental and published values of K_t was - 14.5%. This error is probably due mainly to the inaccuracy of the load cell in the low range.

7. REFERENCES

7.1 ETA 324 Handout, Stress Concentration, 1974.

7.2 ETA 349 Handout, Photoelasticity, J. W. McKinley, 1972.

7.3 Stress Concentration Design Factors, R. E. Peterson, John Wiley and Sons, Inc., New York, 1953.

Appendix A

MATERIAL PROPERTIES

NOTE: Reference Numbers are for references listed at the end of Chapter 2.

Table A.1 Design Mechanical and Physical Properties of Low-Alloy Steels Ref. (1)

Alloy [For specification see Table 2.3.1.0(a)]	AISI 4130, 8630, and 8735		See steels listed in Table 2.3.0.1(a) for the applicable strength levels			
Form	Sheet, strip, plate, tubing		All wrought forms			
Condition	N		Quenched and tempered[a]			
Thickness or diameter, in.	≤0.187	>0.187	See Table 2.3.0.1(a)			
Basis	S	S	S	S	S	S
Mechanical properties:						
F_{tu}, ksi	95	90	125	150	180	200
F_{ty}, ksi	75	70	103	132	163	176
F_{cy}, ksi	75	70	113	145	173	181
F_{su}, ksi	57	54	75	90	108	120
F_{bru}, ksi:						
(e/D = 1.5)	...	...	194	219	250	272
(e/D = 2.0)	200	190	251	287	326	355
F_{bry}, ksi:						
(e/D = 1.5)	...	...	151	189	230	255
(e/D = 2.0)	129	120	180	218	256	280
e, percent	See Table 2.3.1.0(d)		See Table 2.3.1.0(e)			
E, 10^3 ksi	29.0					
E_c, 10^3 ksi	29.0					
G, 10^3 ksi	11.0					
μ	0.32					
Physical properties:						
ω, lb/in.^{3}	0.283					
C, Btu/(lb)(F)	0.114 (at 32 F)					
K, Btu/[(hr)(ft^2)(F)/ft]	22.0 (at 32 F)					
α, 10^{-6} in./in./F	6.3 (0 to 200 F)					

[a]Values in these columns are applicable only to steels for which the indicated F_{tu} has been substantiated through adequate quality-control inspection testing.

Table A.2 Availability of Shapes, Plates and Bars according to ASTM Structural Steel Specifications Ref. (2)

Steel Type	ASTM Designation	F_y Minimum Yield Stress (ksi)	Shapes: Group per ASTM A6 [a]1	2	3	4	5	Plates and Bars: To ½" Incl.	Over ½" to ¾" Incl.	Over ¾" to 1" Incl.	Over 1" to 1½" Incl.	Over 1½" to 2½" Incl.	Over 2½" to 4" Incl.	Over 4" to 5" Incl.	Over 5" to 8" Incl.	Over 8"
Carbon	A36	32														▨
		36	▨	▨	▨	▨	▨	▨	▨	▨	▨	▨	▨	▨	▨	
	A529	42	▨					▨								
High-Strength	A440	42				▨	▨					▨	▨			
		46			▨					▨	▨					
		50	▨	▨				▨	▨							
High-Strength Low-Alloy	A441	40												▨	▨	
		42				▨	▨					▨	▨			
		46			▨					▨	▨					
		50	▨	▨				▨	▨							
	[b]A572 Gr. 42	42	▨	▨	▨	▨		▨	▨	▨	▨	▨	▨			
	[b]A572 Gr. 45	45	▨	▨	▨	▨		▨	▨	▨	▨					
	[b]A572 Gr. 50	50	▨	▨	▨	▨		▨	▨	▨	▨					
	[b]A572 Gr. 55	55	▨	▨	▨	▨		▨	▨	▨	▨					
	[b]A572 Gr. 60	60	▨	▨				▨	▨	▨						
	[b]A572 Gr. 65	65	▨					▨								
Corrosion-Resistant High-Strength Low-Alloy	A242	42				▨	▨					▨	▨			
		46			▨					▨	▨					
		50	▨	▨				▨	▨							
	A588	42													▨	
		46					▨							▨		
		50	▨	▨	▨	▨		▨	▨	▨	▨	▨	▨			
Quenched & Tempered Alloy	A514	90											▨			
		100						▨	▨	▨	▨	▨				

[a] Includes bar size shapes.

[b] To W 14 × 426 only.

▨ Available.

□ Not available.

Table A.3 Ref. (2)

$F_y = 36$ ksi

BEAMS
W shapes

Allowable uniform loads in kips for beams laterally supported

For beams laterally unsupported, see page 2 - 84

W 36

Designation	W36					Deflection Inches
Weight per Foot	300	280	260	245	230	
Flange Width	16⅝	16⅝	16½	16½	16½	
L_c	17.6	17.5	17.5	17.4	17.4	
L_u	35.3	33.1	30.5	28.6	26.8	
$F_y = 36$ ksi, Span in Feet: 16					792	0.18
17	1006	937	884	839	788	0.20
18	987	916	846	795	744	0.22
19	935	867	802	753	705	0.25
20	888	824	762	715	670	0.28
21	846	785	725	681	638	0.30
22	807	749	692	650	609	0.33
23	772	717	662	622	582	0.36
24	740	687	635	596	558	0.40
25	710	659	609	572	536	0.43
26	683	634	586	550	515	0.47
27	658	610	564	530	496	0.50
28	634	589	544	511	478	0.54
29	612	568	525	493	462	0.58
30	592	549	508	477	446	0.62
31	573	532	491	461	432	0.66
32	555	515	476	447	419	0.71
34	522	485	448	421	394	0.80
36	493	458	423	397	372	0.89
38	467	434	401	376	352	1.00
40	444	412	381	358	335	1.10
42	423	392	363	341	319	1.22
44	404	375	346	325	304	1.34
46	386	358	331	311	291	1.46
48	370	343	317	298	279	1.59
50	355	330	305	286	268	1.72
52	342	317	293	275	258	1.86
54	329	305	282	265	248	2.01
56	317	294	272	255	239	2.16
58	306	284	263	247	231	2.32
60	296	275	254	238	223	2.48
62	286	266	246	231	216	2.65
64	278	258	238	224	209	2.82
66	269	250	231	217	203	3.00
68	261	242	224	210	197	3.19
70	254	235	218	204	191	3.38
72	247	229	212	199	186	3.58
Properties and Reaction Values						
S in.^{3}	1110	1030	952	894	837	For explanation of deflection see page 2 - 23
V kips	503	468	442	419	396	
R kips	161	148	138	129	121	
R_i kips	25.5	23.9	22.7	21.7	20.5	
N_e in.	16.9	16.9	16.9	16.9	16.9	

Load above heavy line is limited by maximum allowable web shear.

Table A.3 Continued–Sample Page - ANSI/ASTM Materials Specification Ref. (4)

Standard Specification for
AUSTENITIC STAINLESS STEEL, SHEET, STRIP, PLATE, AND FLAT BAR FOR STRUCTURAL APPLICATIONS[1]

This Standard is issued under the fixed designation A 666; the number immediately following the designation indicates the year of original adoption or, in the case of revision, the year of last revision. A number in parentheses indicates the year of last reapproval.

1. Scope

1.1 This specification covers six types of austenitic stainless steels in four strength levels primarily for use in architectural structural applications. Sheet, strip, plate, and flat bar forms are included.

NOTE 1—The values stated in U.S. customary units are to be regarded as the standard.

2. Applicable Documents

2.1 *ASTM Standards:*

A 370 Methods and Definitions for Mechanical Testing of Steel Products[2]

A 480 Specification for General Requirements for Flat-Rolled Stainless and Heat-Resisting Steel Plate, Sheet, and Strip[3]

A 484 Specification for General Requirements for Stainless and Heat-Resisting Wrought Steel Products (Except Wire)[3]

3. Description of Terms

3.1 Plate, sheet, and strip as used in this specification are described as follows:

3.1.1 *Plate*—Material 3/16 in. (4.76 mm) and over in thickness and over 10 in. (254 mm) in width.

3.1.2 *Sheet*—Material under 3/16 in. in thickness and 24 in. (609.6 mm) and over in width.

3.1.3 *Strip*—Material under 3/16 in. in thickness and under 24 in. in width.

3.1.4 *Flat Bar*—Hot-finished flats: 1/4 to 10 in. (6.35 to 254 mm) inclusive, in width and 1/8 in. (3.17 mm) and over in thickness. Cold-finished flats: 3/8 in. (9.53 mm) and over in width, 1/8 in. and over in thickness.

4. General Requirements for Delivery

4.1.2 Sheet, strip, and plate material furnished under this specification shall conform to applicable requirements of the current edition of ASTM Specification A 480.

4.2 Flat bar material furnished under this specification shall conform to the applicable requirements of the current edition of ASTM Specification A 484.

5. Basis of Purchase

5.1 Orders for material under this specification shall include the following information:

5.1.1 Quantity (weight or number of pieces),

5.1.2 Name of material (stainless steel),

5.1.3 Form (plate, sheet, strip, or flat bar),

5.1.4 Dimensions,

5.1.5 Type (Table 1),

5.1.6 Grade: A (30 000 psi (205 MPa) minimum yield strength), B (40 000 or 45 000 psi (275 or 310 MPa) minimum yield strength), C (75 000 psi (515 MPa) minimum yield strength), or D (100 000 or 110 000 psi (690 or 760 MPa) minimum yield strength),

NOTE 2—All plate thicknesses and sizes of all types are not available in Grades B, C, and D. Producers should be consulted for information on availability of specific sizes and thicknesses.

5.1.7 Edge, strip only (see Specification A 480),

---

[1] This specification is under the jurisdiction of ASTM Committee A-1 on Steel, Stainless Steel and Related Alloys, and is the direct responsibility of Subcommittee A01.17 on Flat Stainless Steel Products.

Current edition approved Aug. 29, 1972. Published November 1972.

[2] *Annual Book of ASTM Standards*, Parts 1, 2, 3, 4, 5, and 10.

[3] *Annual Book of ASTM Standards*, Part 3.

Table A.3 Continued—Typical Properties for Stainless Steels Ref. (4)

Tensile Strength Requirements

Type	Product	Tensile Strength, min, psi (MPa)	Yield Strength, 0.2 percent Offset, psi (MPa)	Elongation in 2 in. or 50 mm, min, percent		
				Thickness, in. (mm)		
				Up to 0.015 (0.38)	Over 0.015 to 0.030 (0.38 to 0.76)	Over 0.030 (0.76)
			Grade A			
301	Plate, Sheet, Strip	75 000 (515)	30 000 (205)	40	40	40
302	Plate, Sheet, Strip	75 000 (515)	30 000 (205)	40	40	40
	Flat Bar	75 000 (515)	30 000 (205)	...	...	40
304	Plate, Sheet, Strip	75 000 (515)	30 000 (205)	40	40	40
	Flat Bar	75 000 (515)	30 000 (205)	...	...	40
316	Plate, Sheet, Strip	75 000 (515)	30 000 (205)	40	40	40
	Flat Bar	75 000 (515)	30 000 (205)	...	...	40
			Grade B			
201	Plate, Sheet, Strip	95 000 (655)	45 000 (310)	40	40	40
	Flat Bar	75 000 (515)	40 000 (275)	...	...	40
202	Plate, Sheet, Strip	90 000 (620)	45 000 (310)	40	40	40
	Flat Bar	75 000 (515)	40 000 (275)	...	...	40
301	Plate, Sheet, Strip	90 000 (620)	45 000 (310)	40	40	40
302	Plate, Sheet, Strip	85 000 (585)	45 000 (310)	40	40	40
	Flat Bar	90 000 (620)	45 000 (310)	...	...	40
304	Plate, Sheet, Strip	80 000 (550)	45 000 (310)	35	35	35
	Flat Bar	90 000 (620)	45 000 (310)	...	...	40
316	Plate, Sheet, Strip	85 000 (585)	45 000 (310)	35	35	35
	Flat Bar	90 000 (620)	45 000 (310)	...	...	40
			Grade C			
201	Plate, Sheet, Strip	125 000 (860)	75 000 (515)	20	20	20
202	Plate, Sheet, Strip	125 000 (860)	75 000 (515)	12	12	12
	Flat Bar (1 in. (25.4 mm) max)	115 000 (790)	75 000 (515)	...	...	15
301	Plate, Sheet, Strip	125 000 (860)	75 000 (515)	25	25	25
302	Plate, Sheet, Strip	125 000 (860)	75 000 (515)	10	10	12
	Flat Bar (1 in. (25.4 mm) max)	115 000 (790)	75 000 (515)	...	...	15
304	Plate, Sheet, Strip	125 000 (860)	75 000 (515)	10	10	12
	Flat Bar	115 000 (790)	75 000 (515)	...	...	15
316	Plate, Sheet, Strip	125 000 (860)	75 000 (515)	10	10	10
	Flat Bar	115 000 (790)	75 000 (515)	...	...	15
			Grade D			
201	Plate, Sheet, Strip	150 000 (1035)	110 000 (760)	9	10	10
202	Plate, Sheet, Strip	150 000 (1035)	110 000 (760)	9	10	10
	Flat Bar (3/4 in. (19 mm) max)	125 000 (860)	100 000 (690)	...	...	12
301	Plate, Sheet, Strip	150 000 (1035)	110 000 (760)	15	18	18
302	Plate, Sheet, Strip	150 000 (1035)	110 000 (760)	9	10	10
	Flat Bar (3/4 in. (19 mm) max)	125 000 (860)	100 000 (690)	...	...	12
304	Plate, Sheet, Strip	150 000 (1035)	110 000 (760)	6	7	7
	Flat Bar	125 000 (860)	100 000 (690)	...	...	7
316	Plate, Sheet, Strip	150 000 (1035)	110 000 (760)	6	7	7
	Flat Bar	125 000 (860)	100 000 (690)	...	...	7

Table A.4 Mechanical Properties of Titanium and Titanium Alloys at Room Temperature Ref. (5)

Material	*Nominal composition* (%)	*Condition*	0·1% *Proof stress* (*tons/in*²)	*U.T.S.* (*tons/in*²)	*Elongation* (%)		*Reduction in area* (%)	*Specification bend radius* (180° *bend*)	*Modulus of elasticity* (*lb/in*²) (×10⁻⁶)‡
					on 2 in	*on* $4\sqrt{A}$			
Ti	Pure iodide 60 V.P.N.	—	6·7 (0·2% PS)	15·6	55		80	—	—
Ti 115†	Commercially pure	Annealed sheet Annealed rod	16·0 15·0	21·0 24·0	36 —	— 40	— 70	16 S.W.G. or thinner 1*t* max	—
Ti 125	Commercially pure	Annealed sheet	25	30	28	—	—	16 S.W.G. or thinner 1*t* max	—
Ti 130	Commercially pure	Annealed sheet Annealed rolled bar and rod	28 26	35 34	25 —	— 29	— 54	16 S.W.G. or thinner 1*t* max	15·9
Ti 150	Commercially pure	Annealed sheet	32	39	24	—	—	16 S.W.G. or thinner 1*t* max	—
Ti 160	Commercially pure	Annealed sheet	38	45	23	—	—	16 S.W.G. or thinner 2*t* max	—
Ti–Al–V	Al 6·0 V 4·0	Annealed sheet Annealed rod	63 58	66 63	12 —	— 15	— 40	16 S.W.G. or thinner 4*t* max	15·3
Ti–Al–Mn	Al 2·0 Mn 2·0	Annealed rod and bar	38	48	—	25	50	—	17·0
Ti–Al–Mn	Al 4·0 Mn 4·0	Annealed rod and bar	64	68	—	19	45	—	17·3
Ti–Al–Sn	Al 5·0 Sn 2·5	Annealed rod and bar Annealed sheet	51 52	58 55	— 15	17 —	30 —	— 16 S.W.G. or thinner 4*t* max	16·8
Ti–Sn–Al–Zr–Mo–Si	Sn 11·00 Al 2·5 Zr 5·00 Mo 1·00 Si 0·20	Fully heat-treated rolled rod and bar	64	71	—	18	41	—	15·6
Ti–Sn–Al–Mo–Si	Sn 11·00 Al 2·25 Mo 4·0 Si 0·20	Fully heat-treated rod and bar	76	84	—	13	35	—	15·3
Ti–Mo	Ti–15 Mo	Sheet beta solution treated Sheet beta solution treated and aged Rod alpha/beta solution treated Rod alpha/beta solution treated and aged	37 70 56 70	50 78 61 76	23 6 — —	— — 16 19	— — 66 66	1*t* max — — —	16·9
Ti–Cu	Cu 2·0	Solution heat-treated and stress relieved sheet Solution heat treated rod Solution treated and duplex aged rod	33 30 45	38 38 51	23 — 15	— 28 20	— 45 41	16–24 S.W.G. 2*t* max Above 16–10 S.W.G. 3*t* max — —	15·8
Ti–Sn–Al–Mo–Si	Sn 4·0 Al 4·0 Mo 2·0 Si 0·5	Wrought, normalized and aged	71·5	80	—	15	40·5	—	16·8
Ti–Sn–Al–Mo–Si	Sn 4·0 Al 4·0 Mo 4·0 Si 0·5	Bar alpha/beta solution treated and aged	80·1	90	—	16	37	—	15·8

Table A.5 Typical Mechanical Properties of Titanium and Titanium Alloys at Elevated and Subnormal Temperatures (Short-time tests) Ref. (5)

Material	*Nominal composition* (%)	*Condition*	*Test temp.* (°C)	0·1% *Proof stress* (*tons/in*²)	*U.T.S.* (*tons/in*²)	*Elongation* (%) *on 2 in*	*Elongation* (%) *on* $4\sqrt{A}$	*Reduction in area* (%)	*Modulus of elasticity* (*lb/in*²) (×10⁻⁶)	*Transformation temp.*
Ti 115	Commercially	Annealed	−196	27·9	41·5	34	—	—	—	—
	pure	sheet	−78	17·8	26·6	34	—	—	—	—
			20	13·4	20·6	40	—	—	—	—
			100	10·9	16·1	45	—	—	—	—
			210	6·4	10·4	61	—	—	—	—
			250	4·1	9·0	60	—	—	—	—
			300	3·4	8·0	58	—	—	—	—
			400	2·5	6·3	58	—	—	—	—
Ti 125	Commercially	Annealed	−196	48·4	55·9	51	—	—	—	—
	pure	sheet	−78	34·3	39·5	28	—	—	—	—
			20	26·9	30·2	29	—	—	—	—
			100	20·8	23·6	40	—	—	—	—
			200	15·4	16·2	56	—	—	—	—
			300	11·4	11·9	50	—	—	—	—
			400	9·6	10·0	39	—	—	—	—
Ti 130	Commercially	Annealed	20	25·7	35·4	28	—	—	—	Alpha to
	pure	sheet	100	21·3	29·9	29	—	—	14·3	alpha +
			200	—	—	—	—	—	13·2	beta: 865°C
			250	10·4	17·9	39	—	—	—	
			300	—	—	—	—	—	12·1	
			400	—	—	—	—	—	9·4	Alpha +
			450	6·6	11·8	34	—	—	—	beta to beta:
			500	—	—	—	—	—	6·6	915°C
Ti 150	Commercially	Annealed	20	29·5	40·5	25	—	—	—	—
	pure	sheet	100	22·2	34·8	26	—	—	—	—
			250	11·8	20·7	34	—	—	—	—
			450	7·6	13·1	33	—	—	—	—
Ti 160	Commercially	Annealed	20	33·7	44·2	23	—	—	—	—
	pure	sheet	100	29·4	39·8	25	—	—	—	—
			250	12·0	21·9	34	—	—	—	—
			450	7·0	12·9	31	—	—	—	—
Ti–Al–Sn	Al 5·0	Annealed	20	53·2	59·5	16	22	39	16·3	Alpha to
	Sn 2·5	rod	100	44·8	51·7	17	24	40	15·8	alpha +
			200	32·0	41·3	17	23	44	15·3	beta: 950°C
			300	26·9	37·3	16	24	42	12·9	Alpha +
			400	22·4	33·8	18	23	41	12·4	beta to beta:
			500	22·4	31·5	18	26	57	11·8	1025°C
Ti–Al–Mn	Al 2·0	Annealed	20	43·0	51·0	23	28	43	16·0	Alpha +
	Mn 2·0	rod	100	35·6	44·6	25	28	43	15·5	beta to
			200	27·6	37·2	27	30	49	13·9	beta: 915°C
			300	22·9	33·2	20	24	51	12·9	± 20°
			400	22·0	31·2	18	22	58	12·2	
			500	14·2	22·9	30	36	76	8·9	

Table A.6 Wrought Aluminum 2021 Ref. (6)

Typical chemical composition, percent by weight: Mn, 0.20–0.40; Fe, 0.40; Si, 0.40; Al, balance; Zr, 0.10–0.25; Cu, 5.8–6.8; Cd, 0.05–0.20; Zn, 0.25

Physical constants and thermal properties
Density, lb/in.^{3}: 0.103
Coefficient of thermal expansion, (70–200°F) in./in./°F × 10^{-6}: 12.6
Modulus of elasticity, psi: tension, 10.7 × 10^6
Melting range, °F: 997–1,195

Heat treatments
Annealing temperature (start) 775°F; solution temperature 985°F; aging temperature 325°F.

Tensile Properties

Temperature, °F	T.S., psi	Y.S., psi, 0.2% offset	Elong., in 2 in., %	Hardness, Brinell
		Annealed		
75	24,000	10,000	23	44
		T81 Temper		
75	73,000	63,000	9	137

Table A.7 Wrought Aluminum 2024 Ref. (6)

Typical chemical composition, percent by weight: Mn, 0.30–0.90; Fe, 0.5; Si, 0.5; Cr, 0.10; Al, balance; Zr, 0.10–0.25; Cu, 3.8–4.9; Cd, 0.05–0.20; Mg, 1.2–1.8; Zn, 0.25

Physical constants and thermal properties
Density, lb/in.^{3}: 0.100
Coefficient of thermal expansion, (70–200°F) in./in./°F × 10^{-6}: 12.9
Modulus of elasticity, psi: tension, 10.6 × 10^6
Melting range, °F: 935–1,180
Specific heat, Btu/lb/°F, 70°F: 0.22
Thermal conductivity, Btu/ft^2/hr/in./°F, 70°F: 1,310
Electrical resistivity, ohms/cmil/ft, 70°F: 20.7 (O); 34.50 (T4)

Heat treatments
Annealing temperature (start) 775°F; solution temperature 930°F; aging temperature 375°F, 11–13 hr for T6.

Tensile Properties

Temperature, °F	T.S., psi	Y.S., psi, 0.2% offset	Elong., in 2 in., %	Hardness, Brinell
		Annealed		
75	27,000	11,000	20	47
		T3 Temper		
75	65,000	45,000	18	120
300	55,000	50,000	11	
400	29,000	22,000	23	
500	12,000	9,000	55	
600	8,000	6,000	75	
700	5,500	4,000	100	

Table A.8 Wrought Aluminum 2024 Ref. (6)

Fatigue Strength

Test temperature, °F	Stress, psi	Cycles to failure
75	13,000	5×10^8 (annealed)
75	20,000	5×10^8 (T3)

Specifications

ASTM: CG42A; SAE-AMS: 24; 24S

Table A.9 Wrought Aluminum 2219 Ref. (6)

Typical chemical composition, percent by weight: Mn, 0.20–0.40; Fe, 0.30; Si, 0.20; Ti, 0.02–0.10; Al, balance; Zr, 0.10–0.25; Cu, 5.8–6.8; V, 0.05–0.15; Mg, 0.02; Zn, 0.10

Physical constants and thermal properties

Density, lb/in.^{3}: 0.103

Coefficient of thermal expansion, (70–200°F) in./in./°F $\times 10^{-6}$: 12.4

Modulus of elasticity, psi: tension, 10.6×10^6

Melting range, °F: 1,010–1,190

Thermal conductivity, Btu/ft^2/hr/in./°F, 70°F: 1,200

Electrical resistivity, ohms/cmil/ft, 70°F: 23.4 (O); 34.2 (T6)

Heat treatments

Annealing temperature (start) 775°F; solution temperature 960°F; aging temperature 340°F.

Tensile Properties

Temperature, °F	T.S., psi	Y.S., psi, 0.2% offset	Elong., in 2 in., %	Hardness, Brinell
Annealed				
75	25,000	11,000	18	130 (T87 temper)
T6 Temper				
212	54,000	35,000	16	
300	46,000	31,000	20	
400	38,000	27,000	23	
500	31,000	22,000	25	
600	7,000	6,000	55	
700	4,500	4,000	60	

Fatigue Strength

Heat Treated T87

Test temperature, °F	Stress, psi	Cycles to failure
70	15,000	5×10^8

Table A.10 Monel Alloy 400 Ref. (6)

Chemical composition, percent by weight: C, 0.15; Mn, 1.00; Fe, 1.25; S, 0.012; Si, 0.25; Ni, 66.5; Cu, 31.5

Physical constants and thermal properties

Density, $lb/in.^3$: 0.319

Coefficient of thermal expansion, (70–200°F) in./in./°F × 10^{-6}: 7.7

Modulus of elasticity, psi: tension, 26.0 × 10^6; torsion, 9.5 × 10^6

Poisson's ratio: 0.32

Melting range, °F: 2,370–2,460

Specific heat, Btu/lb/°F, 70°F: 0.105

Thermal conductivity, $Btu/ft^2/hr/in./°F$, 70°F: 151

Electrical resistivity, ohms/cmil/ft, 70°F: 307

Curie temperature, °F: annealed, 20–50

Heat treatments

Generally used in annealed condition.

Tensile Properties

Annealed

Temperature, °F	T.S., psi	Y.S., psi, 0.2% offset	Elong., in 2 in., %	Hardness, Brinell
70	79,000	30,000	48	–
600	75,000	21,500	50	–
800	63,500	21,000	50	–
1,000	45,500	20,000	26	–
1,200	26,500	14,500	36	–
1,400	17,500	11,000	44	–
1,600	9,000	6,500	52	–
1,800	5,000	2,500	60	–

Rupture Strength, 1,000 hr

Cold Drawn, Annealed

Test temperature, °F	Strength, psi	Elong., in 2 in., %	Reduction of area, %
700	70,000	–	–
900	42,000	–	–
1,100	17,000	–	–

Creep Strength (Stress, psi, to Produce 1% Creep)

Cold Drawn, Annealed

Test temperature, °F	10,000 hr	100,000 hr
750	30,000	–
800	24,000	–
900	16,000	–
1,000	9,500	–

Table A.11 Monel Alloy 400 Ref. (6)

Fatigue Strength

Annealed

Test temperature, °F	Stress, psi	Cycles to failure
70	33,500	10^8

Impact Strength

Cold Drawn, Annealed

Test temperature, °F	Type test	Strength, ft-lb
-310	Charpy–V-notched	212
-112	Charpy–V-notched	219
75	Charpy–V-notched	216

Specifications

ASTM: B127, B163, B164, B165, B395; Federal QQ-N-281 Class A

Producers and Tradenames

Huntington Alloy Products Division, INCO: Monel alloy 400
G. O. Carlson Co.: Carlson 400
Bridgeport Brass: Bronel® 400
H. M. Harper: HMH 400

Table A.12 Fatigue Properties of Titanium and Titanium Alloys at Room Temperature Ref. (5)

Material	*Nominal composition* (%)	*Condition*	*Temp.* (°C)	*Details of test*	*Endurance limit* (*tons/in^2*) (*for 10^8 cycles unless otherwise stated*)
Ti 130	Commercially pure	Annealed rod	Room	Rotating bending	
				Kt = 1·0	± 17·5
				Kt = 2·0	± 11·0
				Kt = 3·0	± 11·0
Ti–Al–Sn	Al 5·0 Sn 2·5	Annealed rod	Room	Rotating bending	
				Kt = 1·0	± 24·0
				Kt = 2·0	± 17·0
				Kt = 3·3	± 15·5
				Direct stress	
				Kt = 1·0	± 28·0
				Kt = 1·5	± 18·0
				Kt = 2·0	± 13·0
				Kt = 3·3	± 10·0
Ti–Al–Mn	Al 2·0 Mn 2·0	Annealed rod	Room	Rotating bending	(10^7 cycles)
				Kt = 1·0	± 30·2
				Kt = 2·0	± 18·2
				Kt = 3·3	± 15·4
			200	Rotating bending	(10^7 cycles)
				Kt = 1·0	± 23·0
			400	Rotating bending	(10^7 cycles)
				Kt = 1·0	± 17·5
				Kt = 2·0	± 8·8
				Kt = 3·3	± 7·8
			Room	Direct stress	(10^7 cycles)
				Kt = 1·0	± 29·5
				Kt = 1·5	± 18·7
				Kt = 2·0	± 14·3
Ti–Al–Mn	Al 4·0 Mn 4·0	Annealed rod	Room	Rotating bending	
				Kt = 1·0	± 41·0
				Kt = 2·0	± 21·0
				Kt = 3·3	± 19·0
				Direct stress	
				Kt = 1·0	± 37·0
				Kt = 1·5	± 26·5
				Kt = 2·0	± 16·0
				Kt = 3·3	± 10·0
Ti–Sn–Al–Zr–Mo–Si	Sn 11·0 Al 2·25 Zr 5·0 Mo 1·0 Si 0·20	Fully heat-treated rod	Room	Rotating bending Kt = 1·0	± 41·5
			200	Kt = 1·0	± 33·0
			400	Kt = 1·0	± 33·0
			450	Kt = 1·0	± 36·0 (10^7 cycles)
			500	Kt = 1·0	± 32·0 (10^7 cycles)
Ti–Sn–Al–Mo–Si	Sn 11·0 Al 2·25 Mo 4·0 Si 0·20	Fully heat-treated rod	Room	Rotating bending Kt = 1·0	± 42·0
			200	Kt = 1·0	± 32·0
			400	Kt = 1·0	± 31·0
			Room	Direct stress	(10^7 cycles)
				Kt = 1·0	± 30·0
				Kt = 3·7	± 22·3
Ti–Mo	Mo 15·0	Beta solution-treated and aged sheet	Room	Reversed plane bending	± 26·0 (10^7 cycles)
		Alpha/beta solution-treated aged rod	Room	Rotating bending	(10^7 cycles)
				Kt = 1·0	± 35·0
				Kt = 2·0	± 26·0
			200	Rotating bending	(10^7 cycles)
				Kt = 1·0	± 23·0
			400	Kt = 1·0	± 23·0

Table A.13 IZOD Impact Properties of Titanium Alloys Ref. (5)

Material	*Nominal composition* (%)	*Condition*	*Izod value* (ft lb)*							
			−196 °C	−78°C	20°C	100°C	200°C	300°C	400°C	500°C
Ti 130†	Commercially pure	Annealed rod	—	46	45	46	53	60½	62	60½
Ti–Al–Mn†	Al 2·0 Mn 2·0	Annealed rod	—	—	16	26	39½	47½	52½	58
Ti–Al–Mn	Al 4·0 Mn 4·0	Annealed rod	5	12	16	25	46	59	65	66
Ti–Al–Sn	Sn 5·0 Al 2·5	Annealed rod	13	15	20	26	39	47	52	53
Ti–Sn–Al–Zr–Mo–Si	Sn 11·0 Al 2·25 Zr 5·0 Mo 1·0 Si 0·20	Fully heat-treated rod	8	8	10	—	10½	15½	24	27
Ti–Sn–Al–Mo–Si	Sn 11·0 Al 2·25 Mo 4·0 Si 0·20	Fully heat-treated rod	6	6½	8	9	11	13	15	19
Ti–Mo‡	Mo 15·0	Alpha/beta solution-treated and aged	4½	8½	15	22	28	31	40	—

* B.S.S. 131(1) 0·45 in diameter straight notched test pieces.

† Izod values of commercial purity titanium are appreciably, and of Titanium 314C are moderately, affected by variation in hydrogen content within commercial limits (0·008% maximum in Titanium 130 rod above 0·5 in diameter and 0·010% maximum in Titanium 314C).

‡ At 80 tons/in² tensile strength level. Room temperature values increase with lowered strength to 21 ft/lb at 70 tons/in² and 56 ft/lb at 60 tons/in² tensile strength.

Table A.14 Working Stresses for Stress-grade Lumber under Normal Loading Conditions Ref. (7)

Species and commercial grade‡	Uses‡	Rules under which graded	Extreme fiber in bending f and tension parallel to grain t	Horizontal shear H	Compression perpendicular to grain $c\perp$	Compression parallel to grain c	Modulus of elasticity E
Cypress, southern:							
1700 f grade	J&P-B&S	Southern Cypress Manufacturers' Association	1,700	145		1,425	
1300 f grade	J&P-B&S		1,300	120	360	1,125	1,200,000
1450 c grade	P&T		.....	...		1,450	
1200 c grade	P&T		.....	...		1,200	
Douglas fir, coast region:							
Dense select structural	J&P-B&S		2,150	145	455	1,550	
Select structural	J&P-B&S		1,900	120	415	1,450	
1700 f—dense No. 1	J&P-B&S	West Coast Bureau of Lumber Grades and Inspection	1,700	145	455	1,325	
1450 f—No. 1	J&P-B&S		1,450	120	390	1,200	
1100 f—No. 2	J&P		1,100	110	390	1,075	1,600,000
Dense select structural	P&T		.....	...	455	1,550	
Select structural	P&T		.....	...	415	1,450	
Dense No. 1	P&T		.....	...	455	1,400	
No. 1	P&T		.....	...	390	1,200	
Douglas fir, inland region:							
Select structural	J&P		2,150	145	455	1,750	1,600,000
Structural	J&P		1,900	100	400	1,400	1,500,000
Common structural	J&P	Western Pine Association	1,450	95	380	1,250	1,500,000
Select structural	P&T		.....	...	455	1,750	1,600,000
Structural	P&T		.....	...	400	1,400	1,500,000
Common structural	P&T		.....	...	380	1,250	1,500,000
Hemlock, eastern:							
Select structural	J&P-B&S	Northern Hemlock and Hardwood Manufacturers Association	1,300	85		850	
Prime structural	J&P		1,200	60		775	
Common structural	J&P		1,100	60	360	650	1,100,000
Utility structural	J&P		950	60		600	
Select structural	P&T		.....	...		850	
Hemlock, west coast:							
1600 f—select structural	J&P	West Coast Bureau of Lumber Grades and Inspection	1,600	100		1,100	
1450 f—No. 1	J&P-B&S		1,450	100	360	1,075	1,400,000
1100 f—No. 2	J&P		1,100	90		850	
No. 1 hemlock timbers	P&T		.....	...		1,075	
Larch:							
Select structural	J&P		2,150	145	455	1,750	
Structural	J&P	Western Pine Association	1,900	120	415	1,450	
Common structural	J&P		1,450	120	390	1,325	1,500,000
Select structural	P&T		.....	...	455	1,750	
Structural	P&T		.....	...	415	1,450	
Common structural	P&T		.....	...	390	1,325	
Oak, red and white:							
2150 f grade	J&P		2,150	145		1,550	
1900 f grade	J&P-B&S	National Hardwood Lumber Association	1,900	145		1,375	
1700 f grade	J&P-B&S		1,700	145		1,200	
1450 f grade	J&P-B&S		1,450	120	600	1,050	1,500,000
1300 f grade	B&S		1,300	120		950	
1325 c grade	P&T		.....	...		1,325	
1200 c grade	P&T		.....	...		1,200	
1075 c grade	P&T		.....	...		1,075	
Pine, southern:							
Dense select structural	J&P-B&S		2,400	120	455	1,750	
Dense structural	J&P-B&S	Southern Pine Inspection Bureau	2,000	120	455	1,400	
Dense structural SE&S	J&P-B&S		1,800	120	455	1,300	
Dense No. 1 structural	J&P-B&S		1,600	120	455	1,150	
No. 1 dense 1400f	J&P-B&S		1,400	140	455	1,400	

WROUGHT MAGNESIUM

Table A.15 Room-temperature Mechanical Properties of Extruded Bars, Rods, and Shapes Ref. (7)

Alloy	Condition[1]	Diam of thickness in least dimension, in.	TS, 1,000 psi		TYS, 1,000 psi		Elongation, % in 2 in.		CYS, 1,000 psi		Bearing strength, 1,000 psi: Ultimate		Bearing strength, 1,000 psi: Yield		Shear, 1,000 psi		Hardness	
			Typ.	Min.	Typ.	Min.	Typ.	Min.	Typ.	Min.	Typ.	Min.	Typ.	Min.	Typ.	Min.	Brinell	Rockwell E
AZ31B	-F	Under 0.250	38	35	28	21	14	10	15	...	56	45	33	23	19	17		
	-F	0.250–1.499	38	35	29	22	15	10	14	12	56	45	33	23	19	17	49	57
	-F	1.500–2.499	38	34	28	22	14	10	14	12	56	45	33	23	19	17		
	-F	2.500–5.000	38	32	28	20	15	7	14	10	56	45	33	23				
AZ61A	-F	Under 0.250	46	38	33	20	17	8	...	...	65	53	38	30				
	-F	0.250–1.499	45	40	33	24	16	10	19	14	68	55	41	32	20	18	60	72
	-F	1.500–2.499	45	40	33	24	16	10	19	14	68	55	41	32	20	18	60	72
	-F	2.500–5.000	45	40	31	22	15	7	21	14	68	55						
M1A	-F	Under 0.250	37	30	26	...	12	2	12	...	51	45	28	23	...	...	44	45
	-F	0.250–1.499	37	32	26	...	11	3	12	...	51	45	28	23	18	15	44	45
	-F	1.500–2.499	39	32	...	...	9	2	...	...	51	45	28	23	18	15	44	45
	-F	2.500–5.000	38	29	...	...	5	2	...	...	51	45	28	23	...	...	44	45
AZ80A	-F	Under 0.250	49	43	36	28	12	9	...	...	68	56	48	40	22	19	60	77
	-F	0.250–1.499	49	43	36	28	11	9	...	17	68	56	48	40	22	19	60	77
	-F	1.500–2.499	49	43	35	28	11	8	...	17	68	56	48	40	22	19	60	77
	-F	2.500–5.000	48	42	36	27	9	4	...	17	68	56	48	40	...	...	60	77
	-T5	Under 0.250	55	47	38	30	8	5	34	...	...	...	...	...	...	...	82	88
	-T5	0.250–1.499	55	48	40	33	7	4	35	28	...	...	...	...	24	21	82	88
	-T5	1.500–2.499	53	48	39	33	6	4	32	27	...	...	...	...	24	21	82	88
	-T5	2.500–5.000	50	45	38	30	6	2	31	26	...	...	...	...	...	...	82	88
		Cross-sect. area, sq in.																
ZK60A	-F	Less than 2	49	43	38	31	14	5	33	27	80	70	55	45	27	22	75	84
	-F	2–3	49	43	37	31	14	5	28	26	80	70	55	45	27	22	75	84
	-F	3–5	49	43	36	31	14	5	27	25	80	70	55	45	27	22	75	84
	-F	5–40	48	43	37	31	9	4	23	20	80	70	55	45	27	22	75	84
	-T5	Less than 2	53	45	44	36	11	4	36	30	85	71	59	47	26	22	82	88
	-T5	2–3	52	45	43	36	12	4	31	28	85	71	59	47	26	22	82	88
	-T5	3–5	51	45	42	36	14	4	30	25	85	71	59	47	24	21	82	88

[1] -F = as extruded; -T5 = artificially aged.

Table A.16 Typical Properties of AISI 8640 Steel Ref. (8)

Mechanical Properties

Form or Condition	Draw Temp	Tensile Strength M psi	Elong 2 in. %	Yield Point M psi	Red Area %	Hard BHN	NBT Ft Lbs
Bars[1]							
Annealed	–	–	–	–	–	197	–
Normalized	–	–	–	–	–	302	–
Aver. Prop.[4] 1 in. rds. Mean chemistry — Cold drawn, unannealed	–	140	11	120	38	277[5]	–
Cold drawn, annealed	–	107	14	90	45	217[5]	–
Turned and Polished, unannealed	–	124	17	94	42	255[5]	–
Turned and Polished, annealed	–	99	24	58	46	183[5]	–
Effect of Mass[4] OQ at 1525F and tempered at 1000F; 0.505 in. tensile specimens — 1 in. diam. ctr.	–	155	17	137	56	331[5]	–
2 in. diam. ½r	–	132	18	112	57	293[5]	–
3 in. diam. ½r	–	125	19	103	58	277[5]	–
Effect of Mass Interpreted from End-Quench Hardenability curve. Quenched in oil at 1550F, drawn as shown[2] — 1 in. rd. ctr.	1000F	174	15.5	154	47.3	352	–
1 in. rd. ctr.	1100	156	17.4	135	51.0	316	–
1 in. rd. ctr.	1200	143	19.0	120	53.5	285	–
2 in. rd. ½r.	1000	156	17.4	135	51.0	316	–
2 in. rd. ½r.	1100	145	18.7	123	53.0	293	–
2 in. rd. ½r.	1200	130	20.6	106	56.3	262	–
3 in. rd. ½r.	1000	141	19.2	118	54.0	285	–
3 in. rd. ½r.	1100	130	20.6	106	56.3	262	–
3 in. rd. ½r.	1200	120	21.8	94	58.2	241	–
4 in. rd. ½r.	1000	131	20.4	107	56.0	269	–
4 in. rd. ½r.	1100	122	21.6	96	57.9	244	–
4 in. rd. ½r.	1200	111	22.7	85	60.0	223	–
Effect of Temp[4] Oil quenched at 1525F and tempered as shown 0.505 in. tensile specimens cut from center of 1 in. rds.	800F	200.5	10	179	46	415[5]	–
	900	181	13	162	51	388[5]	–
	1000	155	17	137	56	331[5]	–
	1100	148.5	16	132	57	302[5]	–
	1200	125.5	20	110.5	61	269[5]	–
	Test Temp						
Low Temp Notched Bar Tests[3] 0.420 in. sq. bars quenched and drawn to 255 BHN. Tested as Charpy bars with keyhole notch.	70F	–	–	–	–	–	45
	0	–	–	–	–	–	40
	-50	–	–	–	–	–	36
	-100	–	–	–	–	–	31
	-200	–	–	–	–	–	22
	-270	–	–	–	–	–	15
	-300	–	–	–	–	–	11

[1]Republic Steel Corp. Approx values.
[2]Ryerson. Aver values.
[3]H. W. Gillett and F. T. McGuire. "Behavior of Ferritic Steels at Low Temperatures."
[4]Bliss & Laughlin
[5]Surface of bar

Table A.17 Minimum Mechanical Properties of Wrought Copper Alloys Ref. (9)

Metric (SI) Units

UNS No.[ee]	Form	Temper	Size Section, mm	Tensile Strength, MPa		Yield Strength, Min MPa	Elongation, Min %[c]	Hardness				Grain Size, mm	
			Over/Thru	Min	Max	0.5% Ext Under Load	In 4 x Dia or Thickness of Specimen	Min	Max	Min	Max	Min	Max
								RF[b]		R30T[b]			
C10200 C11000 C11100 C11300 C11400 C11500 C11600 C12000 C12200	Plate, Sheet, Strip, and Rolled Bar	Soft Anneal	—	—	—	—	—	—	65	—	—	Note a	—
		Deep-Drawing Anneal	—	—	—	—	—	30	75	—	—	Note a	0.050
		Light Cold Rolled	—	220	275	—	—	40	82	—	49	—	—
		1/2 Hard[r]	—	225	315	—	—	77	80	43	57	—	—
		Hard[r]	—	285	360	—	—	86	93	54	62	—	—
		Spring[r]	—	345	400	—	—	91	97	60	66	—	—
		Extra Spring[r]	—	360	—	—	—	92	—	61	—	—	—
		Hot Rolled	—	205	260	—	—	—	75	—	41	—	—
		Hot Rolled and Annealed	—	205	260	—	—	—	65	—	31	—	—
				Type B Matl[u]			Type B[u]	Type A[s,u]					
C10200 C11000 C12000 C12200	Rod, Bar and Shapes	Soft Anneal	All Sizes[m]	—	255	—	25	—	65	—	—	—	—
		Hard	—/6.5[w]	345	—	—	—	68	95	—	—	—	—
			6.5/9.5[w]	310	—	—	10	68	95	—	—	—	—
			9.5/25[w]	275	—	—	12	68	95	—	—	—	—
			25/50[w]	240	—	—	15	68	95	—	—	—	—
			50/75[w]	230	—	—	15	68	95	—	—	—	—
			4.8/9.5[x]	290	—	—	12	68	95	—	—	—	—
			9.5/13[x]	275	—	—	12	68	95	—	—	—	—
			13/50[x]	230	—	—	15	68	95	—	—	—	—
			50/100[x]	220	—	—	15	68	95	—	—	—	—
			All Sizes[y]	220	—	—	15	—	—	—	—	—	—

			OD	Wall				In 50 mm			R15T[d]			
C10200 C12000 C12200	Tube	Soft Anneal	All Sizes	0.4/0.9	—	—	—	40[e]	—	—	—	60	0.040	—
			All Sizes	0.9/—	—	—	—	—	—	50[d]	—	—	0.040	—
		Light Anneal	All Sizes	0.4/0.9	—	—	—	—	—	—	—	65	—	0.040
			All Sizes	0.9/—	—	—	—	—	—	55[d]	—	—	—	0.040
											R30T[d]			
		Light Drawn	All Sizes	All Sizes	250	325	—	—	—	—	30	60	—	—
		Drawn	All Sizes	All Sizes	250	—	—	—	—	—	30	—	—	—
		Hard Drawn	—/25	0.5/3.0	310	—	—	—	—	—	55	—	—	—
			25/50	0.4/4.5	310	—	—	—	—	—	55	—	—	—
			50/100	1.5/6.5	310	—	—	—	—	—	55	—	—	—
			Size Section, mm Over/Thru					In 250 mm	RF[b]		R30T[b]			
C10200 C11000	Wire	Annealed	0.08/0.25		—	—	—	15	—	—	—	—	—	—
			0.25/0.50		—	—	—	20	—	—	—	—	—	—
			0.50/2.5		—	—	—	25	—	—	—	—	—	—
			2.5/7.5		—	—	—	30	—	—	—	—	—	—
			7.5/12		—	—	—	35	—	—	—	—	—	—
								In 4 x Dia						
C14500	Rod	1/2 Hard[f]	1.6/6.5		260	—	205	8	—	—	—	—	—	—
			6.5/65		260	—	205	12	—	—	—	—	—	—
		Hard	1.6/6.5		330	—	275	4	—	—	—	—	—	—
			6.5/30		305	—	260	8	—	—	—	—	—	—
			30/50		275	—	240	8	—	—	—	—	—	—
C14700	Rod	1/2 Hard[f]	1.6/6.5			—	205	8	—	—	—	—	—	—
			6.5/65			—	205	12	—	—	—	—	—	—
		Hard[g]	1.6/6.5			—	275	4	—	—	—	—	—	—
			6.5/30			—	260	8	—	—	—	—	—	—
			30/50			—	240	8	—	—	—	—	—	—
C15000	Round Rod													

Table A.18 Design Mechanical and Physical Properties of 7075 Aluminum Alloy (Sheet and Plate) Ref. (1)

Specification	QQ-A-250/12																				
Form	Sheet							Plate[a]													
Temper	T6 and T62[a]							T651 and T62[a]													
Thickness, in.	0.008-0.011	0.012-0.039		0.040-0.125		0.126-0.249		0.250-0.499		0.500-1.000		1.001-2.000		2.001-2.500		2.501-3.000		3.001-3.500		3.501-4.000	
Basis	S	A	B	A	B	A	B	A	B	A	B	A	B	A	B	A	B	A	B	A	B
Mechanical properties:																					
F_{tu}, ksi:																					
L	...	76	78	78	80	78	80	77	79	77	79	76	78	75	77	71	73	70	72	66	68
LT	74	76	78	78	80	78	80	78	80	78	80	77	79	76	78	72	74	71.	73	67	69
ST	...	...	...	...	...	...	...	...	...	...	...	...	...	68	70	65	67	64	66	60	62
F_{ty}, ksi:																					
L	...	69	72	70	72	71	73	69	71	70	72	69	71	66	68	63	65	60	62	56	58
LT	63	67	70	68	70	69	71	67	69	68	70	67	69	64	66	61	63	58	60	54	56
ST	...	...	...	...	...	...	...	...	...	...	...	...	...	57	59	54	56	52	54	48	50
F_{cy}, ksi:																					
L	...	68	71	69	71	70	72	68	70	68	70	66	68	62	64	59	61	56	57	51	53
LT	...	71	74	72	74	73	75	71	73	72	74	71	73	68	70	65	67	61	64	57	59
ST	...	...	...	...	...	...	...	...	...	...	...	...	...	65	67	62	64	59	61	55	57
F_{su}, ksi	...	46	47	47	48	47	48	44	45	44	45	45	46	45	46	43	44	43	44	41	42
F_{bru}[b], ksi:																					
(e/D=1.5)	...	118	121	121	124	121	124	118	121	118	121	117	120	115	118	109	112	108	111	102	105
(e/D=2.0)	...	152	156	156	160	156	160	146	150	146	150	144	148	142	146	135	138	133	136	125	129
F_{bry}[b], ksi:																					
(e/D=1.5)	...	100	105	102	105	103	106	98	101	101	104	102	105	99	102	96	99	92	96	87	90
(e/D=2.0)	...	117	122	119	122	121	124	115	119	118	122	119	122	116	119	112	115	107	111	101	105
e, percent:																					
L	...	7	...	8	...	8	...	9	...	7	...	6	...	5	...	5	...	5	...	...	...
LT	5	7	...	8	...	8	...	9	...	7	...	6	...	5	...	5	...	5	...	3	...
ST	...	...	...	...	...	...	...	...	...	...	...	...	...	1	...	1	...	1	...	...	...
E, 10^3 ksi	10.3							10.3													
E_c, 10^3 ksi	10.5							10.6													
G, 10^3 ksi	3.9							3.9													
μ	0.33							0.33													
Physical properties:																					
ω, lb/in.^{3}	0.101																				
C, Btu/(lb)(F)	0.23 (at 212 F)																				
K, Btu/[(hr)(ft^2)(F)/ft]	76 (at 77 F)																				
α, 10^{-6} in./in./F	12.9 (68-212 F)																				

[a]The allowables shown for these tempers are based on and have been determined from the results obtained on testing of T6 and T651 temper material and on the testing of T62 temper samples for specification conformance. These allowables also apply when samples of material supplied in the O or F temper are heat treated to demonstrate response to heat treatment. Properties obtained by the user however, may be lower than those listed if the material has been formed or otherwise cold or hot worked, particularly in the annealed temper, prior to solution heat treatment.

[b]See Table 3.1.2.1.1. Bearing values are "dry pin" values per Section 1.4.7.1.

Table A.19 Design Mechanical and Physical Properties of 2024 Aluminum Alloy (Sheet and Plate) Ref. (1)

Specification	QQ-A-250/4																			
Form	Sheet					Plate												Sheet		Plate
Temper	T3					T351												T361		
Thickness, in.	0.008-0.009	0.010-0.128		0.129-0.249		0.250-0.499		0.500-1.000		1.001-1.500		1.501-2.000		2.001-3.000		3.001-4.000		0.020-0.062	0.063-0.249	0.250-0.500
Basis	S	A	B	A	B	A	B	A	B	A	B	A	B	A	B	A	B	S	S	S
Mechanical properties:																				
F_{tu}, ksi:																				
L	64	64[d]	65[d]	65	66	65	67	63	65	62	64	62	64	60	62	57	59	68	69	67
LT	63	63[d]	64[d]	64	65	64	66	63	65	62	64	62	64	60	62	57	59	67	68	66
F_{ty}, ksi:																				
L	47	47	48	47	48	47	49	47	49	47	49	47	49	47	49	46	48	56	56	54
LT	42	42	43	42	43	42	44	42	44	42	44	42	44	42	44	41	43	50	51	49
F_{cy}, ksi:																				
L	39	39	40	39	40	40	42	39	41	39	41	39	40	38	40	36	38	47	48	46
LT	45	45	46	45	46	45	47	45	47	45	47	44	46	44	46	42	44	53	54	52
F_{su}, ksi	39	39[d]	40[d]	40	41	38	40	38	39	37	38	37	38	36	37	34	35	42	42	41
F_{bru}^{ab}, ksi:																				
(e/D = 1.5)	104	104[d]	106[d]	106	107	97	100	95	98	94	97	94	97	91	94	86	89	111	112	109
(e/D = 2.0)	129	129[d]	131[d]	131	133	119	122	117	121	115	119	115	119	111	115	106	109	137	139	135
F_{bry}^{ab}, ksi:																				
(e/D = 1.5)	73	73	75	73	75	73	76	73	76	73	76	73	76	73	76	71	74	82	84	81
(e/D = 2.0)	88	88	90	88	90	87	91	87	91	87	91	87	91	87	91	85	89	97	99	96
e, percent:																				
LT	10	c	...	c	...	12	...	8	...	7	...	6	...	4	...	4	...	8	9	c
E, 10^3 ksi	10.5					10.7												10.5		10.7
E_c, 10^3 ksi	10.7					10.9												10.7		10.9
G, 10^3 ksi	4.0					4.0												4.0		4.0
μ	0.33					0.33												0.33		0.33
Physical properties:																				
ω, lb/in.^{3}	0.100																			
C, Btu/(lb)(F)	See Figure 3.2.3.0																			
K, Btu/[(hr)(ft^2)(F)/ft]	70 (at 77 F). See Figure 3.2.3.0																			
α, 10^{-6} in./in./F	See Figure 3.2.3.0																			

Table A.20 Design Mechanical and Physical Properties of AISI 1025 Carbon Steel Ref. (1)

Specification	MIL-S-7952, 1025	MIL-T-5066	MIL-S-7097, Comp. 3
Form	Sheet and strip	Tubing	Bars
Condition	Cold rolled	Normalized	All
Thickness, in.	...	...	...
Basis	S	S	S[a]
Mechanical properties:			
F_{tu}, ksi:			
L	55	55	55
LT	55	55	55
ST	...	...	55
F_{ty}, ksi:			
L	36	36	36
LT	36	36	36
ST	...	...	36
F_{cy}, ksi:			
L	36	36	36
LT	36	36	36
ST	...	...	36
F_{su}, ksi	35	35	35
F_{bru}, ksi:			
(e/D = 1.5)	...	...	...
(e/D = 2.0)	90	90	90
F_{bry}, ksi:			
(e/D = 1.5)	...	...	...
(e/D = 2.0)	...	...	...
e, percent:			
L	...	b	ab
LT	b	...	...
ST	...	...	...
E, 10^6 psi	29.0		
E_c, 10^6 psi	29.0		
G, 10^3 ksi	11.0		
μ	0.32		
Physical properties:			
ω, lb/in.^{3}	0.284		
C, Btu/(lb)(F)	0.116 (122 to 212 F)		
K, Btu/[(hr)(ft^2)(F)/ft]	30.0 (at 32 F)		
α, 10^{-6} in./in./F	6.78 (68 to 212 F); 7.00 (68 to 392 F)		

[a]Grain direction not specified.

[b]See applicable specification for variation in minimum elongation with ultimate strength.

Appendix B

MATRIX ALGEBRA

1. General Comments

Matrices are arrays of elements. The elements may be scalars (numbers), operators, or variables. When the definition is written down, the resulting array does not look very impressive. Yet when the knowledge of matrix algebra and calculus becomes known, matrices are appreciated as one of the most powerful mathematical tools available to the engineer.

Matrices permit an orderly, compact representation of physical quantities such as equations of motion, force-displacement relations, coordinate transformations and simultaneous linear equations--just to mention a few. They are ideally suited for digital computer operations.

Mastery of skills in matrix algebra is essential, not only for this course, but also because of their use in other areas of engineering and science.

2. Definition of a Matrix

A Matrix $\underline{A}$ is a rectangular array of elements a_{ij} arranged in m rows and n columns.

$$A = \begin{bmatrix} a_{11} & a_{12} & a_{13} & \cdots & a_{1n} \\ a_{21} & a_{22} & a_{23} & \cdots & a_{2n} \\ \cdot & \cdot & \cdot & \cdots & \cdot \\ a_{m1} & a_{m2} & a_{m3} & \cdots & a_{mn} \end{bmatrix}_{mxn}$$

Do not confuse a matrix with a determinant. A matrix is an array. It has no value. A determinant has a numerical value.

The matrix is said to be of "size" or "order" mxn, where m is the number of rows, and n is the number of columns. The matrix $\underline{A}$ is denoted in various ways.

$\underline{A}_{mxn}$ or $[A]_{mxn}$ or a boldface letter in books or papers

$\underline{a}_{mxn}$ or $[a]_{mxn}$

The size designation may be omitted.

By the typical element a_{ij} we denote the element found the intersection of the ith row and j^{th} column. We must understand that a matrix is an array, but an element is a simple scalar, variable or operator.

Examples:

$$\underline{A}_{3 \times 3} = \begin{bmatrix} 2 & 1 & 4 \\ 0 & 2 & 5 \\ 1 & 3 & 1 \end{bmatrix} ; \; a_{22} = 2$$

$$\underline{A}_{3 \times 1} = \begin{Bmatrix} x \\ y \\ z \end{Bmatrix}$$

$$\underline{A}_{2 \times 1} = \begin{Bmatrix} \frac{\partial}{\partial x} \\ \frac{\partial}{\partial y} \end{Bmatrix}$$

The matrix with one column is called a column matrix, or vector. It is usually enclosed in braces and may be denoted by

$$\{R\} = \begin{Bmatrix} R_1 \\ R_2 \\ R_3 \\ R_4 \end{Bmatrix}$$

or

$$\underline{r} = \begin{bmatrix} r_1 \\ r_2 \\ r_3 \\ r_4 \end{bmatrix} \quad \text{etc.}$$

Braces are not used in some publications. We may also see it written

$$\{r\} = \{r_1, r_2, r_3, r_4\}$$

in some publications. This is done to save space.

Other special forms of matrices are:

a. The square matrix; $m = n$ See $\underline{A}_{3 \times 3}$ above.

b. The row matrix; $m = 1$ usually denoted

$$\lfloor A \rfloor \text{ or } A_{1 \times n} = \lfloor 1 \;\; 5 \;\; -6 \;\; 2 \rfloor$$

c. The diagonal matrix, which must be square

$$\left[\diagdown A \diagdown\right]_{mxm} \text{ or } \underset{\sim}{A}_{mxm}$$

$$\left[\diagdown A \diagdown\right]_{3 \times 3} = \begin{bmatrix} 3 & 0 & 0 \\ 0 & 2 & 0 \\ 0 & 0 & 6 \end{bmatrix}_{3 \times 3} \qquad a_{ij} = 0;\ i \neq j$$

d. The null matrix; all elements are zero

$$\left[0\right]_{3 \times 4} = \begin{bmatrix} 0 & 0 & 0 & 0 \\ 0 & 0 & 0 & 0 \\ 0 & 0 & 0 & 0 \end{bmatrix}_{3 \times 4}$$

e. The unit, or identity matrix, must be square

$$\left[I\right]_{2 \times 2} = \begin{bmatrix} 1 & 0 \\ 0 & 1 \end{bmatrix} \qquad \begin{aligned} a_{ij} &= 1;\ i = j \\ &= 0;\ i \neq j \end{aligned}$$

f. A symmetric matrix has elements a_{ij} = elements a_{ji}.
It must be square.

$$\left[\underset{\sim}{A}\right]_{3 \times 3} = \begin{bmatrix} 1 & 4 & 0 \\ 4 & 5 & -8 \\ 0 & -8 & 7 \end{bmatrix}$$

3. Matrix Algebra

Scalar Algebra, Matrix Algebra and Vector Algebra have many things in common as we will see. The following algebraic properties give matrices their meaning.

a. The determinant of a matrix is the determinant formed by taking the elements of the matrix in order.

$$\text{If } \underline{A} = \begin{bmatrix} 0 & 1 & 2 \\ 2 & 1 & 2 \\ -1 & 1 & 1 \end{bmatrix}$$

Then

$$|\underline{A}| = \begin{vmatrix} 0 & 1 & 2 \\ 2 & 1 & 2 \\ -1 & 1 & 1 \end{vmatrix} = 0 \begin{vmatrix} 1 & 2 \\ 1 & 1 \end{vmatrix} - 1 \begin{vmatrix} 2 & 2 \\ -1 & 1 \end{vmatrix} + 2 \begin{vmatrix} 2 & 1 \\ -1 & 1 \end{vmatrix}$$

$$= -1(4) + 2(3) = -1$$

Note that $\underline{A}$ must be square and that $|\underline{A}|$ is a <u>scalar</u>.

b. Matrices $\underline{A}$ and $\underline{B}$ are <u>equal</u> if they are of the same order and if $a_{ij} = b_{ij}$ for all i and j.

c. Matrix <u>addition</u>. To get Matrix $\underline{C}$ the sum of matrices $\underline{A}$ and $\underline{B}$, we form

$$c_{ij} = a_{ij} + b_{ij} \text{ for all i and j}$$

$$\text{If } [A] = \begin{bmatrix} 1 & 2 \\ 3 & 4 \end{bmatrix} \text{ and } [B] = \begin{bmatrix} 10 & 9 \\ 8 & 7 \end{bmatrix} \text{ then}$$

$$[C] = [A] + [B] = \begin{bmatrix} 11 & 11 \\ 11 & 11 \end{bmatrix};$$ $\underline{A}$ and $\underline{B}$ must be of the same order (not necessarily square).

d. Matrix <u>subtraction</u>. The negative of a matrix - A is formed by negating the a_{ij} for all i and j

$$\text{If } [A] = \begin{bmatrix} -1 & 2 & -7 \\ 3 & 5 & 8 \end{bmatrix}$$

$$\text{then } [-A] = \begin{bmatrix} 1 & -2 & 7 \\ -3 & -5 & -8 \end{bmatrix}$$

Then

$$[A] - [B] = [A] + (-[B])$$

or if

$$[A] = \begin{bmatrix} 4 & 5 \\ 2 & -1 \end{bmatrix} \text{and } [B] = \begin{bmatrix} -1 & 3 \\ 2 & -7 \end{bmatrix}$$

$$\text{Then } [C] = [A] - [B] = \begin{bmatrix} 6 & 2 \\ 0 & 6 \end{bmatrix}$$

e. Matrix <u>transpose</u>. If $\underline{A}$ is a general matrix, then $\underline{A}^T$ ($\underline{A}$ transpose) is formed by taking $a^T_{ij} = a_{ji}$ or in plain words, by interchanging rows and columns.

$$\text{If } [A] = \begin{bmatrix} 1 & 2 \\ 3 & 4 \end{bmatrix}, \text{ then } [A]^T = \begin{bmatrix} 1 & 3 \\ 2 & 4 \end{bmatrix}$$

$\underline{A}^T$ is often written $\underline{A}'$.

If $\underline{A}$ is SYMMETRIC, then $\underline{A}^T = \underline{A}$.

f. Multiplication of a matrix by a scalar.
If a matrix $\underline{A}$ is multiplied by a scalar a, then each element in $\underline{A}$ is multiplied by a

$$\underline{A} = \begin{bmatrix} 2 & 4 & 1 \\ 3 & 5 & 7 \\ 4 & 6 & 0 \end{bmatrix}$$

Then

$$3\ \underline{A} = 3 \begin{bmatrix} 2 & 4 & 1 \\ 3 & 5 & 7 \\ 4 & -6 & 0 \end{bmatrix} = \begin{bmatrix} 6 & 12 & 3 \\ 9 & 15 & 21 \\ 12 & -18 & 0 \end{bmatrix}$$

g. Matrix Multiplication.
Two matrices may be multiplied if they are conformable for multiplication

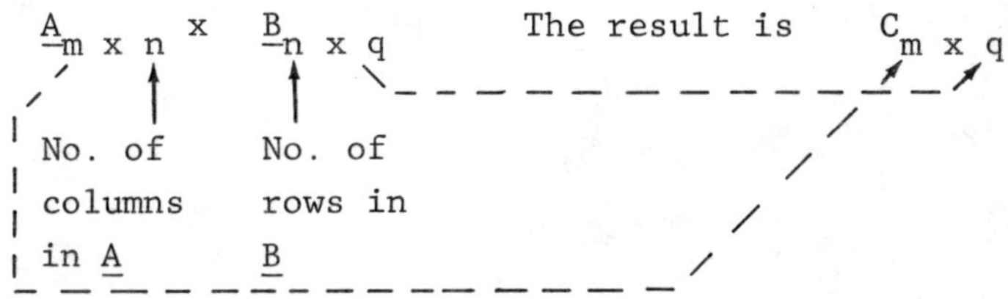

Matrix C is formed in the following way, the ij^{th} element is

$$C_{ij} = \sum_{k=1}^{n} a_{ik} \, b_{kj}$$

For example,

$$C_{12} = \sum_{k=1}^{n} a_{1k} \, b_{k2} = a_{11} \, b_{12} + a_{12} \, b_{22} + \text{---}$$

The easy way to remember this is to work across the row and down the column. Example

$$\underline{A} = \begin{bmatrix} 1 & 2 \\ 3 & 4 \end{bmatrix} \quad \underline{B} = \begin{bmatrix} 1 & 2 & 3 \\ 4 & 5 & 6 \end{bmatrix}$$

$\underline{C} = \underline{A} \times \underline{B}$

$$C_{11} = \sum_{k=1}^{2} a_{1k} \, b_{k1} = a_{11} \, b_{11} + a_{12} \, b_{21} = 1 \times 1 + 2 \times 4$$

etc.

$$\underline{C} \begin{bmatrix} 9 & 12 & 15 \\ 19 & 26 & 33 \end{bmatrix}$$

NOTE: The order of matrix multiplication is important. This is because matrix multiplication is not commutative.

$$\underline{A} \times \underline{B} \neq \underline{B} \times \underline{A}$$

In the above example, $\underline{B} \times \underline{A}$ is not even defined. Also: $\underline{A} \times \underline{B} = \underline{0}$ does not imply that either $\underline{A} = \underline{0}$ or $\underline{B} = \underline{0}$. However, matrix multiplication is distributive

$$\underline{A} \times (\underline{B} + \underline{C}) = \underline{A} \times \underline{B} + \underline{B} \times \underline{C}$$

and also associative

$$\underline{A}\ \underline{B}\ \underline{C} = (\underline{A}\ \underline{B})\underline{C} = \underline{A}(\underline{B}\ \underline{C})$$

The transpose of a matrix product has the following properties

$$(AB)^T = B^T A^T$$

$$(ABC)^T = C^T B^T A^T$$

etc.

The identity matrix $\underline{I}$ has the same property as the scalar unity, i.e.

$$\underline{A}\ \underline{I} = \underline{I}\ \underline{A} = \underline{A}$$

We also denote

$$\underline{A}\ \underline{A} = \underline{A}^2 \quad \text{etc.}$$

4. The Matrix Inverse

In scalar operations. If ab = 1, then $b = \frac{1}{a}$ or $b = a^{-1}$.

We have a similar situation with respect to matrices. If

$$\underline{A}\ \underline{X} = \underline{I},$$

Then $\underline{X} = \underline{A}^{-1}$

The inverse of $\underline{A}$ is denoted by $\underline{A}^{-1}$ and has the property

$$\underline{A}\ \underline{A}^{-1} = \underline{I}$$

Also

$$\underline{A}^{-1} A = I$$

Also

$$(ABC)^{-1} = C^{-1}B^{-1}A^{-1}$$

Consider the specific example of a 3 x 3 matrix. All definitions and logic apply to the general case as well.

$$\underline{A}_{3 \times 3} = \begin{bmatrix} a_{11} & a_{12} & a_{13} \\ a_{21} & a_{22} & a_{23} \\ a_{31} & a_{32} & a_{33} \end{bmatrix}$$

The determinant of $\underline{A}$ is given by

$$\det A = \begin{bmatrix} a_{11} & a_{12} & a_{13} \\ a_{21} & a_{22} & a_{23} \\ a_{31} & a_{32} & a_{33} \end{bmatrix}$$

Recall the definition of the minor of a determinant.

M_{ij} = det A with i^{th} row and j^{th} column deleted

The cofactor is the signed minor

$$C_{ij} = (-1)^{i + j} M_{ij}$$

To understand the inverse of A, let us apply the basic definition.

$$\underline{A}\ \underline{X} = \underline{I}$$

or, for the specific 3 x 3 example

$$\begin{bmatrix} a_{11} & a_{12} & a_{13} \\ a_{21} & a_{22} & a_{23} \\ a_{31} & a_{32} & a_{33} \end{bmatrix} \begin{bmatrix} x_{11} & x_{12} & x_{13} \\ x_{21} & x_{22} & x_{23} \\ x_{31} & x_{32} & x_{33} \end{bmatrix} = \begin{bmatrix} 1 & 0 & 0 \\ 0 & 1 & 0 \\ 0 & 0 & 1 \end{bmatrix}$$

Multiplying out, we have

(3 eq. in 3 ukn.)

$$a_{11}\,x_{11} + a_{12}\,x_{21} + a_{13}\,x_{31} = 1$$
$$a_{21}\,x_{11} + a_{22}\,x_{21} + a_{23}\,x_{31} = 0$$
$$a_{31}\,x_{11} + a_{32}\,x_{21} + a_{33}\,x_{31} = 0$$

(3 eq. in 3 ukn.)

$$a_{11}\,x_{12} + a_{12}\,x_{22} + a_{13}\,x_{32} = 0$$
$$a_{21}\,x_{12} + a_{22}\,x_{22} + a_{23}\,x_{32} = 1$$
$$a_{31}\,x_{12} + a_{32}\,x_{22} + a_{33}\,x_{32} = 0$$

(3 eq. in 3 ukn.)

$$a_{11}\,x_{13} + a_{12}\,x_{23} + a_{13}\,x_{33} = 0$$
$$a_{21}\,x_{23} + a_{22}\,x_{23} + a_{23}\,x_{33} = 0$$
$$a_{31}\,x_{13} + a_{32}\,x_{23} + a_{33}\,x_{33} = 1$$

These equations can be solved for all X_{ij}.

$$X_{11} = \frac{\begin{vmatrix} 1 & a_{12} & a_{13} \\ 0 & a_{22} & a_{23} \\ 0 & a_{32} & a_{33} \end{vmatrix}}{\det \underline{A}} = \frac{a_{22}\,a_{33} - a_{32}\,a_{23}}{\det A} = \frac{C_{11}}{\det A}$$

$$X_{12} = \frac{\begin{vmatrix} 0 & a_{12} & a_{13} \\ 1 & a_{22} & a_{23} \\ 0 & a_{32} & a_{33} \end{vmatrix}}{\det \underline{A}} = \frac{-1\ (a_{12}\ a_{33} - a_{32}\ a_{13}}{\det \underline{A}} = \frac{C_{21}}{\det A}$$

$$X_{23} = \frac{\begin{vmatrix} a_{11} & 0 & a_{13} \\ a_{21} & 0 & a_{23} \\ a_{31} & 1 & a_{33} \end{vmatrix}}{\det \underline{A}} = \frac{-1\ (a_{11}\ a_{23} - a_{21}\ a_{13}}{\det A} = \frac{C_{32}}{\det A}$$

$$X_{ij} = \frac{C_{ij}}{\det A}$$

$$\underline{X} = \frac{1}{\det A} \begin{bmatrix} C_{11} & C_{21} & C_{31} \\ C_{12} & C_{22} & C_{32} \\ C_{13} & C_{23} & C_{33} \end{bmatrix} = \frac{\mathbf{A}}{\det A}$$

where **A** is the adjoint matrix, which is formed by taking the cofactors of $\underline{A}^T$.

Example

$$\underline{A} = \begin{bmatrix} 2 & 2 & 0 \\ 0 & 3 & 1 \\ 1 & 0 & 1 \end{bmatrix}$$

$$\underline{A}^T = \begin{bmatrix} 2 & - & 1 \\ 2 & 3 & 0 \\ 0 & 1 & 1 \end{bmatrix}$$

$$\mathbf{A} = \begin{bmatrix} 3 & -2 & 2 \\ 1 & 2 & -2 \\ -3 & 2 & 6 \end{bmatrix}$$

$|A| = 8$

$$\underline{A}^{-1} = \frac{1}{8}\begin{bmatrix} 3 & -2 & 2 \\ 1 & 2 & -2 \\ -3 & 2 & 6 \end{bmatrix}$$

Problems

Refer to the following matrices for questions 1 through 12:

$$E = \begin{bmatrix} 1 & 0 & 2 \\ 2 & 1 & 3 \\ 0 & 1 & 2 \end{bmatrix} \quad L = \begin{bmatrix} 4 & 2 & 2 \\ 3 & 1 & 5 \\ 2 & 1 & 1 \end{bmatrix} \quad S = \begin{bmatrix} 1 & 6 & 2 \\ 0 & 0 & 0 \\ 2 & 12 & 4 \end{bmatrix}$$

$$F = \begin{Bmatrix} 1 \\ 0 \\ 2 \end{Bmatrix} \quad M = \begin{bmatrix} 2 & 6 & 2 \\ 6 & 5 & 4 \\ 2 & 4 & 3 \end{bmatrix} \quad T = \begin{bmatrix} 5 \end{bmatrix}$$

$$G = \begin{bmatrix} 5 & 1 & 4 \end{bmatrix} \quad N = \begin{bmatrix} 1 & 6 & 2 \end{bmatrix} \quad U = \begin{bmatrix} 1 & 2 & 0 \\ 0 & 1 & 0 \\ 0 & 0 & 0 \end{bmatrix}$$

$$H = \begin{bmatrix} 1 & 0 & 0 \\ 0 & 1 & 0 \\ 0 & 0 & 0 \end{bmatrix} \quad P = \begin{bmatrix} 1 & 2 \\ 0 & 1 \end{bmatrix}$$

$$J = \begin{bmatrix} 0 & 0 & 1 \\ 0 & 1 & 0 \\ 1 & 0 & 0 \end{bmatrix} \quad Q = \begin{bmatrix} 1 & 0 \\ 0 & 1 \end{bmatrix}$$

$$K = \begin{bmatrix} 3 & 1 & 1 \\ 1 & 0 & 2 \\ 1 & 0 & 2 \end{bmatrix} \quad R = \begin{bmatrix} 0 & 0 & 0 \\ 0 & 0 & 0 \\ 0 & 0 & 0 \end{bmatrix}$$

1. The identity matrix is given by matrix
   A. R
   B. H
   C. J
   D. Q
   E. H and Q
2. The null matrix is given by matrix
   A. R
   B. H

C. J

D. K

E. H and Q

3. An example(s) of a singular matrix is given by matrix

A. S and U

B. H and Q

C. J

D. K

E. S, K, U and L

4. Which of the following matrices are symmetric?

A. H

B. J

C. H & Q

D. M & T

E. All of the above

5. Which of the following matrix multiplications are defined?

A. E • K

B. F • K

C. G • F

D. A & C only

E. A & B only

6. The determinate of matrix L, (det L), is equal to

A. 60

B. -30

C. 4

D. 0

E. 10

7. Which of the following matrix additions is not permissible

A. E + J

B. P + Q

C. G + N

D. F + G

E. Both C & D

8. through 20. are TRUE-FALSE type

8. _____ The transpose of matrix H is matrix J.

9. _____ The matrix subtration denoted by P - Q is equal to

$$\begin{bmatrix} 0 & 2 \\ 0 & 0 \end{bmatrix}$$

10. _____ Matrix P is equal to matrix U.
11. _____ N · F = T.
12. _____ Matrices F and G are examples of column and row matrices respectively.
13. _____ The product of any square matrix and the identity matrix yields that matrix.
14. _____ The sum of any square matrix and the null matrix yields that matrix.
15. _____ The determinate of a singular matrix is non-zero.
16. _____ Every non-singular square matrix has an inverse.
17. _____ The product of any non-zero matrix and its inverse is the null matrix.
18. _____ The transpose of a symmetric matrix is symmetric.
19. _____ In general, two matrices A and B of the same order, A · B = B · A.
20. _____ If A^T denotes transpose of matrix A then, $(A\ B)^T = B^T A^T$.

Identify all the characteristics that describe the following matrices.

21.

a.

$$\begin{bmatrix} 1 & 2 & 3 & 4 \\ 2 & 5 & 1 & -7 \\ 3 & 1 & 6 & 8 \\ 4 & -7 & 8 & 1 \end{bmatrix}$$

b.

$$\begin{bmatrix} 1 & 2 & 3 \\ 4 & 5 & 6 \end{bmatrix}$$

c.

$$\begin{Bmatrix} 0 \\ 0 \\ 0 \end{Bmatrix}$$

d.

$$\begin{bmatrix} 1 & 0 & 0 \\ 0 & 1 & 0 \\ 0 & 0 & 1 \end{bmatrix}$$

e.

$$\lfloor 1 \quad 3 \quad -7 \quad 8 \rfloor$$

22. Given the following matrices:

$$[A] = \begin{bmatrix} 1 & 2 & 1 \\ 0 & +3 & -2 \\ 1 & -1 & 2 \end{bmatrix}_{3 \times 3} \qquad [C] = \begin{bmatrix} 1 & 3 \\ -1 & 2 \end{bmatrix}_{2 \times 2}$$

$$[B] = \begin{bmatrix} 1 & 3 & 0 \\ 2 & -2 & 1 \\ -2 & 1 & 2 \end{bmatrix}_{3 \times 3} \qquad [D] = \begin{bmatrix} 1 & 2 \\ 3 & 4 \end{bmatrix}_{2 \times 2}$$

Find:

$[A]^{-1}$

$[B]^{-1}$

$[C]^{-1}$

$[D]^{-1}$

23. Perform the following operations

a. $[A]^{-1} \times [A]$

b. $[A] \times [B]$

c. $[B]^T$

24. Solve the following set of simultaneous linear equations for x, y and z.

$$x + 3y + 2z = 5$$
$$-2x \_ 4y - z = 7$$
$$2x + y + 3z = 1$$

25. If

$$A = \begin{bmatrix} 3 & 0 & 2 \\ 4 & -1 & 5 \end{bmatrix} \qquad B = \begin{bmatrix} 2 & 1 & 0 & -5 \\ 4 & 3 & -1 & 0 \\ 0 & 1 & -7 & 4 \end{bmatrix}$$

Find the product $\underline{A} \cdot \underline{B}$ (matrix product).

26. Invert the matrix

$$[C] = \begin{bmatrix} 2 & 4 & 2 \\ 3 & 1 & 1 \\ 1 & 0 & 1 \end{bmatrix}$$

a. By application of the definition.

b. By Gauss Elimination (row operation).

27. Invert the following matrices:

$$A = \begin{bmatrix} 3 & 1 \\ -2 & -4 \end{bmatrix} \qquad \text{ans.} \quad A^{-1} = -\frac{1}{10}\begin{bmatrix} -4 & -1 \\ 2 & 3 \end{bmatrix}$$

$$B = \begin{bmatrix} 4 & -1 & 2 \\ 3 & 4 & 1 \\ -2 & -2 & 4 \end{bmatrix} \qquad \text{ans.} \quad B^{-1} = \frac{1}{90}\begin{bmatrix} 18 & 0 & -9 \\ -14 & 20 & 2 \\ 2 & 10 & 19 \end{bmatrix}$$

$$C = \begin{bmatrix} 1 & 2 & 5 \\ 3 & 1 & 4 \\ 1 & 1 & 2 \end{bmatrix} \qquad \text{ans.} \quad C^{-1} = \frac{1}{4}\begin{bmatrix} -2 & 1 & 3 \\ -2 & -3 & 11 \\ 2 & 1 & -5 \end{bmatrix}$$

Solve the set of equations:

$$3x_1 - x_2 = 7 \qquad \text{ans.} \quad x_1 = \frac{19}{11}$$

$$2x_1 + 3x_2 = -2 \qquad \text{ans.} \quad x_2 = \frac{20}{-11}$$

Appendix C

EXCERPT FROM ASTM STANDARD E8-57 T

Table C.1 Tension Testing of Metallic Materials (E8–57T)

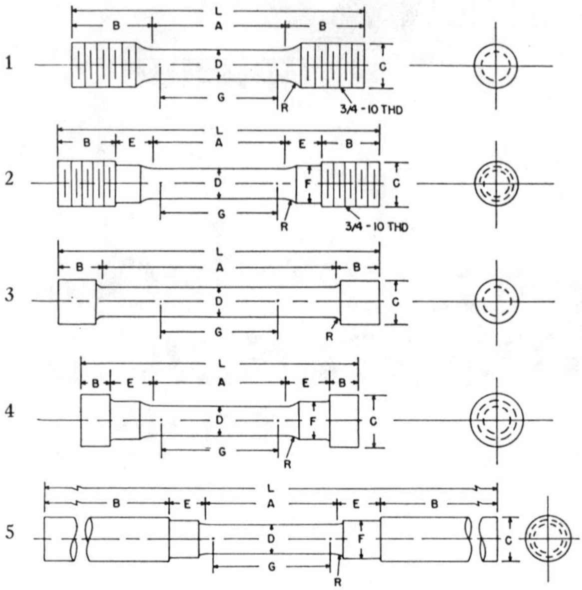

DIMENSIONS

	Specimen				
	1	2	3	4	5
G—Gage length, in.	2.000 ± 0.005	2.000 ± 0.005	2.000 ± 0.005	2.000 ± 0.005	2.000 ± 0.005
D—Diameter (Note 1), in.	0.500 ± 0.010	0.500 ± 0.010	0.500 ± 0.010	0.500 ± 0.010	0.500 ± 0.010
R—Radius of fillet, in.	$\frac{3}{8}$, min	$\frac{3}{8}$, min	$\frac{1}{16}$, min	$\frac{3}{8}$, min	$\frac{3}{8}$, min
A—Length of reduced section, in.	$2\frac{1}{4}$, min	$2\frac{1}{4}$, min	4, approx	$2\frac{1}{4}$, min	$2\frac{1}{4}$, min
L—Overall length, in.	5, approx	$5\frac{1}{2}$, approx	$5\frac{1}{2}$, approx	$4\frac{3}{4}$, approx	$9\frac{1}{2}$, approx
B—Length of end section (Note 2), in.	$1\frac{3}{8}$, approx	1, approx	$\frac{3}{4}$, approx	$\frac{1}{2}$, approx	3, min
C—Diameter of end section	$\frac{3}{4}$ in.	$\frac{3}{4}$ in.	$\frac{23}{32}$ in.	$\frac{7}{8}$ in.	$\frac{3}{4}$ in.
E—Length of shoulder and fillet section, in.	....	$\frac{5}{8}$, approx	....	$\frac{3}{4}$, approx	$\frac{5}{8}$, approx
F—Diameter of shoulder	....	$\frac{5}{8}$ in.	....	$\frac{5}{8}$ in.	$\frac{19}{32}$ in.

NOTE 1.—The reduced section may have a gradual taper from the ends toward the center with the ends not more than 0.005 in. larger in diameter than the center.

NOTE 2.—On specimen (5) it is desirable, if possible, to make the length of the grip section great enough to allow the specimen to extend into the grips a distance equal to two thirds or more of the length of the grips.

Appendix D

AREA PROPERTIES

Formulas for common elements of area

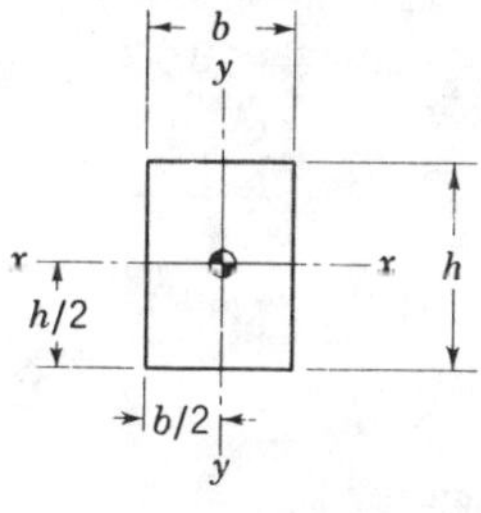

Rectangle:

$$A = bh$$

$$I_x = \tfrac{1}{12}bh^3 \qquad I_y = \tfrac{1}{12}b^3h$$

$$\rho_x = \frac{h}{\sqrt{12}} = 0.289h$$

$$\rho_y = \frac{b}{\sqrt{12}} = 0.289b$$

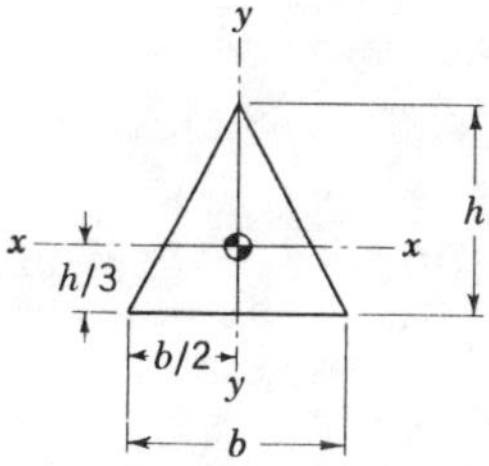

Isosceles triangle:

$$A = \tfrac{1}{2}bh$$

$$I_x = \tfrac{1}{36}bh^3 \qquad I_y = \tfrac{1}{48}b^3h$$

$$\rho_x = \frac{h}{\sqrt{18}} = 0.236h$$

$$\rho_y = \frac{b}{\sqrt{24}} = 0.204b$$

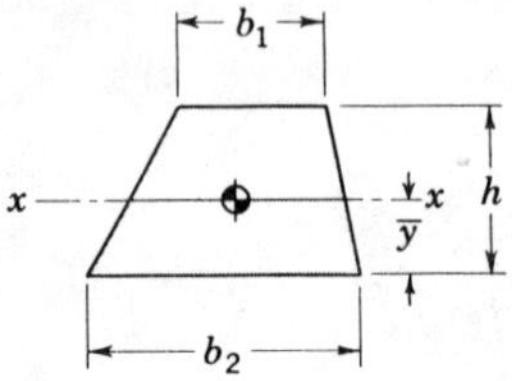

Trapezoid:

$$A = \tfrac{1}{2}(b_1 + b_2)h$$

$$\bar{y} = \frac{1}{3}\,\frac{(2b_1 + b_2)h}{b_1 + b_2}$$

$$I_x = \frac{h^3}{36}\,\frac{b_1^2 + 4b_1b_2 + b_2^2}{b_1 + b_2}$$

$$\rho_x = \frac{h}{6}\,\frac{\sqrt{2(b_1^2 + 4b_1b_2 + b_2^2)}}{b_1 + b_2}$$

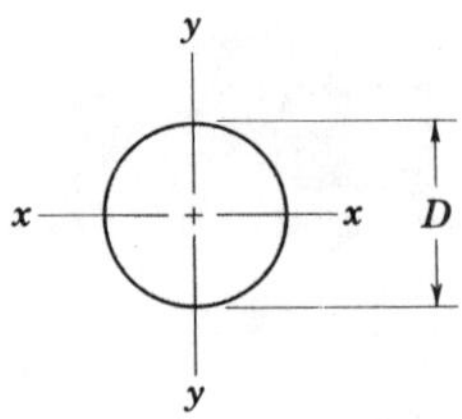

Solid circle:

$$A = \frac{\pi}{4} D^2 = 0.7854D^2$$

$$I_x = I_y = \frac{\pi}{64} D^4 = 0.0491D^4$$

$$\rho_x = \rho_y = \frac{D}{4} = 0.250D$$

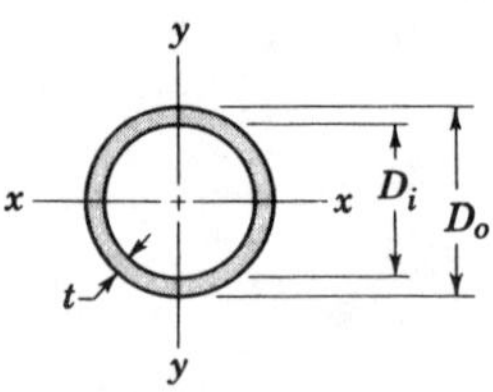

Hollow circle:

$$A = \frac{\pi}{4} (D_0^2 - D_i^2)$$

$$I_x = I_y = \frac{\pi}{64} (D_0^4 - D_i^4)$$

$$\rho_x = \rho_y = \tfrac{1}{4}\sqrt{D_0^2 + D_i^2}$$

Thin-wall approximation:

D_m = mean diameter

$$A = \pi D_m t$$

$$I_x = I_y = \frac{\pi}{8} D_m^3 t$$

$$\rho_x = \rho_y = \sqrt{\tfrac{1}{8}}\, D_m = 0.353D_m$$

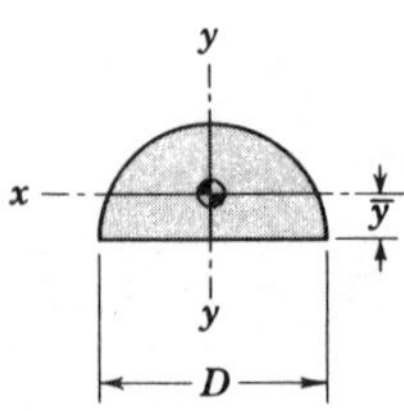

Solid semicircle:

$$A = \frac{\pi D^2}{8}$$

$$\bar{y} = \frac{2D}{3\pi}$$

$$I_x = \frac{D^4(9\pi^2 - 64)}{1{,}152\pi} = 0.00686D^4$$

$$I_y = \frac{\pi D^4}{128} = 0.0245D^4$$

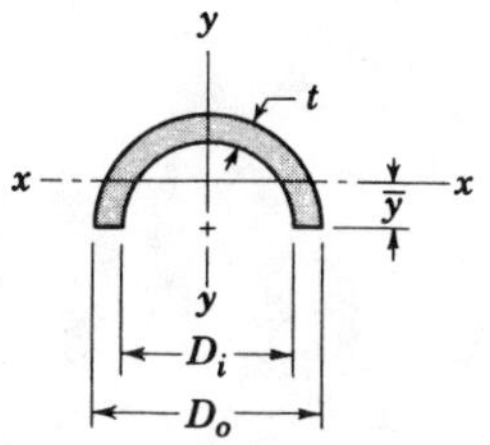

Hollow semicircle:

$$A = \frac{\pi}{8}(D_o^2 - D_i^2)$$

$$\bar{y} = \frac{2}{3\pi}\frac{D_o^3 - D_i^3}{D_o^2 - D_i^2}$$

$$I_x = \frac{\pi}{128}(D_o^4 - D_i^4) - \frac{(D_o^3 - D_i^3)^2}{18\pi(D_o^2 - D_i^2)}$$

$$I_y = \frac{\pi}{128}(D_o^4 - D_i^4)$$

Thin-wall approximation:

D_m = mean diameter

$$A = \frac{\pi}{2}D_m t$$

$$\bar{y} = \frac{D_m}{\pi}$$

$$I_x = \left(\frac{\pi}{16} - \frac{1}{2\pi}\right)D_m^3 t$$

$$I_y = \frac{\pi}{16}D_m^3 t$$

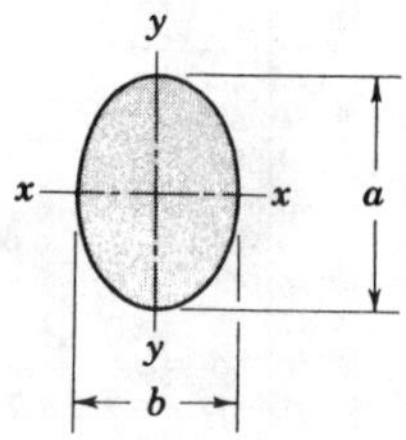

Solid ellipse:

$$A = \frac{\pi}{4}ab$$

$$I_x = \frac{\pi}{64}a^3 b$$

$$I_y = \frac{\pi}{64}b^3 a$$

$$\rho_x = \frac{a}{4}$$

$$\rho_y = \frac{b}{4}$$

Section properties of wide-flange beams

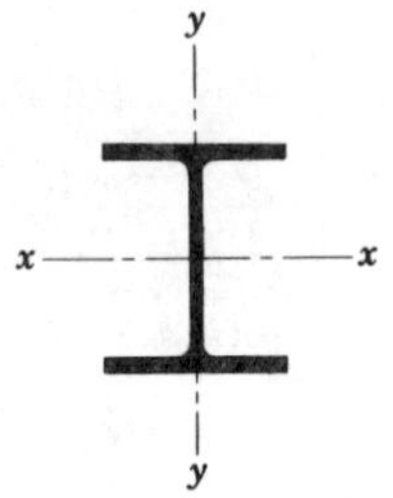

Nominal size, in.	Weight per ft, lb	Area of section, in.²	Depth of section, in.	Flange		Web thickness, in.	Axis xx			Axis yy		
				Width, in.	Thickness, in.		I_x, in.⁴	Z_x, in.³	ρ_x, in.	I_y, in.⁴	Z_y, in.³	ρ_y, in.
14 × 16	193	56.73	15.50	15.71	1.438	0.890	2,402	310.0	6.51	930	118.4	4.05
	184	54.07	15.38	15.66	1.378	0.840	2,274	295.8	6.49	882	112.7	4.04
	176	51.73	15.25	15.64	1.313	0.820	2,149	281.9	6.45	837	107.1	4.02
	167	49.09	15.12	15.60	1.248	0.780	2,020	267.3	6.42	790	101.3	4.01
	158	46.47	15.00	15.55	1.188	0.730	1,900	253.4	6.40	745	95.8	4.00
	150	44.08	14.88	15.51	1.128	0.695	1,786	240.2	6.37	702	90.6	3.99
	142	41.85	14.75	15.50	1.063	0.680	1,672	226.7	6.32	660	85.2	3.97
14 × 14½	136	39.98	14.75	14.74	1.063	0.660	1,593	216.0	6.31	567	77.0	3.77
	127	37.33	14.62	14.69	0.998	0.610	1,476	202.0	6.29	527	71.8	3.76
	119	34.99	14.50	14.65	0.938	0.570	1,373	189.4	6.26	491	67.1	3.75
	111	32.65	14.37	14.62	0.873	0.540	1,266	176.3	6.23	454	62.2	3.73
	103	30.26	14.25	14.57	0.813	0.495	1,165	163.6	6.21	419	57.6	3.72
	95	27.94	14.12	14.54	0.748	0.465	1,063	150.6	6.17	383	52.8	3.71
	87	25.56	14.00	14.50	0.688	0.420	966	138.1	6.15	349	48.2	3.70
14 × 12	84	24.71	14.18	12.02	0.778	0.451	928	130.9	6.13	225.5	37.5	3.02
	78	22.94	14.06	12.00	0.718	0.428	851	121.1	6.09	206.9	34.5	3.00
14 × 10	74	21.76	14.19	10.07	0.783	0.450	796	112.3	6.05	133.5	26.5	2.48
	68	20.00	14.06	10.04	0.718	0.418	724	103.0	6.02	121.2	24.1	2.46
	61	17.94	13.91	10.00	0.643	0.378	641	92.2	5.98	107.3	21.5	2.45
14 × 8	53	15.59	13.94	8.06	0.658	0.370	542	77.8	5.90	57.5	14.3	1.92
	48	14.11	13.81	8.03	0.593	0.339	484	70.2	5.86	51.3	12.8	1.91
	43	12.65	13.68	8.00	0.528	0.308	429	62.7	5.82	45.1	11.3	1.89
14 × 6¾	38	11.17	14.12	6.77	0.513	0.313	385	54.6	5.87	24.6	7.3	1.49
	34	10.00	14.00	6.75	0.453	0.287	339	48.5	5.83	21.3	6.3	1.46
	30	8.81	13.86	6.73	0.383	0.270	289	41.8	5.73	17.5	5.2	1.41
12 × 12	190	55.86	14.38	12.67	1.736	1.060	1,892	263.2	5.82	589.7	93.1	3.25
	161	47.38	13.88	12.51	1.486	0.905	1,541	222.2	5.70	486.2	77.7	3.20
	133	39.11	13.38	12.36	1.236	0.755	1,221	182.5	5.59	389.9	63.1	3.16
	120	35.31	13.12	12.32	1.106	0.710	1,071	163.4	5.51	345.1	56.0	3.13
	106	31.19	12.88	12.23	0.986	0.620	930	144.5	5.46	300.9	49.2	3.11
	99	29.09	12.75	12.19	0.921	0.580	858	134.7	5.43	278.2	45.7	3.09
	92	27.06	12.62	12.15	0.856	0.545	788	125.0	5.40	256.4	42.2	3.08
	85	24.98	12.50	12.10	0.796	0.495.	723	115.7	5.38	235.5	38.9	3.07
	79	23.22	12.38	12.08	0.736	0.470	663	107.1	5.34	216.4	35.8	3.05
	72	21.16	12.25	12.04	0.671	0.430	597	97.5	5.31	195.3	32.4	3.04
	65	19.11	12.12	12.00	0.606	0.390	533	88.0	5.28	174.6	29.1	3.02
12 × 10	58	17.06	12.19	10.01	0.641	0.359	476	78.1	5.28	107.4	21.4	2.51
	53	15.59	12.06	10.00	0.576	0.345	426	70.7	5.23	96.1	19.2	2.48

Section properties of standard channels

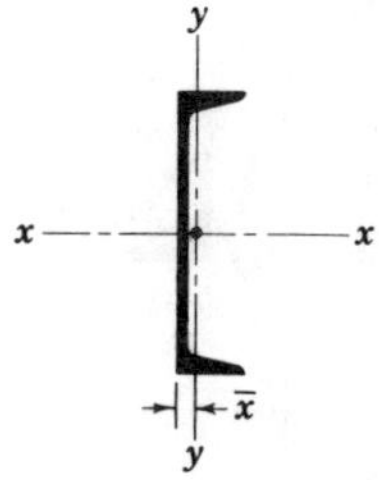

Depth of channel, in.	Weight per ft, lb	Area of section, in.²	Width of flange, in.	Thickness of web, in.	Axis xx			Axis yy	$\bar{x}$, in.
					I, in.⁴	Z, in.³	ρ_x, in.	ρ_y, in.	
15	50.0	14.64	3.716	0.716	401.4	53.6	5.24	0.87	0.80
	40.0	11.70	3.520	0.520	346.3	46.2	5.44	0.89	0.78
	33.9	9.90	3.400	0.400	312.6	41.7	5.62	0.91	0.79
12	30.0	8.79	3.170	0.510	161.2	26.9	4.28	0.77	0.68
	25.0	7.32	3.047	0.387	143.5	23.9	4.43	0.79	0.68
	20.7	6.03	2.940	0.280	128.1	21.4	4.61	0.81	0.70
10	30.0	8.80	3.033	0.673	103.0	20.6	3.42	0.67	0.65
	25.0	7.33	2.886	0.526	90.7	18.1	3.52	0.68	0.62
	20.0	5.86	2.739	0.379	78.5	15.7	3.66	0.70	0.61
	15.3	4.47	2.600	0.240	66.9	13.4	3.87	0.72	0.64
9	20.0	5.86	2.648	0.448	60.6	13.5	3.22	0.65	0.59
	15.0	4.39	2.485	0.285	50.7	11.3	3.40	0.67	0.59
	13.4	3.89	2.430	0.230	47.3	10.5	3.49	0.67	0.61
8	18.75	5.49	2.527	0.487	43.7	10.9	2.82	0.60	0.57
	13.75	4.02	2.343	0.303	35.8	9.0	2.99	0.62	0.56
	11.5	3.36	2.260	0.220	32.3	8.1	3.10	0.63	0.58
7	14.75	4.32	2.299	0.419	27.1	7.7	2.51	0.57	0.53
	12.25	3.58	2.194	0.314	24.1	6.9	2.59	0.58	0.53
	9.8	2.85	2.090	0.210	21.1	6.0	2.72	0.59	0.55
6	13.0	3.81	2.157	0.437	17.3	5.8	2.13	0.53	0.52
	10.5	3.07	2.034	0.314	15.1	5.0	2.22	0.53	0.50
	8.2	2.39	1.920	0.200	13.0	4.3	2.34	0.54	0.52
5	9.0	2.63	1.885	0.325	8.8	3.5	1.83	0.49	0.48
	6.7	1.95	1.750	0.190	7.4	3.0	1.95	0.50	0.49
4	7.25	2.12	1.720	0.320	4.5	2.3	1.47	0.46	0.46
	5.4	1.56	1.580	0.180	3.8	1.9	1.56	0.45	0.46
3	6.0	1.75	1.596	0.356	2.1	1.4	1.08	0.42	0.46
	5.0	1.46	1.498	0.258	1.8	1.2	1.12	0.41	0.44
	4.1	1.19	1.410	0.170	1.6	1.1	1.17	0.41	0.44

Section properties of standard equal angles

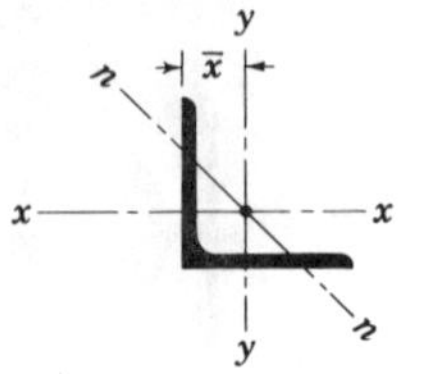

Size, in.	Thick-ness, in.	Weight per ft, lb	Area of section, in.^{2}	Axis xx and axis yy				Axis nn
				I, in.^{4}	Z, in.^{3}	ρ, in.	$\bar{x}$, in.	ρ_{min}, in.
8 × 8	$1\frac{1}{8}$	56.9	16.73	98.0	17.5	2.42	2.41	1.56
	1	51.0	15.00	89.0	15.8	2.44	2.37	1.56
	$\frac{7}{8}$	45.0	13.23	79.6	14.0	2.45	2.32	1.57
	$\frac{3}{4}$	38.9	11.44	69.7	12.2	2.47	2.28	1.57
	$\frac{5}{8}$	32.7	9.61	59.4	10.3	2.49	2.23	1.58
	$\frac{1}{2}$	26.4	7.75	48.6	8.4	2.50	2.19	1.59
6 × 6	1	37.4	11.00	35.5	8.6	1.80	1.86	1.17
	$\frac{7}{8}$	33.1	9.73	31.9	7.6	1.81	1.82	1.17
	$\frac{3}{4}$	28.7	8.44	28.2	6.7	1.83	1.78	1.17
	$\frac{5}{8}$	24.2	7.11	24.2	5.7	1.84	1.73	1.18
	$\frac{1}{2}$	19.6	5.75	19.9	4.6	1.86	1.68	1.18
	$\frac{3}{8}$	14.9	4.36	15.4	3.5	1.88	1.64	1.19
5 × 5	$\frac{7}{8}$	27.2	7.98	17.8	5.2	1.49	1.57	0.97
	$\frac{3}{4}$	23.6	6.94	15.7	4.5	1.51	1.52	0.97
	$\frac{5}{8}$	20.0	5.86	13.6	3.9	1.52	1.48	0.98
	$\frac{1}{2}$	16.2	4.75	11.3	3.2	1.54	1.43	0.98
	$\frac{3}{8}$	12.3	3.61	8.7	2.4	1.56	1.39	0.99
4 × 4	$\frac{3}{4}$	18.5	5.44	7.7	2.8	1.19	1.27	0.78
	$\frac{5}{8}$	15.7	4.61	6.7	2.4	1.20	1.23	0.78
	$\frac{1}{2}$	12.8	3.75	5.6	2.0	1.22	1.18	0.78
	$\frac{3}{8}$	9.8	2.86	4.4	1.5	1.23	1.14	0.79
	$\frac{1}{4}$	6.6	1.94	3.0	1.1	1.25	1.09	0.80
$3\frac{1}{2}$ × $3\frac{1}{2}$	$\frac{1}{2}$	11.1	3.25	3.6	1.5	1.06	1.06	0.68
	$\frac{3}{8}$	8.5	2.48	2.9	1.2	1.07	1.01	0.69
	$\frac{1}{4}$	5.8	1.69	2.0	0.79	1.09	0.97	0.69
3 × 3	$\frac{1}{2}$	9.4	2.75	2.2	1.1	0.90	0.93	0.58
	$\frac{3}{8}$	7.2	2.11	1.8	0.83	0.91	0.89	0.58
	$\frac{1}{4}$	4.9	1.44	1.2	0.58	0.93	0.84	0.59
$2\frac{1}{2}$ × $2\frac{1}{2}$	$\frac{1}{2}$	7.7	2.25	1.2	0.74	0.74	0.81	0.49
	$\frac{3}{8}$	5.9	1.73	0.98	0.57	0.75	0.76	0.49
	$\frac{1}{4}$	4.1	1.19	0.70	0.39	0.77	0.72	0.49
2 × 2	$\frac{3}{8}$	4.7	1.36	0.48	0.35	0.59	0.64	0.39
	$\frac{1}{4}$	3.19	0.94	0.35	0.25	0.61	0.59	0.39
	$\frac{1}{8}$	1.65	0.48	0.19	0.13	0.63	0.55	0.40

Appendix E

CONVERSIONS TO SI UNITS

To convert from	To	Multiply by
Acceleration		
foot/second^2 (ft/s^2)	meter/second^2 (m/s^2)	3.048×10^{-1}
inch/second^2 (in/s^2)	meter/second^2 (m/s^2)	2.54×10^{-2}
Area		
foot^2 (ft^2)	meter^2 (m^2)	9.2903×10^{-2}
inch^2 (in^2)	meter^2 (m^2)	6.4516×10^{-4}
yard^2 (yd^2)	meter^2 (m^2)	8.3613×10^{-1}
Density		
pound mass/inch^3 (lbm/in^3)	kilogram/meter^3 (kg/m^3)	2.7680×10^4
pound mass/foot^3 (lbm/ft^3)	kilogram/meter^3 (kg/m^3)	1.6018×10
Energy, Work		
British thermal unit (Btu)	joule (J)	1.0551×10^3
foot-pound force (ft·lbf)	joule (J)	1.3558
kilowatt-hour (kW·h)	joule (J)	3.60×10^5
Force		
kip (1000 lbf)	newton (N)	4.4482×10^3
pound force (lbf)	newton (N)	4.4482
ounce force	newton (N)	2.7801×10^{-1}

Length		
foot (ft)	meter (m)	3.048×10^{-1}
inch (in)	meter (m)	2.54×10^{-2}
mile (mi) (U.S. statute)	meter (m)	1.6093×10^{3}
mile (mi) (international nautical)	meter (m)	1.852×10^{3}
yard (yd)	meter (m)	9.144×10^{-1}
Mass		
pound mass (lbm)	kilogram (kg)	4.5359×10^{-1}
slug ($lbf \cdot s^2/ft$)	kilogram (kg)	1.4594×10
ton (2000 lbm)	kilogram (kg)	9.0718×10^{2}
Power		
foot-pound/minute ($ft \cdot lbf/min$)	watt (W)	2.2597×10^{-2}
horsepower (550 $ft \cdot lbf/s$)	watt (W)	7.4570×10^{2}
Pressure, stress		
atmosphere(std)(14.71bf/in^2)	newton/meter^2(N/m^{2}or Pa)	1.0133×10^{5}
pound/foot^2(lbf/ft^2)	newton/meter^2(N/m^{2}or Pa)	4.7880×10
pound/inch^2(lbf/in^2 or psi)	newton/meter^2(N/m^{2}or Pa)	6.8948×10^{3}
Velocity		
foot/minute (ft/min)	meter/second (m/s)	5.08×10^{-3}
foot/second (ft/s)	meter/second (m/s)	3.048×10^{-1}
knot (nautical mi/h)	meter/second (m/s)	5.1444×10^{-1}
mile/hour (mi/h)	meter/second (m/s)	4.4704×10^{-1}
mile/hour (mi/h)	kilometer/hour (km/h)	1.6093
mile/second (mi/s)	kilometer/second (km/s)	1.6093
Volume		
foot^3(ft^3)	meter^3 (m^3)	2.8317×10^{-2}
inch^3(in^3)	meter^3 (m^3)	1.6387×10^{-5}

Additional Acceptable Units and their SI Equivalents

(Frequency)	1 hertz (Hz)	= 1 cycle/second (s^{-1})
(Mass)	1 metric ton (t)	= 1000 kilogram (10^3 kg)
(Pressure or stress)	1 pascal (Pa)	= 1 newton/meter^2 (N/m^2)
(Viscosity, absolute)	1 centipoise	= 1 millipascal-second ($mPa \cdot s$ or 10^{-3} $N \cdot s/m^2$)

INDEX